D0084796

Fundamental Mathematics

Fundamental Mathematics

Thomas L. Wade
PROFESSOR OF MATHEMATICS
FLORIDA STATE UNIVERSITY

Howard E. Taylor
CALLAWAY PROFESSOR OF MATHEMATICS
WEST GEORGIA COLLEGE

Fourth Edition

McGraw-Hill Book Company
NEW YORK ST. LOUIS SAN FRANCISCO DÜSSELDORF JOHANNESBURG
KUALA LUMPUR LONDON MEXICO MONTREAL NEW DELHI
PANAMA PARIS SÃO PAULO SINGAPORE SYDNEY TOKYO TORONTO

FUNDAMENTAL MATHEMATICS

2 3 4 5 6 7 8 9 0 KPKP 7 9 8 7 6 5 4

This book was set in News Gothic by York Graphic Services, Inc.
The editors were Jack L. Farnsworth and Carol First;
the designer was Edward A. Butler;
the production supervisor was Thomas J. LoPinto.
New drawings were done by John Cordes, J & R Technical Services, Inc.
Kingsport Press, Inc., was printer and binder.

Library of Congress Cataloging in Publication Data

Wade, Thomas Leonard, date
Fundamental mathematics.

1. Mathematics—1961- I. Taylor, Howard Edward, date joint author. II. Title.
QA39.2.W3 1974 512'.1 74-992
ISBN 0-07-067652-6

Contents

Preface

This book is based on the authors' experiences and efforts over a period of two decades in developing and testing suitable material to be used with the following groups of students:

1 Students whose background in, or mastery of, basic mathematics is insufficient for the study of a sequence of mathematics courses beginning with a strong college algebra, or finite mathematics, or precalculus course, but who need to remove this deficiency in order to study mathematics in connection with their chosen curricula.

2 Students who are not prepared for a standard sequence of mathematics courses and who are undecided about a major. It seems clear that these students should take during the first year of college a course that will prepare them for further courses in mathematics or for courses in other fields which require basic mathematical techniques. To do otherwise will reduce considerably their choice of a major field.

3 Students who have chosen a major that does not require specific courses in mathematics but who need a course in basic mathematics as a general education subject or to provide basic mathematical skills needed for courses in general physical or biological sciences, the social sciences, or education, or the like.

The growing number of colleges with an open door policy of admissions and the growing number of "developmental" programs are bringing into college an increasing number of students who have a need for instruction in mathematics which has the threefold function of strengthening foundations, improving manipulative skills, and providing knowledge of basic topics in mathematics for general education.

For increased motivation, as well as for economy of time, a study of the basic structure of arithmetic is interwoven with an introduction to the basic structure and uses of algebra. A fundamental premise is that there should be a balance between teaching for understanding and teaching for skill in manipulative techniques. The basic structure of arithmetic and algebra is presented and used as a foundation on which to build increased skill and understanding.

Some ways in which the fourth edition differs from the previous edition are:

Some of the longer sections dealing with operations on natural numbers and on rational numbers have been divided into shorter sections with appropriate exercises.

Some of the more formal treatments of conjunction and disjunction have been abbreviated.

The material on relations, functions, and graphs has been gathered into a single chapter with a considerable extension of the material on graphing linear functions, quadratic functions, and lineal inequalities.

There has been some updating of the applications to business.

There has been an elaboration of some parts of the material on trigonometry.

The chapter on probability has been completely rewritten and cast in a more contemporary mold.

The material on statistics, the binomial formula, and the binomial distribution has been reorganized into a more coherent unit.

Beyond a reasonable familiarity with the usual manipulations with whole numbers there are no formal prerequisites.

THOMAS L. WADE
HOWARD E. TAYLOR

SUGGESTED LESSON ASSIGNMENTS TO SERVE COURSES DESIGNED FOR VARIOUS PURPOSES

COURSE I A one-term (3 semester hours or 5 quarter hours) course with 45–50 class meetings. The emphasis is on the *remedial aspect*. The course can be used as a preparation for college algebra or for COURSE III, which follows.

Lesson Number	Material
1	Secs. 1.1–1.3
2	Secs. 1.4–1.5
3	Secs. 1.6–1.7
4	Secs. 1.8–1.10
5	Secs. 2.1–2.2
6	Sec. 2.3
7	Sec. 2.4
8	Secs. 2.5–2.6
9	Secs. 3.1–3.2
10	Review
11	Test 1
12, 13, 14	Secs. 3.3–3.4
15	Sec. 3.5
16, 17, 18	Secs. 3.6–3.7
19, 20, 21	Secs. 3.8–3.10
22	Test 2
23, 24	Secs. 4.1–4.2
25	Sec. 4.3
26, 27	Sec. 4.4
28, 29	Sec. 4.5
30	Sec. 4.7
31	Sec. 4.8
32	Review
33	Test 3
34, 35	Secs. 5.1–5.3
36	Sec. 5.6
37, 38	Secs. 6.2–6.3 with brief mention of Sec. 6.1
39, 40	Secs. 6.4–6.5
41, 42, 43, 44	Secs. 7.1–7.4
45	Test 4
46, 47, 48, 49, 50	Secs. 7.5–7.7

COURSE II A one-term (3 semester hours or 5 quarter hours) course with 45–50 class meetings which contains more algebraic detail than COURSE I. The course can be used as preparation for courses in trigonometry, analytic geometry, or finite mathematics, or as the first term of a two-term sequence to cover the entire book (see COURSE III, which follows). The course assumes reasonable facility with operations with integers (the material in Chapters 1 and 2), and covers topics not included in COURSE I.

Lesson Number	Material
1	Chapter 1, especially
2	Sec. 1.13
3	Chapter 2, especially
4	Sec. 2.6
5	Secs. 3.1–3.2
6	Secs. 3.3–3.4
7	Secs. 3.5–3.6
8	Sec. 3.7
9	Secs. 3.8–3.9
10	Sec. 3.10
11	Sec. 3.12
12	Test 1
13	Secs. 4.1–4.2
14.	Sec. 4.3
15	
16	Secs. 4.4–4.5
17	
18	
19	Secs. 4.7–4.8
20	
21	Sec. 4.9
22	Review
23	Test 2
24	Sec. 5.1
25	Secs. 5.2–5.3

Lesson Number	Material
26	Sec. 5.4
27	Sec. 5.5
28	Sec. 5.6
29	Secs. 5.7–5.8
30	
31	Secs. 6.1–6.4
32	
33	Sec. 6.5
34	Review
35	Test 3
36	Sec. 6.6
37	Sec. 6.7
38	Sec. 6.8
39	Sec. 6.9
40	Secs. 7.1–7.2
41	Sec. 7.3
42–43	Sec. 7.4
44	Review
45	Test 4
46	
47	
48	Secs. 7.5–7.7
49	
50	

COURSE III A one-term (3 semester hours or 5 quarter hours) course with 45–50 class meetings. For students who have reasonable familiarity with the topics in Chapters 1 to 6; for example, students who have successfully completed course I or COURSE II. This course constitutes with COURSE I or COURSE II a *two-term sequence* for students in business, economics, social sciences, etc.

Lesson Number	Material
1–2	Chapter 8
3	
4	
5	
6	Secs. 9.1–9.6
7	Sec. 9.8
8	
9	
10	
11	Test 1
12	Secs. 10.1–10.2

Lesson Number	Material
13	Secs. 10.3–10.4
14	Secs. 10.5–10.6
15	Sec. 10.7
16	Secs. 11.1–11.2 (recall
17	Sec. 4.5 on simple interest)
18	Secs. 11.3–11.4
19	Sec. 11.5
20	Review
21	Test 2
22	
23	Secs. 12.1–12.3
24	

Lesson Number	Material
25	Sec. 12.4
26	Sec. 12.5
27	Sec. 12.6
28 29	Sec. 12.7
30 31	Sec. 12.8
32	Sec. 12.9
33	Test 3
34 35	Secs. 13.1–13.3
36	Secs. 13.4–13.5
37	Sec. 13.6

Lesson Number	Material
38	Sec. 13.7
39 40	Sec. 13.8
41 42 43 44	Secs. 13.9–13.10
45	Test 4
46 47 48 49 50	Review or supplementary topics

COURSE IV A one-term course with 45–50 class meetings intended as a "general education" or "liberal studies" course in fundamental mathematics. The course assumes reasonable familiarity with the arithmetic and algebra of integers and rational expressions.

Lesson Number	Material
1 2	Chapter 1
3 4	Chapter 2
5	Secs. 3.1–3.2
6	Secs. 3.3–3.4
7	Secs. 3.5–3.6
8	Secs. 3.7–3.8
9	Secs. 3.9–3.10
10	Sec. 3.12
11	Test 1
12 13	Secs. 4.1–4.3
14 15	Sec. 4.5
16	Sec. 4.6
17	Sec. 4.8
18 19	Secs. 5.1–5.3
20	Secs. 5.7–5.8
21	Review
22	Test 2
23	Sec. 7.1
24	Sec. 7.2
25 26	Secs. 7.3–7.4

Lesson Number	Material
27 28	Secs. 7.6–7.7
29 30	Chapter 8
31	Review
32	Test 3
33 34 35	Secs. 11.1–11.2
36 37 38	Secs. 12.1–12.3
39	Sec. 12.4
40	Sec. 12.5
41 42	Sec. 12.7
43	Sec. 12.9
44	Review
45	Test 4
46 47 48 49 50	Secs. 13.1–13.6

Fundamental Mathematics

1

Sets and Natural Numbers

1.1 MATHEMATICS AS A LANGUAGE

People exchange ideas through the means of some language, a language being simply an agreed-upon set of symbols or sounds. A person with the usual faculties of seeing, hearing, and talking grows up in a group or society with a mother tongue. For many of us this is English, which has both a spoken form based upon sounds and a written form based upon alphabetical symbols.

You may be accustomed to thinking of a language as English, French, German, Russian, and so forth. To be sure, these make up one sort of language; we say each of them is a "qualitative" language, since each is concerned largely with describing things in terms of qualities. In each of them things are given names and are described as long or short, sweet or sour, present or absent, or are assigned some general quality. There are, however, other important languages besides qualitative ones. Some of these other languages are the language of music, in which rhythmic patterns and tones are used to communicate certain thoughts and feelings; the language of graphic arts, in which form and color provide a basis for communication; and the language of mathematics, which is a language of size and order, of quantities and relationships among quantities.

Among the characteristics distinguishing men from lower animals are two that stand out conspicuously: the development of a written qualitative language, and the development of a mathematical language, starting with the basic concept of number.

The first essential in developing a written verbal language is an adequate alphabet. As a result of a long development, with considerable borrowing from the Egyptians and Phoenicians, the Greeks produced by approximately the fifth century B.C. an alphabet which came to be a model for all the modern languages of Europe. Note that the very name "alphabet" is derived from a combination of the two Greek letters *alpha* and *beta*. Having developed an adequate alphabet, the Greeks developed a written qualitative language which most scholars consider has never been excelled. Through the medium of this language the Greeks produced a literature which has been the inspiration for many later writers and has commanded the admiration of all succeeding generations.

Although the Greeks had remarkable success in developing an alphabet, a qualitative language, and a literature, they did not produce a numeral system that could be widely used. It seems that as far as the Greeks got in this matter was to use letters of the alphabet to stand for numbers. To illustrate, with our present-day numerals written directly above the Greek letters, their scheme was as follows:

1	2	3	4	5	...
α	β	γ	δ	ϵ	...

The limitations of such a crude numeral system were severe. There was no zero, with its dual role of indicating both place value and the number of members in a set which has no members. Computation of even moderate complexity was impossible. Hence, the mathematical language of the Greeks was limited to geometry. You might ask: What difference did it make that the ancient Greeks could do only trivial arithmetic computations and had no algebra? Suffice it to remark, as one illustration, that the fundamental principles of physics, biology, and other basic sciences are the results of many measurements and much computation. This large-scale computation is possible only with an efficient computational system such as we now have available. Without the principles of physics one could not construct microscopes and x-ray machines, and modern medicine could not exist.

As the British writer Prof. Lancelot Hogben has remarked, "The most brilliant intellect is a prisoner within his own social inheritance." Even the intellectual giants among the ancient Greeks, Aristotle and Archimedes, were such prisoners. In this elementary mathematics course of the twentieth century you will do many things that they could not have done in their time.

Mathematics deals (among other things) with numbers, with counting, and with computing. In the operation of a retail store or factory much large-scale figuring is required. Such simple operations as measuring, weighing, and counting are fundamentally of a mathematical nature. Similarly the use of ratios, proportions, and percentages is mathematics. The techniques of counting, of computing, and of measurement have followed the trade routes and the business houses over the centuries.

Some persons consider "practical mathematics" to consist only in skill in arithmetic. For many individuals and for society as a whole, this surely is not true. For example, a knowledge of simple algebra, with some skill in its use, is essential to the study of many college subjects and to the holding of a considerable number of jobs in commerce, industry, and government.

Probably the word "equation" is more closely associated in people's minds with mathematics than any other word. In the language of mathematics the word equation plays a role similar to that played by the word "sentence" in a spoken and written language. Frequently an equation in a special form is spoken of as a "functional relationship," and sometimes as a "formula." Equations, functional relationships, and formulas occur frequently in courses in general physical science, physics, chemistry, and economics, and increasingly often in sociology and psychology. Watch for such instances, and note that each is a compact and highly useful summarization or digest of some important piece of knowledge. A statement is usually much more usable when expressed in a compact and concise formula or equation.

1.2
MATHEMATICS AS AN ABSTRACT SCIENCE

Any written language is constructed with the use of abstract symbols. To illustrate, the English language makes use of the 26 letters *a*, *b*, *c*, ... One of these letters by itself has no particular meaning. We could just as well write ¡ as *i*. Through usage and agreement a specific combination of these abstract alphabetic symbols may have come to stand for an idea, an object, or an action. Thus, the combination *c-h-a-i-r* conveys to our mind a familiar object. Once we have agreed to use the combination, or word, "chair" to stand for a well-known object, we must stick to this agreement. Further there are rules of grammar for combining words in acceptable patterns. What if a person forgets the meaning of a word or a rule of grammar? Such a person is not considered odd, but he is considered lazy or indifferent if he does not consult a dictionary to refresh his memory.

MATHEMATICAL DEFINITION

A *mathematical definition* is simply an arbitrary agreement on the meaning of a mathematical symbol or term or operation. In general when we define a mathematical term, that term will be repeated in the margin adjacent to the definition. If later on you forget a mathematical definition, with the aid of the index and the marginal notes you will be able to locate it easily and refresh yourself on it. Get into the habit of using this book on elementary mathematics in the same way that you would use a dictionary. The practice of refreshing yourself on an idea, a topic, or a procedure will pay dividends in helping you to accumulate knowledge of, appreciation of, and skills in the fundamentals of the language of mathematics.

Too frequently a person will interpret the statement "Mathematics is an abstract science" to mean that mathematics is a conglomeration of meaningless symbols which exist apart from any worthwhile human activity. But, in the basic sense as used here, we mean that mathematics is abstract in the same way any written language must be. In the English language the symbol *chair* has no meaning in itself. In mathematics the symbols 1, 2, 3, and so forth, of arithmetic and the x, y, z, and so forth, of algebra have no meaning in themselves. Just as we learn the meaning of the abstract symbols that form words in the English language, we learn the meaning of these symbols in mathematics; and in the same manner as we learn rules of grammar, we learn the rules, or laws, of operation with the numbers of arithmetic and with the literal symbols of algebra.

In addition to being abstract in the sense that any language is inherently so, mathematics is abstract in a different sense. Basic to the counting process, and hence basic to the development of the simplest "numbers," is the recognition that the sets illustrated below have a common characteristic.

We express this common characteristic by use of the number "four." We have come upon the notion of number as a *noun* rather than an *adjective;* the *number* four, rather than "four dots," or "four apples," or "four people," or "four houses." This process is the first level or lowest order abstraction in the language of mathematics, and mathematics is made up of a chain of abstractions of increasing complexity. For example, the arithmetic equality

$$3 + 5 = 8$$

may be thought of as relating to marbles, bananas, trees, people, and many other objects. The existence of equalities, such as $3 + 5 = 8$, which may apply to a variety of material objects, is of immense significance. Suppose we had to learn a different addition table for each and every kind of material object! We shall return to this abstraction that leads to the idea of number in Sec. 1.4.

It is the study of numbers and symbols apart from physical objects which they may represent that frequently gives the layman the impression that mathematics is weird stuff. But return to the arithmetic equality $3 + 5 = 8$. Note that the very power and significance of this equality is that it does not have any *necessary* connection with particular physical objects, and thereby its range of applicability is unlimited. Once we have straight in our minds a general (abstract) idea or relation, which can be applied to many things, we have an important tool at our command. Our thinking here leads us to this somewhat startling conclusion: *Mathematics is an important tool because it is a human activity concerned with abstract ideas and statements*. This has been aptly stated in these words: "*Only by an abstract statement can the field of application be completely unrestricted.*"

In the twentieth century we have come to understand that mathematics is a study of abstract systems and neither these systems nor the means provided by logic for studying their structural properties have any *necessary* connection with the physical world.

Let us consider what we mean by an *abstract mathematical system*. The indispensable ingredients of such a system are undefined terms, unproved propositions (postulates) that make statements about the undefined terms, definitions, and propositions (theorems) that are logical consequences of the postulates. The need for undefined (or primitive) terms in any language is easily recognized; it is apparent that a project that undertakes to define every word in a language is doomed to failure. Similarly, it is impossible to establish a chain of logical arguments in which *every* proposition is proved on the basis of preceding propositions. We sometimes hear a person say "Mathematics is the science in which everything is proved." This is most certainly *not so,* for in addition to the necessary presence of unproved propositions in a mathematical system, there are many terms and concepts that are matters of definition. One source of great difficulty and frustration for some students is an attempt to "prove" a statement that in reality is a definition. For example, just as we are not dismayed that we cannot "prove" that the word *dog* represents a relatively small four-legged animal that usually has a tail and barks, we should not become flustered when we are told that

$$4^3 = 4 \cdot 4 \cdot 4 \qquad \text{and} \qquad \frac{1}{2} \cdot \frac{3}{7} = \frac{3}{14}$$

are true "by definition" and that these statements are not subject to proof. As you study this book you should notice how the topics fit into systems or structures of the form:

Undefined (primitive) terms

Unproved propositions (postulates) that make statements about the undefined terms

Definitions

Propositions (theorems) that are logical consequences of the postulates

1.3 SETS AND SET OPERATIONS

In the preceding section we pointed out that in any language or in any abstract system it is necessary to begin with some undefined terms. These undefined terms may be either: (*i*) the *primitive* concepts of the language or system (concepts that are intuitively "understood" by the group using the system), or (*ii*) concepts that have little intuitive appeal or commonly "understood" meanings, the meanings of the terms being contained in what the postulates imply about the terms. The undefined terms that we shall encounter in this book will be mainly of the first type.

An example of an undefined term that is of the intuitive or primitive concept type and that is basic to all mathematics is the term "set." Any attempt to define the word set becomes involved in great difficulties and introduces concepts that are clearly neither so primitive nor so intuitive as that of set. So, **set** is taken as one of the fundamental undefined terms of mathematics. Words that are used synonymously with set are "collection," "aggregate," and "class."

SET

We shall consider that a set has been *specified* or *determined* when we are able to tell whether any given object is or is not in the set. Each of the following phrases specifies a set: the set of all chairs in a room; the set of all points on a line; the set of all Presidents of the United States taller than 7 feet; the set of all numbers whose square is 16; the set of all numbers that are one-half of 12.

MEMBER OF A SET

The individual objects that constitute a set are called **members** or elements of the set. If A is a specified set, then any object is either a member of A or not a member of A. To indicate that the set B has a, b, c, d for members, we write

$$B = \{a, b, c, d\}$$

The use of braces $\{\cdots\}$ will always indicate a set whose members either are tabulated or are clearly indicated between the braces. To indicate that an object a is a member of a set A, we write

$$a \in A$$

and to indicate that an object b is not a member of a set A, we write

$$b \notin A$$

We read these symbols, respectively, as "a is a member of A" and "b is not a member of A." To illustrate, for the set $B = \{a, b, c, d\}$ given above, we have

$$a \in B \qquad b \in B \qquad c \in B \qquad d \in B \qquad e \notin B \qquad f \notin B$$

In general, the order in which the members of a set are listed is not significant; further, it is understood that the members of a set are distinct. That is, the sets $\{a, c, b\}$ and $\{c, a, b\}$ are considered to be the same set; the collection a, b, c, d, b, c, a does not constitute a set with seven members, but rather a set with four members, namely, $\{a, b, c, d\}$.

ONE-TO-ONE CORRESPONDENCE

Perhaps one of the earliest uses of the concept of a set grew from noticing that sets could be compared in "size" by use of the concept of **one-to-one correspondence** between the members of two sets. This concept can be taken as another of the undefined terms that is basic to mathematics. We have an intuitive understanding of its meaning, and we generally describe this concept in the following manner: The members of two sets are in *one-to-one correspondence* if the members of the sets can be paired in such a way that each pair consists of one member from each set, all members of both sets are used, and no member of either set is used twice.

Suppose that a goatherd of long ago wanted to keep a record of his flock of goats. He could accomplish this in a most basic way if he transferred a pebble from a large pile to a bag every time a goat passed from the pen in the morning. The pebbles in the bag then would constitute a set whose members were in one-to-one correspondence with the members of the herd of goats. This set of pebbles could be put to several uses. For example, if the goatherd wished to compare the size of his herd with the size of another herd across the valley, he could do this without having to move either herd—all he would have to do would be to move a bag of pebbles. Also, in the evening he could remove a pebble from the bag every time a goat reentered the pen, and if, when the flock was in the pen, every pebble had been removed from the bag it would be apparent that no goat had been lost. In effect our mythical goatherd "counted" his goats. Note that this counting does not depend on the size or color of the individual goats, nor does it depend on the order in which they left or reentered the pen or on the way in which they were grouped as they left or reentered the pen.

In the illustration of the preceding paragraph we had a pairing of goats and pebbles. A pairing of the goats with sticks, notches, or scratches in the sand would work equally well. If the members of two sets are in one-to-one correspondence the two sets are said to have the *same number of members*. This number may be represented by pebbles, by sticks, by the fingers, by the toes, or simply by tally marks, |||. . . . The last was one of the earlier ways of representing the number of objects in a set, and like the method now prevalent, it facilitated (to some extent) the numerical comparison of sets of objects. The prevailing method of representing the number of things in a set differs from the primitive one merely in the substitution of the symbols

1, 2, 3, 4, 5, . . .

for the pebbles

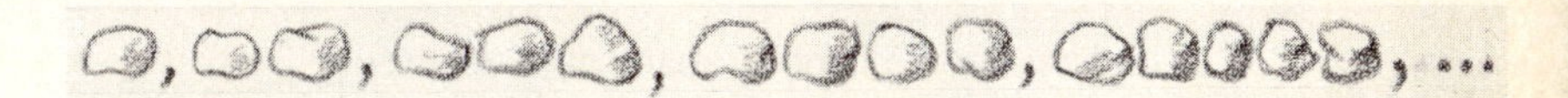

or for the tally marks

|, ||, |||, ||||, ~~||||~~, . . .

In these three representations the set of three dots $\cdots$ is read "and so forth" or "and so on," and is sometimes written "etc."

In addition to comparing the number of members in different sets by using the concept of one-to-one correspondence, we frequently compare sets in another way. If sets A and B have precisely the same members, the sets are said to be **equal** and we write

EQUAL SETS

$A = B$

If the sets are *not equal,* we write

$A \neq B$

To illustrate,

$\{a, b, c\} = \{b, c, a\} \qquad \{a, b, d\} \neq \{a, b, c\}$

If we determine that every member of set A is also a member of set B, we say that A is a **subset** of B, and we indicate this by writing

SUBSET

$A \subseteq B$

If A is *not* a subset of B, we write

$A \not\subseteq B$

To illustrate,

$\{a, b, c\} \subseteq \{a, b, c, d\} \qquad \{a, b, e\} \not\subseteq \{a, b, c, d\}$
$\{6, 8\} \subseteq \{2, 4, 6, 8, 10\} \qquad \{4, 5, 6, 7\} \not\subseteq \{4, 5, 6\}$
$\{4, 5, 6\} \subseteq \{4, 5, 6\} \qquad \{r, u\} \subseteq \{r, u\}$

In connection with the last two illustrations note that

$A \subseteq A$

for any set A, that is, every set is a subset of itself. Also it should be clear that

if $A \subseteq B$ and $B \subseteq A$, then $A = B$

If A is a subset of B and if there is at least one member of B that is *not* a member of A, then we say that A is a **proper subset** of B,

PROPER SUBSET

and we indicate this by writing

$A \subset B$

If A is *not* a proper subset of B we write

$A \not\subset B$

To illustrate,

$\{a, b, c\} \subset \{a, b, c, d\}$
$\{6, 8\} \subset \{4, 6, 8, 10\}$
$\{4, 5, 6\} \not\subset \{4, 5, 6\}$

We note that if A is a proper subset of B, then surely A is a subset of B.

Earlier in this section we mentioned as an example the set of all Presidents of the United States taller than 7 feet. This set is well determined, since if we consider any person, living or dead, we can tell whether or not the person is in this set. This particular set, of course, has no members and is an illustration of the empty set;

EMPTY SET

the **empty set** is the set that contains no members. The symbol $\varnothing$ will be used to denote the empty set. We follow the convention that the empty set is a subset of every set and it is a proper subset of every set except itself. That is, if A is a set, then

$\varnothing \subseteq A$

and

$\varnothing \subset A \qquad \text{unless} \qquad A = \varnothing$

Frequently from two given sets A and B new sets are constructed as described in the next two paragraphs.

If C is the set composed of the members of A together with the members of B, we say that C is the union of sets A and B; that

UNION OF SETS

is, the **union** of sets A and B is the set of objects that are members of *at least one* of the sets A and B. We indicate the union of sets A and B by writing

$A \cup B$

and we read these symbols as "the union of A and B" or "A union B." To illustrate,

$\{a, b, c\} \cup \{d, e\} = \{a, b, c, d, e\}$
$\{1, 3, 5\} \cup \{2, 4, 6, 8\} = \{1, 2, 3, 4, 5, 6, 8\}$
$\{2, 4, 6\} \cup \{1, 2, 3, 4\} = \{1, 2, 3, 4, 6\}$

In connection with the last illustration, note that if an object is a member of both sets A and B, it appears only once in the set $A \cup B$; so the number of members in the union of A and B is not necessarily the sum of the number of members of A and the number of members of B.

If D is the set composed of the members of A that are also members of B, we say that D is the "intersection" of A and B; that is, the **intersection** of sets A and B is the set of objects that are members of *both* A and B. We indicate the intersection of sets A and B by writing

INTERSECTION OF SETS

$$A \cap B$$

and we read these symbols as "the intersection of A and B" or "A intersection B." To illustrate,

$$\{a, b, c\} \cap \{b, c, e\} = \{b, c\}$$
$$\{1, 3, 5\} \cap \{1, 5, 9, 13\} = \{1, 5\}$$
$$\{1, 3, 5\} \cap \{2, 4, 6\} = \varnothing$$

It should be noted that, as in the last illustration, it is possible for the intersection of two sets to be the empty set. If this happens, that is, if $A \cap B = \varnothing$, we say that A and B are **disjoint sets.**

DISJOINT SETS

UNIVERSE

The set composed of all the objects which we wish to consider in a particular discussion is called the *universal set* or **universe.** Thus, each of the sets used in a particular discussion will be a subset of the universal set. For any set A, which is a subset of a given universal set U, we can consider the set A' composed of all the members of U that are not members of A. The set A' is the **complement of** A (*relative to* U). To illustrate, for $U = \{1, 2, 3, 4, 5, 6\}$ and $A = \{2, 4\}$ we have $A' = \{1, 3, 5, 6\}$; for $U =$ the set of all citizens of Alaska and $A =$ the set of persons less than 5 feet tall, we find that $A' =$ the set of all citizens of Alaska who are 5 feet tall or taller. Note that the complement of a set cannot be determined unless the universal set is known. As we shall see, the idea of universal set plays a very important role when we study the topics of variables and open sentences. The relationship between the universe U, a set A and its complement A', and the empty set $\varnothing$ is expressed by the following statements:

COMPLEMENT OF A SET

$$A \cup A' = U \qquad A \cap A' = \varnothing$$
if $A = U$, then $A' = \varnothing$
if $A = \varnothing$, then $A' = U$

EXERCISES

1 Determine which of the following statements are true and which are false:

(*a*) $3 \in \{1, 2, 3, 4\}$ (*b*) $3 \in \{4, 5, 6\}$
(*c*) $8 \notin \{1, 2, 3, 4\}$ (*d*) $a \in \{1, 2, 3, 4\}$
(*e*) $a \in \{a\}$ (*f*) $\{a\} \in a$

2 (*a*) Name four sets each of which is an illustration of the empty set.
(*b*) Is it correct to denote each of these sets by $\{0\}$, or by $\varnothing$?
3 (*a*) List the subsets of $\{1, 2\}$.
(*b*) Which of these are proper subsets of $\{1, 2\}$?
4 (*a*) List the subsets of $\{1, 2, 3\}$.
(*b*) Which of these are proper subsets of $\{1, 2, 3\}$?
5 List the subsets of $\{1, 2, 3, 4, 5\}$ which have exactly two members.
6 If $A = \{1, 2, 3, 4\}$, what are the sets B such that $\{1, 2\} \subset B$ and $B \subset A$?
7 Complete each of the following statements to give a correct answer:

(*a*) $\{1, 2, 3, 4\} \cup \{2, 3, 5\} =$ _______
(*b*) $\{1, 2, 3, 4\} \cup \{3, 4\} =$ _______
(*c*) $\{1, 2, 3, 4\} \cap \{2, 3, 5\} =$ _______
(*d*) $\{1, 2, 3, 4\} \cap \{5, 6, 7\} =$ _______

8 Complete each of the following statements to give a correct answer:

(*a*) $\{1, 3, 5, 7\} \cap \{2, 3, 4, 5, 6\} = \{\qquad\}$
(*b*) $\{1, 3, 5, 7\} \cup \{2, 3, 4, 5, 6\} = \{\qquad\}$
(*c*) $\{2, 6, 10, 14\} \cup \{2, 4, 5\} = \{\qquad\}$
(*d*) $\{2, 4, 6\} \cap \{6, 10, 14\} = \{\qquad\}$

9 For the sets $A = \{a, b, c, d\}$ and $B = \{b, d, e\}$, find $A \cap B$, $A \cup B$, $A \cap A$, $A \cup A$, $A \cap \varnothing$, $A \cup \varnothing$.
10 Give an argument intended to convince someone that for any sets A and B it is true that $A \cup B = B \cup A$, $A \cap B = B \cap A$, and $(A \cup B) \cup C = A \cup (B \cup C)$.
11 If $U = \{1, 2, 3, 4, 5, 6\}$ and $A = \{1, 2, 3\}$, $B = \{2, 3\}$, and $C = \{3\}$, give A', B', and C'.
12 If $U = \{1, 2, 3, 4, 5, 6, 7, 8, 9\}$, give the complement of each of the following sets: $A = \{1, 2, 3, 4\}$, $B = \{1, 3, 5, 7, 9\}$, and $C = \{2, 3, 6\} \cup \{1, 2, 5, 7\}$

1.4
COUNTING AND THE NATURAL NUMBERS

In the previous section we mentioned that basic to the "counting" of the objects in a set were the concept of one-to-one correspondence and the concept of two sets having the same number of members. We agree that if the members of two sets can be put in one-to-one correspondence, then the sets have the **same number of members.** Let us now regard (in our mind's eye) the collection of all sets that have the same number of members as the set of tally marks shown in Fig. 1.1. The *common property* that is shared

SAME NUMBER OF MEMBERS

FIGURE 1.1

FIGURE 1.2

FIGURE 1.3

FIGURE 1.4

ONE

by all the sets in this collection is called the *number one;* we say that the tally mark in Fig. 1.1 *represents* this number. We may think also of the collection of all sets that have the same number of members as the set of tally marks shown in Fig. 1.2. The *common property* of the sets in this collection is called the *number two;* we say that the tally marks in Fig. 1.2 *represent* this number. Similarly, the common property of all the sets with the same number of members as the set shown in Fig. 1.3 is called the *number three;* the common property of the sets with the same number of members as the set shown in Fig. 1.4 is called the *number four*. The tally marks in Fig. 1.3 and Fig. 1.4 *represent,* respectively, the numbers three and four. This process can be continued in the obvious way.

TWO

THREE

FOUR

In the procedure just described we have achieved an abstraction of the first level that was mentioned in Sec. 1.2, that is, we have considered a collection of sets (all those sets having the same number of objects as a specified nonempty set) and have "abstracted" the common property of these sets. This common property is called a **number,** and we give a name to each number that is associated with each of the collections of sets.

A NUMBER

NUMBER OF MEMBERS IN A SET

It is natural of course to say that the **number of members in a** (*nonempty*) **set** A is the number associated with the collection of all sets that have the same number of members as A. We say that we have *counted* the members of a set A when we have determined the number of members in A; that is, *counting* is the determination of the number of members in a set. Therefore, it is natural to call the numbers obtained as described in the preceding paragraph the *counting numbers*. Another name given to these numbers is the **natural numbers,*** and this is the name we shall use throughout this

COUNTING NUMBERS

NATURAL NUMBERS

*Some authors include the number zero (which we shall discuss later) in the set of natural numbers.

POSITIVE INTEGERS

chapter. Still another name for these numbers is the *positive integers.**

We said above that the tally mark in Fig. 1.1 represents the number one, the tally marks in Fig. 1.2 represent the number two, and so on. The prevailing method in most of the world for representing or denoting the natural numbers is by use of the symbols

$$1, 2, 3, 4, 5, 6, \ldots \tag{1.1}$$

NUMERALS

The written symbols used to represent or denote numbers are called **numerals.** The symbols given in (1.1) that we use to denote the natural numbers are called Hindu-Arabic numerals.

It should be clearly understood that numbers are inherently abstract mathematical ideas, and their existence does not depend on any particular choice of the numerals to represent them. We know from history that the Egyptians and Babylonians in the East developed their own systems of numerals, and archaeology has established the fact that the Incas and Mayans in the West did likewise. Very probably each ancient civilization had its own system of numerals. In Sec. 1.1 we mentioned a scheme that the Greeks used for symbolizing the natural numbers. Of all these ancient numeral systems, the only one that is used to any extent at present is the Roman numeral system. In case you have forgotten the Roman system, you may review it in Appendix A.

The fact that there are several systems of symbols, or systems of numerals, that can be used to represent numbers points up the fact that there is a distinction between number and numeral. The symbol, or numeral, that is used to represent, or denote, a number is *not* the number, just as a name used to denote a person is not the person. So, for example, 5 is not a number but rather a numeral which denotes a number. However, we do not insist on making this distinction explicit every time we use numerals to represent numbers, and in line with common usage we shall refer to the symbol as a number. That is, we speak of "the number 5" rather than "the numeral 5" or "the number represented by 5."

It is worth noting that in this section we have not given a definition of "a number." We have defined "same number of members"; "the number one," "the number two," etc.; "the number of members in a set"; the "counting numbers" or the "natural numbers"; but we have not said what we mean by "a number" in the most general sense. Such a definition will be given later after we have examined in detail some specific types of numbers.

* A conceptual distinction can be made between the counting numbers and the positive integers, but we shall not make such a distinction in this book.

1.5 ADDITION AND MULTIPLICATION OF NATURAL NUMBERS

What do we do with natural numbers besides using them to count with? For one thing we *add* natural numbers. Addition is almost as instinctive as counting; indeed it is essentially a counting process. Basically, addition of natural numbers is the determination of the number of members in the union of two disjoint sets when the number of elements in each of these sets is known.

Suppose that A and B are disjoint nonempty sets, and let a denote the number of members in A and b denote the number of members in B. Then the number of members in the set $A \cup B$ is called the **sum** of a and b and is denoted by $a + b$. Each of the numbers in the sum $a + b$ is called an **addend,** and the process of determining the sum $a + b$ is called **addition.** Addition is an example of an *operation* on the set of natural numbers (an operation on a set N associates a member of N with each ordered pair of members of N). In the case of addition, $a + b$ is the natural number* associated with the ordered pair a, b of natural numbers.

SUM

ADDEND

ADDITION

Addition of natural numbers is one of the earliest skills learned in elementary school, and during that period one usually masters the "addition table" for natural numbers. This addition table, together with certain rules relating to our place-value system of numeration (which we shall consider in Sec. 1.11), enables us to find the sum of any two natural numbers without counting the members in the union of two sets.

We are accustomed to speaking about two natural numbers being "equal" and about two numbers being "unequal." Let us agree on what we mean by these statements. Suppose that a and b are natural numbers, a being the number of members in set A and b being the number of members in set B. If the members of set A can be put in one-to-one correspondence with the members of set B, we say that a and b are **equal.** When a and b are equal we write

EQUAL NATURAL NUMBERS

$$a = b$$

and call this statement an equality, in which a is the *left side* and b is the *right side*. If no one-to-one correspondence can be established between the members of set A and the members of set B, we say that a and b are **unequal** or not equal. When a and b are not equal we write

UNEQUAL NATURAL NUMBERS

$$a \neq b$$

and call this statement an *inequality*.

*It should be clear that since the union of two nonempty sets is itself nonempty, the sum of two natural numbers is a natural number.

From our long experience with the natural numbers and the addition of natural numbers we are aware (perhaps almost subconsciously) of several properties possessed by the operation of addition on the set of natural numbers. One of these properties is expressed by saying that the sum of two natural numbers a and b is *unique.* The fact that the sum of two natural numbers is unique is formally expressed by the following statement in which a, b, c, d represent natural numbers:

UNIQUENESS PROPERTY OF ADDITION

If $a = b$ and $c = d$,
then $a + c = b + d$ (1.2)

This property is frequently described by the statement: *If equals are added to equals, the results are equal.*

Another familiar property of addition of natural numbers is illustrated by such statements as

$2 + 3 = 3 + 2$ and $16 + 29 = 29 + 16$

If a is the number of members in set A and b is the number of members in set B, we have from the definition of natural numbers that if A and B are disjoint,

$a + b$ is the number of members in $A \cup B$

and

$b + a$ is the number of members in $B \cup A$

It should be clear from the definition of the union of sets in Sec. 1.3 that $A \cup B = B \cup A$ and hence that $A \cup B$ and $B \cup A$ have the same number of members. So,

for any natural numbers a, b,
$a + b = b + a$ (1.3)

COMMUTATIVE PROPERTY OF ADDITION

The property* given in (1.3) is called the **commutative property of addition** (of natural numbers).

It is to be understood that a letter symbol appearing in an equality such as (1.3) must represent the same number wherever the symbol appears in that equality.

From experience we are aware of still another property of addition. We are aware that if we wish to add three or more natural numbers,†

*We do not propose to give a rigorous logical proof of the commutative property of addition. Rather we have given an argument intended to show the *plausibility* of the property. The associative property for addition, the commutative and associative properties for multiplication, and the distributive property will be dealt with similarly.

†The sum of three or more natural numbers may be defined in a manner analogous to that used to define the sum of two natural numbers.

we may group these numbers in any way that is most convenient. For example, if we wish to add 4, 3, and 2, we can add 4 and 3, getting 7, and then add 2 to 7, obtaining 9; or we can add 3 and 2, getting 5, and then add 4 to 5, obtaining 9. That is,

$$(4 + 3) + 2 = 4 + (3 + 2)$$

In general,

for any natural numbers a, b, c,
$$(a + b) + c = a + (b + c) \tag{1.4}$$

ASSOCIATIVE PROPERTY OF ADDITION

The property expressed by (1.4) is called the **associative property of addition** (of natural numbers). The fact that this property holds for the natural numbers follows immediately when we notice that the definition of the union of sets given in Sec. 1.3 implies that for any sets A, B, C,

$$(A \cup B) \cup C = A \cup (B \cup C)$$

There is another basic operation that we perform with natural numbers—the operation of *multiplication*. Multiplication was probably first explained to us in terms of addition. If we wish to determine the number of members in the union of five sets when each of the five sets contains seven members, we may use *addition,* $7 + 7 + 7 + 7 + 7 = 35$, or we may use *multiplication,* $5 \times 7 = 35$. In other words, the symbol 5×7, called the *product* of 5 and 7, is to be identified with the sum, $7 + 7 + 7 + 7 + 7$, of five sevens. We may extend this idea to make the following definition. Suppose that a and b are natural numbers, then the sum in which b is used as an addend a times is called the **product** of a and b and is denoted* by ab or $a \cdot b$ or $(a)(b)$; that is,

PRODUCT

$a \cdot b$ is the sum of a b's

or

$$a \cdot b = b + b + b + \cdots + b$$
where there are a addends in the sum (1.5)

In the product $a \cdot b$, the natural number a is called the *multiplier,* the natural number b is called the *multiplicand,* and the numbers a and b are called **factors** of the product. The process of determining the product $a \cdot b$ is called **multiplication.** In speaking of $a \cdot b$, we frequently say "a times b." Multiplication is of course an operation

FACTORS

MULTIPLICATION

*When letters are used to denote numbers, it is convenient to use the symbol ab for the product. However, when Hindu-Arabic numerals are being used, then we must use the symbolism $a \cdot b$ or $(a)(b)$; for example, the product of 5 and 7 must be written $5 \cdot 7$ or $(5)(7)$ and *not* as 57.

on the set of natural numbers, for it associates, with each ordered pair a,b of natural numbers, the natural number* $a \cdot b$.

Skill in multiplication of natural numbers is ordinarily attained in elementary school, and during that period one usually masters the "multiplication table" for natural numbers. This multiplication table, together with certain rules relating to our place-value system of numeration, enables us to find the product of any two natural numbers without making repeated additions or counting the members in the union of sets.

Just as our experience with natural numbers has made us aware of some basic properties of addition, so we are aware of similar properties possessed by the operation of multiplication. We express one of these properties by saying that the product of two natural numbers a and b is *unique*. This property is formally expressed by the following statement in which a, b, c, d represent natural numbers:

UNIQUENESS PROPERTY OF MULTIPLICATION

If $a = b$ and $c = d$,
then $a \cdot c = b \cdot d$ (1.6)

This property is frequently described by the words: *If equals are multiplied by equals, the results are equal.*

A second property of multiplication is illustrated by such statements as

$4 \cdot 5 = 5 \cdot 4$ and $16 \cdot 9 = 9 \cdot 16$

In thinking about the statement $4 \cdot 5 = 5 \cdot 4$, we could think $4 \cdot 5 = 5 + 5 + 5 + 5 = 20$ and $5 \cdot 4 = 4 + 4 + 4 + 4 + 4 = 20$, or we could consider the diagram shown in Fig. 1.5 in which there are four rows of five dots each. The problem of determining the number of dots in the figure can be looked at in two ways. One way the number would be $5 \cdot 4 = 4 + 4 + 4 + 4 + 4 = 20$. This procedure would be equivalent to counting the dots in the figure, beginning with the dots in the first *column* and continuing by columns until the dots in all the columns have been counted. But since counting the members of a set is independent of the order in which the members are taken, we will obtain the same number of dots if we count by *rows*. The latter procedure is symbolized by $4 \cdot 5 = 5 + 5 + 5 + 5 = 20$. So, $5 \cdot 4 = 4 \cdot 5$. Now if we let a and b be natural numbers, we can imagine a figure with a rows of b dots each, which is also a figure with b columns of a dots each. Since the order of

*Since the sum of natural numbers is a natural number, the definition (1.5) assures us that the product $a \cdot b$ of natural numbers is itself a natural number.

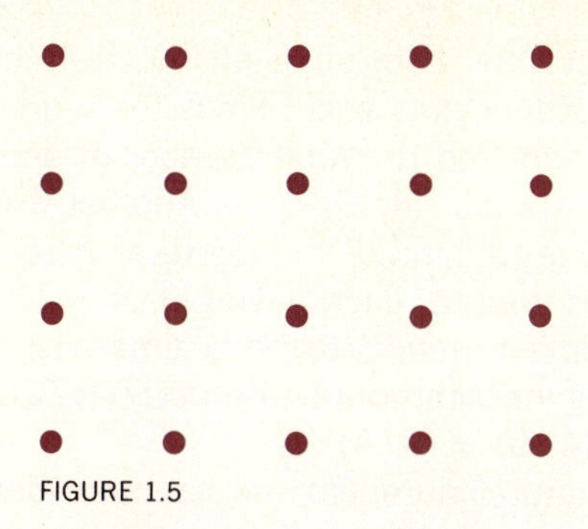

FIGURE 1.5

counting the dots does not affect the number of dots, we have a times b is equal to b times a. We indicate this result as follows:

For any natural numbers a, b,
$$a \cdot b = b \cdot a \tag{1.7}$$

COMMUTATIVE PROPERTY OF MULTIPLICATION

The property given in (1.7) is called the **commutative property of multiplication** (of natural numbers).

The definition of the operation of multiplication on the set of natural numbers given in (1.5) does not provide a meaning for the product of three or more natural numbers. What meaning can be given to the product $a \cdot b \cdot c$ of three natural numbers? We are accustomed to considering the product $3 \cdot 4 \cdot 5$ to be $(3 \cdot 4) \cdot 5$ or $3 \cdot (4 \cdot 5)$, and we are aware that we obtain the same result in either case: $(3 \cdot 4) \cdot 5 = 12 \cdot 5 = 60$ and $3 \cdot (4 \cdot 5) = 3 \cdot 20 = 60$. One way of thinking about the result $3 \cdot (4 \cdot 5) = (3 \cdot 4) \cdot 5$ is to consider the diagram given in Fig. 1.6. In this figure there are three tiers, or

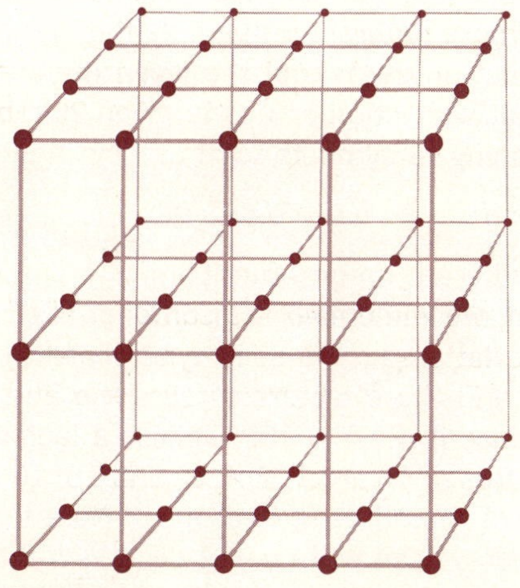

FIGURE 1.6

layers, of dots, with each tier consisting of four rows of five dots each, or five columns of four dots each. Since there are $4 \cdot 5 = 20$ dots in each layer, we can find the total number of dots by multiplying 3 times 20. This gives $3 \cdot (4 \cdot 5) = 60$. Another way of counting the dots in Fig. 1.6 is to think of the figure as made up of five slices (moving from left to right) each having $3 \cdot 4 = 12$ dots; then the total number of dots is found to be 5 times 12. This gives $5 \cdot (3 \cdot 4) = 60$, or, using the commutative property (1.7), $5 \cdot (3 \cdot 4) = (3 \cdot 4) \cdot 5 = 60$. So $3 \cdot (4 \cdot 5) = (3 \cdot 4) \cdot 5$.

Now if a, b, c are natural numbers, we can consider a similar figure with a tiers of dots, each tier consisting of b rows of c dots each, and so conclude:

For any natural numbers a, b, c,

$$a \cdot (b \cdot c) = (a \cdot b) \cdot c \tag{1.8}$$

ASSOCIATIVE PROPERTY OF MULTIPLICATION

The property expressed by (1.8) is called the **associative property of multiplication** (of natural numbers).

Notice that by using both the commutative and associative properties of multiplication we find

$$a \cdot (b \cdot c) = a \cdot (c \cdot b) = (a \cdot c) \cdot b$$

and so the associative property could be written as

$$a \cdot (b \cdot c) = (a \cdot b) \cdot c = (a \cdot c) \cdot b$$

Like results hold in case there are more than three factors; for example,

$$a \cdot [b \cdot (c \cdot d)] = a \cdot [(b \cdot c) \cdot d] = [(a \cdot b) \cdot c] \cdot d$$

The associative property of multiplication allows us to define the product of three or more natural numbers as the number obtained by finding the product of any two of the given numbers, then the product of this result with another of the given numbers, and so on, thus giving meaning to symbols such as

$$a \cdot b \cdot c \qquad a \cdot b \cdot c \cdot d \qquad a \cdot b \cdot c \cdot d \cdot e \quad \text{etc.}$$

where a, b, c, d, and e are natural numbers.

FACTORS

MULTIPLE

The numbers that are multiplied in a product are called **factors** of the product, and the product is said to be a **multiple** of each of its factors. Thus, 8 and 2 are factors of 16 since $2 \cdot 8 = 16$; also, 4 is a factor of 16 since $2 \cdot 2 \cdot 4 = 16$; 1 is also a factor of 16 since $1 \cdot 16 = 16$. The factors of 16 are 1, 2, 4, 8, and 16; 16 is a multiple of 1, and of 2, and of 4, and of 8, and of 16. Note that, since for n any natural number we can write $n = 1 \cdot n$, 1 *is a factor of every natural number* and every natural number is a multiple of 1.

PRIME NUMBER

If a natural number *different from* 1 has no factor except itself and 1, it is called a **prime number.** For example, the only factors of 2 are 2 and 1, so 2 is a prime number. Likewise, 3, 5, 7, 11, 13, 17, and 19 are prime numbers. You should satisfy yourself that this is so. Make careful note of the fact that, according to the definition of prime number, the number 1 is not prime. It is frequently important that we be able to find the prime factors of a given natural number. To illustrate, the prime factors of 15 are 5 and 3, for each of the numbers 5 and 3 is a factor of 15 and each is a prime number; 2 is the only prime factor of 16 since, as we saw above, the factors of 16 are 1, 2, 4, 8, and 16, and of these numbers only 2 is prime.

The factors of 20 are 1, 2, 4, and 5 since $20 = 4 \cdot 5 = 1 \cdot 2 \cdot 2 \cdot 5$ and 20 is a multiple of each of 1, 2, 4, and 5. The only prime factors of 20 are 2 and 5, and 20 can be written as a product involving *only* these prime factors: $20 = 2 \cdot 2 \cdot 5$. When a natural number is written as the product of *prime* factors it is said to be **factored completely.**

FACTORED COMPLETELY

EXAMPLE Factor each of the following numbers completely: (*a*) 90, (*b*) 16.

SOLUTION (*a*) We can write

$$90 = 2 \cdot 45 = 2 \cdot 5 \cdot 9 = 2 \cdot 5 \cdot 3 \cdot 3$$

and observe that the prime factors of 90 are 2, 5, and 3. Using these prime factors, 90 may be written as $90 = 2 \cdot 3 \cdot 3 \cdot 5$, and we have the complete factorization of 90. Notice that 90 is a multiple of each of 2, 3, 5, and 9.

(*b*) We noted above that the only prime factor of 16 is 2. Hence to factor 16 completely, we write $16 = 2 \cdot 2 \cdot 2 \cdot 2$.

It is true that except for the order of the factors, the complete factorization of any given natural number except 1 is unique; that is, except for the order of the factors a given natural number (other than 1) can be written as the product of prime factors in only one way. This statement would not be true if 1 were included in the set of prime numbers.

EVEN NATURAL NUMBER

If a natural number is a multiple of 2, or in other words, if 2 is a factor of a natural number, we say that the number is **even.** Thus, 2, 4, 6, 8, 10, and 12 are even, but 1, 3, 5, 7, 9, and 11 are not even. If a natural number is *not* a multiple of 2, or in other words, if 2 is *not* a factor of a natural number, we say that the number is **odd.** For example, 1, 3, 5, 7, 9, and 11 are odd. Notice that if n is any natural number, then $2n$ is even and $2n + 1$ is odd.

ODD NATURAL NUMBER

EXERCISES

In each of Exercises 1 and 2, verify the associative property of multiplication, $a(bc) = (ab)c$, for the given values of a, b, and c.

1 $a = 4$, $b = 5$, $c = 6$.

2 $a = 8$, $b = 7$, $c = 3$.

3 List all the prime natural numbers which are members of the set $\{1, 2, 3, \ldots, 30\}$.

4 List all the prime natural numbers which are members of the set $\{30, 31, 32, \ldots, 50\}$.

5 Factor each of the following into prime factors: 8; 11; 14; 18; 24; 29.

6 Factor each of the following completely: 15; 17; 21; 45; 39; 42.

7 What prime factors do the numbers 21, 42, and 56 have in common?

8 What prime factors do the numbers 30, 210, and 770 have in common?

In each of Exercises 9 to 12, the given number is a multiple of what natural numbers?

9 93.

10 210.

11 315.

12 267.

1.6 THE DISTRIBUTIVE PROPERTY

In addition to the commutative and associative properties possessed separately by the operations of addition and multiplication in the set of natural numbers, there is a property that involves both addition and multiplication and which gives emphasis to the fact that one must use care in combining these two operations. For example,

$$3 \cdot (4 + 5) = 3 \cdot 9 = 27$$

and

$$3 \cdot 4 + 3 \cdot 5 = 12 + 15 = 27$$

so

$$3 \cdot (4 + 5) = 3 \cdot 4 + 3 \cdot 5 \tag{1.9}$$

On the other hand,

$$3 + (4 \cdot 5) = 3 + 20 = 23$$

and

$$(3 + 4) \cdot (3 + 5) = 7 \cdot 8 = 56$$

so

$$3 + (4 \cdot 5) \neq (3 + 4) \cdot (3 + 5) \tag{1.10}$$

The result in (1.9) illustrates the exceedingly important property possessed by addition and multiplication in the set of natural numbers that can be stated as follows:

$$\textit{For any natural numbers } a, b, c, \quad a(b + c) = ab + ac \tag{1.11}$$

DISTRIBUTIVE PROPERTY

The property expressed by (1.11) is called the **distributive property of multiplication with respect to addition** (in the set of natural numbers). The multiplication in $a(b + c)$ can be "distributed," or spread out, over the two numbers that are to be added. A geometrical interpretation of the property (1.11) can be obtained by considering Fig. 1.7. Recall that the number of square units of area in a rectangle is equal to the product of the number of units of length and the number of the same units of width. Let the number of units in the dimensions of the rectangles in Fig. 1.7 be as indicated. Then the areas of the two smaller rectangles are ab and ac, and the area of the larger rectangle is $a(b + c)$. But the area of the larger one is equal to the sum of the areas of the smaller ones; that is, $a(b + c) = ab + ac$.

The result in (1.10) illustrates the fact that addition is *not* distributive with respect to multiplication.

Because the equations in (1.3), (1.4), (1.7), (1.8), and (1.11) are true when the letters in the equations represent *any* natural numbers, we frequently say that these equations are *identities* in the set of natural numbers. By using the distributive property and the commutative and associative properties of addition and multiplication, other identities in the set of natural numbers can be proved. Let us prove that

$$\textit{for any natural numbers } a, b, c, \quad (b + c)a = ba + ca \tag{1.12}$$

This theorem is almost like the distributive property (1.11), for it shows that we can "multiply out" from the right as well as from

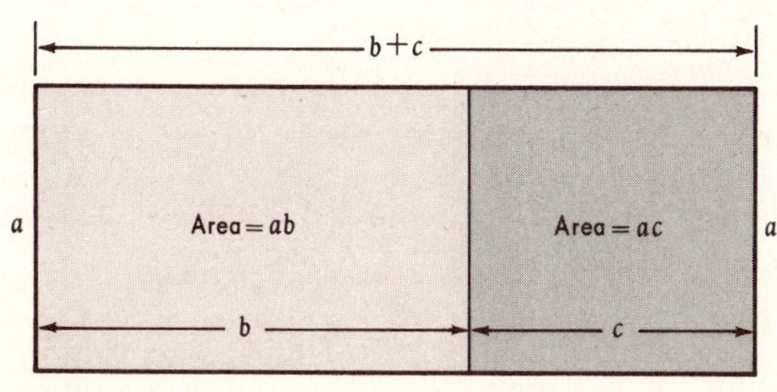

FIGURE 1.7

the left. Now for its proof. We may arrange the steps as follows, with the justification for each step indicated in the column on the right. We have, for any natural numbers a, b, c

$(b + c)a = a(b + c)$	*commutative property of multiplication*
$= ab + ac$	*distributive property*
$= ba + ca$	*commutative property of multiplication*

Thus (1.12) is proved.

The properties (or identities) given in (1.3), (1.4), (1.7), (1.8), (1.11), and (1.12) are true for *any* natural numbers a, b, c. As a consequence of this fact we can obtain other identities from these properties by replacing the symbols which appear in them by combinations of symbols that represent natural numbers. To illustrate, since for any natural numbers a and b, the sum $a + b$ is a natural number, we have, by the use of the distributive property, that

for any natural numbers a, b, c, d,
$(a + b)(c + d) = (a + b)c + (a + b)d$

Also, by (1.12)

$$(a + b)c + (a + b)d = ac + bc + ad + bd$$

So we have

$$(a + b)(c + d) = ac + bc + ad + bd$$

It is very important that we understand the distributive property well, for it justifies several mechanical processes in arithmetic and is a basis for much of algebra. Some of its uses in algebra will be discussed in the next section, and some of the mechanical processes in arithmetic for which it is a justification will be illustrated in Sec. 1.12.

Relative to problems that involve addition and multiplication, we adhere to the convention that in such a problem the *multiplications will be performed before the additions except where otherwise indicated by parentheses.* To illustrate,

$$2 + 4 \cdot 5 + 3 = 2 + 20 + 3 = 25$$

EXERCISES

In each of Exercises 1 to 4, verify the distributive property $a(b + c) = ab + ac$ for the given values of a, b, and c.

1 $a = 3$, $b = 5$, $c = 6$.
2 $a = 10$, $b = 9$, $c = 8$.
3 $a = 21$, $b = 3$, $c = 2$.
4 $a = 20$, $b = 30$, $c = 40$.

In each of Exercises 5 to 12, find the value of the given expression.

5 $5(6 + 3)$.
6 $(4 + 3)(2 + 3)$.
7 $3 + 5 \cdot 4 + 6$.
8 $6 \cdot 5 + 6 \cdot 9$.
9 $4 + 3(7 + 2)$.
10 $4 \cdot 5 + 6$.
11 $6(5 + 8)$.
12 $(4 + 7)5$.

13 Prove that for all natural numbers a and b, $(a + b)(a + b) = aa + 2ab + bb$.

14 Prove that for all natural numbers a and b, $(a + b)(a + b)(a + b) = aaa + 3aab + 3abb + bbb$.

15 Give a definition of the sum of three natural numbers a, b, and c, and give a definition of the sum of four natural numbers a, b, c, and d.

1.7 ALGEBRAIC SYMBOLS AND ADDITION

As we have pointed out, mathematics is concerned with the *abstracting,* or *generalizing,* of ideas. The development of the notion of *natural numbers* and the fundamental operations with these numbers as used in arithmetic are first steps in that generalization process. In Chaps. 2 and 3 we shall study how the notion of number is extended to include zero, then the negative integers, and then the fractions. All these make up a system of numbers, called *rational numbers,* which is more general than the system of natural numbers.

Another type of generalization in mathematics is algebra. In algebra the letters of our alphabet are used to represent numbers: when we use them in this manner we call them *letter symbols.* Elementary, or classical, algebra is that branch of mathematics which deals with the properties of numbers by using letter symbols, signs of operation, and other special symbols. Algebra developed slowly over a long period, as did arithmetic. This is not surprising, since algebra encompasses arithmetic and goes beyond it in the study of the properties of numbers. It was not until the seventeenth century that algebra reached a stage nearing perfection in its use of letter symbols and other special symbols.

LETTER SYMBOLS

We emphasize that the letter symbols of algebra represent numbers, just as the symbols 1, 2, 3, and so on, represent numbers. However, algebra uses letter symbols for numbers in a sense quite different from that used by the Greeks. For the Greeks used a letter of their alphabet to stand for a particular number, whereas algebra uses a letter symbol to stand for *any* one of an entire set of numbers. In fact, we have already used some algebra. When we symbolized the commutative property of addition of natural numbers by writing

$$a + b = b + a$$

for any natural numbers a and b, we made use of algebra.

The algebraic statement $a + b = b + a$ is an example of an identity, since it is true for any natural numbers. Of course not all algebraic statements are identities; for example, the equation

$$n + 5 = 14$$

where n represents a natural number, is *not* an identity. This equation is called a *conditional* equation because it is true only under a certain condition, namely, when 9 is put in place of n.

We emphasize that in algebra a letter symbol is used in such a way that it may represent any number in a given set of numbers. In the statement $a + b = b + a$, a and b may represent *any* natural numbers, and the statement is true when a is replaced by any natural number and b is replaced by any natural number. In the statement $n + 5 = 14$, n may represent *any* natural number, although the statement is true only when 9 is put in place of n. Because of this property of a letter symbol, it is frequently called

VARIABLE

a variable. A **variable** is a symbol that represents any member of a given set. The given set is called the *universal set* of the variable

UNIVERSE

or the **universe** of the variable, and each member of the universe is a *value* of the variable. For example, if x is a variable whose universe is the set $\{1, 3, 5, 7, 9, 11, 13\}$, then x has the values 1, 3, 5, 7, 9, 11, 13. In other words, x may be replaced by any odd natural number less than 14. For this reason it is often said that a variable is a *placeholder* for any member of its universe.

When we express the commutative property of addition of natural numbers by writing the equation

$$a + b = b + a$$

a and b are variables each having the set of natural numbers as its universe. This equation is *true* for all members of the universe of the variables. If the set of natural numbers is the universe of the variable n, the equation

$$n + 5 = 14$$

is *true* for only one member of the universe, namely, the natural number 9.

If a set has only one member, a symbol for that member is called

CONSTANT

a **constant**; in other words, a constant is a symbol for a fixed element of the universal set. For example, 3 is a constant, since it is a symbol

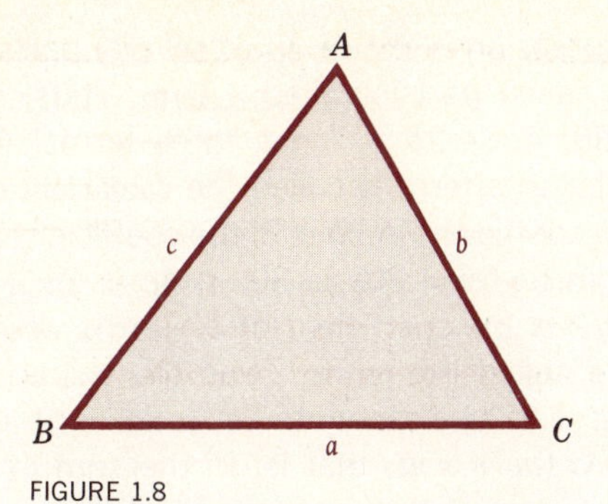

FIGURE 1.8

for the member of the set whose only member is the number 3.

Recall the definition: *The perimeter of a triangle is the sum of the lengths of the three sides of the triangle.* Suppose that we wish to translate this verbal statement into algebraic language by use of letter symbols, or variables. Let (see Fig. 1.8)

p be the perimeter of the triangle

a be the length of one side

b be the length of a second side

c be the length of the third side

Then

$$p = a + b + c$$

Obviously, this algebraic statement is more concise than the verbal statement to which it refers. This may be an advantage in itself, but it is not all there is to algebra. An important use of algebra results from the fact that an equation or an inequality that contains a variable can be transformed or operated on so that the value or values of the variable which make the equation or inequality true can be determined. The mere act of writing a statement in algebraic form may enable us to solve a problem which would be humanly impossible otherwise.

If letter symbols and numbers are combined by the use of the four fundamental operations of algebra (addition and multiplication, which were introduced in Sec. 1.5, and subtraction and division which will be introduced in Sec. 1.9), we have an **algebraic expression.** Thus

ALGEBRAIC EXPRESSION

$$a + 3b \qquad 4a + 2b + 7 \qquad \text{and} \qquad 6x + 5$$

are illustrations of algebraic expressions obtained by the use of the operations of addition and multiplication.

TERM

COEFFICIENT

LIKE TERMS

If an algebraic expression is made up of several parts connected by the plus sign +, each part is called a **term.** Thus, $a + 3b$ has two terms, a and $3b$; $4a + 2b + 7$ has three terms, $4a$, $2b$, and 7. The constant factor in a term is called the **coefficient** of the letter part of the term. The coefficient in $3b$ is 3; the coefficient in $1 \cdot a = a$ is 1; the coefficient in $5 \cdot 6x = 30x$ is 30; in $4a + 2b + 7$, 4 is the coefficient of a and 2 is the coefficient of b. Terms which have the same letter part are called **like terms.** Thus the terms in $2a + 3a$ are like terms. Using (1.12) we can write $2a + 3a = (2 + 3)a = 5a$. Similarly, $ma + na = (m + n)a$; that is, in the sum of like terms, the coefficient is the sum of the coefficients in the addends, and the letter part is the same as the letter part in each addend. For example,

$4a + 3a = (4 + 3)a = 7a$ *by the distributive property* (1.12)

Also

$$\begin{aligned}(2a + 6b + c) + (5a + b) &= 2a + 5a + 6b + b + c \\ &\quad \textit{by the commutative and associative} \\ &\quad \textit{properties of addition} \\ &= (2 + 5)a + (6 + 1)b + c \\ &\quad \textit{by the distributive property } (1.12) \\ &= 7a + 7b + c\end{aligned}$$

In an expression such as $(2a + 6b + c) + (5a + b)$ only like terms can be added, giving $7a + 7b + c$ as the result. The combination of unlike terms can only be indicated. Notice the similarity between an indicated algebraic sum, such as $6a + 4b$, and the problem of finding the "sum" of 6 feet and 4 yards.

MULTINOMIAL

BINOMIAL

TRINOMIAL

MONOMIAL

A sum of two or more terms in algebra is called a **multinomial.** When there are just two terms in a sum, as in $a + b$ or $x + 5y$, we say it is a **binomial.** An expression with three terms, as $a + 4b + 2c$, is called a **trinomial.** When an expression is made up of only one term, for example, $3xy$ or $4b$, it is called a **monomial.**

EXERCISES

In the following exercises we assume that the letter symbols represent natural numbers. In each of Exercises 1 to 10, add the terms that are alike, and indicate the sum of unlike terms.

1 $\begin{array}{r}4a\\ \underline{7a}\end{array}$

2 $\begin{array}{r}c\\ \underline{c}\end{array}$

3 $\begin{array}{r}3b\\ \underline{2a}\end{array}$

4 $\begin{array}{r}y\\ \underline{10y}\end{array}$

5
$$\begin{array}{r} 6a + b + 3c \\ \underline{2a + 5b + c} \end{array}$$

6
$$\begin{array}{l} 2a + 3b + 4c \\ \underline{2a + 5b \qquad + 2d} \end{array}$$

7 $3a + 7a + 2a$

8 $16z + 4z + z$

9 $3aa + 7a + 2a$

10 $16aa + 2aa + 7b$

The *perimeter* of a closed plane figure is defined to be the sum of the lengths of its sides. In Exercises 11 to 15, draw a figure, and write a formula for the perimeter p for each of the figures, using as few terms as possible.

11 A square of side s.

12 A rectangle of length l and width w.

13 An equilateral triangle (a triangle with sides of equal length) with side s.

14 A regular pentagon (a plane figure with five sides of equal length) with side d.

15 A regular octagon (a plane figure with eight sides of equal length) with side c.

16 The lengths of the sides of a lot, which is in the shape of a quadrilateral (a figure with four sides), when expressed in feet are $2a + 3$, $a + 7$, $2a + 9$, and $5a + 6$. Write a formula for the perimeter p of the lot.

17 Obtain a formula for the distance d above the ground of a flag that is on a pole $4a$ feet high, which is set on a building $3a + 7$ feet high (see Fig. 1.9).

18 A rectangle is three times as long as it is wide. Express the perimeter p of the rectangle in terms of its width w.

19 What is the perimeter of a triangle, the lengths of whose sides are $4x$, $3x$, and $6x$?

20 What is the perimeter of a rectangle whose length and width are $7x$ and $2x + y$, respectively?

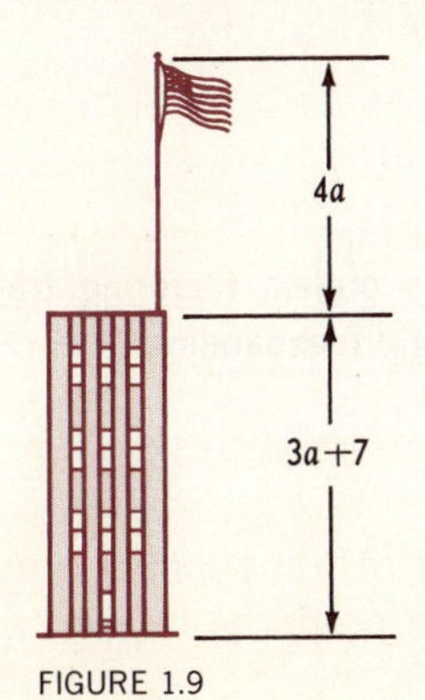

FIGURE 1.9

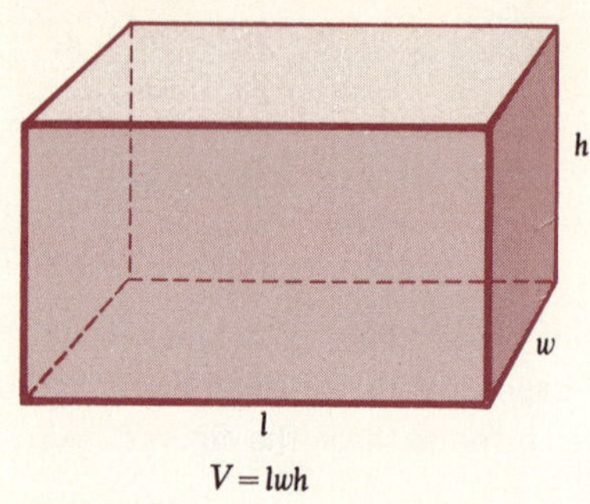

FIGURE 1.10

21 What are the terms in the expression $a + 3b + 6c$? In the expression $7a + 2cd$? In the expression $3abc$?

22 What is the coefficient in each of the following expressions: x, $3x$, $2 \cdot 4x$, and $(5 + 6)x$?

23 What are the terms in the expression $7x + 3y + 2z$? What is the coefficient in each term?

24 What are the terms in the expression $2xy + 7y + 13yz + 6z$? What is the coefficient in each term?

25 If $a = 4$, find the value of $6a + 5$; of $2a + 6$.

26 If $a = 2$ and $b = 5$, find the value of $5a + 2b$; of $2a + 3b$.

27 If $a = 3$, $b = 5$, and $c = 6$, what is the value of $4abc$?

28 The number of cubic units of volume V of a rectangular box is given by the formula $V = lwh$, where l is the number of units of length of the box and w and h are, respectively, the number of units of width and height (see Fig. 1.10). Here l, w, and h are all measured in the same unit. If l, w, and h are measured in inches, the volume will be in cubic inches. If feet or yards are used, then the volume will be expressed in cubic feet or cubic yards, respectively. Find V when $l = 3$ inches, $w = 2$ inches, and $h = 9$ inches; $l = 7$ feet, $w = 9$ feet, and $h = 13$ feet; $l = 6$ yards, $w = 8$ yards, and $h = 7$ yards; $l = 24$ inches, $w = 4$ feet, $h = 2$ yards (express V in cubic feet).

1.8 MULTIPLICATION IN ALGEBRA

FACTORING

EXPANDING

Expressing a sum as a product is called **factoring** the sum, and writing a product as a sum is called **expanding** the product. The distributive property, in the form

$$a(b + c) = ab + ac$$

gives a rule for writing the product of a monomial and a binomial as a sum. Similarly, the distributive property, in the form

$$ab + ac = a(b + c)$$

gives a procedure for factoring a suitable sum into the product of a monomial and a binomial. The distributive property is the basis for a large number of rules and identities that are concerned with "removing parentheses," "factoring," and "combining like terms." You should learn to think in terms of the distributive property, rather than memorize a host of rules and formulas for doing the things just mentioned.

The following are some illustrations for the use of the distributive property:

In factoring,

$$3x + 3y = 3(x + y)$$
$$2ax + 4ay = (2a)x + (2a)2y = 2a(x + 2y)$$

In expanding,

$$7(a + x) = 7a + 7x$$
$$4a(r + 5s) = 4ar + 20as$$

In removing parentheses,

$$6a + 4(x + 2a) = 6a + 4x + 4(2a) = 6a + 4x + 8a = 14a + 4x$$
$$2a(x + 1) + 3x(a + 2) = 2ax + 2a + 3xa + 6x = 5ax + 2a + 6x$$

Observe that the identity

$$(a + b)(c + d) = ac + bc + ad + bd \tag{1.13}$$

which was established in Sec. 1.6, gives a procedure for writing the product of two binomials as a sum, and also a procedure for writing an appropriate sum as the product of two binomials. A form of (1.13) that is of frequent occurrence is

$$(x + a)(y + b) = xy + ay + bx + ab \tag{1.14}$$

For example,

$$(x + 2)(y + 5) = xy + 2y + 5x + 10$$

If y is replaced by x in (1.14), we have $(x + a)(x + b) = x \cdot x + ax + bx + ab = x \cdot x + (a + b)x + ab$. It is customary of course to write $x \cdot x$ as x^2, and then this last identity becomes

$$(x + a)(x + b) = x^2 + (a + b)x + ab \tag{1.15}$$

You should check through this in detail and note the various properties that were used in order to obtain the final form.

Either by use of (1.15) or by direct multiplication we have

$$(x + 2)(x + 3) = x^2 + 5x + 6$$

Suppose that we were faced with the problem of factoring $x^2 + 6x + 8$ into the product of two binomials. The first terms of the binomials are factors of x^2, so they are x and x. We write

$$x^2 + 6x + 8 = (x + \quad)(x + \quad)$$

Now, according to the identity (1.15), the second terms with which we are to fill the parentheses must be such that their sum is 6 and their product is 8. Clearly they must be 2 and 4, and so we conclude that

$$x^2 + 6x + 8 = (x + 2)(x + 4)$$

as can be checked by finding the product on the right side.

In the discussion of (1.15) we replaced the product $x \cdot x$ by the symbol x^2. This is an example of the use of the notation of *exponents*. In case the factors of a product are all the same, the factor is written with a superscript integer to the right. This integer tells the number of times the factor appears, and it is called an **exponent.** Then this one factor together with its exponent is used in place of the product of the repeated factors. Thus

EXPONENT

$$a \cdot a = a^2 \qquad a \cdot a \cdot a = a^3 \qquad 2^4 = 2 \cdot 2 \cdot 2 \cdot 2$$

In general,

if n is a natural number,

$$a^n = a \cdot a \cdot a \cdots a \qquad n \text{ factors } a \tag{1.16}$$

that is, a is used as a factor n times. We call a^n the **nth power of** a. Thus a^2 is the second power of a; this is often read "the square of a" or "a squared." Similarly, a^3 is the third power of a; it is often read "the cube of a" or "a cubed." In a^n, a is the **base,** and n is the exponent. If the exponent is 1, it is usually understood and not written; thus $a^1 = a$ and $3^1 = 3$. Different powers of the same number constitute unlike terms; hence we can only indicate their sum as, for example, $x + x^2$. For reasons that will be discussed in Chap. 3, it is very convenient to give a meaning to the use of the symbol 0 as an exponent. At this point we shall simply make the **definition** that

nth POWER

BASE

if a is a natural number, then

$$a^0 = 1$$

A discussion of exponents that are not natural numbers will be given in Chap. 3. In the meantime we shall have occasion to use the definition $a^0 = 1$ in Sec. 1.11.

EXERCISES

1 Expand the following by using the distributive property: $2(b + c)$; $a(3c + 4d)$; $(2 + a)(4 + b)$.

2 Prove the following by use of the identities previously established:

$$a(b + c + d) = ab + ac + ad$$
$$a(b + c + d + e) = ab + ac + ad + ae$$
$$(a + b)(c + d + e) = ac + bc + ad + bd + ae + be$$

3 Factor each of the following sums into the product of a monomial and a binomial: $7by + 21bx$; $6ab + b$; $20xy + 15$.

4 By use of the identity (1.14), find the following products: $(x + 3)(y + 4)$; $(x + 5)(y + 6)$; $(x + 3a)(y + 2b)$.

5 By use of the identity (1.14), factor each of the following sums into the product of two binomials: $xy + 3x + 2y + 6$; $xy + 5x + 4y + 20$; $xy + 3sx + 2ry + 6sr$.

6 A field in the shape of a rectangle is 205 rods long and 91 rods wide. How many rods of fence will be required to enclose the field?

7 When one man has worked 1 hour, we say that one man-hour of work has been performed. It takes 13 men 85 hours working together to complete a job. If the rate of pay is $3 per man-hour, what is the total labor bill?

8 A room is 18 feet wide and 24 feet long. Find the cost of covering the floor of the room with linoleum which costs $3 per square yard.

9 On the floor of a room which is 20 feet long and 15 feet wide, there is a rug whose dimensions are 12 feet by 10 feet. How many square feet of floor space are not covered by the rug?

10 A cellar for a new building is to be 72 feet long and 38 feet wide, and the dirt is to be removed to a depth of 12 feet. How many cubic feet of dirt are to be removed?

In Exercises 11 to 18, remove the parentheses in the given expression, and simplify the result by combining any like terms.

11 $3b + 3(2)$

12 $3(b + 2)$

13 $2(x + 3) + x$

14 $x(2 + y) + 4x$

15 $a + (bc) + d$

16 $(a + b)c + d$

17 $2(x + y) + 3(x + y)$

18 $a + 3(a + b)$

In Exercises 19 to 26, find the value of the given expression if $a = 2$, $b = 3$, and $c = 4$.

19 $2a + 3b$

20 $2(a + b) + 3$

21 $(2)(3)(a + b)$

22 $a + bc$

23 abc

24 $a + b + c$

25 $ab + bc + ac$

26 $a(b + c)$

In Exercises 27 to 30, find an expression for the area of the rectangle whose length l and width w are given.

27 $l = 2x,\ w = 3y$
28 $l = 2 + x,\ w = 3 + y$
29 $l = 2a + b,\ w = 3c$
30 $l = 6a,\ w = 2b + 3c$

In Exercises 31 to 34, find an expression for the volume of a rectangular box whose length l, width w, and height h are given. Expand all products.

31 $l = 2,\ w = 3a,\ h = 4b$
32 $l = 2c,\ w = 3a,\ h = 4b$
33 $l = 2 + c,\ w = 3a,\ h = 4b$
34 $l = 2 + c,\ w = 3 + a,\ h = 4 + b$

In Exercises 35 to 40, factor the given expression, and check your work by finding the product of the factors.

35 $ab + ac + a$
36 $6x + 6ax + axy$
37 $7x + 7$
38 $3xy + 6ax + 18xu$
39 $9a + 3ab$
40 $15x + 5y + 10z$

In Exercises 41 to 43, write a formula for the quantity described.

41 The volume V, in cubic feet, of a rectangular box which is a feet long, 60 inches wide, and b feet high.
42 The value V, in cents, of x quarters and y dimes.
43 The distance D a person travels who goes by airplane for x hours at 275 miles per hour and by car for y hours at 45 miles per hour.

In Exercises 44 to 47, translate the given sentence into an algebraic equation.

44 Seven times a number x is equal to x plus three times y.
45 Five times the sum of the two numbers x and y is equal to 4.
46 Three times a number x plus four times the same number is equal to 14.
47 The sum of three consecutive integers, the first of which is x, is equal to 60.

48 Write each of the following products in brief form by using exponents: aaa; $2xxy$; $3 \cdot 3xyyy$; $rssttt$; $4 \cdot 4 \cdot 4$.
49 Express each of the following as a product, without exponents, of the variables involved: x^2y^3; a^4b^5; $2ay^2$; $(x^2y^3)(x^3y^2)$.
50 Simplify by combining like terms: $2x + 3x^2 + x$; $4y^3 + y^3 + y$; $2a^3 + a^2 + a + 4a^2$.
51 Multiply and combine like terms when possible: $(x + 1)(x + 2)$; $(x + 2)(x + 5)$; $(2x + 3)(x + 1)$; $(x + y)(x + 2y)$.
52 Factor each of the following expressions into the product of two binomials. In each case check your work by finding the product of the factors. $x^2 + 7x + 12$; $x^2 + 6x + 9$; $y^2 + 11y + 24$; $a^2 + 6a + 5$.

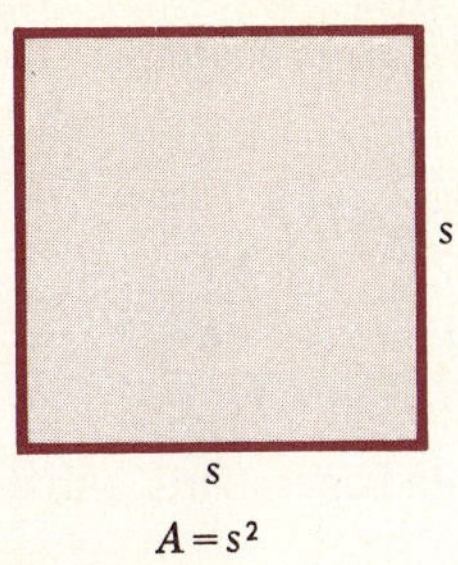

$A = s^2$

FIGURE 1.11

53 If s is the number of units of length in the side of a square, the number of square units of area A in the square is given by the formula $A = s \cdot s$, or $A = s^2$ (see Fig. 1.11). Find the area of the square whose side has the length of 7 inches; 15 feet; 29 yards; 19 miles.

54 If the length of the side of a square is doubled, what is the effect upon its area? Illustrate, using numerical values.

55 If e is the number of units of length in the edge of a cube (see Fig. 1.12), the number of cubic units of volume V of the cube is given by the formula $V = e \cdot e \cdot e$ or $V = e^3$. Find the volume of the cube whose edge has the length of 5 inches; 17 feet; 21 yards; 4 miles.

56 If the length of the edge of a cube is doubled, what is the effect upon its volume? Illustrate, using numerical values.

57 The cube with edge e units long (see Fig. 1.12) has six plane faces, each of which is a square. What is the area of each of these squares? What is the sum of the areas of these six plane faces?

58 Calculate the sum of the areas of two squares if the side of one square is 7 inches long and the side of the other square is 9 inches long.

59 Calculate the sum of the volumes of two cubes if the edge of one cube is 2 inches long and the edge of the other is 3 inches long.

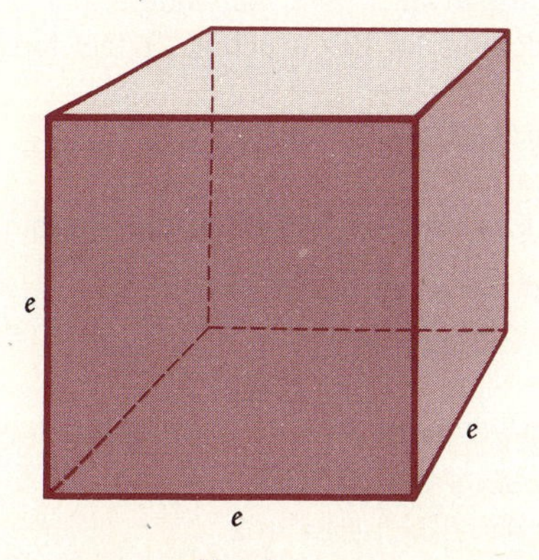

$V = e^3$

FIGURE 1.12

In Exercises 60 to 67, find the value of the given expression if $a = 2$, $b = 3$, and $c = 4$.

60 ab^2
61 $(ab)^2$
62 $(ab)^2c$
63 $(a + b)^2$
64 $a^2 + b^2$
65 $a(b + c)^2$
66 $ab^2 + c$
67 $(abc)^2$

In Exercises 68 to 73, perform the indicated operations, and write the result with as few terms as possible.

68 $2x(3x + 5) + 7x$
69 $a(b + 2c) + 3a(b + c)$
70 $2a(a^2 + 4)$
71 $a(b + c) + b(a + c) + c(a + b)$
72 $a^2(b + 2a) + 4a^3$
73 $3x^2 + 2x(x + 4)$

In Exercises 74 to 77, find an expression for the area of a rectangle whose length l and width w are as given. Expand all products.

74 $l = 3x$, $w = 2x + 1$
75 $l = x + 5$, $w = 4x + 2$
76 $l = x^2$, $w = 1 + x$
77 $l = 6x$, $w = 2x + 3y$

78 Express as a sum of powers of x the volume of a rectangular box whose length is $2x$, whose width is $x + 3$, and whose height is $x + 4$.

In Exercises 79 to 84, factor the given expression, and check your work by finding the product of the factors.

79 $2x^2 + x$
80 $3x^2 + 4xy + 2x$
81 $5a^3 + ax$
82 $x^2 + 10x + 21$
83 $y^2 + 9y + 14$
84 $x^2 + 4x + 4$

In Exercises 85 to 89, write a formula for the quantity described.

85 The area A of a rectangle which is x feet wide and $x + 3$ feet long.
86 The volume V of a rectangular box which is $2x$ feet long, $x + 2$ feet wide, and $x + 1$ feet high.
87 The sum of the squares of the numbers a and b.
88 The square of the sum of the numbers a and b.
89 Are your answers to Exercises 87 and 88 equal? Explain.

In Exercises 90 to 95, add the given multinomials.

90 $3x + 4$; $4x + 8$
91 $5x^2 + 6x$; $8x + 9$
92 $3x^2 + 4x + 5$; $x^2 + 2$
93 $6x^2 + 2x + 4$; $2x^2 + 3x + 7$
94 $7x^3 + 4x^2 + 6x + 2$; $4x + 3x^2 + 8x + 2$
95 $5x^3 + 4$; $3x^2 + 9$

96 What are the base, the exponent, and the value of the power in each of the following: 3^2; 4^3; 2^4; 5^2?
97 What are the base of the power, the exponent, and the coefficient in each of the following terms: $2x^3$; $3x^4$; $4x^2$; $7x^3$?
98 What are the terms in the expression $4x^3 + 3x^2 + 4x + 9$? What is the coefficient in each of the terms in which x appears?
99 Complete the following table:

WHEN $x =$	1	2	3	4	5	6	7	8
THEN $x^2 =$	1	4	9					
AND $x^3 =$		8						

1.9
SUBTRACTION AND DIVISION OF NATURAL NUMBERS

In addition to being familiar with the operations of addition and multiplication on the set of natural numbers, we are also aware of the processes called *subtraction* of natural numbers and *division* of natural numbers. In this section we shall recall and make precise the meanings and properties of subtraction and division of natural numbers.

We have all used such language as "When 4 is subtracted from 9, the result is 5," or "9 minus 4 is 5," and we have written this as $9 - 4 = 5$. Similarly we "know" that $7 - 3 = 4$, $16 - 9 = 7$, $4 - 1 = 3$, etc. We say that 5 is the difference when 4 is subtracted from 9, and we note that $4 + 5 = 9$, that 7 is the difference when 9 is subtracted from 16, and $9 + 7 = 16$, and so on.

If a and b are natural numbers and if there exists a natural number c with the property that

$$b + c = a$$

DIFFERENCE

then c is said to be the **difference** that results when we *subtract* b from a. If such a natural number c exists, we write $a - b = c$. The expression $a - b$ is usually read "a *minus* b." In other words, we have *defined* $a - b$ to be the natural number c, if such a natural number exists, with the property that $b + c = a$.

a MINUS b

Thus $9 - 4$ is found by asking "$4 +$ what number $= 9$?" and using our knowledge of addition to give the answer. The proviso in the definition as to the existence of $c = a - b$ is necessary because in the set of natural numbers the difference does not exist in all cases. There is no natural number c with the property that $9 + c = 2$, and hence we cannot subtract 9 from 2 in the set of natural numbers, and we say that $2 - 9$ does not represent a natural number.

Contrast the fact that if a and b are natural numbers, then $a + b$ is always a natural number, with the fact just mentioned that $a - b$ is *not* always a natural number. We express these properties of the natural numbers with respect to addition and subtraction by saying that the set of natural numbers is *closed under addition* but is *not closed under subtraction.* Recalling the meaning we gave to the word "operation" in Sec. 1.5, we see that, strictly speaking, subtraction is not an operation on the set of natural numbers because subtraction does not associate a natural number with each ordered pair of natural numbers. However, in spite of this fact, it is very common to find references to the "operation" of subtraction in the set of natural numbers. Since it is not our purpose in this book to study the concept of operation in detail, we shall make use of this common form of expression and speak of the operation of subtraction on the set of natural numbers, recalling as we do so that the set of natural numbers is *not closed* under subtraction.

The nonclosure of the set of natural numbers under subtraction means that if we desire subtraction to be a true operation and if we desire to remove the rather "unaesthetic" consequence that $8 - 3$ exists but $3 - 8$ does not, we need a "larger" set of "numbers." This is indeed one of the strong motives for extending the set of numbers by introducing zero and the negative integers. Such an extension is made in the following chapter.

We note that subtraction is neither commutative nor associative since

$$8 - 3 \neq 3 - 8$$

and

$$(9 - 4) - 2 \neq 9 - (4 - 2)$$

In the difference $a - b$, a is called the *minuend* and b is called the *subtrahend.* According to the definition of subtraction, $a - b = c$ if and only if $b + c = a$; that is, in a subtraction problem the subtrahend plus the difference must be equal to the minuend. We shall use this principle to prove that

for any natural numbers a, b,

$$(a + b) - b = a \tag{1.17}$$

In order to prove (1.17) we see that on the basis of the definition of subtraction it will be sufficient to show that the subtrahend b plus the difference a is equal to the minuend $(a + b)$. Now

$$b + a = a + b$$

by the commutative property of addition. So (1.17) is proved. The result stated in (1.17) shows that subtraction "undoes" addition. Because of this we say that *subtraction is the inverse of addition*. Evaluation of a difference in a subtraction problem is based on the inverse relation between addition and subtraction. If you have mastered the addition table and understand the place-value concept in our system of writing numerals, you should have no difficulty with subtraction. We shall review the idea of place-value notation in Sec. 1.11.

We have commented that subtraction is the inverse process to addition, since subtraction undoes addition. There is also a process that is the inverse of multiplication, a process that "undoes" multiplication; this process is called *division*. Division probably had its origin in the physical process of dividing a set of objects among several people. Thus if a farmer had 12 cows to divide equally among 4 sons, each would get 3 cows. We say that 12 divided by 4 is 3 and note that the product of 4 and 3 is 12.

If a and b are natural numbers, and if there exists a unique natural number c with the property that

$$bc = a$$

QUOTIENT

then c is said to be the **quotient** which results when a is *divided* by b. If such a natural number c exists, we write $a \div b = c$. In other words, we have *defined* $a \div b$ to be the natural number c, if such a natural number exists, with the property that $bc = a$. We call a the **dividend** and b the **divisor.**

DIVIDEND

DIVISOR

Thus $8 \div 4$ is found by asking "4 times what number equals 8?" and observing from our knowledge of multiplication that 2 is the number required. The proviso in the definition as to the existence of $c = a \div b$ is necessary because (as was the case for subtraction of natural numbers) in the set of natural numbers the quotient does not always exist. There is no natural number c with the property that $3c = 7$, and hence we cannot divide 7 by 3 in the set of natural numbers, and we say that $7 \div 3$ does not represent a natural number.

Contrast the fact that if a and b are natural numbers, then ab is always a natural number, with the fact just mentioned that $a \div b$ is *not* always a natural number. We express these contrasting properties of the natural numbers with respect to multiplication and division by saying that the set of natural numbers is *closed under multiplication* but is *not closed under division.* As was the case for subtraction, division is not, strictly speaking, an operation on the set of natural numbers, because division does not associate a natural number with each ordered pair of natural numbers. However, it is very common to find references to the "operation" of division on the set of natural numbers. We shall use this common form of expression and speak of the operation of division, recalling as we do so that the set of natural numbers is *not closed* under division.

The nonclosure of the set of natural numbers under division means that if we desire to remove the rather "unaesthetic" consequence that $9 \div 3$ exists but $3 \div 9$ does not exist, we shall need a "larger" set of numbers. This is one strong motive for the introduction of the objects we call "rational numbers" in Chap. 3.

We note that division is neither commutative nor associative since

$$8 \div 2 \neq 2 \div 8$$

and

$$(16 \div 4) \div 2 \neq 16 \div (4 \div 2)$$

According to the definition of division, $a \div b = c$ if and only if $bc = a$; therefore to prove that c is the quotient of a divided by b, all that is necessary is to show that $bc = a$. We shall use this principle to prove that

$$(ab) \div b = a \tag{1.18}$$

To prove (1.18) it is sufficient to show that b multiplied by a is equal to ab. Now

$$ba = ab$$

by the commutative property of multiplication. So (1.18) is proved, and we see that division "undoes" multiplication. We say that *division is the inverse of multiplication.*

FACTOR

MULTIPLE

If a and b are natural numbers and if $a \div b = c$ is a natural number, we say that b is a *divisor* or **factor** of a; further we say that a is a **multiple** of b (see Sec. 1.5). Thus 4 is a divisor of, or a factor of, 8, and 8 is a multiple of 4 since $8 \div 4 = 2$. In general, *if the dividend is a multiple of the divisor,* there is an equality of the form

$$(\textit{divisor})\cdot(\textit{quotient}) = \textit{dividend} \tag{1.19}$$

corresponding to the equality

$$(\textit{dividend}) \div (\textit{divisor}) = \textit{quotient} \tag{1.20}$$

Evaluation of a quotient in a division problem is based on the inverse relation between multiplication and division. If you have mastered the multiplication table and understand the place-value concept used in writing numerals, you should have no difficulty with division. For example, to divide 2,226 by 53, the work may be arranged as follows:

$$\begin{array}{r} 42 \\ 53\overline{)2226} \\ \underline{212} \\ 106 \\ \underline{106} \end{array}$$

Here 2,226 is the dividend, 53 is the divisor, and 42 is the quotient. We write

$$2{,}226 \div 53 = 42 \tag{1.21}$$

and as a check we note

$$53 \cdot 42 = 2{,}226 \tag{1.22}$$

Compare statements (1.21) and (1.22) with statements (1.20) and (1.19), respectively.

Relative to problems that involve addition, subtraction, multiplication, and division, we adhere to the convention that the *multiplications and divisions* (if any) *will be performed before the additions and subtractions* (if any) *except where otherwise indicated by parentheses.* To illustrate,

$$2 + 4 \cdot 5 - 4 \div 2 = 2 + 20 - 2 = 20$$
$$(2 + 4) \cdot 5 - (4 + 2) \div 2 = 6 \cdot 5 - 6 \div 2 = 30 - 3 = 27$$

EXERCISES

In Exercises 1 to 12, perform the subtraction.

1 $15a^2b - 9a^2b$
2 $(6a + b) - (2a + b)$
3 $(7x + 3y) - (4x + y)$
4 $(2x^2 + 3x) - 2x$
5 $(9x^2 + 2x + 7) - (2x^2 + x + 5)$
6 $(3x + 4y + 5z) - (2y + 3z)$
7 $(2ab + 3a) - (ab)$
8 $(19xy + 4xyz) - (15xy + xyz)$
9 $(95a + 17b) - (90a + 17b)$
10 $(4ab + 5ac) - (3ac + ab)$
11 $(9a + 10b + 11c + 12d) - (4a + b + 10c + 9d)$
12 $(17a^2b + 9ab) - (8ab)$

In Exercises 13 to 21, tell whether the given expression represents a natural number. If a quotient exists in the set of natural numbers, give its value.

13 $12 \div 3$ 14 $7 \div 4$ 15 $8 \div 1$
16 $3 \div 3$ 17 $9 \div 7$ 18 $1 \div 4$
19 $(9 + 12) \div 3$ 20 $(8 + 5) \div 2$ 21 $6 \div (2 + 5)$

22 Assuming that a and b are natural numbers, tell, by answering "yes" or "no" with reference to each, which of the following expressions *always* represents a natural number: $a + b$; $a - b$; ab; $a \div b$.
23 If your answer to a part of Exercise 22 is "no," what restriction must be placed on a and b for the expression in question to represent a natural number?

In Exercises 24 to 27, perform the indicated operations.

24 $6 + 4 \cdot 5 - 8 \div 4$ 25 $10 \cdot 5 - 2(4 \cdot 2) - 5$
26 $(6 + 4) \cdot 5 - (8 + 4) \div 4$ 27 $(4 + 6)2 - 8$

28 Is 3 a divisor of 15? Why?
29 Is 5 a divisor of 14? Why?
30 Is 16 a multiple of 3? Why?
31 Is 12 a multiple of 3? Why?

Find the quotient in Exercises 32 to 35. Check your answer by showing that the product of the divisor and the quotient is equal to the dividend.

32 $27\overline{)4{,}320}$ 33 $36\overline{)2{,}484}$
34 $56{,}448 \div 504$ 35 $49{,}928 \div 79$

36 If fence posts are placed with centers 12 feet apart, how many posts will be needed for 1 mile (5,280 feet) of fence? It is understood that a post is to be used at each end of the mile.
37 If 16 pairs of shoes can be packed in a carton, how many cartons are needed for 608 pairs of shoes?
38 A coal bin is 19 feet long, 15 feet wide, and 10 feet deep. If a ton of coal fills 38 cubic feet, how many tons will fill the bin?
39 The sum of $1,024 was divided equally among 16 people. How much did each one receive?

1.10
ORDER IN THE SET OF NATURAL NUMBERS

The fact that the natural numbers are first used in our experience in counting leads to a very natural concept of "order" in the set of natural numbers. We say that 1 comes "before" 2, that 4 comes "before" 9, that 5 comes "before" 61, and so on. Further, if the natural number b comes "before" the natural number a, we say that b is "less than" a or a is "greater than" b; thus, 1 is less than 2 and 2 is greater than 1; also 4 is less than 9 and 9 is greater than 4, and so on. This common way of thinking about natural numbers leads to the following definitions.

If a and b are natural numbers and if there exists a natural number c with the property that $a - b = c$ (or, equivalently, $b + c = a$), we say that a is **greater than** b and we write

GREATER THAN

$$a > b$$

LESS THAN

In the same instance we say that b is **less than** a and we write

$$b < a$$

For example, $7 > 4$, or $4 < 7$, because $7 - 4 = 3$.

We use the symbol $a \geq b$ to mean that either a is greater than b or a is equal to b, and we read $a \geq b$ as "a is greater than or equal to b." Similarly, $a \leq b$ means that either a is less than b or a is equal to b, and is read "a is less than or equal to b."

With the definitions given in the preceding two paragraphs we say that we have defined an order relation in the set of natural numbers. Let us list some properties of this order relation.

Suppose that a and b are any natural numbers with $a \neq b$; then either $a - b$ is a natural number or $b - a$ is a natural number. That is,

for any natural numbers a and b,
either $a > b$ or $a < b$ or $a = b$ (1.23)

Other properties of order in the set of natural numbers are

for any natural numbers a, b, and c,
if $a > b$ and $b > c$, then $a > c$ (1.24)
if $a > b$, then $a + c > b + c$ (1.25)
if $a > b$, then $ac > bc$ (1.26)

Note carefully that the property (1.26) supposes that c is a *natural number,* and do not try to apply this property in any other situation.

A useful way of visualizing the set of natural numbers, the addition of natural numbers, the subtraction of natural numbers, and the order relation in the set of natural numbers is by making use of the *number scale* for natural numbers. To construct such a number scale, we draw a horizontal straight line upon which we designate a reference point called the *origin,* represented by O in Fig. 1.13. We choose a segment of fixed length, called the *unit length.*

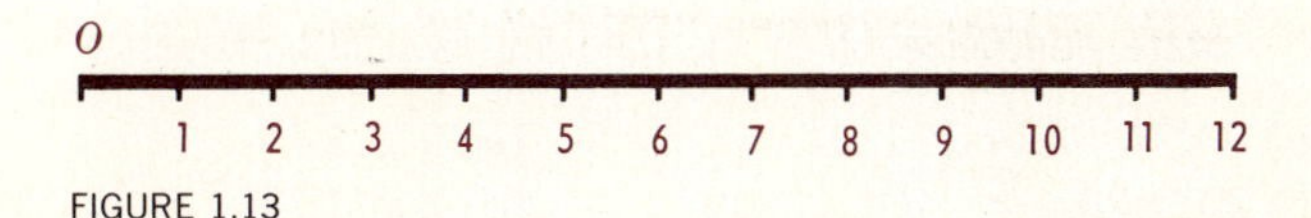

FIGURE 1.13

Next we place points to the right of the origin O at distances equal to the unit length, twice the unit length, three times the unit length, and so on. These points to the right of O then have a natural one-to-one correspondence with the natural numbers, as indicated in Fig. 1.13.

This number scale provides us with a geometric interpretation (or model) of the set of natural numbers. Also the number scale provides a geometric interpretation of the process of addition of two natural numbers. For example, the addition $3 + 5$ can be interpreted as follows: Start at the point 3 and move 5 units to the *right* from 3; this brings us to the point 8, so $3 + 5 = 8$. For an interpretation of subtraction, say the difference $9 - 7$, we start at the point 9 and move 7 units to the *left* from 9; this brings us to the point 2, so $9 - 7 = 2$.

We note of course that if the natural number a is greater than the natural number b, that is, if $a > b$, then the point corresponding to a lies to the right of the point corresponding to b. For example, consider $4 > 2$, $7 > 1$, $18 > 9$, etc.

EXERCISES

1 Given that A is the set of natural numbers less than 13, indicate which of the following statements are true and which are false:

(*a*) $10 \in A$ (*b*) $15 \in A$ (*c*) $16 \in A$
(*d*) $7 \in A$ (*e*) $8 \in A$ (*f*) $6 \notin A$

In Exercises 2 to 5, translate the given equality into two statements, one that uses "greater than" and the other that uses "less than."

2 $20 - 6 = 14$
3 $9 = 5 + 4$
4 $40 = 26 + 14$
5 $26 - 9 = 17$

In Exercises 6 to 9, translate the given symbolic statement into a verbal statement which uses either "greater than" or "less than."

6 $5 < 8$
7 $6 > 4$
8 $3 > 1$
9 $10 < 20$

10 State whether each of the following is true or false:

(*a*) $6 < 9$ (*b*) $13 > 15$
(*c*) $10 < 6$ (*d*) $12 > 6$

11 State whether each of the following is true or false:

(*a*) $4 < 4$ (*b*) $9 > 3$
(*c*) $20 > 20$ (*d*) $5 < 15$

1.11 NUMERATION SYSTEMS AND PLACE VALUE

We have stated that the written symbols used to represent numbers are called *numerals*. The symbols used together with the rules for their use in forming numerals are known as a numeral system or a numeration system. The numeral system that we use is called the *Hindu-Arabic numeral system*. This system makes use of the symbols 0, 1, 2, 3, 4, 5, 6, 7, 8, 9, known as **digits**, and the place-value principle. The two most important features of the system is its use of the place-value idea and its use of the symbol 0.

DIGITS

PLACE VALUE

Place value means the value given to a digit by virtue of the place it occupies in a numeral relative to the units or ones place. To illustrate, in the numeral 22, the right-hand digit is in the units place and designates two ones; the left-hand digit is in the tens place and designates two tens. The various places occupied by the digits in a numeral in the Hindu-Arabic system, reading from right to left, are units or ones place, tens place, hundreds place, thousands place, ten-thousands place, hundred-thousands place, and so forth. Thus in the numeral 4,856, we say

4 is the thousands digit   8 is the hundreds digit
5 is the tens digit   6 is the units digit

and the numeral 4,856 represents

4 thousands + 8 hundreds + 5 tens + 6 ones

or

$4{,}000 + 800 + 50 + 6$

Since (see Sec. 1.8) $1{,}000 = 10 \cdot 10 \cdot 10 = 10^3$, $100 = 10 \cdot 10 = 10^2$, $10 = 10^1$, and $1 = 10^0$, we can write

$$4{,}856 = 4(10)^3 + 8(10)^2 + 5(10)^1 + 6(10)^0$$

Note, by the way, that the symbol 10 is *not* a digit but that it represents "one ten and no ones." Similarly,

$$\begin{aligned} 62{,}973 &= 6(10)^4 + 2(10)^3 + 9(10)^2 + 7(10)^1 + 3(10)^0 \\ 203 &= 2(10)^2 + 3(10)^0 \end{aligned} \qquad (1.27)$$

A comment on the role played by the symbol 0 (zero) in a numeral system is in order. Without zero the place-value principle would not be possible for a number such as two hundreds + no tens + three ones—that is, for 203. The zero is needed to push the 2 into the hundreds place. Without the zero (or an equivalent symbol) we could not distinguish clearly between the numerals for two-hundred and three and twenty-three.

Because, in the Hindu-Arabic numeral system, any natural number can be written in terms of powers of ten, as illustrated by (1.27), that system is said to be a numeral system with *base ten.* Any natural number other than one could play the role of a base of a numeral system. For example, by analogy with equations (1.27), if we let the base of a numeral system be five and use the digits 0, 1, 2, 3, 4, we find that the numeral 241 in the base-five system represents the number "2 twenty-fives plus 4 fives plus 1 one" in the base-ten system, and we write

241 = 2 twenty-fives + 4 fives + 1 one

or

$$241 = 2(\text{five})^2 + 4(\text{five}) + 1(\text{one})$$

In order to avoid confusion, whenever we write a numeral in a system *with base other than ten,* we shall agree to indicate the base as a subscript on the numeral. A numeral without a subscript will always be a base-ten numeral. For example, in the illustration above we should write

$$241_5 = 2(5)^2 + 4(5)^1 + 1(5)^0$$

Similarly,

$$623_7 = 6(7)^2 + 2(7)^1 + 3(7)^0$$

In general, if n is any natural number other than one, and a, b, c, and d are natural numbers, each less than n,

$$abcd_n = a(n)^3 + b(n)^2 + c(n)^1 + d(n)^0 \qquad (1.28)$$

BASE OF A NUMERAL SYSTEM

The number n in this scheme is called the **base of the numeral system.**

If we are given a numeral with base other than ten, we can determine a numeral with base ten that represents the same number. Consider the following examples.

EXAMPLE 1 Given the numeral 243_5 in a base-five system, find the numeral in the base-ten system that represents the same number.

SOLUTION According to the definition (1.28),

$$\begin{aligned} 243_5 &= 2(5)^2 + 4(5)^1 + 3(5)^0 \\ &= 2(25) + 4(5) + 3 = 50 + 20 + 3 \\ &= 73 \end{aligned}$$

Therefore the numerals 243_5 and 73 represent the same number.

EXAMPLE 2 Find the numeral in the base-ten system that represents the same number as the numeral 101101_2 in the base-two system.

SOLUTION Using an extension of the idea expressed by the definition (1.28), we find

$$101101_2 = 1(2)^5 + 1(2)^3 + 1(2)^2 + 1(2)^0$$

Now $2^0 = 1$, $2^1 = 2$, $2^2 = 4$, $2^3 = 8$, $2^4 = 16$, $2^5 = 32$, so that

$$\begin{aligned} 101101_2 &= 1(32) + 1(8) + 1(4) + 1 \\ &= 45 \end{aligned}$$

Suppose that we are given a numeral with base ten and we wish to determine a numeral with base four that represents the same number. How shall we proceed? Let 573 be a base-ten numeral; we then wish to determine digits a, b, c, d, and e (each of these digits being one of the digits 0, 1, 2, 3) so that

$$\begin{aligned} 573 &= a(4)^4 + b(4)^3 + c(4)^2 + d(4)^1 + e(4)^0 \\ &= a(256) + b(64) + c(16) + d(4) + e \end{aligned}$$

We see that the first problem is to determine how many 256's are in 573. We have

$$\begin{array}{r} 2 \\ 256\overline{)573} \\ \underline{512} \\ 61 \end{array}$$

So as a first step we write

$$573 = 2(256) + 61$$

Now we look for digits b, c, d, and e, so that

$$61 = b(64) + c(16) + d(4) + e$$

We ask how many 64's are in 61. Obviously there are none, and so b is the digit 0. Next we determine the number of 16's in 61. We have

$$\begin{array}{r} 3 \\ 16\overline{)61} \\ \underline{48} \\ 13 \end{array}$$

Therefore $c = 3$, and we now can write

$$573 = 2(256) + 3(16) + 13$$

Now we ask how many 4's are in 13. We see that

$$\begin{array}{r} 3 \\ 4\overline{)13} \\ \underline{12} \\ 1 \end{array}$$

So finally we have

$$573 = 2(256) + 3(16) + 3(4) + 1(1)$$

or

$$573 = 20331_4$$

EXAMPLE 3 Determine the numeral written in a base-seven system that represents the same number as 2,198.

SOLUTION First we determine the values of the first five places in a base-seven numeral. We find

$$7^0 = 1 \qquad 7^1 = 7 \qquad 7^2 = 49 \qquad 7^3 = 343 \qquad 7^4 = 2{,}401$$

From this computation we can see that four digits will be required in the base-seven numeral; that is,

$$\begin{aligned} 2{,}198 &= a(7)^3 + b(7)^2 + c(7)^1 + d(7)^0 \\ &= a(343) + b(49) + c(7) + d(1) \end{aligned}$$

By the long division shown below we see that 2,198 can be written as

$$2{,}198 = 6(343) + 140$$

Next we find that

$$140 = 2(49) + 42$$

Also

$$42 = 6(7)$$

Therefore,

$$\begin{aligned} 2{,}198 &= 6(343) + 2(49) + 6(7) \\ &= 6260_7 \end{aligned}$$

$$\begin{array}{r} 6 \\ 343\overline{)2{,}198} \\ \underline{2{,}058} \\ 140 \end{array} \qquad \begin{array}{r} 2 \\ 49\overline{)140} \\ \underline{98} \\ 42 \end{array} \qquad \begin{array}{r} 6 \\ 7\overline{)42} \\ \underline{42} \\ 0 \end{array}$$

Note that a numeral system with base ten uses ten digits, a numeral system with base five uses five digits, a numeral system with base two uses two digits, and in general a numeration system with base n uses n digits. If the base of a numeral system is less than ten, the digits used are usually taken to be a subset of the ten Hindu-Arabic digits; for example, for a base-five system, the digits used are 0, 1, 2, 3, 4. If the base is greater than ten, then new symbols for some digits are required. A system with base twelve will be considered later in Sec. 1.12.

EXERCISES

1 (*a*) What digits would you employ in a numeral system of base seven? (*b*) Write the numerals in this system which correspond to the numerals 1 to 31 in the decimal system.

2 Construct a calendar for the current month using the numeral system of Exercise 1.

3 (*a*) What digits would you employ in a numeral system of base thirteen? (*b*) Write the numerals in this system which correspond to the numerals 1 to 31 in the decimal system.

4 The following are numerals in the binary (base-two) system; find their equivalents in the decimal system: 10010; 10011; 100011; 1001011.

5 The following are numerals in the decimal system; find their equivalents in the binary (base-two) system: 20; 15; 43; 37.

6 Determine the numerals in a base-two system that represent, respectively, the same numbers as the following numerals in the decimal system: 32; 35; 40; 50; 80; 128.

7 Determine the numerals in the base-three system that represent, respectively, the same numbers as the numerals 3 to 20 in the decimal system.

8 Determine the numerals in the base-six system that represent, respectively, the same numbers as the numerals 3 to 20 in the decimal system.

9 Determine the numeral in each of the base-four, base-eight, and base-nine systems that represents the same number as the numeral 532 in the decimal system.

10 Show that $2101_4 = 265_7$.

1.12 PLACE VALUE AND ADDITION AND MULTIPLICATION

We have commented earlier upon the superiority for computational purposes of the Hindu-Arabic numeral system over the numeration systems used by the Greeks and Romans. This superiority does not depend on the fact that the base of the numeration system is ten; the choice of base ten was probably made because we have ten fingers. The superiority of our numeration system in facilitating computations involving addition, multiplication, subtraction, and division is due to *the presence of the zero symbol and the use of the place-value principle.* When we use the processes of "carrying" in addition or multiplication and "borrowing" in subtraction or the device known as "long division," we are making use of the place-value principle. Let us examine some of these computations.

Suppose we wish to add the three numbers 3,974, 885, and 4,566.

We may display the process as follows:

3,974 is 3 thousands + 9 hundreds + 7 tens + 4 ones
885 is 8 hundreds + 8 tens + 5 ones
4,566 is 4 thousands + 5 hundreds + 6 tens + 6 ones

The sum is

7 thousands + 22 hundreds + 21 tens + 15 ones

or, since 15 ones is 1 ten and 5 ones, the sum is

7 thousands + 22 hundreds + 22 tens + 5 ones

or, since 22 tens is 2 hundreds and 2 tens, the sum is

7 thousands + 24 hundreds + 2 tens + 5 ones

or, since 24 hundreds is 2 thousands and 4 hundreds, the sum is

9 thousands + 4 hundreds + 2 tens + 5 ones

or 9,425

CARRYING

The process of changing the number of ones in the sum (15 in the illustration above) to tens and ones (1 and 5), and then adding the tens in with the tens column, and doing a similar thing with the number of tens, and so on, is called *carrying* from one column to the next column on the left.

How do we compute the product of 3 and 241? We could proceed as follows:

$3 \cdot 241 = 3(200 + 40 + 1)$	*base-ten notation*
$= 3 \cdot 200 + 3 \cdot 40 + 3 \cdot 1$	*distributive property*
$= 600 + 120 + 3$	
$= 600 + (100 + 20) + 3$	*base-ten notation*
$= (600 + 100) + 20 + 3$	*associative property of addition*
$= 700 + 20 + 3$	
$= 723$	*base-ten notation*

Customarily we arrange the work as shown below:

$$\begin{array}{r} 241 \\ 3 \\ \hline 723 \end{array}$$

Notice that, in the mechanical process represented, the reduction of 12 tens to 1 hundred and 2 tens is performed during the multiplication by carrying from one step to another. This carrying in multiplication is different from the carrying in addition. In multiplication, a digit carried must be retained mentally until after the succeeding product is obtained; in addition, a digit carried may be added immediately.

Next let us consider multiplication in which each of the factors contains at least two digits. For example,

$$
\begin{aligned}
53 \cdot 42 &= 53(40 + 2) && \textit{base-ten notation} \\
&= 53 \cdot 40 + 53 \cdot 2 && \textit{distributive property} \\
&= (50 + 3)40 + (50 + 3)2 && \textit{base-ten notation} \\
&= (50 \cdot 40 + 3 \cdot 40) + (50 \cdot 2 + 3 \cdot 2) && \textit{equation } (1.12) \\
&= (2{,}000 + 120) + (100 + 6) && \\
&= [2{,}000 + (100 + 20)] + 100 + 6 && \textit{base-ten notation} \\
&= 2{,}000 + (100 + 100) + 20 + 6 && \textit{associative and commutative properties of addition} \\
&= 2{,}000 + 200 + 20 + 6 && \textit{associative property of addition} \\
&= 2{,}226 && \textit{base-ten notation}
\end{aligned}
$$

As usually arranged the multiplication appears as

$$
\begin{array}{rl}
53 & \\
\underline{42} & \\
106 & \quad (2 \cdot 53 = 106) \\
\underline{2{,}120} & \quad (40 \cdot 53 = 2{,}120) \\
2{,}226 &
\end{array}
$$

The student should identify the corresponding steps in the two computations.

With a knowledge of the addition and multiplication tables for the digits 0, 1, 2, 3, 4, 5, 6, 7, 8, and 9 and an understanding of the place-value principle, it is possible to perform computations involving addition or multiplication with relative ease. In order that we may see more clearly that this ease of computation depends only on these two things, we shall consider some computations using numerals with bases other than ten.

First let us exhibit the numerals with bases two, five, ten, and twelve that represent the natural numbers from one through thirty. The student should verify the entries in the table.

	BASE			
NUMBER	TWO	FIVE	TEN	TWELVE
ZERO	0	0	0	0
ONE	1	1	1	1
TWO	10	2	2	2
THREE	11	3	3	3
FOUR	100	4	4	4
FIVE	101	10	5	5
SIX	110	11	6	6
SEVEN	111	12	7	7
EIGHT	1000	13	8	8
NINE	1001	14	9	9
TEN	1010	20	10	α
ELEVEN	1011	21	11	β
TWELVE	1100	22	12	10
THIRTEEN	1101	23	13	11
FOURTEEN	1110	24	14	12
FIFTEEN	1111	30	15	13
SIXTEEN	10000	31	16	14
SEVENTEEN	10001	32	17	15
EIGHTEEN	10010	33	18	16
NINETEEN	10011	34	19	17
TWENTY	10100	40	20	18
TWENTY-ONE	10101	41	21	19
TWENTY-TWO	10110	42	22	1α
TWENTY-THREE	10111	43	23	1β
TWENTY-FOUR	11000	44	24	20
TWENTY-FIVE	11001	100	25	21
TWENTY-SIX	11010	101	26	22
TWENTY-SEVEN	11011	102	27	23
TWENTY-EIGHT	11100	103	28	24
TWENTY-NINE	11101	104	29	25
THIRTY	11110	110	30	26

Now, assuming for the purposes of this section that we have defined the number zero and the operations of addition and multiplication involving this number so that $a + 0 = a$ and $a \cdot 0 = 0$, let us consider the addition and multiplication tables in two of these systems.

The numeral system with base two surely provides the simplest addition and multiplication tables possible:

For addition

+	0	1
0	0	1
1	1	10

For multiplication

·	0	1
0	0	0
1	0	1

For the numeral system with base five the tables are as follows:

For addition

+	0	1	2	3	4
0	0	1	2	3	4
1	1	2	3	4	10
2	2	3	4	10	11
3	3	4	10	11	12
4	4	10	11	12	13

For multiplication

·	0	1	2	3	4
0	0	0	0	0	0
1	0	1	2	3	4
2	0	2	4	11	13
3	0	3	11	14	22
4	0	4	13	22	31

The student should use the results in the table on page 52 to verify the entries in these addition and multiplication tables.

We may use these addition and multiplication tables for base-two and base-five numerals together with the place-value idea to do computations. In the computations in the following examples, the student should notice that the mechanical procedures are the same as those employed in computation with base-ten numerals.

EXAMPLE 4 Perform the following additions:

$$(a)\quad \begin{array}{r} 10111_2 \\ +\ 1010_2 \\ \hline \end{array} \qquad (b)\quad \begin{array}{r} 4303_5 \\ +\ 244_5 \\ \hline \end{array}$$

SOLUTION (*a*)

$$\begin{array}{r} {\scriptstyle 111} \\ 10111_2 \\ +\ 1010_2 \\ \hline 100001_2 \end{array}$$

We explain the solution in the following way: 1 + 0 = 1; 1 + 1 = 10, put down 0, carry 1; 1 + 1 + 0 = 10, put down 0, carry 1; 1 + 0 + 1 = 10, put down 0, carry 1; 1 + 1 = 10.

(*b*)

$$\begin{array}{r} {\scriptstyle 111} \\ 4303_5 \\ +\ 244_5 \\ \hline 10102_5 \end{array}$$

We explain the solution in the following way: 3 + 4 = 12, put down 2, carry 1; 1 + 4 = 10, put down 0, carry 1; 1 + 3 + 2 = 11, put down 1, carry 1; 1 + 4 = 10.

EXAMPLE 5 Perform the following multiplications:

(*a*) $\begin{array}{r} 1101_2 \\ \times\ 101_2 \\ \hline \end{array}$ (*b*) $\begin{array}{r} 4403_5 \\ \times\ 213_5 \\ \hline \end{array}$

SOLUTION (*a*)

$$\begin{array}{r} 1101_2 \\ \times\ 101_2 \\ \hline 1101 \\ 110100 \\ \hline 1000001_2 \end{array}$$

The result is obtained by multiplying 1 × 1101 and 100 × 1101 and adding these products.

(*b*)

$$\begin{array}{r} 4403_5 \\ \times\ 213_5 \\ \hline 24214 \\ 44030 \\ 1431100 \\ \hline 2104344_5 \end{array}$$

First we multiply 4403 by 3:

$$\begin{array}{r} {\scriptstyle 2\ 1\ \ } \\ 4403 \\ \times\ \ \ 3 \\ \hline 24214 \end{array}$$

3 × 3 = 14, put down 4, carry 1
3 × 0 = 0, add 1 and get 1
3 × 4 = 22, put down 2, carry 2
3 × 4 = 22, add 2 and get 24

Next we multiply 4403 by 10, getting 44030

Next we multiply 4403 by 200:

$$\begin{array}{r} {\scriptstyle 1\ 1\ \ } \\ 4403 \\ \times\ \ 200 \\ \hline 1431100 \end{array}$$

2 × 3 = 11, put down 1, carry 1

$2 \times 0 = 0$, add 1 and get 1
$2 \times 4 = 13$, put down 3, carry 1
$2 \times 4 = 13$, add 1 and get 14

Then we add 24214, 44030, and 1431100.

EXERCISES

1 Write the addition table for the numeral system with base seven.
2 Write the multiplication table for the numeral system with base seven.
3 Using the addition table of Exercise 1, find the sum of each of the following pairs of base-seven numerals: 4 and 3; 5 and 4; 6 and 5.
4 The following are numerals in the duodecimal (base-twelve) system; find their equivalents in the decimal system: 72; 105; 76; 23α.
5 The following are numerals in the decimal system; find their equivalents in the duodecimal system: 144; 99; 77; 678.
6 Perform the following multiplications in the binary numeral system; in each case, check your result by translating the given numerals and the product into the decimal system: (101)(111); (11)(101); (101)(101).
7 A ternary system employs a base of three. In this system the symbols 0, 1, and 2 are adequate to represent any number. Give the addition and multiplication tables for the ternary numeral system.
8 Using the addition and multiplication tables which you constructed in Exercise 7, perform the following calculations in the ternary system:

$$\begin{array}{r} 201 \\ +\ 12 \\ \hline \end{array} \qquad \begin{array}{r} 122 \\ +222 \\ \hline \end{array} \qquad \begin{array}{r} 12 \\ \times\ 2 \\ \hline \end{array} \qquad \begin{array}{r} 12 \\ \times 21 \\ \hline \end{array}$$

9 The ancient Thracians used a quaternary numeral system which employed a base of four. Using the symbols 0, 1, 2, 3, construct the addition and multiplication tables for this system.
10 Using the addition and multiplication tables for the quaternary system which you constructed in Exercise 9, perform the following calculations in the quaternary system:

$$\begin{array}{r} 11 \\ +12 \\ \hline \end{array} \qquad \begin{array}{r} 23 \\ +\ 2 \\ \hline \end{array} \qquad \begin{array}{r} 11 \\ \times\ 2 \\ \hline \end{array} \qquad \begin{array}{r} 12 \\ \times\ 2 \\ \hline \end{array} \qquad \begin{array}{r} 13 \\ \times\ 2 \\ \hline \end{array}$$

1.13 SUMMARY

In this chapter we have discussed the set of natural numbers and some of the things we do with natural numbers. A natural number is obtained by considering the common property shared by all sets that have the same number of members. Let us agree that the set of natural numbers will be denoted by N.

We have seen that it is possible to define what it means to say that two natural numbers a and b are equal. If the natural numbers a and b are equal we write $a = b$. From our common experience we are aware that "equality" has the following properties:

For every $a, b, c \in N$,
E_1: $a = a$
E_2: if $a = b$, then $b = a$
E_3: if $a = b$ and $b = c$, then $a = c$

These three properties can be proved by appealing to the definition of a natural number and the definition of equality.

Also we have seen that two operations, called addition and multiplication, can be defined on the set of natural numbers. That is, we can give meanings to addition (+) and multiplication (·) so that, if $a, b \in N$, then $a + b \in N$ and $a \cdot b \in N$. These two conclusions are called, respectively, the **closure property of addition** and the **closure property of multiplication** in the set of natural numbers. These operations separately and collectively have other properties that we have called the uniqueness, commutative, associative, and distributive properties.

CLOSURE PROPERTIES

Let us summarize the facts we have recalled in the two preceding paragraphs.

The set N of natural numbers is a set in which a concept of equality (=) has been defined and on which two operations, addition (+) and multiplication (·), have been defined in such a way that the properties listed below hold; the properties pertaining to equality are labeled E_1, E_2, E_3, and the other properties are labeled P_1, P_2, P_3, P_4.

For every $a, b, c \in N$,
E_1: $a = a$
E_2: if $a = b$, then $b = a$
E_3: if $a = b$ and $b = c$, then $a = c$
P_1: $a + b \in N$ and $a + b$ is unique
$a \cdot b \in N$ and $a \cdot b$ is unique
P_2: $a + b = b + a$
$a \cdot b = b \cdot a$
P_3: $a + (b + c) = (a + b) + c$
$a \cdot (b \cdot c) = (a \cdot b) \cdot c$
P_4: $a \cdot (b + c) = a \cdot b + a \cdot c$

If for a set S, there exist a concept of equality and two operations, denoted by =, +, ·, respectively, and if the equality and the operations possess the seven properties listed above, then the set S,

NUMBER SYSTEM

together with the equality and the two operations, is called a **number system.** Each member of the set S in a number system is called a number; that is, **a number** is a member of the set S in a number system.

A NUMBER

The set N, together with the equality and the addition and multiplication we have defined, is called the **natural number system.**

NATURAL NUMBER SYSTEM

There is another property possessed by the natural number system that deserves mention. The set N of natural numbers contains a member, namely 1, with the property that

for any natural number a,
$a \cdot 1 = 1 \cdot a = a$

IDENTITY UNDER MULTIPLICATION

We call the number that possesses this property, the* **identity under multiplication.**

If the set S in a number system has this property, we write the following:

P_5: there exists a member $e \in S$ with the property that for any $a \in S$, $a \cdot e = e \cdot a = a$.

(For the set N of natural numbers, the member e is the number 1.)

In Sec. 1.10 we introduced the concept of order in the set of natural numbers. In that section we defined what it means to say, for $a, b \in N$, "a is greater than b," which is symbolized $a > b$. We restate that definition here: *If a, $b \in N$ and if there exists a natural number c with the property that $a - b = c$, then $a > b$.* The order concept possesses the four properties listed below, which are labeled O_1, O_2, O_3, O_4:

O_1: either $a > b$ or $b > a$, or $a = b$
O_2: if $a > b$ and $b > c$, then $a > c$
O_3: if $a > b$, then $a + c > b + c$
O_4: if $a > b$, then $a \cdot c > b \cdot c$

We have also discussed subtraction $(-)$ and division $(\div)$ in the set N of natural numbers. Recall the following definitions:

For $a, b \in N$, $a - b$ is the natural number c, *if such a natural number exists,* with the property that $b + c = a$; that is,

$$a - b = c \quad \text{if and only if} \quad b + c = a, \quad \text{where } a, b, c \in N \tag{1.29}$$

*It can be proved that if a set of numbers has an identity under an operation, then there is only one identity.

For $a, b \in N$, $a \div b$ is the natural number c, *if such a natural number exists,* with the property that $b \cdot c = a$; that is,

$$a \div b = c \quad \text{if and only if} \quad b \cdot c = a, \quad \text{where } a, b, c \in N \tag{1.30}$$

A very significant fact concerning subtraction and division in the set of natural numbers is that the set of natural numbers is not closed under subtraction or under division. That is,

if $a, b \in N$, then
$a - b$ may not represent a natural number
$a \div b$ may not represent a natural number

Clearly then, if we wish to have a set of numbers that is closed under subtraction, so that we may find the difference of any two numbers, we need to augment the set of natural numbers. Likewise, if we wish to have a set of numbers that is closed under division, so that we can find the quotient of any two numbers, we need to augment the set of natural numbers.

We have then two motives for wishing to extend the set of natural numbers. We shall discuss how these extensions can be made in the next two chapters. We shall produce a set of numbers which is closed under subtraction and a set of numbers in which the quotient of two numbers exists except when the divisor is zero.

2

The Arithmetic and Algebra of Integers

2.1
OPEN SENTENCES AND SOLUTION SETS

In Chap. 1 we discussed the concept of set as being a basic undefined concept in mathematics. Also we defined a *variable* as a symbol that represents any member of a given set, the given set being called the *universal set* or *universe* of the variable. Frequently we wish to consider certain subsets of a given universal set; one of the most common and useful ways of designating a subset of a given universal set is by use of a sentence containing a variable. To indicate how this is done, we shall first consider the concept of an open sentence.

Suppose that A is a given set and that x is a variable whose universe is A. Let S_x denote a sentence that contains the variable x and which has the property that when we replace x by any member of A, we obtain a statement which is either true or false, but not both. Some examples of such sentences are given below, where the specified set A is the universe of the variable.

$$x + 3 = 9 \qquad A = \{1, 2, 3, 4, 5, 6, 7, 8\} \tag{2.1}$$
$$x + 3 > 5 \qquad A = \{2, 4, 6, 8, 10\} \tag{2.2}$$
$$x \div 4 = 1 \qquad A = \{1, 2, 4, 8, 16\} \tag{2.3}$$
$$2x > 4 \text{ and } 3x < 15 \qquad A = \{1, 2, 3, 4, 5\} \tag{2.4}$$
$$x^2 + 2x = 1 \qquad A = \{3, 4, 5\} \tag{2.5}$$
$$x^2 > 3 \qquad A = \{2, 3, 4, 5, 6, 7, 8, 9\} \tag{2.6}$$

Observe that the set of numbers with which we are working is the set N of natural numbers. In each of the above examples the universe A of the variable must be a subset of N, and any constants that appear in the sentences must be members of N.

There are several significant features exhibited by these sentences.

Notice that each of the sentences in (2.1) to (2.6) is neither true nor false; that is, we cannot meaningfully ask whether $x + 3 = 9$ is true or false, or whether $x \div 4 = 1$ is true or false, etc. Because of this fact a sentence that contains a variable is called an **open sentence** (that is, the question of its truth or falsity is an "open" question). In each case, however, as the variable is replaced in turn by the members of the universal set, the sentence becomes a statement about which it is meaningful to ask whether it is true or false.

OPEN SENTENCE

Let us consider (2.2). If in the sentence $x + 3 > 5$, we replace x by the number 2 from the set A, we obtain the statement $2 + 3 > 5$, and this statement is false. If we replace x by the number 4 from the set A, we obtain the statement $4 + 3 > 5$, and this statement is true. By replacing x by 6, we obtain $6 + 3 > 5$; by replacing x by 8 we obtain $8 + 3 > 5$; and by replacing x by 10 we obtain $10 + 3 > 5$. These three statements also are true. We see that the sentence $x + 3 > 5$ becomes a true statement when x is replaced by any member of A *except* the member 2; we say that $x + 3 > 5$ is *true* for the members 4, 6, 8, 10 of A and *false* for the member 2 of A.

If we consider the sentence $x + 3 = 9$ in (2.1), we see that by replacing x in turn by each of the eight members of the set A we obtain seven false statements:

$$1 + 3 = 9 \quad 2 + 3 = 9 \quad 3 + 3 = 9 \quad 4 + 3 = 9$$
$$5 + 3 = 9 \quad 7 + 3 = 9 \quad 8 + 3 = 9$$

and one *true* statement:

$$6 + 3 = 9$$

In this case the sentence is true for only one member of the universe, namely 6, and false for all the other members of the universe.

The open sentence in (2.6) exhibits a different type of behavior. If in the sentence $x^2 > 3$, we put each of the members of the universal set A, in turn, in place of x, we obtain eight *true* statements:

$$4 > 3 \quad 9 > 3 \quad 16 > 3 \quad 25 > 3$$
$$36 > 3 \quad 49 > 3 \quad 64 > 3 \quad 81 > 3$$

In this example, the sentence $x^2 > 3$ is *true* for *every* member of the universal set.

On the other hand, notice that the sentence $x^2 + 2x = 1$, in (2.5), is *false* for *every* member of the universal set $A = \{3, 4, 5\}$.

IDENTITY

When, as in (2.6), an open sentence is true for every member of the universe of the variable, the sentence is called an **identity.** When, as in (2.1), (2.2), (2.3), (2.4), and (2.5), an open sentence is false for at least one member of the universe of the variable, the sentence is called a **conditional sentence.** If a conditional sentence has the form of an equation, as in (2.1), (2.3), and (2.5), it is called a **conditional equation;** if a conditional sentence has the form of an inequality, as in (2.2) and (2.4), it is called a **conditional inequality.**

CONDITIONAL SENTENCE

CONDITIONAL EQUATION

CONDITIONAL INEQUALITY

An open sentence S_x containing a variable x whose universe is A can be used to determine a subset of A. Let us see how. If a is a member of the universal set A and if when we put a in place of x in S_x, we obtain a *true* statement, we say that a **satisfies** the sentence S_x. Then, being given a sentence S_x, we may consider the subset of A consisting of those members of A that satisfy S_x; we

a SATISFIES A SENTENCE S_x

use the symbols $\{x \in A \mid S_x\}$ to denote this subset of A. That is,

$$\{x \in A \mid S_x\} \tag{2.7}$$

is *the set of all the members of the universal set A that satisfy the sentence* S_x.

For example, if $A = \{2, 4, 6, 8, 10\}$, then $\{x \in A \mid x + 3 > 5\}$ is the set of all members of A for which $x + 3 > 5$ becomes a true statement. As we have seen above,

$$\{x \in A \mid x + 3 > 5\} = \{4, 6, 8, 10\}$$

Also from our preceding discussion we see that:

If $A = \{1, 2, 3, 4, 5, 6, 7, 8\}$, then $\{x \in A \mid x + 3 = 9\} = \{6\}$
If $A = \{3, 4, 5\}$, then $\{x \in A \mid x^2 + 2x = 1\} = \varnothing$
If $A = \{2, 3, 4, 5, 6, 7, 8, 9\}$, then $\{x \in A \mid x^2 > 3\} = A$

The set denoted by

$$\{x \in A \mid S_x\}$$

SOLUTION SET
SOLUTION

is called the *solution set,* or simply the **solution** (in the universe A), of the sentence S_x. To illustrate, $\{4, 6, 8, 10\}$ is the solution of $x + 3 > 5$ in the universe $\{2, 4, 6, 8, 10\}$; $\{6\}$ is the solution of $x + 3 = 9$ in the universe $\{1, 2, 3, 4, 5, 6, 7, 8\}$; $\varnothing$ is the solution of $x^2 + 2x = 1$ in the universe $\{3, 4, 5\}$; the universe itself is the solution of $x^2 > 3$ in the universe $\{2, 3, 4, 5, 6, 7, 8, 9\}$.

If the universe of the variable is clearly indicated in the context or is clearly understood, we abbreviate (2.7) as

$$\{x \mid S_x\} \tag{2.8}$$

and we read this as "the set of all members of the universe that satisfy the sentence S_x" or, more concisely, as "the set of all x (in the universe) for which S_x is true."

EXAMPLE 1 Find the solution of $x \div 4 = 1$ in the universe $A = \{1, 2, 4, 8, 16\}$; in other words, tabulate $\{x \in A \mid x \div 4 = 1\}$, where $A = \{1, 2, 4, 8, 16\}$.

SOLUTION Replacing x by the members of A in turn, we obtain the following statements:

$1 \div 4 = 1 \qquad 2 \div 4 = 1 \qquad 4 \div 4 = 1 \qquad 8 \div 4 = 1 \qquad 16 \div 4 = 1$

Since, by the definition of division in the set of natural numbers, $a \div b = c$ if and only if $b \cdot c = a$, we see that only the third statement is true. Therefore

$$\{x \in A \mid x \div 4 = 1\} = \{4\}$$

EXAMPLE 2 If $A = \{1, 2, 3, 4, 5\}$, tabulate $\{x \in A \mid 2x > 4 \text{ and } 3x < 15\}$.

SOLUTION Replacing x, in turn, by 1, 2, 3, 4, 5, we obtain

$2 \cdot 1 > 4$ and $3 \cdot 1 < 15$
$2 \cdot 2 > 4$ and $3 \cdot 2 < 15$
$2 \cdot 3 > 4$ and $3 \cdot 3 < 15$
$2 \cdot 4 > 4$ and $3 \cdot 4 < 15$
$2 \cdot 5 > 4$ and $3 \cdot 5 < 15$

The third and fourth of these statements are true and the others are false. Therefore

$$\{x \in A \mid 2x > 4 \text{ and } 3x < 15\} = \{3, 4\}$$

EXAMPLE 3 If the universe of the variable is the set N of natural numbers, tabulate $\{x \mid x + 5 = 14\}$.

SOLUTION From our knowledge of the addition table for natural numbers we see that the only natural number we can put in place of x in the sentence $x + 5 = 14$ to produce a true statement is the number 9. Therefore

$$\{x \mid x + 5 = 14\} = \{9\}$$

EXAMPLE 4 Find the solution of $4x = 8$ in the universe of natural numbers.

SOLUTION By definition, the solution of $4x = 8$ is $\{x \in N \mid 4x = 8\}$. From our knowledge of the multiplication table for natural numbers we see that the only natural number we can put in place of x in the sentence $4x = 8$ to produce a true statement is the number 2. Therefore, $\{2\}$ is the solution of $4x = 8$ in the universe of natural numbers.

EXAMPLE 5 Tabulate $\{x \in N \mid x + 11 = 5\}$.

SOLUTION Observe that if any natural number is added to 11, the result is greater than 11. Therefore, there is no natural number that can be put in place of x in $x + 11 = 5$ to produce a true statement; hence

$$\{x \in N \mid x + 11 = 5\} = \varnothing$$

EXAMPLE 6 If the universe of the variable is the set N of natural numbers, tabulate $\{x \mid 8x = 4\}$.

SOLUTION Observe that if any natural number is multiplied by 8, the result cannot be less than 8. Therefore, there is no natural number that satisfies $8x = 4$; hence $\{x \mid 8x = 4\} = \varnothing$ in the universe of natural numbers.

EXERCISES

In each of Exercises 1 to 14, tabulate the solution set. Wherever N appears, it designates the set of natural numbers.

1 $\{x \in A \mid x - 8 = 1\}$ $A = \{1, 3, 6, 9, 12\}$

2 $\{x \in A \mid 6x = 24\}$ $A = \{1, 2, 3, 4, 5, 6, 7\}$

3 $\{x \in A \mid x + 2 = 8\}$ $A = \{2, 4, 6, 8\}$

4 $\{x \in B \mid x - 3 = 6\}$ $B = \{2, 4, 6, 8, 10\}$

5 $\{x \in C \mid 5x + 2 = 7\}$ $C = \{1, 2, 3, 4, 5\}$

6 $\{x \in A \mid 3x - 5 = 4\}$ $A = \{1, 2, 3, 4, 5\}$

7 $\{x \in N \mid 2x - 1 = 5\}$

8 $\{x \in N \mid 4x + 2 = 22\}$

9 $\{x \in N \mid 2x = 3\}$

10 $\{x \in N \mid 2x - 1 = 3\}$

11 $\{x \in A \mid 3x > 20\}$ $A = \{1, 3, 5, 7, 9, 11\}$

12 $\{x \in B \mid x + 4 < 11\}$ $B = \{2, 4, 6, 8, 10\}$

13 $\{x \in N \mid 4x < 12\}$

14 $\{x \in N \mid 2x + 5 < 14\}$

In each of Exercises 15 to 26, find the solution of the given sentence for the given universal set A. N designates the set of natural numbers.

15 $x + 3 = 5$ $A = \{1, 2, 3, 4, 5\}$

16 $x - 3 = 4$ $A = \{1, 2, 3, 4, 5\}$

17 $x - 3 = 4$ $A = \{6, 7, 8, 9, 10\}$

18 $x + 3 > 6$ $A = \{1, 3, 5, 7, 9\}$

19 $x + 3 > 6$ $A = \{1, 2, 3, 4, 5\}$

20 $2x + 3 = 13$ $A = \{2, 4, 6, 8, 10\}$

21 $2x + 3 = 13$ $A = \{1, 3, 5, 7, 9\}$

22 $3x = 7$ $A = N$

23 $x + 4 = 4$ $A = N$

24 $x + 15 = 5x$ $A = N$

25 $x > 3$ and $x < 10$ $A = \{1, 3, 5, 7, 9, 11, 13\}$

26 $2x > 4$ and $3x < 7$ $A = \{1, 2, 3, 4\}$

2.2 ZERO AND THE NEGATIVE INTEGERS

In Sec. 1.13 we pointed out that the set of natural numbers is not closed under subtraction. Indeed, recalling the definition (1.29) of subtraction in Sec. 1.13, and the definition of the statement $a > b$ in Sec. 1.10, we see that if $a, b \in N$, then $a - b$ is a natural number if and only if $a > b$. In the preceding section we have seen that in the universe of natural numbers, the solution set of some sentences of the form $x + b = a$ is the empty set. For example,

$\{x \in N \mid x + 5 = 3\} = \varnothing$, $\{x \in N \mid x + 7 = 7\} = \varnothing$. In general $\{x \in N \mid x + b = a\}$ is the empty set *unless* $a > b$; if $a > b$, then we see that $\{x \in N \mid x + b = a\} = \{a - b\}$. Therefore, for natural numbers a and b, the requirement that $a - b$ be a natural number is equivalent to the requirement that $\{x \in N \mid x + b = a\}$ be non-empty.

In this section we shall discuss a way of enlarging or extending the set of natural numbers to produce a set I of numbers which is closed under subtraction, or equivalently, in which $\{x \in I \mid x + b = a\}$ is always nonempty for $a, b \in I$.

We have seen that for $a \in N$

$$\{x \in N \mid x + a = a\} = \varnothing \tag{2.9}$$

or equivalently, $a - a$ is not a natural number. In order that $\{x \mid x + a = a\}$ not be empty, we define a number **zero**, symbolized by 0, by the statement

ZERO

$$0 + a = a \qquad \text{for every natural number } a \tag{2.10}$$

In other words, "0" is a new object that is adjoined to N, with addition with 0 being defined by (2.10). Zero could also be defined as the number of members in the empty set. It is apparent that $0 \notin N$.

We mentioned in Sec. 1.4 that the set of natural numbers, $\{1, 2, 3, 4, 5, \ldots\}$, is also called the set of **positive integers.** We shall adopt this terminology for use throughout the remainder of this book.

POSITIVE INTEGERS

The set consisting of the positive integers and zero, that is, $\{0, 1, 2, 3, 4, 5, \ldots\}$, is called the set of **nonnegative integers** and is denoted by N_0.

NONNEGATIVE INTEGERS

In order that the nonnegative integers can properly be called *numbers,* we must define equality and the operations of addition and multiplication on the set, and we must be assured that the commutative, associative, and distributive properties hold (see Sec. 1.13).

If a and b are members of N_0, we say that a and b are **equal** if and only if a and b are equal positive integers (see the definition on page 15 in Sec. 1.5) or both a and b are zero.

EQUAL NONNEGATIVE INTEGERS

We *define addition and multiplication on the set* N_0 as follows: If $a, b \in N_0$ and neither a nor b is zero, then $a, b \in N$ and both $a + b$ and $a \cdot b$ are defined as in Sec. 1.5; if $a \in N_0$, then we define

$$a + 0 = 0 + a = a \tag{2.11}$$

and

$$a \cdot 0 = 0 \cdot a = 0 \tag{2.12}$$

Definition (2.11) of course is suggested by the definition (2.10) of zero and by the wish to have addition be commutative. Definition (2.12) is suggested by the notion that the union of a sets, each with no members, is a set with no members; that is,

$$a \cdot 0 = 0 + 0 + 0 + \cdots + 0 = 0$$

where there are a addends in the sum. In order to have the commutative property hold for multiplication, we of course require that $a \cdot 0 = 0 \cdot a$.

Using the definitions given above, it can be proved that the operations of addition and multiplication on the set N_0 of nonnegative integers are commutative and associative and that multiplication is distributive over addition. We shall not carry out the details of these proofs here.

Subtraction in the set N_0 of nonnegative integers is defined in the same way as in the set N of positive integers, namely, if $a, b \in N_0$, then $a - b$ is the nonnegative integer c, if such a nonnegative integer exists, with the property that $b + c = a$; or briefly,

$$a - b = c \quad \text{if and only if} \quad b + c = a \tag{2.13}$$

From definition (2.13) and definition (2.11) it follows that

$$\begin{aligned} &\text{for } a \in N_0, \\ &a - a = 0 \\ &a - 0 = a \end{aligned} \tag{2.14}$$

As special cases of (2.11) and (2.14) we have, respectively,

$$0 + 0 = 0 \qquad 0 - 0 = 0$$

We now have a set of numbers in which the solution set of such equations as $x + 4 = 4$ is not empty. Let us consider the following set:

$$\{x \in N_0 \mid x + 4 = 0\}$$

From the definition of addition in the set N_0 of nonnegative integers we see that this set is the empty set. In general,

$$\begin{aligned} &\text{if } a \in N, \\ &\{x \in N_0 \mid x + a = 0\} = \varnothing \end{aligned} \tag{2.15}$$

If we wish $\{x \in A \mid x + a = 0\}$ to be nonempty for $a \in N$, then the universal set A must contain numbers other than the nonnegative integers. Such numbers are introduced by the following definition.

Corresponding to each positive integer a, a number *negative* a, symbolized by $-a$, is *defined* by the statement

$$(-a) + a = 0 \qquad \text{for every natural number } a \tag{2.16}$$

That is,

$$\text{for each positive integer } a, \quad \{x \mid x + a = 0\} = \{-a\} \tag{2.17}$$

SET OF NEGATIVE INTEGERS

Thus we have defined a set whose members are $-1, -2, -3, -4, -5, \ldots$. This set is called the *set of* **negative integers.**

EQUAL NEGATIVE INTEGERS

If a and b are *negative integers* we say that a and b are **equal** if and only if the positive integers corresponding to a and b are equal.

The set composed of the positive integers, zero, and the negative integers,

$$\{\ldots, -5, -4, -3, -2, -1, 0, 1, 2, 3, 4, 5, \ldots\}$$

INTEGERS

is called the **set of integers** and is frequently denoted by I. Sometimes the plus sign (+) is prefixed to each of the positive integers; however, when no sign appears before an integer, it is understood to be a positive integer.

EQUAL INTEGERS

If a and b are *integers,* we say that a and b are **equal** if and only if they are equal positive integers or equal negative integers or both zero. If a and b are equal integers, we write

$$a = b$$

If a and b are not equal, we write

$$a \neq b$$

Recall the number scale in Fig. 1.13 and the way in which we marked off points to the right of the origin O corresponding to the positive integers. Similarly we mark off points to the *left* of O corresponding to the negative integers. Further, we assign the number 0 to the origin O. We then have the set of integers represented on a number scale, as shown in Fig. 2.1. On this extended number scale, for each point on the right side of O there is a corresponding point on the left side of O. These corresponding points are at equal distances from O but on opposite sides. For example, the distance between 4 and O is equal to the distance between -4 and O; both 4 and -4 are four units distance from O.

When an integer a is represented by a point on a number scale,

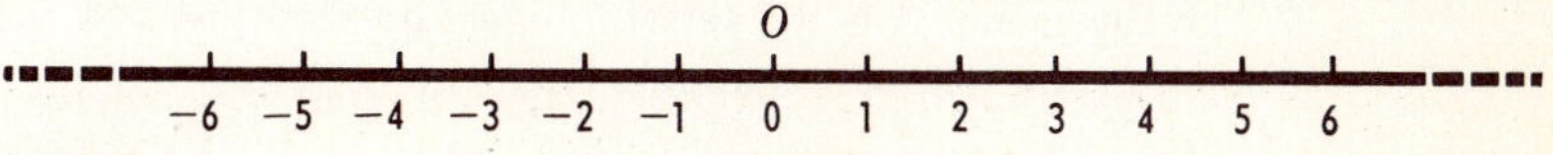

FIGURE 2.1

NUMERICAL VALUE

the distance between the point and the origin O is called the **numerical value** of a. Thus, the numerical value of 4 is 4, the numerical value of -4 is 4, the numerical value of -10 is 10, the numerical value of 0 is 0.

In Sec. 1.10 we discussed an ordering of the set N of positive integers. For a, $b \in N$, we agreed that $a > b$ means that there is a positive integer c with the property that $a - b = c$. In that section we related this definition to the number scale for positive integers, and saw that $a > b$ whenever the point representing a lies to the right of the point representing b. We shall extend this idea to the set of integers.

If a and b are integers, there are points on the number scale for integers (Fig. 2.1) that correspond to a and b, respectively. For convenience we shall speak of the point corresponding to a as "the point a" and the point corresponding to b as "the point b," and so on.

GREATER THAN

LESS THAN

We say that the integer a is **greater than** the integer b if the point a is to the right of the point b on the number scale, and for this we write $a > b$. On each of the number scales shown below, a is greater than b. If a is greater than b, then b is **less than** a, which

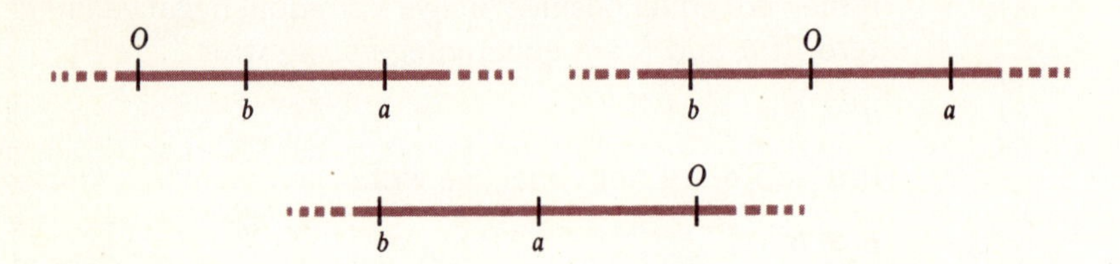

is written $b < a$. To illustrate, $2 > -3$ and $-3 < 2$; $-4 > -6$ and $-6 < -4$; $0 > -2$ and $-2 < 0$. Any positive integer is greater than zero, that is, if a is a positive integer, than $a > 0$; for this reason we read $a > 0$ as "a is positive." Similarly, $a < 0$ is read "a is negative."

Before the integers can properly be called numbers, we must define addition and multiplication on the set of integers in such a way that the commutative, associative, and distributive properties will hold. This will be done in the next section. Then in Sec. 2.4 we shall show that the set I of integers is closed under subtraction.

EXERCISES

1 Given that A is the set of integers greater than -4, indicate whether each of the following statements is true or false:

(a) $7 \in A$ (b) $-3 \in A$ (c) $-5 \in A$

(d) $0 \in A$ (e) $8 \in A$ (f) $-5 \notin A$

In Exercises 2 to 5, insert the symbol $<$ or $>$ between the given pair of numbers to make a true statement.

2 2, −4
3 −6, −8
4 −3, 2
5 −4, 0

In each of Exercises 6 to 11, give the numerical value of the integer.

6 8
7 −13
8 −2
9 −5
10 2
11 5

In each of Exercises 12 to 17, tabulate the solution set. N denotes the set of positive integers, and I denotes the set of integers.

12 $\{x \in N \mid x < 4\}$

13 $\{x \in N \mid x < 6\}$

14 $\{x \in A \mid x < -2\}$ $A = \{-4, -3, -2, -1\}$

15 $\{x \in A \mid x < 2\}$ $A = \{-3, -2, -1, 0, 1, 2, 3\}$

16 $\{x \in I \mid x + 1 = 1\}$

17 $\{x \in I \mid x + 4 = 4\}$

18 The noon temperature on a certain day in Augusta, Maine, was 18° above zero and was listed in the weather report as +18°. At 10 P.M. that night the temperature was 5° below zero and was listed in the weather report as −5°. How many degrees had the temperature changed from noon to 10 P.M.?

19 What temperatures are indicated by the reports +32°, −35°, +100°, −15°?

20 In a certain locality the lowest temperature during a certain day was −15° at 4 A.M., and the highest temperature was +29° at 3 P.M. How much did the temperature rise from 4 A.M. to 3 P.M.?

21 If the temperature in a certain locality is −8° at 9 P.M. and if the temperature had fallen 24° during the preceding 3 hours, what was the temperature at 6 P.M.?

22 Longitude is expressed in degrees east or west of a meridian near London, England. If a point 38° west longitude is written as −38°, how should you write 52° east longitude? 65° west longitude? 80° east longitude?

2.3 ADDITION AND MULTIPLICATION OF INTEGERS

We shall now define two operations, called addition and multiplication, on the set

$$I = \{\ldots, -5, -4, -3, -2, -1, 0, 1, 2, 3, 4, 5, \ldots\}$$

of integers. What we select as the meaning for the *sum* of two integers and the *product* of two integers is a matter of choice. In making these decisions we are guided, of course, by the definitions already made of addition and multiplication in the set of nonnegative integers—a subset of the set of integers. In other words, whatever definitions we select for the sum $a + b$ and the product $a \cdot b$ of integers a and b, we want those definitions to agree with the definitions previously given (in Sec. 1.5) for $a + b$ and $a \cdot b$ whenever a and b are *positive integers*. To this end it is useful to observe that the operation of addition in the set of positive integers can be given an interpretation on the number scale (see Sec. 1.10).

The sum $5 + 3 = 8$ can be interpreted on the number scale as follows: We fix our attention on the point 5 and move three units to the *right* of 5; we get to the point 8; so

$$5 + 3 = 8$$

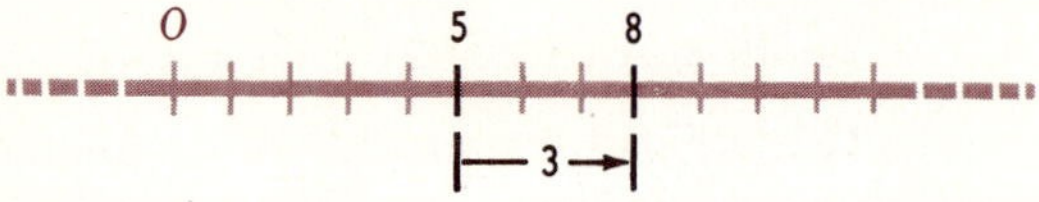

In general, if $a, b \in N$, the sum $a + b = c$ can be interpreted as follows: Start at the point a and move b units to the *right* to arrive at point c,

$$a + b = c$$

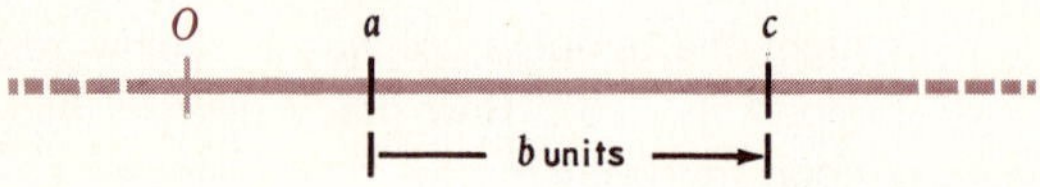

Hence it is natural to interpret the addition of a *positive* integer to a given integer a as a movement to the *right* on the number scale. Similarly it seems natural to interpret the addition of a *negative* integer to a given integer a as a movement to the *left* on the number scale. Let us look at some examples.

To add -5 and $+3$, fix attention on the point -5 and move 3 units to the *right*. We get to the point -2:

$$(-5) + (+3) = (-2)$$

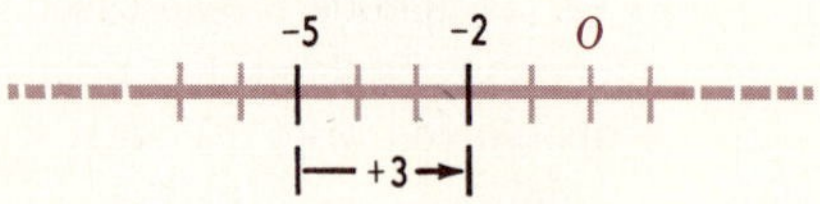

To add $+6$ and -5, start at the point 6 and move 5 units to the *left*. We get to the point 1:

$$(+6) + (-5) = (+1)$$

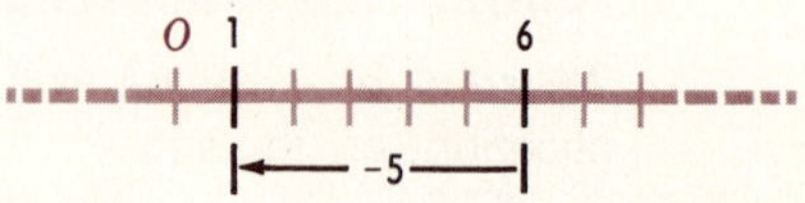

To add -5 and -3, start at the point -5 and move 3 units to the *left.* We get to the point -8:

$(-5) + (-3) = (-8)$

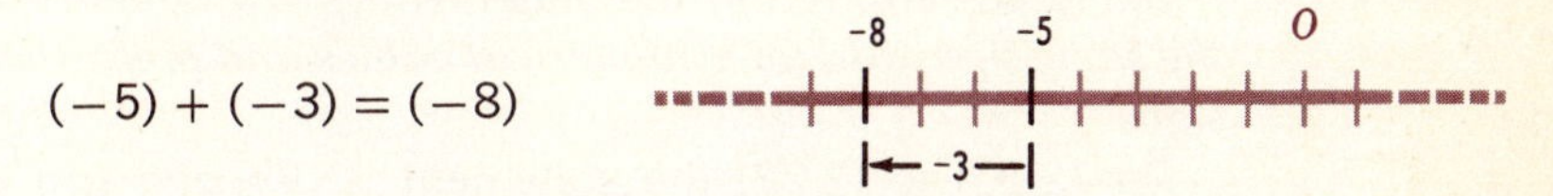

To add -3 and $+3$, start at the point -3 and move 3 units to the *right.* We arrive at the point 0:

$(-3) + (+3) = 0$

To add -3 and 0, start at the point -3 and do not move:

$(-3) + 0 = (-3)$

To add 0 and -3, start at the point 0 and move 3 units to the *left.* We arrive at the point -3:

$0 + (-3) = (-3)$

We elect to define addition in the set of integers so that it will agree with the examples given above. This is done formally in the following definition of **addition in the set of integers.**

ADDITION OF INTEGERS

For a and b any positive integers,

$$(+a) + (+b) = +(a + b) \tag{2.18}$$

$$(-a) + (-b) = -(a + b) \tag{2.19}$$

$$(+a) + (-b) = (-b) + (+a) = +(a - b) \quad \text{if } a > b \tag{2.20}$$

$$(+a) + (-b) = (-b) + (+a) = -(b - a) \quad \text{if } a < b \tag{2.21}$$

$$(+a) + (-b) = (-b) + (+a) = 0 \quad \text{if } a = b \tag{2.22}$$

$$(+a) + 0 = 0 + (+a) = +a \tag{2.23}$$

$$(-a) + 0 = 0 + (-a) = -a \tag{2.24}$$

$$0 + 0 = 0 \tag{2.25}$$

Notice that these definitions are given in terms of addition and subtraction and order in the set of *positive integers.* Notice also that definitions (2.23) and (2.25) agree with definition (2.11) given in Sec. 2.2, and that definition (2.22) agrees with definition (2.16) in Sec. 2.2. It should be clear that (2.18) to (2.25) enable us to add *any two integers;* in other words, addition is an *operation* on the set of integers. This is also expressed by saying that the set I of integers is *closed under addition.*

The definitions (2.18) to (2.25) may conveniently be summarized in the following statements.

For (2.18) and (2.19) the statement is: *To add two integers of the same sign, add their numerical values and prefix their common sign.*

For (2.20) to (2.22) the statement is: *To add two integers of unlike signs, subtract the smaller numerical value from the larger numerical value and prefix the sign of the number with the larger numerical value.*

For (2.23) to (2.25) the statement is: *The sum of any given integer and zero is the given integer.*

The sum of two integers is unique, that is, for a, b, c, $d \in I$,

if $a = b$ and $c = d$,
then $a + c = b + d$

Also, the definition of addition on the set of integers as given in (2.18) to (2.25) has been made in such a way that *addition is both commutative and associative; that is,*

for any integers a, b, c,

$$a + b = b + a \tag{2.26}$$

$$a + (b + c) = (a + b) + c \tag{2.27}$$

We shall not give the details of the proofs of these statements. The student, by a careful examination of the various parts of the definition, should be able to convince himself of the truth of (2.26) and (2.27).

We recall (Sec. 1.5) that $3 \cdot 4$ means $4 + 4 + 4$. It is natural to extend this definition of multiplication on the set of positive integers so that

$$3 \cdot (-4) = (-4) + (-4) + (-4) = -12$$

Such a definition agrees with experience, for if a business loses $4 per day, it will lose $12 in three days. However, we cannot define $(-4) \cdot 3$ in a similar way as the result of using 3 as an addend -4 times. But since we want the commutative property to be true for multiplication, we shall agree that

$$(-4) \cdot 3 = 3 \cdot (-4) = -12$$

Another consideration that leads to the agreement that $3(-4) = -12$ is the following. If we wish the distributive property $a(b + c) = ab + ac$ to hold for all integers, we must have

$$3[4 + (-4)] = 3 \cdot 4 + 3(-4)$$

The left member of this equality is $3(0) = 0$; for the distributive property to hold, the right member must be 0. In order for

$$3 \cdot 4 + 3(-4) = 0$$

to be true we must agree that

$$3(-4) = -(3 \cdot 4) = -12$$

PRODUCT OF TWO INTEGERS OF DIFFERENT SIGNS

These considerations suggest the following definition: *The product of two integers of different signs is always a negative integer, its numerical value being the product of the numerical values of the two integers.*

If a and b are any two positive integers, then

$$(+a)(-b) = -(ab) \tag{2.28}$$

$$(-a)(+b) = -(ab) \tag{2.29}$$

Recall that in Sec. 2.2 we had the definition

$$a \cdot 0 = 0 \cdot a = 0 \tag{2.30}$$

for any nonnegative integer a. Similarly, we *define*

$$(-a)0 = 0(-a) = 0 \qquad \text{for any negative integer } -a \tag{2.31}$$

We have yet to consider products like $(-3)(-4)$. Consider the expression $(-3)[4 + (-4)]$. We want the distributive property

$$a(b + c) = ab + ac$$

to hold for all integers, so that

$$(-3)[4 + (-4)] = (-3)(4) + (-3)(-4)$$

The left member of this equality is $(-3)0 = 0$; for the distributive property to hold, the right member must be 0. We have previously defined $(-3)(4)$ to be -12. In order to have

$$-12 + (-3)(-4) = 0$$

$(-3)(-4)$ must be 12. This argument does *not* prove that

$$(-3)(-4) = 12$$

It only shows that if we want the distributive property and the previously agreed-upon definitions to hold, we must agree that $(-3)(-4) = 12$.

PRODUCT OF TWO POSITIVE OR TWO NEGATIVE INTEGERS

We state the following definition: *The product of two positive integers or of two negative integers is the product of their numerical values.*

If a and b are any two positive integers, then

$$(+a)(+b) = +(ab) \tag{2.32}$$

$$(-a)(-b) = +(ab) \tag{2.33}$$

It should be clear that (2.28) to (2.33) enable us to find the product of *any two integers;* in other words, multiplication is an *operation* on the set of integers. Equivalently we say that the set I of integers is *closed under multiplication.*

Notice that definitions (2.28), (2.29), (2.32), and (2.33) are given in terms of multiplication on the set of positive integers and the definition of the set of negative integers. Notice also that definition (2.30) agrees with (2.12) in Sec. 2.2.

We assume the existence of another very useful property of zero: *If* $ab = 0$, *then either* $a = 0$ *or* $b = 0$ *or both* a *and* b *are equal to zero.* More generally, if the product of any number of numbers is zero, then at least one of the numbers must be zero.

LAW OF SIGNS FOR MULTIPLICATION

That part of the definitions (2.28) to (2.33) pertaining to the sign of the product of any two integers is frequently referred to as the *law of signs for the multiplication of integers* and is stated as follows: *The product of two integers of like signs is a positive integer, the product of two integers of unlike signs is a negative integer, and the product of any integer and zero is zero.*

The product of two integers is unique, that is, for $a, b, c, d \in I$,

if $a = b$ and $c = d$,
then $ac = bd$

From definitions (2.28) to (2.33) the student should see that multiplication on the set of integers is *commutative;* that is,

for any integers a, b,
$$a \cdot b = b \cdot a \tag{2.34}$$

From these definitions it can also be proved that multiplication on the set of integers is *associative;* that is,

for any integers a, b, c,
$$a \cdot (b \cdot c) = (a \cdot b) \cdot c \tag{2.35}$$

We shall not give the details of this proof here.

Considering the definitions of addition and multiplication given in this section, it can be shown that multiplication is distributive with respect to addition; that is,

for any integers a, b, c,
$$a(b + c) = ab + ac \tag{2.36}$$

Since (2.26), (2.27), (2.34), (2.35), and (2.36) are true for the set of integers, we can now say that the set of integers is a set of *numbers* (see Sec. 1.13). In the next section we shall define subtraction on the set I of integers, and we shall show that I is closed under subtraction.

EXERCISES

In Exercises 1 to 14, use the number scale, as in the illustrations on pages 70–71, to find each of the sums.

1 $(+7) + (-4)$
2 $(-4) + (+7)$
3 $(+8) + (+10)$
4 $(-8) + (-6)$
5 $0 + (-7)$
6 $(-9) + 0$
7 $(-7) + (+7)$
8 $(+7) + (+2) + (-4)$
9 $(-8) + (-3) + (+5)$
10 $(-1) + (-2) + (-3)$
11 $(+7) + (+3) + (-5)$
12 $(-8) + (+4) + (-4)$
13 $(+2) + (+3) + (+4)$
14 $0 + (-4) + (-5)$

In Exercises 15 to 32, find the sums by using the appropriate definition.

15 $9 + (-5)$
16 $(-1) + (-3)$
17 $(-1) + 7$
18 $14 + (-5)$
19 $(-14) + (-5)$
20 $[(+16) + (+15)] + (-19)$
21 $(-14) + 5$
22 $7 + [(-3) + 4]$
23 $[8 + (-9)] + (-3)$
24 $63 + (-71)$
25 $18 + [(-81) + 6]$
26 $[19 + (-31)] + (-9)$
27 $15 + (-15)$
28 $9 + 18 + (-27)$
29 $(-21) + (-3) + (-9) + 16$
30 $(-14) + 3 + (-9)$
31 $21 + (-9) + (-36) + 44$
32 $(-17) + (-3) + 4 + 10$

In Exercises 33 to 42, find the result by using the appropriate definition.

33 $(+6)(-3)$
34 $(-6)(-3)$
35 $(-7)(+8)$
36 $(-7)(-8)$
37 $(-9)(0)$
38 $7(-42)$
39 $10(-10)$
40 $(-1)(-2) + (-4)(-5)$
41 $(+3)(-5)(-2)(-3)$
42 $(-3)(-5) - (-2)(-5)$

In Exercises 43 to 50, find the value of the given expression if $a = 2$, $b = -3$, and $c = -4$.

43 $2a + bc$
44 $bc - 2a$
45 $2(a + c)$
46 $a^2 + b^2 + c^2$
47 $a^2 + c^2 - 2b^2$
48 ab^2c^2
49 $a^2b + c$
50 $2a^2 + 3b^2 - c^2$

2.4 SUBTRACTION AND DIVISION OF INTEGERS

SUBTRACTION OF INTEGERS

The definition of subtraction on the set of integers has the same form as the definition of subtraction on the set of positive integers. If a and b are any integers, we *define* $a - b$ to be the integer c, if such an integer exists, with the property that $b + c = a$; that is,

$a - b = c$ if and only if c is an integer with the property $b + c = a$

DIFFERENCE

If $a - b = c$, we call c the **difference** which results when we *subtract* b from a.

To illustrate, $(+3) - (+8)$ is an integer c such that $(+8) + c = (+3)$, or $8 + c = 3$. Here $c = -5$, since $(+8) + (-5) = (+3)$. We may visualize this subtraction with the aid of the number scale as follows: Start at the point $(+8)$; to get to the point $(+3)$, we go five units to the left.

$$(+8) + (-5) = (+3)$$
So $(+3) - (+8) = (-5)$

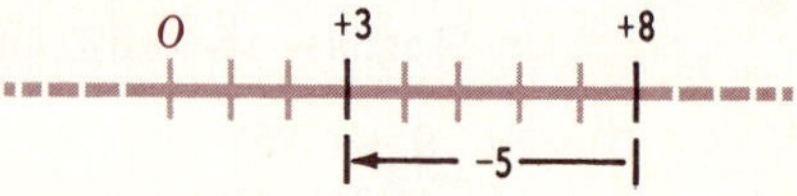

Observe that we would get the same result if we added (-8) to $(+3)$, for

$$(+3) + (-8) = (-5)$$

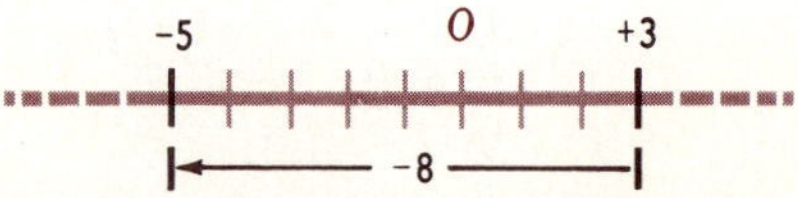

For another example, $(-5) - (-2)$ is an integer c such that $(-2) + c = (-5)$. Here $c = (-3)$, since $(-2) + (-3) = (-5)$. If we start at (-2), we must go three units to the left to get to the point (-5).

$$(-2) + (-3) = (-5)$$
So $(-5) - (-2) = (-3)$

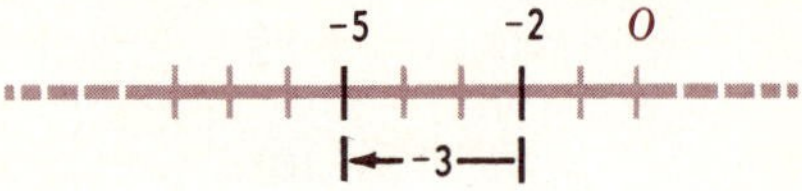

The same result is obtained by adding $(+2)$ to (-5):

$$(-5) + (+2) = (-3)$$

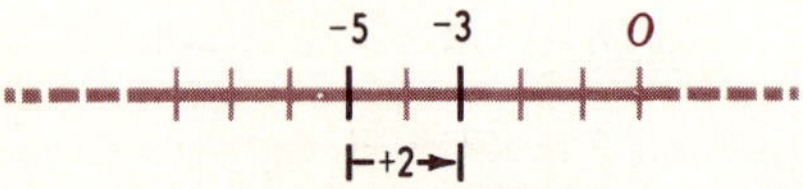

The concept of the negative of an integer can be used to great advantage in connection with subtraction in the set of integers. Let us define this concept.

NEGATIVE OF AN INTEGER

The **negative of an integer** a (positive, negative, or zero) is the number $(-a)$ which when added to a gives the sum 0. Some illustrations are:

The negative of 4 is -4, for $4 + (-4) = 0$.

The negative of -5 is 5, because $(-5) + 5 = 0$.

The negative of 0 is $-0 = 0$, since $0 + 0 = 0$.

From the above definition of the negative of an integer, it follows that we have

for any integer a,
$a + (-a) = 0$ (2.37)

Since the minus sign before any integer signifies the negative of that integer, it follows that:

$-a$ is negative if a is positive, and conversely.

$-a$ is positive if a is negative, and conversely.

Thus, if $a = 6$, then $-a = -6$; if $a = -8$, then $-a = 8$.

In connection with the two illustrations of subtraction given above, we observed that

$(+3) - (+8) = (-5)$ gives the same result as $(+3) + (-8) = (-5)$.
$(-5) - (-2) = (-3)$ gives the same result as $(-5) + (+2) = (-3)$.

That is, in these cases we see that subtracting a given integer gives the same result as adding its negative. This is true in general, as we shall now prove.

We want to prove that

if a and b are *any integers*,
then $a - b = a + (-b)$ (2.38)

By the definition of subtraction we have only to show that b plus the right side of the equation in (2.38) is equal to a. Now

$b + [a + (-b)] = b + [(-b) + a]$	*addition is commutative*
$= [b + (-b)] + a$	*addition is associative*
$= 0 + a$	*definition (2.37) of a negative*
$= a$	*use of (2.23) to (2.25)*

So the equality in (2.38) is proved. This result is sometimes expressed by this statement: *To subtract one integer from another, change the sign of the subtrahend and add the result to the minuend.* Therefore, we can think of $a - b$ as the sum of a and $(-b)$ rather than as a difference. This obviates the necessity of working out special formulas and rules for subtractions. It is as though negative numbers were created to make subtraction always possible, and as a consequence, subtraction as a necessary operation ceased to exist.

Interpreted geometrically, through the use of a number scale, result (2.38) says that to subtract a positive integer or to add a negative integer involves a movement to the left on a number scale, and to add a positive integer or to subtract a negative integer involves a movement to the right on the number scale.

An examination of the result (2.38) shows that we have attained one of our goals; namely, we have produced a set of numbers that is closed under subtraction. If a and b are *any* integers, result (2.38) tells us that $a - b = a + (-b)$. If b is an integer, then its negative $-b$ is also an integer, and since we know that the set of integers is closed under addition, we know that $a + (-b)$ is an integer and hence that $a - b$ is an integer.

In Sec. 2.2 we pointed out that the requirement that the set I be closed under subtraction is equivalent to the requirement that $\{x \in I \mid x + b = a\}$ be nonempty for any integers a and b. We can now prove that $\{x \in I \mid x + b = a\} = \{a - b\}$. Let us see how this is done. To say that $\{x \in I \mid x + b = a\} = \{a - b\}$ is equivalent (see Sec. 2.1) to saying that

$$(a - b) + b = a$$

is a true statement for every $a, b \in I$. To prove this we can proceed as follows: Let $a, b \in I$, then

$$\begin{aligned}(a - b) + b &= [a + (-b)] + b && \textit{by (2.38)}\\ &= a + [(-b) + b] && \textit{addition is associative}\\ &= a + 0 && \textit{definition of a negative}\\ &= a && \textit{use of (2.23) to (2.25)}\end{aligned}$$

In Sec. 2.2 we introduced the concept of the numerical value of an integer in a geometrical way. Another term for the numerical value of an integer is *absolute value,* and it is usually defined in a nongeometrical way. The **absolute value** of an integer a is denoted by $|a|$ and is defined as follows:

ABSOLUTE VALUE

1 The absolute value of a nonnegative integer is the integer itself, $|a| = a$ if $a \geq 0$.
2 The absolute value of a negative integer is the negative of that integer, $|a| = -a$ if $a < 0$.

Recalling that $-a$ is positive if a is negative, we see that the absolute value of an integer is never negative and is zero only when the integer is zero.

If an algebraic expression is made up of several parts connected by plus or minus signs, each part together with the sign that precedes it is called a **term.** In $+3x - 4y + 7$, which we usually write as $3x - 4y + 7$, the terms are $+3x$, $-4y$, and $+7$, or simply $3x$, $-4y$, and 7. In $5x(-y)z - 4xz$, the terms are $5x(-y)z$ and $-4xz$; here, since $5x(-y)z = -5xyz$, we could also say that the terms are $-5xyz$ and $-4xz$. Notice that expressions of this type can be written

TERM

ALGEBRAIC SUM

as sums, and for this reason each is called an **algebraic sum.** For example,

$$3x - 4y + 7 \qquad \text{or} \qquad 3x + (-4y) + 7$$

is an algebraic sum.

Throughout the study and application of arithmetic and algebra, the distributive property is perhaps the most important of the commonly encountered properties of numbers. We indicate here examples of some uses of this property.

For any integers a, b, and c,

$$a(b - c) = ab - ac \tag{2.39}$$

and

$$(-1)(a + b - c) = (-a) + (-b) + (+c)$$

or simply

$$-(a + b - c) = (-a) + (-b) + (+c) = -a - b + c \tag{2.40}$$

Equation (2.40) states that the negative of the expression $a + b - c$ is the sum of the negatives of the terms in that expression. In a similar way it can be shown that *the negative of the sum of any number of terms is equal to the sum of the negatives of those terms.* This principle is useful in simplifying expressions such as $(5x - 6y) - (-3x + 4y)$. We have

$$\begin{aligned}(5x - 6y) - (-3x + 4y) &= (5x - 6y) + [-(-3x + 4y)] && \textit{by (2.38)}\\ &= (5x - 6y) + (3x - 4y) && \textit{by the above principle}\\ &= 8x - 10y\end{aligned}$$

Similarly

$$(a - b) - (c - d) = (a - b) + [-(c - d)] = a - b - c + d$$

Also

$$-2(7 - 4c) = (-2)\cdot 7 - (-2)(4c) = -14 + 8c$$

$$6x(a - 4b) - 2y(3a + 2b - c) = 6xa - 24xb - 6ya - 4yb + 2yc$$

The definition of division on the set I of integers has the same form as the definition of division on the set of positive integers.

DIVISION OF INTEGERS

If a and b are any integers, we **define** $a \div b$ to be *the* integer c, if such an integer exists, with the property that

$$b \cdot c = a$$

that is, $a \div b = c$, if and only if c is *the* (unique) integer with the property $b \cdot c = a$.

QUOTIENT DIVIDEND DIVISOR If $a \div b = c$, where a, b, and c are integers, we call c the **quotient** which results when a is *divided* by b. We call a the **dividend** and b the **divisor.** Thus $8 \div 4$ is found by asking "4 times what number equals 8?" and observing from our knowledge of multiplication that 2 is the answer. The proviso in the definition as to the existence of c is necessary because in the set of integers, the quotient does not always exist. Thus $7 \div 3$ does not exist within the set of integers, since there is no integer c such that $3c = 7$. Similarly $8 \div (-5)$ does not exist within the set of integers, for there is no integer c such that $(-5)c = 8$. Therefore the set of integers is not closed under division.

We saw in Sec. 2.1 that $\{x \in N \mid bx = a\}$ is sometimes the empty set. This is also true if the universal set is the set I of integers. For example, $\{x \in I \mid 4x = 3\} = \varnothing$ and $\{x \in I \mid 3x = -2\} = \varnothing$. Recalling from Sec. 1.8 the meaning of the words "factor" and "multiple" in the set N of positive integers, we see that $\{x \in N \mid bx = a\} = \varnothing$ *unless* b is a factor of a (or equivalently, a is a multiple of b). We give meaning to the words "factor" and "multiple" in the set I of integers so that a similar statement will be true if the universal set is the set of integers.

FACTOR MULTIPLE If a and b are integers and if $a \div b = c$ is an integer, so that $bc = a$, we say that b is an (integral) **factor** of a and that a is an (integral) **multiple** of b. Thus 4 is a factor of 8, and 8 is a multiple of 4, since $8 \div 4 = 2$.

The following are some illustrations of divisions:

$12 \div 6$ means the number c such that $6 \cdot c = 12$. Here $c = 2$; that is, $12 \div 6 = 2$, and 6 is a factor of 12.

$(-18) \div 3$ means the number c such that $3 \cdot c = -18$. In this case $c = -6$; that is, $-18 \div 3 = -6$, since $3 \cdot (-6) = -18$, and so 3 is a factor of -18.

$(-12) \div (-3)$ means the number c such that $(-3) \cdot c = -12$. Here $c = 4$, since $(-3) \cdot 4 = -12$. So -3 is a factor of -12.

$0 \div 5$ means the number c such that $5 \cdot c = 0$. In this case $c = 0$, since $5 \cdot 0 = 0$; so $0 \div 5 = 0$, and 5 is a factor of 0.

The following situations are of a different nature from those just discussed:

If $7 \div 5$ existed in the set of integers, it would mean the integer c such that $5 \cdot c = 7$. But there is no such integer; 5 is not a factor of 7.

If $3 \div 0$ existed in the set of integers, it would mean the integer c such that $0 \cdot c = 3$. But $0 \cdot c = 0$ for any integer c. So $3 \div 0$ does not exist in the set of integers; 0 is not a factor of 3.

If $0 \div 0$ existed in the set of integers, it would mean the *unique* integer c with the property that $0 \cdot c = 0$. Since this equation is true when c is replaced by *any* integer, there is no unique integer with the required property, and so $0 \div 0$ is not an integer and 0 is not a factor of 0.

From the definition of division of integers we see that

$$\{x \in I \mid bx = a\} = \{c\}$$

if and only if c is an integer with the property that

$$a \div b = c$$

Therefore, $\{x \in I \mid bx = a\} = \varnothing$ *unless* b is a factor of a.

To prove that c is the quotient of a divided by b, all that is required is to show that $bc = a$. We shall use this principle to show that,

for any integers a, b, c,
$(ab) \div b = a$ (2.41)

To prove (2.41), we must show that b multiplied by a is equal to (ab). Now

$$ba = ab$$

by the commutative property for multiplication; so (2.41) is proved. This shows that division "undoes" multiplication. For this reason *division is called the inverse of multiplication.*

Since division is the inverse of multiplication, the law of signs for division of integers is obtained by considering the law of signs for multiplication of integers. Let a and b be positive integers. Then $(+a) \div (+b)$ means an integer c such that $+a = (+b) \cdot c$, and c must be positive (if it exists as an integer).

Similarly, $(+a) \div (-b)$ means an integer c such that $+a = (-b) \cdot c$, and c must be negative; $(-a) \div (-b)$ means an integer c such that $(-a) = (-b) \cdot c$, and c must be positive; $(-a) \div (+b)$ means an integer c such that $(-a) = (+b) \cdot c$, and c must be negative.

LAW OF SIGNS FOR DIVISION

These observations may be summarized in the following **law of signs for division of integers:** *If the quotient of two integers exists, the quotient is positive if the two integers are of like sign and the quotient is negative if the two integers are of unlike sign.*

Further, *the quotient* "0 *divided by* a," *where* a *is any integer except zero, is zero:* $0 \div a = 0$ $(a \neq 0)$.

The following are some illustrations:

$(-8) \div 2 = -4$ $(-8) \div (-2) = 4$
$8 \div (-2) = -4$ $0 \div (-2) = 0$

Just as was the case with division in the set of positive integers, if the dividend is a multiple of the divisor there is an equality of the form

$(divisor) \cdot (quotient) = dividend$

corresponding to the equality

$(dividend) \div (divisor) = quotient$

We again shall use the following convention (see Sec. 1.9): *In a problem which involves addition, subtraction, multiplication, and division, the multiplications and divisions will be performed before the additions and subtractions, except where otherwise indicated by parentheses.* To illustrate,

$$2 + 4 \cdot 5 - 4 \div 2 = 2 + 20 - 2 = 20$$
$$(2 + 4) \cdot 5 - (4 + 2) \div 2 = 6 \cdot 5 - 6 \div 2 = 30 - 3 = 27$$

EXERCISES

In Exercises 1 to 18, perform the indicated subtractions and additions.

1 $(-4) - 3$
2 $(-4) - (-3)$
3 $8 - (-2)$
4 $(-2) - 8$
5 $(-6) - 0$
6 $(-10) - (-16)$
7 $10 + (-5) - (-3)$
8 $(-3) + 16 - 4$
9 $0 - (-5)$
10 $0 - 5$
11 $4 + (-4) - (-4)$
12 $(-6) - (-3) + (-9)$
13 $(-5) + (-6) - (-4)$
14 $15 + (-18) - 3$
15 $20 - (-9) - (-4)$
16 $8 + (-5) - (-3)$
17 $(-8) + (-9) - (-17)$
18 $(-61) + 59 - (-5)$

19 Give the absolute value of each of the following: 7; -3; 4; 0; -13.
20 Give the absolute value of each of the following: 9; -4; 11; $7 - 10$; $5 - 2$.

In Exercises 21 to 24, find the value of the given expression.

21 $|3| + |-3|$
22 $|-3| + |-3|$
23 $|-3| - |-3|$
24 $|7| + |-8| + |-4| - |-1|$

25 Give the negative of each of the following: 9; -2; 0; 25; -15.
26 If $a = 7$ and $b = -4$, give the negative of each of the following: a; b; $-a$; $-b$; $|a|$; $|a + b|$; $|a| + |b|$; $-(-a)$.

Perform the indicated operations. Simplify the results by combining like terms when possible.

27 $(x + y) - (x + 2y)$
28 $(4x - 5y) - (2x - 2y)$
29 $8a - 2a - (-3a)$
30 $-(a - b)$

31 $4x - 2(1 - 3x)$
32 $-3(x - 2)$
33 $1 - c(a + b) + ac$
34 $-2n - (n - 7)$
35 $5 - (2a - 5)$
36 $2a^2 - a(a - 2)$

Factor the following expressions:

37 $3a - 4ac$
38 $-14x - 7y$
39 $-ax + 2ax$
40 $3xy - 3x$

Expand:

41 $(s + a)(r - b)$
42 $(x - 2)(x + 3)$
43 $(x - 2)(x - 2)$
44 $(x + 4)(x + 1)$

Factor the following into the product of two binomials:

45 $x^2 - 5x + 6$
46 $x^2 - 5x - 6$
47 $x^2 - 7x + 10$
48 $x^2 + 3x - 10$

In Exercises 49 to 56, remove the brackets and parentheses, and simplify the results by combining like terms.

49 $(3a + 2b) - [(a - 2b) - 2(3a + b)]$
50 $5x - 2[5y - (7x - 2y)]$
51 $[(3a + b)^2 - 4ab] - 3a[2b(5 - 7)]$
52 $3[2(a + b) - 4(a - b)] - 5(a + 3b)$
53 $7 - [3x - (5 - 2x)]$
54 $a[b(c - 4) + 5] - 2abc$
55 $a - (b - c) + 2[a + 3(b + c)]$
56 $2x^2 - [(3x + 3)x + x^2]$

Which of the following expressions are meaningless within the set of all integers? In case a quotient exists, give its value as an integer.

57 $10 \div 2$
58 $7 \div 3$
59 $4 \div 1$
60 $2 \div 2$
61 $9 \div 4$
62 $1 \div 4$
63 $10 \div (-2)$
64 $(-16) \div (-4)$
65 $(-21) \div 7$
66 $(-12 + 2) \div (-4 - 1)$

67 Is -4 a divisor of -24? Why?
68 Is $(-2 - 3)$ a divisor of -15? Why?
69 Is -7 a divisor of $(20 + 8)$? Why?

2.5 EQUATIONS AND INEQUALITIES INVOLVING INTEGERS

We recall from Sec. 2.1 that the *solution,* in the universe A, of an open sentence S_x is the *set* of all members of the universal set A that satisfy the sentence S_x. The solution of S_x in the universe A

is denoted by

$$\{x \in A \mid S_x\}$$

Since we are working within the set I of integers, the universe of the variable must be a subset of I and any constants that appear in S_x must represent numbers in the set I. The most common types of open sentences that occur in elementary mathematics are equations, inequalities, and compound sentences whose components are equations or inequalities, or both. In this section we shall discuss some examples of open sentences and their solutions. In these examples and throughout the remainder of the book we shall use the following symbolism:

$N = \{1, 2, 3, 4, 5, \ldots\}$, the set of positive integers
$N_0 = \{0, 1, 2, 3, 4, 5, \ldots\}$, the set of nonnegative integers
$I = \{\ldots, -5, -4, -3, -2, -1, 0, 1, 2, 3, 4, 5, \ldots\}$, the set of integers

EXAMPLE 1 Find the solution of the sentence $3x = 12$ in the universe I.

SOLUTION We wish to find the set of all members of I that satisfy $3x = 12$; that is, the set of all members of I that can be put in place of x in the equation $3x = 12$ to produce a true statement. It should be clear that the solution is $\{4\}$. In symbols we write

$$\{x \in I \mid 3x = 12\} = \{4\}$$

EXAMPLE 2 Find the solution of the sentence $3x = 12$ in the universe of negative integers.

SOLUTION If $A =$ the set of negative integers, we wish to determine the members of

$$\{x \in A \mid 3x = 12\}$$

It should be clear that there is no negative integer that satisfies $3x = 12$; therefore

$$\{x \in A \mid 3x = 12\} = \varnothing$$

that is, the solution of $3x = 12$ in the set of negative integers is the empty set.

EXAMPLE 3 Find the solution of $2x + 1 = 6x - 15$ in the universe $A = \{0, 2, 4, 6\}$.

SOLUTION We wish to determine the members of $\{x \in A \mid 2x + 1 = 6x - 15\}$, where $A = \{0, 2, 4, 6\}$. Substituting each member of A in turn in the equation $2x + 1 = 6x - 15$, we find that we obtain a false statement for 0, 2, and 6 and a true statement for 4. Therefore the solution is $\{4\}$.

EXAMPLE 4 Find the solution of $x < 5$ in the universe $A = \{-4, -2, 0, 2, 4, 6, 8, 10\}$.

SOLUTION The members of $\{x \in A \mid x < 5\}$ are $-4, -2, 0, 2, 4$, since each of the points corresponding to each of these numbers on the number scale lies to the left of the point 5, and each of the points 6, 8, and 10 lies to the right of the point 5.

$$\{x \in A \mid x < 5\} = \{-4, -2, 0, 2, 4\}$$

EXAMPLE 5 Find the solution of the sentence "$-3 < x$ and $x < 5$" in the universe I.

SOLUTION To find the solution, we must determine the set consisting of all the members of I that satisfy *both* $-3 < x$ *and* $x < 5$. Note that $\{x \in I \mid -3 < x\} = \{-2, -1, 0, 1, 2, 3, 4, 5, 6, \ldots\}$ and $\{x \in I \mid x < 5\} = \{\ldots, -4, -3, -2, -1, 0, 1, 2, 3, 4\}$. The set of all members of I that satisfy both $-3 < x$ and $x < 5$ will consist of the integers common to these two sets, so, recalling the definition of intersection of sets given in Sec. 1.3, we have

$$\begin{aligned}\{x \in I \mid -3 < x \text{ and } x < 5\} &= \{-2, -1, 0, 1, 2, 3, 4, 5, \ldots\} \\ &\quad \cap \{\ldots, -4, -3, -2, -1, 0, 1, 2, 3, 4\} \\ &= \{-2, -1, 0, 1, 2, 3, 4\}\end{aligned}$$

The sentence whose solution was sought in Example 5 is an example of a compound sentence made up of two sentences joined by the word "and"; such a sentence is called a *conjunction.* In this example we were asked to determine the numbers that could be put in place of x to produce a true statement. Let us consider for a moment what we mean by a "true" conjunction.

CONJUNCTION OF STATEMENTS

Let p designate one statement, and let q designate another statement. The **conjunction** of these two statements is the statement

p and q

This conjunction is true *only* if p and q *both* are true.

Suppose now that P_x and Q_x represent open sentences in the universe A. The open sentence "P_x and Q_x" is the conjunction of the two sentences P_x and Q_x; the solution of this compound open sentence in the universe A is the set of all members of A which satisfy *both* P_x *and* Q_x (as illustrated in Example 5). It follows from the definition of the intersection of sets, given in Sec. 1.3, that

$$\{x \in A \mid P_x \text{ and } Q_x\} = \{x \in A \mid P_x\} \cap \{x \in A \mid Q_x\} \tag{2.42}$$

It is customary to write the compound sentence "$-3 < x$ and $x < 5$" in the condensed form "$-3 < x < 5$," which is read "-3 is less than x and x is less than 5" or "x is greater than -3 and

$a<x<b$ less than 5." In general, the sentence "$a < x < b$" means "*$a < x$ and $x < b$.*"

EXAMPLE 6 Find the solution of $4 < x < 10$ in the universe $A = \{0, 2, 4, 6, 8, 10, 12, 14\}$.

SOLUTION We know that $4 < x < 10$ means "$4 < x$ and $x < 10$." Therefore we wish to determine the members of $\{x \in A \mid 4 < x \text{ and } x < 10\}$, that is, the members of A that are both greater than 4 and less than 10. By the use of (2.42), we have

$$\{x \in A \mid 4 < x \text{ and } x < 10\} = \{x \in A \mid 4 < x\} \cap \{x \in A \mid x < 10\}$$

Now

$$\{x \in A \mid 4 < x\} = \{6, 8, 10, 12, 14\}$$
$$\{x \in A \mid x < 10\} = \{0, 2, 4, 6, 8\}$$

and

$$\{x \in A \mid 4 < x \text{ and } x < 10\} = \{6, 8, 10, 12, 14\} \cap \{0, 2, 4, 6, 8\} = \{6, 8\}$$

EXAMPLE 7 Find the solution of the sentence "$x < 6$ or $x = 9$" in the universe N of positive integers.

SOLUTION Note that $\{x \in N \mid x < 6\} = \{1, 2, 3, 4, 5\}$ and $\{x \in N \mid x = 9\} = \{9\}$. Recalling the definition of union of sets given in Sec. 1.3, we see that the set of all members of N that satisfy "$x < 6$ or $x = 9$" will be the union of the two sets $\{x \in N \mid x < 6\}$ and $\{x \in N \mid x = 9\}$. Therefore

$$\{x \in N \mid x < 6 \text{ or } x = 9\} = \{1, 2, 3, 4, 5\} \cup \{9\} = \{1, 2, 3, 4, 5, 9\}$$

The sentence whose solution was sought in Example 7 is an example of a compound sentence made up of two sentences joined by the word "or." Such a sentence is called a *disjunction*. If we let

DISJUNCTION OF STATEMENTS

p represent a statement and let q represent a statement, the **disjunction** of these statements is the statement

p or q

This disjunction is true if *at least one* of the statements p, q is true. To illustrate:

$4 = 2$ or $2 + 5 = 7$	is true, because $2 + 5 = 7$ is true
$1 - 8 = 7$ or $6 + 6 = 10$	is false, because neither $1 - 8 = 7$ nor $6 + 6 = 10$ is true
$9 \cdot 2 = 18$ or $2 \cdot 1 = 3$	is true, because $9 \cdot 2 = 18$ is true
$9 - 1 = 8$ or $2 \cdot 2 = 4$	is true, because both $9 - 1 = 8$ and $2 \cdot 2 = 4$ are true

We shall always use the word "or" in this sense throughout the book. The student should note that the meaning of "or" in ordinary English discourse may be ambiguous. In some instances "or" is used in the way we have agreed to use it as meaning one or the other or both of two possibilities; in other instances it is used as meaning one or the other but *not* both of two possibilities. An example of this latter use is in the sentence "At noon tomorrow I will be at home or at Bill's house."

If P_x and Q_x represent open sentences in the universe A, then "P_x or Q_x" is the disjunction of these sentences. The solution of the compound open sentence "P_x or Q_x" in the universe A is the set of all members of A that satisfy *at least one* of the sentences P_x, Q_x (as illustrated in Example 7). It follows from the definition of the union of sets, given in Sec. 1.3, that

$$\{x \in A \mid P_x \text{ or } Q_x\} = \{x \in A \mid P_x\} \cup \{x \in A \mid Q_x\} \tag{2.43}$$

EXAMPLE 8 Find the solution of the sentence "$x > 3$ or $x < -2$" in the universe $S = \{-4, -3, -2, -1, 0, 1, 2, 3, 4, 5\}$

SOLUTION By the use of (2.43) we have

$$\begin{aligned}\{x \in S \mid x > 3 \text{ or } x < -2\} &= \{x \in S \mid x > 3\} \cup \{x \in S \mid x < -2\}\\ &= \{4, 5\} \cup \{-4, -3\}\\ &= \{-4, -3, 4, 5\}\end{aligned}$$

Recall that the sentence "$-3 < x$ and $x < 5$," which appeared in Example 5, could be written in the condensed form "$-3 < x < 5$." There is no comparable condensed form for sentences like "$x > 3$ or $x < -2$."

EXERCISES

1 What is the solution of the sentence $4x = 24$ in the universe I? In the universe N? In the universe of negative integers?

2 What is the solution of the sentence $3x = -15$ in the universe I? In the universe N? In the universe of negative integers?

In Exercises 3 to 8, find the solution of the given equality in the specified universe A.

3 $3x - 2 = 2x + 5$ $\quad A = \{3, 5, 7, 9\}$
4 $3x - 2 = 2x + 5$ $\quad A = \{2, 4, 6, 8\}$
5 $5x = 35$ $\quad A = I$
6 $5x = 35$ $\quad A =$ the set of negative integers
7 $2x + 5 = 25$ $\quad A = \{7, 8, 9, 10, 11, 12\}$
8 $4x - 5 = 3(x + 2)$ $\quad A = \{10, 11, 12, 13\}$

In Exercises 9 to 14, find the solution of the given inequality in the specified universe A.

9 $2x > 10$ $A = \{3, 5, 7, 9\}$
10 $5x < 10$ $A = \{2, 4, 6, 8\}$
11 $5x - 5 > -9 + 3x$ $A = \{-3, -2, -1, 0\}$
12 $2x - 9 < 5x - 3$ $A = \{1, 2, 3, 4\}$
13 $4 + 3x < 10$ $A = \{0, 1, 2, 3, 4\}$
14 $x - 7 < -4x - 1$ $A = \{0, 1, 2, 3, 4\}$

15 If $A = \{3, 5, 7, 9\}$, using your results in Exercises 3 and 9, determine the members of

$\{x \in A \mid 3x - 2 = 2x + 5 \text{ and } 2x > 10\}$

16 If $A = \{2, 4, 6, 8\}$, using your results in Exercises 4 and 10, determine the members of

$\{x \in A \mid 3x - 2 = 2x + 5 \text{ and } 5x < 10\}$

In Exercises 17 to 24, find the solution of the given sentence in the specified universe.

17 $x > -4$ and $x < 3$ universe I
18 $x < 7$ and $x > -2$ universe I
19 $x > -4$ or $x < 3$ universe I
20 $x < -4$ or $x > -2$ universe $\{-5, -4, -3, -2, -1\}$
21 $x > -4$ and $x < -2$ universe $\{-5, -4, -3, -2, -1\}$
22 $x < 3$ or $x > 6$ universe $\{0, 1, 2, 3, 4, 5, 6, 7\}$
23 $2 < x < 8$ universe I
24 $x < 5$ or $x = 10$ universe N

25 Indicate whether each of the following statements is true or false, and give a reason for your classification:

(*a*) $5 = 7$ or $3 + 6 = 9$
(*b*) $9 - 4 = 5$ or $8 + 8 = 88$
(*c*) $3 \cdot 4 = 12$ or $7 + 8 = 15$
(*d*) $5 = 7$ and $3 + 6 = 9$
(*e*) $3 \cdot 4 = 12$ and $7 + 8 = 15$

2.6 SUMMARY OF THE PROPERTIES OF THE SET OF INTEGERS

In this chapter we have "enlarged" the set of positive integers by defining the number zero and the set of negative integers. In this way we have produced the set I of integers.

Just as was the case for the set N of positive integers (or natural numbers), we have defined what it means to say that two integers a and b are equal ($a = b$), and from our common experience we are aware that "equality" has the same properties E_1, E_2, and E_3 as were given in Sec. 1.13 for positive integers.

We have defined addition (+) and multiplication (·) on the set of integers in such a way that the set of integers is *closed* under each of these operations. These operations have the properties that we have called the uniqueness, commutative, associative, and distributive properties.

Therefore the set I of integers constitutes a *set of numbers* (see Sec. 1.13), that is:

For every $a, b, c \in I$,

E_1: $a = a$

E_2: if $a = b$, then $b = a$

E_3: if $a = b$ and $b = c$, then $a = c$

P_1: $a + b \in I$ and $a + b$ is unique
$a \cdot b \in I$ and $a \cdot b$ is unique

P_2: $a + b = b + a$
$a \cdot b = b \cdot a$

P_3: $a + (b + c) = (a + b) + c$
$a \cdot (b \cdot c) = (a \cdot b) \cdot c$

P_4: $a(b + c) = ab + ac$

These seven properties of course are possessed by any number system; in particular we have seen that the set N of positive integers and the set N_0 of nonnegative integers have these properties.

IDENTITY UNDER MULTIPLICATION

There is another property that the set I shares with the sets N and N_0; in each of these sets there is an **identity under multiplication.** That is,

P_5: there exists a member $e \in I$ with the property that for any $a \in I$, $a \cdot e = e \cdot a = a$

UNIT ELEMENT

The member $e \in I$ that has this property is the number 1. We call this member the* identity under multiplication or the **unit element** in the set I.

The set I (and also the set N_0) contains a member, namely 0, with the property that

for any integer a,
$a + 0 = 0 + a = a$

IDENTITY UNDER ADDITION

We call the number that possesses this property the* **identity under addition.** We state this property as follows:

P_6: there exists a member $i \in I$ with the property that for any $a \in I$, $a + i = i + a = a$

ZERO ELEMENT

The member of I (and of N_0) that has this property is called the **zero element** in the set, and for I and N_0 it is of course the number 0.

A feature that distinguishes the set I of integers from the sets N and N_0 is that corresponding to any given integer a there is an

*See the footnote on p. 57.

integer, denoted by $-a$, with the property that

$$a + (-a) = 0$$

INVERSE OF a UNDER ADDITION

We call the number $-a$ that possesses this property the* **inverse of a under addition,** or, as in Sec. 2.4, the *negative* of a. We may state this result as follows:

P_7: if i is the identity under addition, and if a is any member of I, then there exists a number $-a \in I$ with the property that $a + (-a) = (-a) + a = i$

We have seen [equation (2.38)] that the subtraction of b from a is the same as the addition of $-b$ to a. Therefore, the requirement that the property P_7 be true in a number system is equivalent to requiring that subtraction always be possible. So, because of P_7, we can say,

for any $a, b \in I$,
$a - b \in I$

GREATER THAN

LESS THAN

In Sec. 2.2 we defined an order relation in the set of integers. The definition given there was stated in geometric terms; it can be restated in algebraic terms as follows: *If* $a, b \in I$ *and if there exists a positive integer* c *with the property that* $a - b = c$, *then we say* a *is* **greater than** b *and write* $a > b$. If $a > b$, then b is **less than** a, which is written $b < a$. This order relation has properties that are analogous to, but not entirely identical with, the properties of the order relation in the set of positive integers as given in Sec. 1.13. These properties are:

For $a, b, c \in I$,
O_1: either $a > b$, or $b > a$, or $a = b$
O_2: if $a > b$ and $b > c$, then $a > c$
O_3: if $a > b$, then $a + c > b + c$
O_4: if $a > b$, then $ac > bc$ *provided* $c > 0$
$ac < bc$ *provided* $c < 0$

We have discussed division in the set I of integers and have pointed out the fact that the set of integers is not closed under division; that is,

if $a, b \in I$,
$a \div b$ may not represent an integer

In the next chapter we shall discuss a set of numbers in which the quotient of two numbers exists except when the divisor is zero.

* It can be proved that if in a set of numbers, a number has an inverse under addition, then there is only one such inverse.

3

The Arithmetic and Algebra of Rational Numbers

3.1
THE SET OF FRACTIONS—EQUALITY OF FRACTIONS

We have pointed out that the set of integers is not closed under division. In Sec. 2.4 we saw that if $a, b \in I$, then $a \div b$ is *not* an integer, or equivalently, $\{x \in I \mid bx = a\} = \varnothing$, *unless* b is a factor of a. Of course, if b is a factor of a, so that $a \div b = c$, where $c \in I$, then

$$\{x \in I \mid bx = a\} = \{c\}$$

As a first step in producing a set of numbers which is closed under division, except for division by zero, we define the set of fractions.

FRACTION

If a and b are integers and $b \neq 0$, we call $\frac{a}{b}$ or a/b a **fraction.**

NUMERATOR

DENOMINATOR

We call a the **numerator** and b the **denominator** of the fraction $\frac{a}{b}$.

Thus $\frac{3}{2}, \frac{7}{15}, \frac{-8}{4}, \frac{6}{-11}, \frac{0}{4}, \frac{5}{1}, \frac{2}{2}$, and $\frac{-1}{-3}$ are fractions.

As we have noted, among the set of fractions are those whose denominators are 1. We agree to identify the fraction $\frac{a}{1}$ with the integer a; that is,

$$\frac{a}{1} = a \tag{3.1}$$

Suppose that the rectangle of Fig. 3.1 is divided into six equal smaller rectangles. Each of these small rectangles has an area which is $\frac{1}{6}$ of the area of the large rectangle. Again, divide the given large rectangle into two equal smaller rectangles. The area of each of the latter rectangles is equal to $\frac{1}{2}$ of the area of the large rectangle and is equal to the area of three of the smaller rectangles. Otherwise stated,

$$\frac{1}{2} = \frac{3}{6}$$

This suggests that we may want to say that two fractions are equal even though they have different symbolic forms.

Our experience with fractions, exemplified by the preceding example, leads us to the conclusion that if $a, b, k \in I$ with $b \neq 0$ and

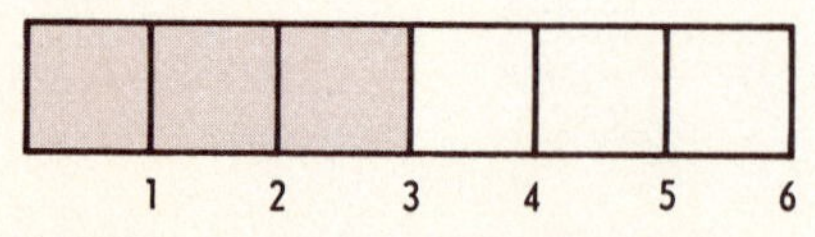

FIGURE 3.1

$k \neq 0$, then we wish to agree that $\frac{a}{b}$ and $\frac{ak}{bk}$ are equal fractions. We shall indeed make a definition of equality in the set of fractions that will ensure that

$$\frac{a}{b} = \frac{ak}{bk} \tag{3.2}$$

whenever $a, b, k \in I$ *with* $b \neq 0$ *and* $k \neq 0$.

Let $\frac{a}{b}$ and $\frac{c}{d}$ be fractions; we say that $\frac{a}{b}$ is *equal* to $\frac{c}{d}$ if and only if $ad = bc$. If $\frac{a}{b}$ is equal to $\frac{c}{d}$, we write $\frac{a}{b} = \frac{c}{d}$. That is,

EQUALITY OF FRACTIONS

$$\frac{a}{b} = \frac{c}{d} \quad \textit{if and only if} \quad ad = bc \tag{3.3}$$

Some illustrations of the use of (3.3) are

$$\frac{1}{7} = \frac{4}{28} \qquad \textit{because } 1 \cdot 28 = 7 \cdot 4$$

$$\frac{2}{3} = \frac{-2}{-3} \qquad \textit{because } 2(-3) = 3(-2)$$

$$\frac{18}{9} = \frac{6}{3} \qquad \textit{because } 18 \cdot 3 = 9 \cdot 6$$

Let us now show that definition (3.3) makes (3.2) true.

According to (3.3), $\frac{a}{b} = \frac{ak}{bk}$ will be true provided $a(bk) = b(ak)$. Now

$$\begin{aligned} a(bk) &= (ab)k && \textit{multiplication of integers is associative} \\ &= (ba)k && \textit{multiplication of integers is commutative} \\ &= b(ak) && \textit{multiplication of integers is associative} \end{aligned}$$

Therefore, making use of the definition (3.3), we have *proved* that

$$\frac{a}{b} = \frac{ak}{bk} \qquad \text{for } a, b, k \in I \text{ with } b \neq 0 \text{ and } k \neq 0$$

The student should note that the symbol = is used in two different ways in (3.3); in $\frac{a}{b} = \frac{c}{d}$ it is used to denote equality of fractions, while in $ad = bc$ it is used to denote equality of integers. Thus we have defined the new concept of equality in the set of fractions in terms of the previously defined concept of equality in the set of integers.

Equality (3.2) tells us that if the numerator and denominator of a given fraction are multiplied by the same nonzero integer, the resulting fraction is equal to the given fraction. To illustrate,

$$\frac{1}{2} = \frac{1 \cdot 4}{2 \cdot 4} = \frac{4}{8} \qquad \frac{-2}{3} = \frac{(-2)(-3)}{3(-3)} = \frac{6}{-9}$$

Equality (3.2) also tells us that if the numerator and denominator of a given fraction are divided by the same nonzero integer that is a factor of both the numerator and denominator, the resulting fraction is equal to the given fraction. To illustrate,

$$\frac{5}{15} = \frac{1 \cdot 5}{3 \cdot 5} = \frac{(1 \cdot 5) \div 5}{(3 \cdot 5) \div 5} = \frac{1}{3} \qquad \frac{-9}{6} = \frac{(-3)(3)}{2 \cdot 3} = \frac{-3}{2}$$

As a special case of equality (3.1) we see that $\frac{1}{1} = 1$. From definition (3.3) of equality of fractions we have

$$\frac{a}{a} = \frac{1}{1} \qquad \textit{since } a \cdot 1 = 1 \cdot a$$

Therefore

$$\frac{a}{a} = \frac{1}{1} = 1 \qquad \textit{for any nonzero integer } a \tag{3.4}$$

From equality (3.2) it follows that if we are given a fraction $\frac{a}{b}$, then there are an infinite number of fractions that are equal to $\frac{a}{b}$; for example,

$$\frac{1}{2} = \frac{2}{4} = \frac{-2}{-4} = \frac{3}{6} = \frac{-3}{-6} = \frac{4}{8} = \frac{-4}{-8} = \frac{5}{10} = \cdots \tag{3.5}$$

$$\frac{7}{3} = \frac{14}{6} = \frac{-14}{-6} = \frac{21}{9} = \frac{-21}{-9} = \frac{28}{12} = \frac{-28}{-12} = \cdots \tag{3.6}$$

$$\frac{-16}{64} = \frac{16}{-64} = \frac{-4}{16} = \frac{1}{-4} = \frac{2}{-8} = \frac{-6}{24} = \cdots \tag{3.7}$$

EXERCISES

State the missing numerator in each of the following:

1 $\frac{?}{24} = \frac{1}{6}$ 2 $\frac{1}{3} = \frac{?}{6}$ 3 $\frac{7}{16} = \frac{?}{48}$

4 $\frac{3}{8} = \frac{?}{288}$ 5 $\frac{3x}{7} = \frac{?}{14}$ 6 $\frac{a}{6} = \frac{?}{24}$

State the missing denominator in each of the following:

7 $\frac{4}{12} = \frac{1}{?}$ 8 $\frac{3}{7} = \frac{6x}{?}$ 9 $\frac{2}{5x} = \frac{16x}{?}$

10 $\frac{7}{2x} = \frac{14x^2}{?}$ 11 $\frac{7}{8} = \frac{14}{?}$ 12 $\frac{3}{4x} = \frac{12xy}{?}$

Find the missing numbers in each of the fractions in the following exercises:

13 $\frac{3}{4} = \frac{?}{8} = \frac{?}{12} = \frac{?}{16} = \frac{?}{20} = \frac{?}{32} = \frac{?}{100}$

14 $\frac{2}{3} = \frac{?}{6} = \frac{?}{9} = \frac{?}{12} = \frac{?}{15} = \frac{?}{30} = \frac{?}{60}$

15 $\frac{3}{5} = \frac{?}{10} = \frac{?}{15} = \frac{?}{20} = \frac{?}{25} = \frac{?}{50} = \frac{?}{100}$

16 $\frac{2}{7} = \frac{?}{14} = \frac{?}{21} = \frac{?}{28} = \frac{?}{35} = \frac{?}{70} = \frac{?}{280}$

17 For each of the following fractions, write a fraction with denominator 12 that is equal to the given fraction:

$\frac{1}{2}, \frac{3}{4}, \frac{5}{-6}, \frac{2}{3}, \frac{-1}{4}$

18 For each of the following fractions, write a fraction with denominator 16 that is equal to the given fraction:

$\frac{1}{2}, \frac{3}{4}, \frac{5}{-8}, \frac{7}{8}, \frac{-48}{32}$

19 For each of the following fractions, write a fraction with numerator 8 that is equal to the given fraction:

$\frac{1}{2}, \frac{4}{3}, \frac{-2}{5}, \frac{1}{-7}, \frac{56}{14}$

20 For each of the following fractions, write a fraction with numerator 12 that is equal to the given fraction:

$\frac{3}{5}, \frac{-4}{7}, \frac{6}{5}, \frac{2}{-3}, \frac{60}{25}$

3.2 RATIONAL NUMBERS

Our experience, which has suggested that we define equality so that $\frac{1}{2} = \frac{3}{6}$, also suggests that we shall wish to say that $\frac{1}{2}$ and $\frac{3}{6}$ represent the same "number." Let us consider the collection [indicated in (3.5)] of all fractions that are equal to $\frac{1}{2}$. The common property (of being equal to $\frac{1}{2}$) that is shared by all the fractions in this collection is called the *rational number that is represented by each of the fractions in* (3.5); this rational number may be denoted by $\left(\frac{p}{q}\right)$, where $\frac{p}{q}$ is any one of the fractions in (3.5). Thus $\left(\frac{1}{2}\right)$, $\left(\frac{3}{6}\right)$, and $\left(\frac{-4}{-8}\right)$ each denote the same rational number. We may consider also the collection [indicated in (3.6)] of all fractions that are equal to $\frac{7}{3}$. The common property (of being equal to $\frac{7}{3}$) that is shared by all the fractions in this collection is called the *rational number that is represented by each of the fractions in* (3.6); this rational number may be denoted by $\left(\frac{r}{s}\right)$, where $\frac{r}{s}$ is any one of the fractions in (3.6). Thus $\left(\frac{14}{6}\right)$, $\left(\frac{-28}{-12}\right)$, and $\left(\frac{21}{9}\right)$ each denote the same rational number.

In general, if $\frac{a}{b}$ is a given fraction, we may consider the collection of all fractions that are equal to $\frac{a}{b}$; the common property $\left(\text{of being equal to } \frac{a}{b}\right)$ shared by all the fractions in this collection is called a **rational number.** We say that each fraction in the collection *represents* the rational number and this rational number may be denoted by $\left(\frac{a}{b}\right)$. Thus $\left(\frac{1}{-4}\right)$ denotes, or as we shall say* in line with common usage, $\left(\frac{1}{-4}\right)$ *is* the rational number represented by each fraction that is equal to $\frac{1}{-4}$; clearly $\left(\frac{1}{-4}\right)$, $\left(\frac{-1}{4}\right)$, $\left(\frac{-16}{64}\right)$ all denote the same rational number since $\frac{1}{-4} = \frac{-1}{4} = \frac{-16}{64}$. It should be clear that we wish to say that $\left(\frac{1}{-4}\right) = \left(\frac{-1}{4}\right)$, $\left(\frac{1}{-4}\right) = \left(\frac{-16}{64}\right)$, and in general if $\left(\frac{a}{b}\right)$ and $\left(\frac{c}{d}\right)$ are rational numbers, that

RATIONAL NUMBER

EQUALITY OF RATIONAL NUMBERS

$$\left(\frac{a}{b}\right) = \left(\frac{c}{d}\right) \quad \textit{if and only if} \quad \frac{a}{b} = \frac{c}{d} \tag{3.8}$$

The procedure just described is similar to that used in Sec. 1.4 to produce the set of natural numbers. At this point the student should review the discussion of natural numbers in Sec. 1.4 and also should reread Sec. 1.2.

In our initial discussion of rational numbers we found it convenient to make use of parentheses to distinguish between the fraction $\frac{a}{b}$ and the rational number $\left(\frac{a}{b}\right)$ represented by the fraction. However, continued use of this notation would be cumbersome and out of line with common usage. Therefore, let us agree that $\frac{a}{b}$, where

*See the discussion of the distinction between numeral and number on page 14.

$a, b \in I$ and $b \neq 0$, is* the *rational number* represented by the fraction $\frac{a}{b}$. Thus, $\frac{6}{5}$ is the rational number that is represented by each fraction equal to $\frac{6}{5}$; $\frac{8}{24}$ is the rational number represented by each fraction equal to $\frac{1}{3}$; and $\frac{6x+y}{2y}$, where $x, y \in I$ and $y \neq 0$, is the rational number represented by each fraction equal to $\frac{6x+y}{2y}$.

An examination of definitions (3.8) and (3.3) indicates that we can understand (3.3) as referring to either fractions or rational numbers. Therefore, since equality (3.2) is a consequence of (3.3), as we showed in Sec. 3.1, we can interpret (3.2) as referring to either fractions or rational numbers. We restate (3.2) in terms of rational numbers as follows: If $\frac{a}{b}$ is a rational number and if k is a nonzero integer, then $\frac{ak}{bk}$ is a rational number and

$$\frac{a}{b} = \frac{ak}{bk} \tag{3.9}$$

Equality (3.9) is fundamental to all work with rational numbers.

Equality (3.9) tells us that if $\frac{a}{b}$ is a rational number, then we may multiply both numerator and denominator by the same nonzero integer and still have the same rational number. Also, equation (3.9) tells us that if the numerator and denominator of a rational number have a common nonzero factor k, then we may divide both the numerator and denominator by this common factor and still have the same rational number. Observe the following examples of the use of (3.9):

$$\frac{32}{20} = \frac{8 \cdot 4}{5 \cdot 4} = \frac{8}{5} \qquad \frac{-2}{5} = \frac{(-2)(-1)}{5(-1)} = \frac{2}{-5}$$

$$\frac{6}{27} = \frac{2 \cdot 3}{9 \cdot 3} = \frac{2}{9} \qquad \frac{5}{-32} = \frac{5(-1)}{(-32)(-1)} = \frac{-5}{32}$$

$$\frac{8}{-3} = \frac{8(-1)}{(-3)(-1)} = \frac{-8}{3}$$

$$\frac{-50}{-75} = \frac{(-10)(5)}{(-15)(5)} = \frac{-10}{-15} = \frac{2(-5)}{3(-5)} = \frac{2}{3}$$

*A conceptual distinction can be made of course between a rational number, which is the common property of a set of fractions, and the symbol or numeral which denotes that rational number. We believe that, while this distinction should be recognized, it is not helpful at this level to belabor such a distinction. Therefore, in line with common usage, we shall say that $\frac{a}{b}$ (or a/b) *is* the rational number.

The results in the preceding paragraph lead us to conclude that any rational number can be written with *a positive denominator and with the numerator and denominator having no common factor other than* 1 *and* -1; when a rational number is so written, it is said to be expressed in **lowest terms** or in *simplest form.* To express a rational number in lowest terms, we need to recognize any factors common to both the numerator and denominator; this recognition is frequently made easier if we first factor both numerator and denominator into prime factors.

EXPRESSED IN LOWEST TERMS

EXAMPLE 1 Express $\frac{90}{105}$ in lowest terms.

SOLUTION Factoring both numerator and denominator into prime factors, we have

$$\frac{90}{105} = \frac{2 \cdot 3 \cdot 3 \cdot 5}{3 \cdot 5 \cdot 7}$$

The factors 3 and 5 appear in both the numerator and denominator; therefore we divide both numerator and denominator by 3 and by 5, and we have

$$\frac{90}{105} = \frac{2 \cdot 3 \cdot 3 \cdot 5}{3 \cdot 5 \cdot 7} = \frac{2 \cdot 3}{7} = \frac{6}{7}$$

EXAMPLE 2 Express $\frac{9}{-27}$ in lowest terms.

SOLUTION Here we see that the numerator 9 is a factor of the denominator; $-27 = (-3)(9)$, so upon dividing both numerator and denominator by 9, we obtain

$$\frac{9}{-27} = \frac{1}{-3}$$

However, $\frac{1}{-3}$ is not expressed in lowest terms since the denominator is negative. We multiply both numerator and denominator by -1, and we find

$$\frac{1}{-3} = \frac{1(-1)}{(-3)(-1)} = \frac{-1}{3}$$

Therefore, expressed in lowest terms,

$$\frac{9}{-27} = \frac{-1}{3}$$

To indicate that both numerator and denominator have been divided by a common factor, frequently a slanting line is drawn through the factor in both numerator and denominator.

EXAMPLE 3 Express $\frac{4x^2y}{2xy}$ in lowest terms, where x and y represent positive integers.

SOLUTION We have $\frac{6x^2y}{2xy} = \frac{\not{2}\cdot 3\cdot \not{x}\cdot x\cdot y}{\not{2}\cdot\not{x}\cdot\not{y}} = \frac{3\cdot x}{1} = 3x.$

Notice that in the solution of Example 3 we have used the fact that equation (3.1) for fractions applies also to rational numbers. That is, we agree that the rational number $\frac{a}{1}$ will be identified with the integer a; for any integer a, $\frac{a}{1}$ is a rational number, and

$$\frac{a}{1} = a \tag{3.10}$$

The procedure illustrated above for expressing a rational number in lowest terms may be combined with the distributive property in working with rational numbers. Some examples of this are

$$\frac{ab+ac}{a} = \frac{\not{a}(b+c)}{\not{a}} = \frac{b+c}{1} = b+c$$

$$\frac{ab+ac+ad}{ac} = \frac{\not{a}(b+c+d)}{\not{a}c} = \frac{b+c+d}{c}$$

EXAMPLE 4 Express $\frac{5x^2-x}{y-5xy}$ in lowest terms, where x and y represent positive integers.

SOLUTION Using the distributive property to factor the numerator and denominator, we have

$$\frac{5x^2-x}{y-5xy} = \frac{x(5x-1)}{y(1-5x)}$$

At first glance it may not seem that there is a factor common to both numerator and denominator. Closer examination however reveals that

$$(5x-1) = (-1)(-5x+1) = (-1)(1-5x)$$

Therefore

$$\frac{5x^2-x}{y-5xy} = \frac{x(5x-1)}{y(1-5x)} = \frac{-x\cancel{(1-5x)}}{y\cancel{(1-5x)}} = \frac{-x}{y}$$

Failure to understand the procedure for expressing a rational number in lowest terms causes serious mistakes in working with rational numbers.

When a careful person divides the numerator and denominator of a rational number by an integer, he makes certain that the integer is a factor of the *entire* numerator and the *entire* denominator. For example, in the rational number $\frac{ax+y}{ay}$ there is no factor common to both numerator and denominator, while in $\dfrac{ax + ay}{ay}$ there is a factor, namely a, common to both numerator and denominator and

$$\frac{ax + ay}{ay} = \frac{\cancel{a}(x + y)}{\cancel{a}y} = \frac{x + y}{y}$$

EXERCISES

Express the following in lowest terms. The letter symbols that appear represent positive integers.

1 $\dfrac{18}{48}$

2 $\dfrac{85}{100}$

3 $\dfrac{56}{-12}$

4 $\dfrac{-91}{49}$

5 $\dfrac{32a^2}{16a}$

6 $\dfrac{45x^3y}{-15y^2}$

7 $\dfrac{abc}{cd}$

8 $\dfrac{2x + 2}{x + 1}$

9 $\dfrac{2a - 1}{b - 2ab}$

10 $\dfrac{ab}{ba}$

11 $\dfrac{6a + 4}{12a}$

12 $\dfrac{a + 1}{ab + b}$

13 $\dfrac{14 - 7x}{21}$

14 $\dfrac{3x - x^2}{x^2 - x} \quad (x \neq 1)$

15 $\dfrac{a^2 + 7a}{a^2}$

16 $\dfrac{x^2 - 3x}{6x - 2x^2} \quad (x \neq 3)$

17 $\dfrac{5a^2 - a}{5a - 1}$

18 $\dfrac{10a^2 - 2a}{10a^2 + 2a}$

Are the following results correct? Answer "yes" or "no" in each case. If the answer is "no", be able to explain why.

19 $\dfrac{2 + \cancel{7}}{\cancel{7}} = 2$

20 $\dfrac{2 + 7}{7} = \dfrac{9}{7}$

21 $\dfrac{\cancel{3} + 13}{\cancel{3} + 17} = \dfrac{13}{17}$

22 $\dfrac{3 + 13}{3 + 17} = \dfrac{16}{20}$

23 $\dfrac{\cancel{d}x + b}{\cancel{d}y + d} = \dfrac{x + b}{y + d}$

24 $\dfrac{y + \cancel{f}}{2z + \cancel{f}} = \dfrac{y}{2z}$

25 $\dfrac{\cancel{a}(x + b)}{\cancel{a}y + \cancel{a}z} = \dfrac{x + b}{y + z}$

26 $\dfrac{\cancel{a}x + b}{\cancel{a}y + \cancel{a}z} = \dfrac{x + b}{y + z}$

3.3
ADDITION OF RATIONAL NUMBERS

In the preceding section we discussed certain properties of the entities we have called rational numbers without justifying the use of the word *number*. In order for the rational numbers to be properly called numbers, there must be defined on the set two operations that satisfy the commutative, associative, and distributive properties (see Sec. 1.13). We have already defined equality in the set of rational numbers by definition (3.8). In this section we shall define addition (+), and in the next section we shall define multiplication (·) on the set of rational numbers.

Since rational numbers have been created by us, we could define the sum of two rational numbers in any one of several ways. For example, we could define addition so that the sum of $\frac{2}{3}$ and $\frac{3}{4}$ would be $\frac{5}{7}$ and the sum of $\frac{3}{8}$ and $\frac{6}{5}$ would be $\frac{9}{13}$. While such a definition would make addition of rational numbers simple, it is neither a useful nor a practical definition. It is desirable that we make a definition of addition that is consistent with previous definitions and with practical experiences.

If you were to divide a pie into six equal parts and if you ate two parts for lunch and three parts for dinner, altogether you would eat five of the six equal parts. That is,

$$\frac{2}{6}+\frac{3}{6}=\frac{5}{6}$$

Note that the numerator on the right side of the above equation is equal to the sum of the numerators on the left side. We adopt the following definition for the sum of two rational numbers expressed with the same denominator:

$$\frac{a}{c}+\frac{b}{c}=\frac{a+b}{c} \tag{3.11}$$

For example,

$$\frac{2}{7}+\frac{3}{7}=\frac{5}{7} \qquad \frac{2}{x}+\frac{3}{x}=\frac{5}{x}$$

$$\frac{a+b}{x+y}+\frac{b+c}{x+y}=\frac{(a+b)+(b+c)}{x+y}=\frac{a+2b+c}{x+y}$$

$$\frac{7}{15}+\frac{-4}{15}=\frac{7+(-4)}{15}=\frac{3}{15}=\frac{1}{5}$$

With reference to the pie problem described in the preceding paragraph, note that two of the six equal pieces of the pie constitute $\frac{1}{3}$ of the pie and three of the six equal pieces constitute $\frac{1}{2}$ of the pie, and so we wish to define addition of rational numbers so that

$$\frac{1}{2}+\frac{1}{3}=\frac{5}{6}$$

This result can be obtained if $\frac{1}{2}$ is expressed with denominator 6 and $\frac{1}{3}$ is expressed with denominator 6 and the results are added in accordance with (3.11). Now if $\frac{a}{b}$ and $\frac{c}{d}$ are rational numbers, we see that by the use of (3.9), $\frac{a}{b}$ can be expressed with denominator bd and $\frac{c}{d}$ can be expressed with denominator bd as follows:

$$\frac{a}{b} = \frac{ad}{bd} \qquad \frac{c}{d} = \frac{bc}{bd}$$

According to (3.11), $\frac{ad}{bd} + \frac{bc}{bd} = \frac{ad + bc}{bd}$. Hence we adopt the following definition for the **sum** *of the rational numbers* $\frac{a}{b}$ and $\frac{c}{d}$:

SUM OF RATIONAL NUMBERS

$$\frac{a}{b} + \frac{c}{d} = \frac{ad}{bd} + \frac{bc}{bd} = \frac{ad + bc}{bd} \qquad (3.12)$$

Since a, b, c, and d are integers and since the set of integers is closed under addition and multiplication, it is true that $ad + bc$ is an integer and bd is an integer. Further, bd is not zero since $b \neq 0$ and $d \neq 0$. Therefore $\frac{ad + bc}{bd}$ is a rational number, and the set of rational numbers is closed under addition as defined by (3.12).

It can be shown that the sum of two rationals is unique, that is,

if $\frac{a}{b} = \frac{c}{d}$ and $\frac{m}{n} = \frac{p}{q}$,

then $\frac{a}{b} + \frac{m}{n} = \frac{c}{d} + \frac{p}{q}$

This result can be established as a consequence of the fact that the sum of two integers is unique (page 72).

Note that if d is a nonzero integer and $\frac{a}{b}$ is any rational number,

$$\frac{a}{b} + \frac{0}{d} = \frac{ad + b \cdot 0}{bd} = \frac{ad}{bd} = \frac{a}{b}$$

Thus, the rational number $\frac{0}{d}$ is the identity under addition, or as we usually say, the *zero element* in the set of rational numbers. In accordance with (3.9) and (3.10), we write $\frac{0}{d} = \frac{0}{1} = 0$.

In (3.12) when we replace $\frac{a}{b}$ by $\frac{ad}{bd}$ and $\frac{c}{d}$ by $\frac{bc}{bd}$, we say that we express $\frac{a}{b}$ and $\frac{c}{d}$, respectively, with a common denominator.

In general, if $\frac{a}{b} = \frac{r}{n}$ and $\frac{c}{d} = \frac{s}{n}$, we say that $\frac{r}{n}$ and $\frac{s}{n}$ express $\frac{a}{b}$ and $\frac{c}{d}$, respectively, with a **common denominator.** An alternative form of the definition of the sum of rational numbers $\frac{a}{b}$ and $\frac{c}{d}$ can be given as follows: *To add* $\frac{a}{b}$ *and* $\frac{c}{d}$, *first express each of* $\frac{a}{b}$ *and* $\frac{c}{d}$ *with a common denominator, and then add in accordance with* (3.11). In using this form of the definition of addition, it is sometimes desirable, although not necessary, to use the least common denominator. The **least common denominator** of several given rational numbers (with positive denominators) is the least common multiple of the denominators. The **least common multiple** of two or more positive integers is the smallest positive integer which is a multiple of each of the integers. Recall that the integer p is a multiple of the integer q if and only if there is a positive integer m with the property that $p = mq$; then q is a factor, or divisor, of p. Also recall that if a positive integer has no factor except itself and 1, it is prime.

COMMON DENOMINATOR

LEAST COMMON DENOMINATOR

LEAST COMMON MULTIPLE

The least common multiple of several positive integers, and hence the least common denominator of several rational numbers, may frequently be found by inspection. To illustrate, the least common multiple of 2, 3, and 4 is 12; the least common multiple of x^2y and xy^2 is x^2y^2.

However, it is sometimes convenient to use the following rule: *The least common multiple of two or more positive integers is found by taking the product of all the different prime factors in these integers, each taken the greatest number of times that it occurs in any of the given integers.* We give the following illustrations, in which least common multiple is abbreviated as LCM:

The LCM of 8 and 12 is 24, since $8 = 2^3$, $12 = 2^2 \cdot 3$, and $2^3 \cdot 3 = 24$.

The LCM of 24 and 63 is 504, since $24 = 2^3 \cdot 3$, $63 = 3^2 \cdot 7$, and $2^3 \cdot 3^2 \cdot 7 = 504$. The LCM of $3a^2b^3$ and $4ab^4$ is $12a^2b^4$.

EXAMPLE 1 Find the sum of $\frac{1}{3}$ and $\frac{5}{12}$, and express the result in lowest terms.

SOLUTION (a) Using definition (3.12), we have

$$\frac{1}{3} + \frac{5}{12} = \frac{1 \cdot 12 + 3 \cdot 5}{3 \cdot 12} = \frac{12 + 15}{36} = \frac{27}{36}$$

To express $\frac{27}{36}$ in lowest terms, we note that 9 is a common factor of both numerator and denominator, so that

$$\frac{1}{3} + \frac{5}{12} = \frac{27}{36} = \frac{3}{4}$$

(*b*) To use the alternative formulation of the definition of addition of rational numbers, we look for a common denominator for $\frac{1}{3}$ and $\frac{5}{12}$. We note that 12 is the least common multiple of 3 and 12, and we can write $\frac{1}{3} + \frac{5}{12} = \frac{4}{12} + \frac{5}{12} = \frac{9}{12}$. We can express $\frac{9}{12}$ in lowest terms by dividing both numerator and denominator by 3, and we have

$$\frac{1}{3} + \frac{5}{12} = \frac{9}{12} = \frac{3}{4}$$

which agrees with the result in part (*a*).

EXAMPLE 2 Add $\frac{6}{13} + \frac{-5}{12}$, and express the result in lowest terms.

SOLUTION We have

$$\frac{6}{13} + \frac{-5}{12} = \frac{6 \cdot 12 + 13(-5)}{13 \cdot 12} = \frac{72 - 65}{156} = \frac{7}{156}$$

Since 7 is prime and it is not a factor of 156, there is no factor common to 7 and 156 except 1 and -1. Therefore $\frac{7}{156}$ is expressed in lowest terms.

EXAMPLE 3 Add $\frac{2}{a} + \frac{3}{b}$, where a and b are positive integers.

SOLUTION By definition (3.12) we have

$$\frac{2}{a} + \frac{3}{b} = \frac{2 \cdot b + a \cdot 3}{ab}$$

$$= \frac{2b + 3a}{ab} \qquad \textit{since multiplication of integers is commutative}$$

This result is expressed in lowest terms since there is no common factor of $2b + 3a$ and ab.

We want the addition of rational numbers to be *commutative*. That is, we want $\frac{a}{b} + \frac{c}{d} = \frac{c}{d} + \frac{a}{b}$ to be true for any rational numbers $\frac{a}{b}$ and $\frac{c}{d}$.

By definition (3.12) we have

$$\frac{a}{b} + \frac{c}{d} = \frac{ad + bc}{bd}$$

and

$$\frac{c}{d} + \frac{a}{b} = \frac{cb + da}{db}$$

Thus we see that addition of rational numbers will be commutative if and only if

$$\frac{ad + bc}{bd} = \frac{cb + da}{db} \tag{3.13}$$

The student should see that because addition and multiplication of integers are commutative, it is true that $ad + bc = cb + da$ and $bd = db$ and hence that (3.13) is true.

Similarly the commutative, associative, and distributive properties of addition and multiplication of integers can be used to show that addition of rational numbers is *associative;* that is,

for any rational numbers $\frac{a}{b}, \frac{c}{d}, \frac{e}{f}$,

$$\frac{a}{b} + \left(\frac{c}{d} + \frac{e}{f}\right) = \left(\frac{a}{b} + \frac{c}{d}\right) + \frac{e}{f}$$

The student is asked to show that this is true in Exercise 27 of this section.

EXAMPLE 4 Add $\frac{1}{3} + \frac{3}{4} + \frac{5}{6}$.

SOLUTION The least common multiple of 3, 4, and 6 is 12. So we express each of the rational numbers with denominator 12. We get

$$\frac{1}{3} + \frac{3}{4} + \frac{5}{6} = \frac{4}{12} + \frac{9}{12} + \frac{10}{12} = \frac{4 + 9 + 10}{12} = \frac{23}{12}$$

The integer 23 is prime and is not a factor of 12; hence $\frac{23}{12}$ is expressed in lowest terms.

EXAMPLE 5 Add $\frac{1}{2} + \frac{5}{6} + \frac{3}{5}$, and express the result in lowest terms.

SOLUTION The least common multiple of 2, 6, and 5 is 30. So

$$\frac{1}{2} + \frac{5}{6} + \frac{3}{5} = \frac{1 \cdot 15 + 5 \cdot 5 + 3 \cdot 6}{30} = \frac{15 + 25 + 18}{30} = \frac{58}{30} = \frac{29}{15}$$

EXERCISES

Perform the following additions. Express each result in lowest terms. The letter symbols which appear in these exercises represent positive integers.

1 $\frac{2}{3} + \frac{5}{3}$

2 $\frac{1}{9} + \frac{4}{9} + \frac{5}{9}$

3 $\frac{2}{7} + \frac{3}{7} + \frac{5}{7}$

4 $\frac{a}{y} + \frac{b}{y} + \frac{c}{y}$

5 $\frac{a}{x} + \frac{3a}{x} + \frac{4a}{x}$

6 $\frac{x}{x + y} + \frac{x}{x + y}$

7 $\frac{a}{a + 1} + \frac{1}{a + 1}$

8 $\frac{2}{3} + \frac{4}{5}$

9 $3 + \frac{7}{6} + \frac{2}{3}$

10 $\frac{y}{3} + \frac{2y}{9}$ 11 $3y + \frac{y}{5}$ 12 $\frac{x}{2} + \frac{x}{3} + \frac{x}{4}$

13 $\frac{4}{a} + \frac{3}{2a}$ 14 $\frac{1}{a} + \frac{1}{b} + \frac{1}{c}$ 15 $bc + \frac{1}{c}$

16 $\frac{x}{x+y} + 7$ 17 $\frac{1}{2m} + \frac{a+3}{4m}$ 18 $\frac{9}{x^2} + \frac{4}{x}$

19 $\frac{a}{x+y} + \frac{b}{x+y}$ 20 $\frac{1}{a} + \frac{3}{ab} + \frac{2}{b}$ 21 $\frac{2x}{a} + \frac{3y}{b}$

22 $\frac{2}{a} + \frac{7}{abc} + \frac{6}{c}$

23 A steam pipe is to be covered with asbestos. If each piece of asbestos is $3\frac{1}{2}$ feet long, how many pieces are needed to cover a pipe $66\frac{1}{2}$ feet long?

24 How many blocks $\frac{4}{5}$ foot long must be laid end to end to make a row 48 feet long?

25 The top of a table is to be made from three pieces of board. One board is $4\frac{5}{8}$ inches wide, and another is $7\frac{2}{5}$ inches wide. How wide should the third board be in order for the top to be $16\frac{1}{2}$ inches wide?

26 A piece of cloth $27\frac{5}{8}$ yards long shrank $1\frac{2}{3}$ yards in bleaching. How long was the cloth after bleaching?

27 Show that addition of rational numbers is associative. First notice that by (3.12)

$$\frac{a}{b} + \left(\frac{c}{d} + \frac{e}{f}\right) = \frac{a}{b} + \frac{cf + de}{df} = \frac{a(df) + b(cf + de)}{b(df)}$$

and

$$\left(\frac{a}{b} + \frac{c}{d}\right) + \frac{e}{f} = \frac{ad + bc}{bd} + \frac{e}{f} = \frac{(ad + bc)f + (bd)e}{(bd)f}$$

Then use the distributive, commutative, and associative properties of addition and multiplication of integers to complete the argument.

3.4 MULTIPLICATION OF RATIONAL NUMBERS

Turning now to multiplication of rational numbers, let us use a physical problem to suggest a possible definition for the product of two rational numbers.

Suppose that we have a square with side 1 unit in length. If we divide one side into equal parts each $\frac{1}{7}$ unit in length and the other into equal parts each $\frac{1}{5}$ unit in length, we may effect a corresponding division of the square into 35 equal rectangles. The area of each of these rectangles is $\frac{1}{35}$ square unit, since each is 1 of 35 equal parts into which the square has been divided. But using the formula $A = lw$, we get $\frac{1}{7} \cdot \frac{1}{5}$ square unit for each of the small rectangles.

If we want the formula for the area of a rectangle to hold for fractional lengths and widths, we must define the product of two fractions so that $\frac{1}{7} \cdot \frac{1}{5} = \frac{1}{35}$.

Again, consider the rectangle made up of six of these small rectangles, with its length $\frac{3}{7}$ (3 of 7 equal parts) unit and its width $\frac{2}{5}$ (2 of 5 equal parts) unit. The area of this rectangle is $\frac{6}{35}$, since it is 6 of 35 equal parts. We want the multiplication of fractions to be defined so that $\frac{3}{7} \cdot \frac{2}{5} = \frac{6}{35}$.

These considerations suggest the following definition for the

PRODUCT OF RATIONAL NUMBERS

product *of the rational numbers* $\frac{a}{b}$ and $\frac{c}{d}$:

$$\frac{a}{b} \cdot \frac{c}{d} = \frac{a \cdot c}{b \cdot d} \tag{3.14}$$

To illustrate,

$$\frac{1}{2} \cdot \frac{6}{7} = \frac{1 \cdot 6}{2 \cdot 7} = \frac{3}{7} \qquad \frac{2a}{3b} \cdot \frac{5b^2}{7} = \frac{2a \cdot 5 \cdot b \cdot b}{3 \cdot b \cdot 7} = \frac{10ab}{21}$$

$$\frac{3a}{a+2} \cdot \frac{ab+2b}{d} = \frac{3a}{a+2} \cdot \frac{b(a+2)}{d} = \frac{3a \cdot b \cdot (a+2)}{d(a+2)} = \frac{3ab}{d}$$

You should observe that the simplification of the product of two rational numbers is facilitated by recognizing a factor which is common to any numerator and any denominator.

Since a, b, c, and d are integers, and since the set of integers is closed under multiplication, it is true that ac and bd are integers. Further, bd is not zero since $b \neq 0$ and $d \neq 0$. Therefore $\frac{ac}{bd}$ is a rational number, and the set of rational numbers is closed under multiplication as defined by (3.14).

It can be shown that the product of two rational numbers is unique; that is,

$$\text{if } \frac{a}{b} = \frac{c}{d} \text{ and } \frac{m}{n} = \frac{p}{q},$$

$$\text{then } \frac{a}{b} \cdot \frac{m}{n} = \frac{c}{d} \cdot \frac{p}{q}$$

This result can be established as a consequence of the fact that the product of two integers is unique (page 74).

We note that the rational number $\frac{1}{1}$ has the property required of the *identity* under multiplication (or the unit element) in the set of rational numbers; indeed $\frac{a}{b} \cdot \frac{1}{1} = \frac{a \cdot 1}{b \cdot 1} = \frac{a}{b}$.

Multiplication of rational numbers is commutative and associative; that is,

$$\frac{a}{b}\cdot\frac{c}{d}=\frac{c}{d}\cdot\frac{a}{b} \tag{3.15}$$

and

$$\frac{a}{b}\left(\frac{c}{d}\cdot\frac{e}{f}\right)=\left(\frac{a}{b}\cdot\frac{c}{d}\right)\frac{e}{f} \tag{3.16}$$

The student should see that (3.15) follows from the commutative property of multiplication of integers and that (3.16) follows from the associative property of multiplication of integers.

Because of property (3.16) we may define the product of more than two rational numbers as follows:

$$\frac{a}{b}\cdot\frac{c}{d}\cdot\frac{e}{f}=\frac{ace}{bdf} \qquad \frac{a}{b}\cdot\frac{c}{d}\cdot\frac{e}{f}\cdot\frac{g}{h}=\frac{aceg}{bdfh} \qquad \text{etc.}$$

By using the definitions of addition and multiplication of rational numbers together with the properties of addition and multiplication of integers it is possible to prove that in the set of rational numbers, multiplication is distributive with respect to addition; that is

for any rational numbers $\frac{a}{b}, \frac{c}{d}, \frac{e}{f}$,

$$\frac{a}{b}\cdot\left(\frac{c}{d}+\frac{e}{f}\right)=\frac{a}{b}\cdot\frac{c}{d}+\frac{a}{b}\cdot\frac{e}{f}$$

We shall not give the details of this proof here.

As a special case of (3.14) we have $\frac{a}{1}\cdot\frac{c}{d}=\frac{a\cdot c}{1\cdot d}=\frac{ac}{d}$, or since

$$\frac{a}{1}=a,$$

$$a\cdot\frac{c}{d}=\frac{ac}{d} \tag{3.17}$$

That is, *to multiply a rational number by a positive integer, multiply the numerator by that integer.* For example,

$$3\cdot\frac{7}{5}=\frac{21}{5} \qquad 5\cdot\frac{2}{11}=\frac{10}{11} \qquad x\cdot\frac{x}{y}=\frac{x^2}{y} \qquad z\cdot\frac{z^2}{w}=\frac{z^3}{w}$$

Also, as a special case of (3.14), we have

$$\frac{a}{b}\cdot c=\frac{ac}{b}$$

Thus

$$\frac{2}{3} \cdot 6 = 4$$

Recall that in order to obtain $\frac{2}{3}$ of anything, we divide that thing into three equal parts and take two of those parts. Now 6 divided by 3 gives 2 as each part, and two of these parts totals 4. We interpret

$$\frac{2}{3} \text{ of } 6 \text{ is } 4 \qquad \text{to mean} \qquad \frac{2}{3} \cdot 6 = 4$$

That is, for any fraction $\frac{a}{b}$ and any number c,

$$\frac{a}{b} \text{ of } c \qquad \text{means} \qquad \frac{a}{b} \cdot c$$

To illustrate,

$$\frac{3}{4} \text{ of } 20 \qquad \text{means} \qquad \frac{3}{4} \cdot 20$$

We are now in a position to give a meaning to the quotient $a \div b$ of integers a and b when b is not a factor of a. Recall that we have defined $a \div b$ as the number c with the property that $b \cdot c = a$. Now we notice that

if $b \neq 0$,

$$b \cdot \frac{a}{b} = \frac{b}{1} \cdot \frac{a}{b} = \frac{ba}{b} = \frac{a}{1} = a$$

so that $\frac{a}{b}$ has the property required of c in the definition of $a \div b = c$. Therefore, if a and b are integers with $b \neq 0$, we can define $a \div b$ to be the *rational number* c with the property that $b \cdot c = a$, and we have

$$a \div b = \frac{a}{b} \tag{3.18}$$

For example,

$$3 \div 4 = \frac{3}{4} \qquad \text{and} \qquad 6 \div 3 = \frac{6}{3} = \frac{2}{1} = 2$$

In accord with (3.18), we shall use the phrases "the rational number with numerator a and denominator b" and "the quotient of a divided by b" to indicate the same number.

RECIPROCAL The **reciprocal** of a nonzero integer a is 1 divided by that integer a. If a' is the reciprocal of a, then

$$a' = \frac{1}{a}$$

For example, $\frac{1}{5}$ is the reciprocal of 5; $\frac{1}{7}$ is the reciprocal of 7. As a special case of (3.14), we have

$$\frac{1}{b} \cdot \frac{1}{d} = \frac{1}{bd}$$

That is, *the product of the reciprocals of two nonnegative integers is the reciprocal of their products.*

If $c = 1$ in (3.17), we have

$$a \cdot \frac{1}{d} = \frac{a \cdot 1}{d} = \frac{a}{d} \tag{3.19}$$

This result shows that any rational number can be expressed as the product of an integer and the reciprocal of a nonnegative integer. For example, $\frac{2}{3} = 2 \cdot \frac{1}{3}$ and $\frac{7}{5} = 7 \cdot \frac{1}{5}$.

If $d = a$ in (3.19), we have $a \cdot \dfrac{1}{a} = 1$. The product of any nonnegative integer and its reciprocal is 1.

MIXED NUMBER The sum of a positive integer and a rational number is called a **mixed number.** Thus $9 + \frac{1}{2}$ is a mixed number. Recall that in arithmetic $9 + \frac{1}{2}$ is written $9\frac{1}{2}$, $5\frac{2}{3}$ means $5 + \frac{2}{3}$, and $6\frac{7}{8}$ means $6 + \frac{7}{8}$, and so forth. You will have to be careful not to confuse $9\frac{1}{2}$, which means $9 + \frac{1}{2}$, with the expression $9 \cdot \frac{1}{2}$ or $9(\frac{1}{2})$, which means 9 times $\frac{1}{2}$.

It is sometimes desirable to express a mixed number in *the form* $\dfrac{a}{b}$ where a and b are integers, with $b \neq 0$. This is done as in the following illustrations:

$$9\tfrac{1}{2} = 9 + \frac{1}{2} = \frac{9}{1} + \frac{1}{2} = \frac{9 \cdot 2 + 1 \cdot 1}{1 \cdot 2} = \frac{18 + 1}{2} = \frac{19}{2}$$

$$7\tfrac{4}{3} = 7 + \frac{4}{3} = \frac{7}{1} + \frac{4}{3} = \frac{7 \cdot 3 + 1 \cdot 4}{1 \cdot 3} = \frac{21 + 4}{3} = \frac{25}{3}$$

$$5\tfrac{2}{3} = 5 + \frac{2}{3} = \frac{5}{1} + \frac{2}{3} = \frac{5 \cdot 3 + 1 \cdot 2}{1 \cdot 3} = \frac{15 + 2}{3} = \frac{17}{3}$$

In general,

$$a + \frac{b}{c} = \frac{a}{1} + \frac{b}{c} = \frac{ac + b}{c} \tag{3.20}$$

By making use of (3.20), we are able to add and multiply mixed numbers.

EXAMPLE 1 Add $6\frac{5}{12} + 13\frac{3}{5}$.

SOLUTION We have

$$6\tfrac{5}{12} + 13\tfrac{3}{5} = \frac{6}{1} + \frac{5}{12} + \frac{13}{1} + \frac{3}{5} = \frac{360 + 25 + 780 + 36}{60} = \frac{1{,}201}{60}$$

EXAMPLE 2 Find the product $(6\frac{5}{12}) \cdot (13\frac{3}{5})$.

SOLUTION Here

$$(6\tfrac{5}{12}) \cdot (13\tfrac{3}{5}) = \left(\frac{6}{1} + \frac{5}{12}\right) \cdot \left(\frac{13}{1} + \frac{3}{5}\right) = \left(\frac{72 + 5}{12}\right) \cdot \left(\frac{65 + 3}{5}\right)$$
$$= \frac{77}{12} \cdot \frac{68}{5} = \frac{77 \cdot 4 \cdot 17}{3 \cdot 4 \cdot 5} = \frac{1{,}309}{15}$$

If $\frac{a}{b}$ is a rational number with a and b positive integers and $a > b$, we sometimes wish to express $\frac{a}{b}$ as a mixed number. This can be accomplished by interpreting $\frac{a}{b}$ as $a \div b$ in accordance with (3.18). To illustrate, suppose we wish to express $\frac{2{,}097}{51}$ as a mixed number. The work may be arranged as follows.

```
     41
51)2097
   204
    57
    51
     6
```

We call 2,097 the dividend, 51 the divisor, 41 the quotient, and 6 the remainder. We write

$$\frac{2{,}097}{51} = 41 + \tfrac{6}{51} = 41\tfrac{6}{51}$$

This result can be checked by expressing the mixed number $41\frac{6}{51}$ in the form $\frac{a}{b}$. Whenever the dividend is not a multiple of the divisor there is an equality of the form

$$\frac{\text{dividend}}{\text{divisor}} = \text{quotient} + \frac{\text{remainder}}{\text{divisor}}$$

EXAMPLE 3 Write $\frac{486}{25}$ as a mixed number.

SOLUTION

$$\begin{array}{r} 19 \\ 25\overline{)486} \\ \underline{25} \\ 236 \\ \underline{225} \\ 11 \end{array}$$

So $\frac{486}{25} = 19\frac{11}{25}$.

EXERCISES

Find the following products. Express each result in lowest terms. The letter symbols represent positive integers.

1 $3 \cdot \frac{4}{9}$

2 $5 \cdot \frac{7}{11}$

3 $\frac{2}{3} \cdot 6$

4 $\frac{5}{8} \cdot \frac{4}{15}$

5 $\frac{91}{119} \cdot \frac{34}{39}$

6 $\frac{a}{b} \cdot \frac{1}{a}$

7 $\frac{3a}{2b} \cdot \frac{5a^2}{9b}$

8 $\frac{a}{b} \cdot \frac{a}{b}$

9 $\frac{1}{x^2} \cdot \frac{2}{3} \cdot \frac{x}{4}$

10 $\frac{m^2}{n} \cdot \frac{n}{m^2}$

11 $\frac{a^2 + ax}{3} \cdot \frac{6}{a + x}$

12 $\frac{2x + 12}{x + 5} \cdot \frac{3x + 15}{x + 6}$

13 $\frac{x + 1}{x} \cdot \frac{x^2}{x^2 + x}$

14 $\frac{4xy}{x + 3} \cdot \frac{3x^2 + 9x}{16y^2}$

15 $\frac{2}{6} \cdot \frac{1}{2} \cdot \frac{3}{4}$

16 $\frac{a}{2} \cdot \frac{4}{a^2} \cdot \frac{3}{5}$

17 $\frac{3x + 12}{x} \cdot \frac{2x^2}{x + 4}$

18 $\frac{y^3}{15x + 6} \cdot \frac{5x + 2}{y}$

19 $\frac{2a + b}{7ab} \cdot \frac{3b^2}{4a + 2b}$

20 $\frac{5}{a^3 + a^2y} \cdot \frac{a^2 + ay}{25}$

21 $\frac{2a^2 + 3a}{b} \cdot \frac{b^3}{a^2}$

22 $\frac{4y + 4}{7} \cdot \frac{14}{2y^2 + 2y}$

23 Give the reciprocal of each of the following: 7; 10; 12; 15; 18.

24 What is the reciprocal of 1?

With the use of the distributive property, perform the following divisions:

25 $\frac{4a + 4}{4}$

26 $\frac{2x + 6}{2}$

27 $\frac{a^2 + a}{a}$

28 $\frac{ax + ay}{a}$

Express each of the following as a mixed number. Verify your results by expressing the mixed number in the form $\frac{a}{b}$.

29 $\frac{31}{4}$ 30 $\frac{89}{7}$ 31 $\frac{243}{14}$

32 $\frac{2,976}{121}$ 33 $\frac{5,276}{7}$ 34 $\frac{10,792}{1,003}$

35 $\frac{8,762}{115}$ 36 $\frac{123,456}{647}$

37 How many pieces of wire each $3\frac{1}{2}$ feet long can be cut from a coil of wire 140 feet long?

38 If an airplane can fly 260 miles in 1 hour, how far can it fly in $11\frac{1}{2}$ hours under the same conditions?

39 If a city block is $\frac{1}{6}$ mile long, how many miles has a man gone when he has walked 7 blocks south and $9\frac{1}{2}$ blocks west?

In Exercises 40 to 44, use $\frac{22}{7}$ as the value for π.

40 If r is the length of the radius of a circle, the circumference of the circle is given by the formula $C = 2\pi r$ (see Fig. 3.2). Find the circumference of the circle for which the radius is: 7 inches; 14 feet; 35 yards; $\frac{9}{2}$ feet.

41 If r is the number of units of length in the radius of a circle, the number of square units of area A of the circle is given by the formula $A = \pi r^2$. Find the area of the circles given in Exercise 40.

42 The number of cubic units of volume V of a right circular cylinder is given by the formula $V = \pi r^2 h$, where r is the number of units of length of the radius and h is the number of units of length of the altitude (see Fig. 3.3). Find the volume of each of the right circular cylinders whose radius and altitude are given in the following table:

	(a)	(b)	(c)	(d)
r	3 inches	$\frac{5}{2}$ inches	4 feet	$\frac{4}{3}$ feet
h	4 inches	$\frac{2}{3}$ inch	7 feet	$\frac{5}{6}$ foot
V				

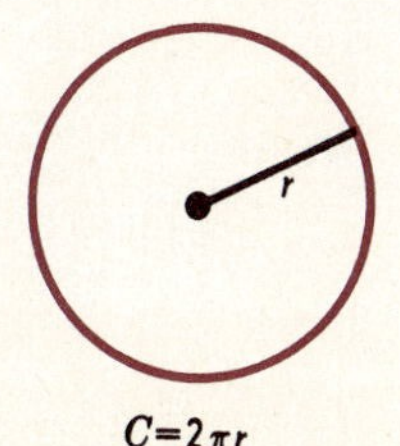

$C=2\pi r$

FIGURE 3.2

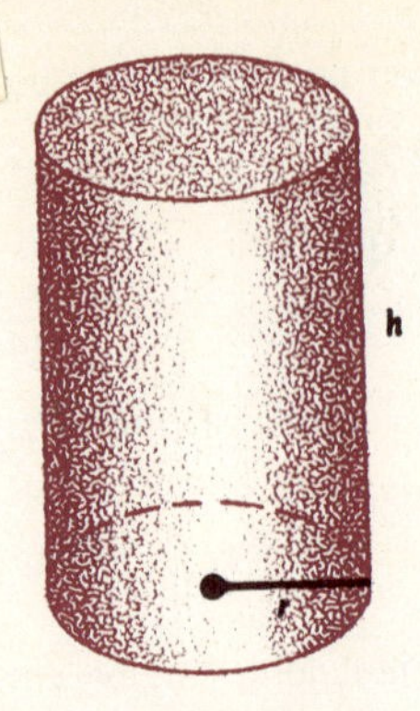

$V = \pi r^2 h$

FIGURE 3.3

43 The number of square units of lateral area S of the right circular cylinder with radius r and altitude h is given by the formula $S = 2\pi rh$. Find the lateral area of each of the cylinders specified in Exercise 42.

44 Let r be the radius of a given sphere. The volume V of the sphere (see Fig. 3.4) is given by the formula $V = \frac{4}{3}\pi r^3$, and the surface area of the sphere is given by the formula $A = 4\pi r^2$. Find the volume V and the surface area A of each of the spheres whose radius is given in the following table:

	(*a*)	(*b*)	(*c*)	(*d*)
r	3 inches	$\frac{7}{3}$ inches	6 feet	$\frac{5}{2}$ feet
V				
A				

45 $\frac{1}{5}$ of a pint is what part of $\frac{4}{5}$ of a pint?

46 $\frac{1}{4}$ of a dollar is what part of $\frac{3}{4}$ of a dollar?

47 $\frac{1}{2}$ of an inch is how many one-tenths of an inch?

48 What part of a foot is 3 inches? 8 inches? 10 inches?

49 What part of a pound is 8 ounces? 12 ounces?

50 What part of an hour is 15 minutes? 20 minutes? 30 minutes?

51 Find the number of square feet in the area of the square whose side has the length of: 6 inches; $\frac{2}{3}$ foot; $\frac{7}{4}$ feet; 9 inches.

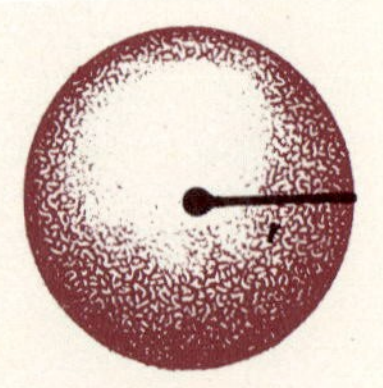

$V = \frac{4}{3}\pi r^3$

FIGURE 3.4

52 Find the number of cubic feet in the volume of the cube whose edge has the length of: 8 inches; $\frac{3}{2}$ feet; $\frac{4}{5}$ foot; 6 inches.

53 Complete the following table:

WHEN $x =$	$\frac{1}{2}$	$\frac{2}{3}$	$\frac{3}{4}$	$\frac{1}{5}$	$\frac{3}{2}$	$\frac{5}{7}$	$\frac{1}{6}$
THEN $x^2 =$		$\frac{4}{9}$					
AND $x^3 =$		$\frac{8}{27}$					

54 Find the area (in square feet) of each of the rectangles whose length l and width w are, respectively, as follows:

$l = 2$ feet, $w = 6$ inches
$l = \frac{3}{5}$ foot, $w = \frac{2}{3}$ foot
$l = 3$ yards, $w = 27$ inches
$l = 20$ inches, $w = 15$ inches

55 Find the volume (in cubic feet) of each of the rectangular boxes whose length l, width w, and height h are, respectively, as follows:

$l = \frac{7}{2}$ feet, $w = \frac{5}{7}$ foot, $h = \frac{1}{3}$ foot
$l = \frac{5}{2}$ feet, $w = 27$ inches, $h = \frac{2}{3}$ yard
$l = \frac{3}{4}$ foot, $w = \frac{2}{9}$ foot, $h = \frac{7}{2}$ feet
$l = \frac{3}{2}$ yards, $w = \frac{2}{3}$ foot, $h = 9$ inches

56 Find an expression for the area A of each rectangle whose length l and width w are given, respectively, below. Express each result in lowest terms.

$$l = \frac{x + 2}{3}, \quad w = \frac{3x}{4}$$

$$l = 2y + 2, \quad w = \frac{x}{4}$$

$$l = \frac{5}{3}, \quad w = \frac{6x}{15}$$

$$l = \frac{a^2 + a}{8}, \quad w = \frac{12}{a}$$

57 Find an expression for the volume V of each rectangular box whose length l, width w, and height h are given, respectively, below. Express each result in lowest terms.

$$l = \frac{x}{3}, \quad w = \frac{x^2 + 3x}{4}, \quad h = \frac{6}{x}$$

$$l = \frac{a^2}{2}, \quad w = \frac{b + 3}{2a}, \quad h = \frac{2b}{3a}$$

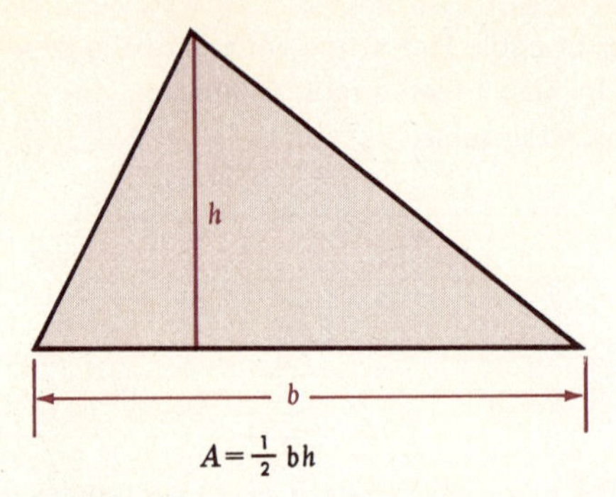

FIGURE 3.5

$$l = \frac{x^2 + 5x + 6}{4}, \; w = \frac{7}{x + 2}, \; h = \frac{2}{21x}$$

$$l = \frac{5a - 10b}{2}, \; w = \frac{3a - 6b}{10}, \; h = \frac{8a}{3}$$

58 The number of square units of area A of a triangle is given by the formula $A = \frac{1}{2}bh$, where b is the number of units of length of the base and h is the number of units of length of the altitude (see Fig. 3.5). Find the area (in square feet) of each triangle whose base b and altitude h are, respectively, as follows:

$b = \frac{3}{2}$ feet, $h = \frac{10}{3}$ feet
$b = 24$ inches, $h = \frac{5}{2}$ feet
$b = \frac{3}{4}$ yard, $h = 27$ inches
$b = \frac{9}{2}$ feet, $h = \frac{18}{3}$ feet

59 Find an expression for the area A of each triangle whose base b and altitude h are given, respectively, below. Express each result in lowest terms.

$$b = \frac{4x}{3}, \; h = \frac{27x + 15}{15}$$

$$b = \frac{5a + 5}{x}, \; h = \frac{10x^2 + 6x}{3}$$

$$b = \frac{8x}{5}, \; h = 5x + 4$$

$$b = \frac{14x + 7}{3}, \; h = x + 2$$

3.5 SUBTRACTION OF RATIONAL NUMBERS

The definition of subtraction on the set of rational numbers has the same form as the definitions of subtraction on the set of integers

(Sec. 2.4) and on the set of natural numbers (Sec. 1.9). If $\frac{a}{b}$ and $\frac{c}{d}$ are rational numbers, we **define** $\frac{a}{b} - \frac{c}{d}$ to be the rational number $\frac{x}{y}$, if such a rational number exists, with the property that $\frac{c}{d} + \frac{x}{y} = \frac{a}{b}$; that is,

SUBTRACTION OF RATIONAL NUMBERS

$$\frac{a}{b} - \frac{c}{d} = \frac{x}{y}$$

if and only if $\frac{x}{y}$ is a rational number with the property that

$$\frac{c}{d} + \frac{x}{y} = \frac{a}{b}$$

DIFFERENCE

If $\frac{a}{b} - \frac{c}{d} = \frac{x}{y}$, we call $\frac{x}{y}$ the **difference** that results when we *subtract* $\frac{c}{d}$ from $\frac{a}{b}$.

Suppose that we wish to determine the rational number $\frac{x}{y}$ so that

$$\frac{1}{2} - \frac{1}{3} = \frac{x}{y} \tag{3.21}$$

Recall that two rational numbers can be added by expressing the rational numbers with a common denominator. In (3.21) let us express $\frac{1}{2}$ and $\frac{1}{3}$ with a common denominator; when this is done we have

$$\frac{3}{6} - \frac{2}{6} = \frac{x}{y}$$

We are seeking a rational number $\frac{x}{y}$ with the property that

$$\frac{2}{6} + \frac{x}{y} = \frac{3}{6}$$

and using our knowledge of addition of rational numbers, we find that $\frac{1}{6}$ is the desired rational number. Observe that in this case

$$\frac{3}{6} - \frac{2}{6} = \frac{1}{6}$$

This example suggests that if we wish to determine the difference $\frac{a}{b} - \frac{c}{d}$, we express $\frac{a}{b}$ and $\frac{c}{d}$ with a common denominator as $\frac{ad}{bd}$

and $\frac{bc}{bd}$ and ask whether it is true that

$$\frac{ad}{bd} - \frac{bc}{bd} = \frac{ad - bc}{bd} \tag{3.22}$$

The definition of subtraction given in the preceding paragraph tells us that in order to prove (3.22) is true, we need to show the truth of

$$\frac{bc}{bd} + \frac{ad - bc}{bd} = \frac{ad}{bd} \tag{3.23}$$

Using definition (3.12), we have

$$\frac{bc}{bd} + \frac{ad - bc}{bd} = \frac{bc + (ad - bc)}{bd}$$

so

$$\frac{bc}{bd} + \frac{ad - bc}{bd} = \frac{bc - bc + ad}{bd}$$

$$= \frac{ad}{bd}$$

Thus (3.23) is true, and we have proved that (3.22) is true.

Using the result in the preceding paragraph, we can say that

for any rational numbers $\frac{a}{b}$ *and* $\frac{c}{d}$*,*

$$\frac{a}{b} - \frac{c}{d} = \frac{ad - bc}{bd} \tag{3.24}$$

Since ad, bc, and bd are integers with $bd \neq 0$, and since the set of integers is closed under subtraction, then $\frac{ad - bc}{bd}$ is a rational number. Thus (3.24) shows that the set of rational numbers is closed under subtraction.

EXAMPLE 1 Perform the subtraction $\frac{7}{9} - \frac{2}{3}$, and express the result in lowest terms.

SOLUTION (a) Using the result (3.24), we write

$$\frac{7}{9} - \frac{2}{3} = \frac{7 \cdot 3 - 9 \cdot 2}{9 \cdot 3} = \frac{21 - 18}{27} = \frac{3}{27} = \frac{1}{9}$$

(b) If we note that the lowest common denominator of $\frac{7}{9}$ and $\frac{2}{3}$ is 9, we can write

$$\frac{7}{9} - \frac{2}{3} = \frac{7}{9} - \frac{6}{9} = \frac{7 - 6}{9} = \frac{1}{9}$$

EXAMPLE 2 Perform the subtraction $\frac{6x - 3y}{2x} - \frac{4x + 2y}{3y}$, where x and y are nonzero integers. Express the result in lowest terms.

SOLUTION By the use of (3.24), we have

$$\frac{6x - 3y}{2x} - \frac{4x + 2y}{3y} = \frac{(6x - 3y)3y - 2x(4x + 2y)}{(2x)(3y)}$$
$$= \frac{18xy - 9y^2 - 8x^2 - 4xy}{6xy}$$
$$= \frac{14xy - 9y^2 - 8x^2}{6xy}$$

The only (integer) factors of the denominator are 1, −1, 2, −2, 3, −3, 6, −6, x, y, and since none of these except 1 and −1 is a factor of the numerator, the result is in lowest terms.

Just as we defined the concept of the negative of an integer (Sec. 2.4), we can define the concept of the negative of a rational number.

NEGATIVE OF A RATIONAL NUMBER

The **negative** *of a rational number* $\frac{a}{b}$ is the number $-\frac{a}{b}$ which when added to $\frac{a}{b}$, gives the sum zero. For example,

$\frac{-3}{5}$ is the negative of $\frac{3}{5}$ since

$$\frac{3}{5} + \frac{-3}{5} = \frac{3 + (-3)}{5} = \frac{0}{5} = 0$$

that is,

$$-\frac{3}{5} = \frac{-3}{5}$$

Of course, since $\frac{-3}{5} = \frac{3}{-5}$ [because $(-3)(-5) = 5 \cdot 3$], $\frac{3}{-5}$ is another way of writing the negative of $\frac{3}{5}$, that is,

$$-\frac{3}{5} = \frac{3}{-5}$$

If $\frac{a}{b}$ is any rational number, let us consider the sum

$$\frac{a}{b} + \frac{-a}{b} = \frac{a + (-a)}{b} = \frac{0}{b} = 0$$

Thus, the rational number $\frac{-a}{b}$ is the negative of the rational number $\frac{a}{b}$, that is,

$$-\frac{a}{b} = \frac{-a}{b} \tag{3.25}$$

Also, since $\frac{-a}{b} = \frac{a}{-b}$, we have

$$-\frac{a}{b} = \frac{a}{-b} \tag{3.26}$$

To illustrate,

$$-\frac{4}{7} = \frac{-4}{7} = \frac{4}{-7}$$

$$-\frac{-6}{13} = \frac{-(-6)}{13} = \frac{6}{13} \qquad \textit{since the negative of } -6 \textit{ is } 6$$

$$-\frac{7}{-5} = \frac{-7}{-5} = \frac{7}{5}$$

We can now prove that

if $\frac{a}{b}$ and $\frac{c}{d}$ are *any rational numbers,* then

$$\frac{a}{b} - \frac{c}{d} = \frac{a}{b} + \left(-\frac{c}{d}\right) \tag{3.27}$$

According to the definition of subtraction, all we have to do to prove that (3.27) is true is to show that

$$\frac{c}{d} + \left[\frac{a}{b} + \left(-\frac{c}{d}\right)\right] = \frac{a}{b}$$

Now

$$\begin{aligned}
\frac{c}{d} + \left[\frac{a}{b} + \left(-\frac{c}{d}\right)\right] &= \frac{c}{d} + \left[\left(-\frac{c}{d}\right) + \frac{a}{b}\right] && \textit{addition is commutative} \\
&= \left[\frac{c}{d} + \left(-\frac{c}{d}\right)\right] + \frac{a}{b} && \textit{addition is associative} \\
&= 0 + \frac{a}{b} && \textit{definition of a negative} \\
&= \frac{a}{b} && \textit{property of zero under addition}
\end{aligned}$$

So (3.27) is proved. The student should compare this proof with the proof of (2.38) in Sec. 2.4.

If a rational number can be represented by $\frac{a}{b}$, where a and b

POSITIVE RATIONAL NUMBER

NEGATIVE RATIONAL NUMBER

are *positive* integers, we call the number a **positive rational number.** If a rational number can be represented by $\frac{-a}{b}$, or by $\frac{a}{-b}$ where a and b are *positive* integers, we call the number a **negative rational number.**

Equations (3.25) and (3.26) show that a negative rational number can always be expressed as the negative of a positive rational number. Thus,

$$\frac{-2}{5} = -\frac{2}{5} \qquad \frac{16}{-9} = -\frac{16}{9}$$

The results of the preceding paragraphs indicate that the set of rational numbers is composed of the *positive rational numbers,* the *negative rational numbers,* and the *zero rational number.*

In the exercises at the end of this section (unless noted otherwise), it is assumed that the letter symbols which appear may be any rational numbers, except that the letter symbols cannot have values which make the denominator of a fraction zero. To illustrate, in Example 3 below, a may be any rational number except -1; in Example 4, x may be any rational number except 4; and, in Example 5, a and b may be any unequal rational numbers.

EXAMPLE 3 Express $\frac{x}{a+1} + \frac{x^2}{a+1} - \frac{2}{a+1}$ as a single fraction in lowest terms.

SOLUTION A common denominator is $a + 1$. So we write

$$\frac{x}{a+1} + \frac{x^2}{a+1} - \frac{2}{a+1} = \frac{x + x^2 - 2}{a+1}$$

EXAMPLE 4 Combine $\frac{x}{x-4} + \frac{-3}{4-x}$ into a single fraction in lowest terms.

SOLUTION Note that the second fraction can be changed to a fraction with the same denominator as the first fraction by multiplying the numerator and denominator by -1. We have then

$$\frac{x}{x-4} + \frac{-3}{4-x} = \frac{x}{x-4} + \frac{3}{x-4} = \frac{x+3}{x-4}$$

EXAMPLE 5 Express $\frac{5a+b}{a-b} - \frac{2a-3b}{a-b}$ as a single fraction in lowest terms.

SOLUTION We have

$$\frac{5a+b}{a-b} - \frac{2a-3b}{a-b} = \frac{5a+b-(2a-3b)}{a-b}$$

$$= \frac{5a+b-2a+3b}{a-b} = \frac{3a+4b}{a-b}$$

Notice that we enclosed in parentheses the numerator of the fraction that we were subtracting.

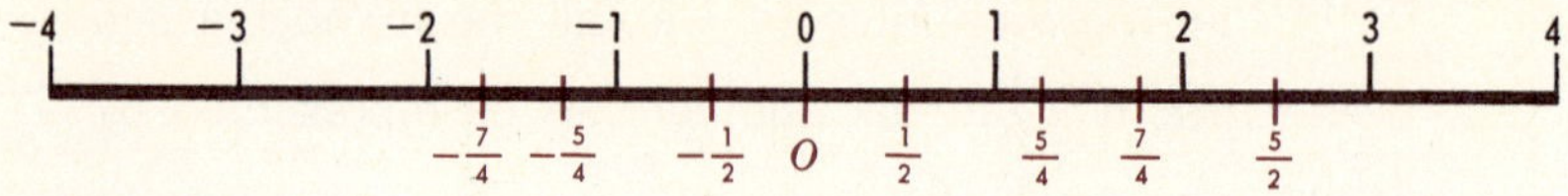

FIGURE 3.6

Similarly to the way in which we marked off points on the number scale representing the integers, we can mark off points representing the rational numbers as shown in Fig. 3.6. On this number scale, for each point that is to the right of 0 and which represents a positive rational number, there corresponds a point that is to the left of 0 and which represents a negative rational number. These two rational numbers are the negatives of each other. For example, the negative of $\frac{7}{4}$ is $-\frac{7}{4}$ (and the negative of $-\frac{7}{4}$ is $\frac{7}{4}$) since

$$\frac{7}{4}+\left(-\frac{7}{4}\right)=\frac{7}{4}-\frac{7}{4}=0$$

and we note that $\frac{7}{4}$ and $-\frac{7}{4}$ are on opposite sides of, but at equal distances from, the origin.

GREATER THAN

LESS THAN

If p and q are rational numbers and if there exists a *positive* rational number r such that $p-q=r$, we say that p is **greater than** q and write $p>q$. Equivalently, we say that q is **less than** p and write $q<p$. For example, $-\frac{13}{42}>-\frac{1}{3}$, because $-\frac{13}{42}-(-\frac{1}{3})=-\frac{13}{42}+\frac{14}{42}=\frac{1}{42}$, which is positive.

If $p-q=r$, where p and q are any rational numbers and r is a *positive* rational number, we say that p is r *more than* q, or that q is r *less than* p. To illustrate, $\frac{1}{2}-\left(-\frac{1}{5}\right)=\frac{1}{2}+\frac{1}{5}=\frac{5+2}{10}=\frac{7}{10}$, which is positive. Therefore, $\frac{1}{2}$ is $\frac{7}{10}$ more than $-\frac{1}{5}$, or $-\frac{1}{5}$ is $\frac{7}{10}$ less than $\frac{1}{2}$. Since any positive rational number is greater than zero, we frequently write $p>0$ for "p is positive." Similarly, $p<0$ means "p is negative."

The student should see that if $p>q$, then the point on the number scale corresponding to p lies to the *right* of the point on the number scale representing q.

ABSOLUTE VALUE

The **absolute value** of a rational number p is denoted by $|p|$ and is defined as follows:

1 The absolute value of a nonnegative rational number is the rational number itself, $|p|=p$ if $p\geqslant 0$.

2 The absolute value of a negative rational number is the negative of that number, $|p|=-p$ if $p<0$. Thus

$$\left|\frac{3}{4}\right|=\frac{3}{4}\qquad\left|-\frac{7}{8}\right|=\frac{7}{8}\qquad\left|\frac{6}{5}\right|=\frac{6}{5}\qquad\left|\frac{-3}{4}\right|=\frac{|-3|}{|4|}=\frac{3}{4}$$

EXERCISES

By the use of one or more of equations (3.9), (3.10), (3.25), and (3.26), show that each of the following results is correct:

1 $\frac{-5}{3} = -\frac{5}{3}$

2 $\frac{5}{-3} = -\frac{5}{3}$

3 $-\frac{a}{b} = -\frac{-a}{-b}$

4 $\frac{-b}{x-y} = \frac{b}{y-x}$

5 $\frac{x-y}{y-x} = -1$

6 $\frac{-ax-ay}{ax+ay} = -1$

7 $\frac{3a-3b}{b-a} = -3$

8 $\frac{3a-3b}{-3} = b-a$

Write the following rational numbers in lowest terms, with both numerator and denominator positive; assume that all letter symbols represent positive integers.

9 $\frac{-5}{15}$

10 $\frac{-21}{-42}$

11 $\frac{3a}{-6a}$

12 $\frac{-13x^2}{26x^2}$

13 $\frac{7y}{-14y}$

14 $\frac{-14}{-42}$

Apply the definition of the subtraction of rational numbers to calculate the following differences. Express results in lowest terms.

15 $\frac{3}{2} - \frac{2}{3}$

16 $\frac{2}{3} - \frac{3}{2}$

17 $\frac{1}{5} - \frac{1}{6}$

18 $\frac{1}{6} - \frac{1}{5}$

19 $\left(\frac{2}{3} + \frac{1}{6}\right) - \frac{3}{5}$

20 $\frac{3}{5} - \left(\frac{2}{3} - \frac{1}{6}\right)$

21 $\left(3 + \frac{1}{11}\right) - \frac{3}{2}$

22 $\frac{3}{2} - \left(3 + \frac{1}{11}\right)$

23 Assuming that x and y are positive rational numbers, tell, by answering "yes" or "no" with reference to each, which of the following expressions always represent a positive rational number: $x + y$; $x - y$; xy.

24 If your answer to a part of Exercise 23 is "no," what restriction must be placed on x and y for the expression in question to represent a positive rational number?

25 Suppose that from a bag of potatoes weighing 120 pounds, a merchant sells lots weighing 10 pounds, $6\frac{3}{4}$ pounds, and $15\frac{1}{2}$ pounds. How many pounds does he sell? How many pounds remain in the bag?

26 If a boy works four consecutive days of $6\frac{1}{2}$ hours, 7 hours, $4\frac{1}{4}$ hours, and 6 hours, respectively, how much money will he earn during the 4-day period if he gets 60 cents per hour?

27 What is the total time you spend in doing homework, if you devote $1\frac{1}{3}$ hours to one subject, $2\frac{1}{4}$ hours to a second, $1\frac{1}{6}$ hours to a third, and

$\frac{7}{12}$ of an hour to a fourth? Express your answer first in hours and then in hours and minutes.

28 A man has a half interest in a business. He gives one-third of his share to his son. What is his son's interest in the business? How much is the son's interest worth if the business is capitalized at $4,800?

29 The distance between the centers of two holes in a metal bar is $7\frac{1}{2}$ inches. The holes are $\frac{9}{16}$ inch in diameter. What is the distance between the holes?

30 From a bolt of cloth containing $85\frac{1}{2}$ yards, a salesman sold four pieces whose lengths were $6\frac{3}{4}$, $9\frac{1}{3}$, $17\frac{3}{4}$, and $22\frac{1}{5}$ yards. How many yards of cloth remained on the bolt?

31 Four cars finished a race in the following respective times: $7\frac{3}{4}$, $8\frac{2}{3}$, $7\frac{3}{5}$, and $6\frac{1}{2}$ hours. Find the difference between the time required for each of the slower cars and the winner's time.

32 If $1\frac{2}{3}$ yards of material are required to make an apron, how many yards are needed to make four aprons? If $7\frac{1}{2}$ yards of material were purchased, how much was left over after making four aprons?

33 How many pieces of rope each $\frac{2}{3}$ yard long can be cut from a rope 20 yards long?

34 Which is the greater, $\frac{13}{42}$ or $\frac{1}{3}$? How much greater?

35 Arrange the fractions $\frac{1}{4}$, $\frac{1}{3}$, $\frac{7}{20}$, $\frac{9}{28}$ in order of size, beginning with the smallest.

36 Which is the greater, -3 or $-\frac{5}{2}$? How much greater?

37 Which is the greater, $-\frac{1}{7}$ or $-\frac{4}{35}$? How much greater?

38 -2 is how much more than $-\frac{21}{10}$?

39 Which is the greater, $-\frac{2}{5}$ or $-\frac{7}{17}$? The larger is how much more than the smaller?

40 If 4 is added to the numerator and also to the denominator of the fraction $\frac{3}{5}$, is the value of the fraction increased or decreased? How much?

41 If p is positive, is $-p$ positive or negative? If p is negative, is $-p$ positive or negative?

42 If p, q, and r are rational numbers, we say that q is between p and r if $p < q < r$. Of considerable interest is the fact that between any two unequal rational numbers there is always another rational number. If $p < r$, it can be shown that

$$p < (\tfrac{1}{2})(p + r) < r$$

This result tells us that the rational number $q = (\frac{1}{2})(p + r)$ is between the rational numbers p and r. For the values of p and r given below, find $q = (\frac{1}{2})(p + r)$, and verify that $p < q < r$:

$p = \frac{1}{5}$, $r = \frac{8}{35}$

$p = -\frac{7}{54}$, $r = -\frac{1}{9}$

$p = \frac{17}{46}$, $r = \frac{9}{23}$

$p = -2\frac{1}{4}$, $r = -2\frac{1}{5}$

3.6 DIVISION OF RATIONAL NUMBERS

The definition of division on the set of rational numbers has the same form as the definition of division on the set of integers (Sec. 2.4) and on the set of natural numbers (Sec. 1.9). If $\frac{a}{b}$ and $\frac{c}{d}$ are rational numbers, we **define** $\frac{a}{b} \div \frac{c}{d}$ to be *the* rational number $\frac{x}{y}$, if such a rational number exists, with the property that $\frac{c}{d} \cdot \frac{x}{y} = \frac{a}{b}$; that is,

DIVISION OF RATIONAL NUMBERS

$$\frac{a}{b} \div \frac{c}{d} = \frac{x}{y}$$

if and only if $\frac{x}{y}$ is the rational number with the property that

$$\frac{c}{d} \cdot \frac{x}{y} = \frac{a}{b}$$

QUOTIENT

DIVIDEND

DIVISOR

If $\frac{a}{b} \div \frac{c}{d} = \frac{x}{y}$, we call $\frac{x}{y}$ the **quotient** which results when $\frac{a}{b}$ is *divided* by $\frac{c}{d}$. We call $\frac{a}{b}$ the **dividend** and $\frac{c}{d}$ the **divisor.**

Suppose that we wish to determine the rational number $\frac{x}{y}$ so that

$$\frac{1}{2} \div \frac{1}{3} = \frac{x}{y}$$

From the definition of division we know that $\frac{x}{y}$ must satisfy the equality

$$\frac{1}{3} \cdot \frac{x}{y} = \frac{1}{2}$$

Notice that if we multiply $\frac{1}{3}$ by $\frac{3}{1}$, we get $\frac{1}{3} \cdot \frac{3}{1} = \frac{1}{1}$; and if we then multiply $\frac{1}{1}$ by $\frac{1}{2}$, we get $\frac{1}{1} \cdot \frac{1}{2} = \frac{1}{2}$. So, if we let $\frac{x}{y} = \frac{3}{1} \cdot \frac{1}{2} = \frac{3}{2}$, we have

$$\frac{1}{3} \cdot \frac{x}{y} = \frac{1}{3} \cdot \frac{3}{2} = \frac{1}{2}$$

and we can write

$$\frac{1}{2} \div \frac{1}{3} = \frac{1}{2} \cdot \frac{3}{1} = \frac{3}{2}$$

Now suppose that we wish to determine the rational number $\frac{x}{y}$

so that

$$\frac{2}{5} \div \frac{3}{7} = \frac{x}{y}$$

For this to be true we must have

$$\frac{3}{7} \cdot \frac{x}{y} = \frac{2}{5}$$

Notice that if we multiply $\frac{3}{7}$ by $\frac{7}{3}$, we get $\frac{3}{7} \cdot \frac{7}{3} = \frac{1}{1}$; and if we then multiply $\frac{1}{1}$ by $\frac{2}{5}$, we get $\frac{1}{1} \cdot \frac{2}{5} = \frac{2}{5}$. So in this case

$$\frac{x}{y} = \frac{7}{3} \cdot \frac{2}{5} = \frac{14}{15}$$

and

$$\frac{2}{5} \div \frac{3}{7} = \frac{2}{5} \cdot \frac{7}{3} = \frac{14}{15}$$

The results of these two illustrations suggest that it may be true that for any two rational numbers $\frac{a}{b}$ and $\frac{c}{d}$,

$$\frac{a}{b} \div \frac{c}{d} = \frac{a}{b} \cdot \frac{d}{c} = \frac{ad}{bc}$$

In order for $\frac{a}{b} \div \frac{c}{d}$ to be equal to $\frac{ad}{bc}$, it must be true that $\frac{ad}{bc}$ is the *rational number* with the property that $\frac{c}{d} \cdot \frac{ad}{bc} = \frac{a}{b}$. Now $\frac{ad}{bc}$ represents a rational number *provided* a, b, c, and d are integers with $b \neq 0$ and $c \neq 0$. Thus $\frac{ad}{bc}$ will represent a rational number if $\frac{a}{b}$ and $\frac{c}{d}$ are rational numbers and *if* $\frac{c}{d}$ *is not the zero rational number* (that is, if $c \neq 0$). If $\frac{ad}{bc}$ is a rational number, we have

$$\frac{c}{d} \cdot \frac{ad}{bc} = \frac{c \cdot a \cdot d}{d \cdot b \cdot c}$$

and, since c and d are factors of both numerator and denominator, we can express $\frac{c \cdot a \cdot d}{d \cdot b \cdot c}$ in lowest terms as $\frac{a}{b}$. Therefore we have proved that for any rational numbers $\frac{a}{b}$ and $\frac{c}{d}$ with $\frac{c}{d} \neq 0$,

$$\frac{a}{b} \div \frac{c}{d} = \frac{a}{b} \cdot \frac{d}{c} = \frac{ad}{bc} \tag{3.28}$$

EXAMPLE 1 Divide $\frac{3}{4}$ by $\frac{5}{12}$ and express the result in lowest terms.

SOLUTION By the use of (3.28) we have

$$\frac{3}{4} \div \frac{5}{12} = \frac{3}{4} \cdot \frac{12}{5} = \frac{3 \cdot 12}{4 \cdot 5} = \frac{3 \cdot 3 \cdot 4}{4 \cdot 5} = \frac{3 \cdot 3}{5} = \frac{9}{5}$$

EXAMPLE 2 Divide 4 by $\frac{-3}{5}$, and express the result in lowest terms.

SOLUTION We can write 4 as $\frac{4}{1}$ and use (3.28) to obtain

$$\frac{4}{1} \div \frac{-3}{5} = \frac{4}{1} \cdot \frac{5}{-3} = \frac{20}{-3}$$

This result is not in lowest terms since the denominator is negative, so we write

$$\frac{20}{-3} = \frac{-20}{3} \qquad \text{or} \qquad \frac{20}{-3} = -\frac{20}{3}$$

and, expressed in lowest terms,

$$4 \div \frac{-3}{5} = \frac{-20}{3} \qquad \text{or} \qquad 4 \div \frac{-3}{5} = -\frac{20}{3}$$

In Sec. 3.4 we defined the reciprocal of a nonzero integer. Here we shall extend the concept of reciprocal. The **reciprocal** of any nonzero number is 1 divided by that number. The reciprocal of the rational number $\frac{a}{b}$, with $a \neq 0$, is

RECIPROCAL

$$1 \div \frac{a}{b} = \frac{1}{1} \div \frac{a}{b} = \frac{1}{1} \cdot \frac{b}{a} = \frac{b}{a}$$

That is, if $\frac{a}{b} \neq 0$,

the rational number $\frac{b}{a}$ *is the reciprocal of the rational number* $\frac{a}{b}$

The result (3.28) concerning the quotient of rational numbers can be stated as follows:

To divide one rational number by another, multiply the dividend by the reciprocal of the divisor.

Observe that

$$a \div \frac{1}{b} = \frac{a}{1} \cdot \frac{b}{1} = ab$$

Thus, dividing by the reciprocal of a nonzero integer is the same as multiplying by that integer.

EXAMPLE 3 Divide 5 by $\frac{1}{7}$.

SOLUTION We write

$$5 \div \frac{1}{7} = 5 \cdot \frac{7}{1} = 5 \cdot 7 = 35$$

EXAMPLE 4 Divide $\frac{2x+10}{y^2}$ by $\frac{x+5}{2y}$, where x and y are positive integers, and express the result in lowest terms.

SOLUTION We have

$$\frac{2x+10}{y^2} \div \frac{x+5}{2y} = \frac{2x+10}{y^2} \cdot \frac{2y}{x+5} = \frac{2(x+5) \cdot 2y}{y \cdot y(x+5)} = \frac{4}{y}$$

In the preceding three sections we saw that the set of rational numbers is closed under addition, multiplication, and subtraction. Further, the result (3.28) shows that if $\frac{a}{b}$ and $\frac{c}{d}$ are rational numbers and *if* $\frac{c}{d}$ *is not zero,* then $\frac{a}{b} \div \frac{c}{d}$ is a rational number. That is, we have shown that the set of rational numbers is closed under division *except* when the divisor is zero.

Let us denote the set of rational numbers by the symbol Ra. Then the results summarized in the preceding paragraph can be written:

If $p \in Ra$ and $q \in Ra$, then
$p + q \in Ra$
$p \cdot q \in Ra$
$p - q \in Ra$
$p \div q \in Ra$ *provided* $q \neq 0$

So far we have shown that if the divisor is not zero, then the quotient of two rational numbers exists and is a uniquely determined rational number. Now let us see that division by zero *must* be excluded; that is, if the divisor is zero, then the quotient cannot exist as a uniquely determined rational number. Recall that to divide the rational number $\frac{a}{b}$ by the rational number $\frac{c}{d}$ means to find the uniquely determined rational number $\frac{x}{y}$ for which $\frac{c}{d} \cdot \frac{x}{y} = \frac{a}{b}$, if such a number $\frac{x}{y}$ exists. Let us try to apply this definition to find $\frac{x}{y}$ in the situation where $\frac{c}{d} = 0$, that is, where $c = 0$. There are two cases.

CASE 1

$\frac{a}{b} \neq 0$. If the division of $\frac{a}{b}$ by $\frac{0}{d}$ could be performed, we would need to find the uniquely determined number $\frac{x}{y}$ with the property that $\frac{0}{d} \cdot \frac{x}{y} = \frac{a}{b}$. But $\frac{0}{d} \cdot \frac{x}{y} = \frac{0 \cdot x}{dy} = 0$, and we cannot find a number $\frac{x}{y}$

so that $\frac{0}{d} \cdot \frac{x}{y} = \frac{a}{b}$ where $\frac{a}{b} \neq 0$. To illustrate, if division by zero were possible, $\frac{1}{2} \div 0$ would represent the rational number p for which $0 \cdot p = \frac{1}{2}$. But there is no number that satisfies this requirement.

CASE 2

$\frac{a}{b} = 0$. In this case the definition of division requires us to find the uniquely determined number $\frac{x}{y}$ with the property that $\frac{0}{d} \cdot \frac{x}{y} = 0$. But *any* rational number will satisfy the equation $0 \cdot \frac{x}{y} = 0$, and therefore $\frac{x}{y}$ cannot be uniquely determined.

We see then that if we want division of one number by another to exist and the quotient to be uniquely determined, we must exclude division by zero.

In Sec. 3.4, in equation (3.18), we identified $\frac{a}{b}$ with $a \div b$, where a and b are integers with $b \neq 0$. We now agree to extend the use of the fraction bar — to indicate division of rational numbers, and we make the following *definition:* if a and b represent rational numbers,* with $b = 0$, the symbol $\frac{a}{b}$ means $a \div b$. That is, for rational numbers a and b, with $b \neq 0$,

$$\frac{a}{b} = a \div b \qquad (3.29)$$

FRACTION

We also extend the meaning of the word **fraction** to include any symbol of the form $\frac{a}{b}$, where a and b represent rational numbers, with $b \neq 0$. In the fraction $\frac{a}{b}$, a is the numerator and b is the denominator.

To illustrate,

$$\frac{x}{a+1} \qquad \frac{5a+b}{a-b} \qquad \frac{x^2+x-2}{x+y}$$

are fractions if a, b, x, and y are rational numbers with $a + 1 \neq 0$, $a - b \neq 0$, and $x + y \neq 0$.

A fraction is said to be in lowest terms if the numerator and denominator have no common factor except 1 and -1 and the denominator is positive.

* In Chap. 5 the identification of $\frac{a}{b}$ with $a \div b$ and the use of the word "fraction" for the symbol $\frac{a}{b}$ will be extended to the situation where a and b are *any* numbers, with $b \neq 0$.

EXERCISES

Perform the following divisions. Express results in lowest terms. The letter symbols represent positive integers.

1 $\frac{7}{8} \div \frac{2}{3}$

2 $4 \div \frac{3}{5}$

3 $\frac{3}{5} \div 4$

4 $\frac{a}{b} \div a$

5 $a \div \frac{a}{b}$

6 $\frac{x+y}{x} \div x$

7 $x \div \frac{x+y}{x}$

8 $\frac{x}{y} \div \frac{y}{x}$

9 $\frac{x}{x+y} \div \frac{x}{y}$

10 $a \div ab$

11 $ab \div a$

12 $\frac{x}{2} \div \frac{5x^2}{8}$

13 $\frac{a+b}{a} \div \frac{b}{a}$

14 $\frac{a+b}{a} \div \frac{a}{b}$

15 $\frac{2a}{3b} \div \frac{4a}{27}$

16 $\frac{a}{a+b} \div \frac{3a}{ac+bc}$

17 $\frac{5x+10}{x^2} \div \frac{5}{x}$

18 $\frac{xy+x}{y} \div \frac{ay+a}{y^2}$

19 $\frac{y^2}{15x+6} \div \frac{y}{5x+2}$

20 $\frac{3x+12}{x} \div \frac{x+4}{2x^2}$

21 $\frac{5}{a^3+a^2y} \div \frac{25}{a^2+ay}$

22 $\frac{2a+b}{7ab} \div \frac{4a+2b}{3b^2}$

23 $\frac{4y+4}{7} \div \frac{2y^2+2y}{14}$

24 $\frac{2a^2+3a^2}{b} \div \frac{a^2}{b^3}$

25 Recall that the number of square units of area A of a rectangle is given by the formula $A = lw$, where l is the number of units of length and w is the number of units of width, it being understood that l and w are measured in the same units. If we divide both sides of the equation $A = lw$ by w, we obtain $l = \frac{A}{w}$. Using this formula, find the length l of each of the rectangles whose area A and width w are, respectively, as follows:

$A = 20$ square feet, $w = \frac{5}{3}$ feet
$A = \frac{22}{5}$ square feet, $w = 2$ feet
$A = \frac{23}{4}$ square feet, $w = \frac{2}{3}$ foot
$A = 30$ square feet, $w = 18$ inches

26 Give the reciprocal of each of the following: $\frac{3}{4}$; $\frac{1}{5}$; $\frac{4}{9}$; $\frac{9}{10}$; $\frac{2}{a}$; $-(\frac{3}{7})$; -2; $2\frac{1}{3}$.

Perform the following calculations. Express results in lowest terms. In exercises in which letter symbols appear, tell what values of the letter symbols must be excluded.

27 $\left(\frac{1}{2}+\frac{1}{3}\right) \div \left(\frac{1}{4}+\frac{1}{5}\right)$

28 $\left(\frac{1}{x}+\frac{1}{y}\right) \div \left(\frac{2}{x}+\frac{2}{y}\right)$

29 $(8\frac{3}{4}) \cdot (\frac{2}{3})$

30 $(4\frac{1}{3}) \cdot (6\frac{1}{2})$

31 $(2\frac{1}{2}) \div (5\frac{1}{3})$

32 $(6\frac{1}{2}) + (14\frac{1}{3})$

33 $(4\frac{2}{7}) \div (3\frac{1}{3})$

34 $(4\frac{2}{7}) \cdot (3\frac{1}{3})$

35 $\left(\frac{1}{x}+3\right) \div \left(4+\frac{2}{x}\right)$

36 $\left(\frac{x+y}{4}\right) \div \left(\frac{2x+2y}{8}\right)$

37 $\left(\frac{1+x}{3}\right) \div \left(\frac{3+3x}{7}\right)$

38 $\left(3-\frac{1}{x^2}\right) \div \left(2+\frac{1}{x}\right)$

39 $\left(\frac{a}{2}+\frac{b}{3}\right) \div \left(\frac{3a+2b}{5}\right)$

40 $\left(\frac{2}{a}+\frac{3}{b}\right) \div \left(\frac{5}{a}+\frac{4}{b}\right)$

Combine the following expressions into a single fraction in lowest terms. Indicate the values of the letter symbols which must be excluded.

41 $\frac{4}{2x} - \frac{3x}{5} + \frac{6}{x}$

42 $\frac{8}{x+3} + 5 + \frac{3}{7}$

43 $\frac{2}{y+2} - \frac{3}{y+2}$

44 $\frac{3}{2x-1} + 4 - \frac{x}{1-2x}$

45 $\frac{6a}{2a-3} - \frac{9}{2a-3}$

46 $\frac{ax}{x+a} \cdot \frac{x+a}{xa}$

47 $\frac{m-4}{12} \cdot \frac{18}{m^2-4m}$

48 $\frac{xy-x}{y} \div \frac{ay-a}{y^2}$

49 $\frac{3a-b}{a+b} - \frac{2a-2b}{a+b}$

50 $\frac{2a}{a-b} + \frac{a}{b-a}$

51 $\frac{b}{3a} - \frac{a-1}{5b}$

52 $\frac{ab+ac}{bm+bn} \div \frac{b+c}{m+n}$

53 $\frac{2x+8}{3x-9} \cdot \frac{3}{x+4}$

54 $\frac{x^2+y^2}{m+n} - \frac{x^2-y^2}{m+n}$

55 $\frac{1}{2} - \frac{a-1}{a} + \frac{a-2}{a^2}$

56 $\frac{3}{x-4} - \frac{4}{x-4}$

57 $\frac{7}{a-b} - \frac{5}{b-a}$

58 $\frac{2x}{2x-3} - \frac{3}{2x-3}$

59 $\frac{7x}{2a+2b} \cdot \frac{a+b}{x^2}$

60 $\frac{x+2}{3x-12} + \frac{2x-1}{4-x}$

61 $ab\left(\frac{1}{a}+\frac{1}{b}\right)$

62 $\frac{x^2}{x-y} - x + y$

63 $\left(x-\frac{1}{y}\right) \div \left(y-\frac{1}{x}\right)$

64 Assuming that p and q are rational numbers, tell, by answering "yes" or "no" with reference to each, which of the following expressions always represent a rational number: $p+q$; $p-q$; pq; $p \div q$.

65 If your answer to a part of Exercise 64 is "no," what restriction must be placed on p or q for the expression in question to represent a rational number?

3.7 SOME ALGEBRAIC IDENTITIES

Recall that the distributive property

$$a(b + c) = ab + ac \tag{3.30}$$

gives a method for writing the *product* of a monomial and a binomial as a *sum*. The distributive property, when read from right to left, gives a procedure for factoring a suitable *sum* into the *product* of a monomial and a binomial. For instance,

$$2am - 6an = 2a(m - 3n)$$

We make frequent use of the distributive property in the form

$$a(b + c + d) = ab + ac + ad \tag{3.31}$$

The identity (3.31) enables us to factor a suitable trinomial into the product of a monomial and another trinomial. To illustrate,

$$3r^2 + 6r - 3 = 3(r^2 + 2r - 1)$$
$$-4m^2x^2 + 16mx - 4x = -4x(m^2x - 4m + 1)$$

The distributive property (3.30) enables us to express the square of a binomial as a sum. For

$$\begin{aligned}(a + b)(a + b) &= (a + b)a + (a + b)b \\ &= a^2 + ba + ab + b^2 \\ &= a^2 + 2ab + b^2\end{aligned}$$

That is,

$$(a + b)^2 = a^2 + 2ab + b^2 \tag{3.32}$$

To illustrate,

$$(3x + 2y)^2 = (3x)^2 + 2(3x)(2y) + (2y)^2 = 9x^2 + 12xy + 4y^2$$

Similarly, using the distributive property, we obtain

$$\begin{aligned}(a - b)(a - b) &= (a - b)a - (a - b)b \\ &= a^2 - ab - ab + b^2 \\ &= a^2 - 2ab + b^2\end{aligned}$$

So

$$(a - b)^2 = a^2 - 2ab + b^2 \tag{3.33}$$

For example,

$$(3x - 2y)^2 = (3x)^2 - 2(3x)(2y) + 4y^2 = 9x^2 - 12xy + 4y^2$$

If we write the identities (3.32) and (3.33) in reverse order, as

$$a^2 + 2ab + b^2 = (a + b)^2 \qquad (3.34)$$
$$a^2 - 2ab + b^2 = (a - b)^2 \qquad (3.35)$$

we see that we have rules for factoring two special types of trinomials.

EXAMPLE 1 Factor $x^2 + 6x + 9$.

SOLUTION Observe that x^2 is the square of x, and 9 is the square of 3; also notice that $6x = 2(x \cdot 3)$. Hence

$$(x + 3)^2 = x^2 + 2 \cdot x \cdot 3 + 9 = x^2 + 6x + 9$$

EXAMPLE 2 Factor $9x^2 - 6x + 1$.

SOLUTION We may write

$$9x^2 - 6x + 1 = (3x)^2 - 2(3x)1 + 1^2 = (3x - 1)^2$$

by use of (3.35).

Let a and b be any two numbers. Then

$$\begin{aligned}(a + b)(a - b) &= (a + b)a - (a + b)b \\ &= a^2 + ab - ab - b^2 \\ &= a^2 - b^2\end{aligned}$$

That is,

$$(a + b)(a - b) = a^2 - b^2 \qquad (3.36)$$

We are especially interested in the identity (3.36) written in the form

$$a^2 - b^2 = (a + b)(a - b) \qquad (3.37)$$

EXAMPLE 3 Factor $25x^2 - 4y^2$.

SOLUTION Recognizing each term as a square, we write

$$25x^2 - 4y^2 = (5x)^2 - (2y)^2 = (5x + 2y)(5x - 2y)$$

Recall from Sec. 1.8 that

$$(x + a)(x + b) = x^2 + x(a + b) + ab \qquad (3.38)$$

To factor a trinomial of the type

$$x^2 + sx + p$$

compare it with the trinomial

$$x^2 + (a + b)x + ab$$

We see that if we can find two numbers a and b whose sum is s and whose product is p, we may write

$$x^2 + sx + p = (x + a)(x + b)$$

EXAMPLE 4 Factor $x^2 - x - 6$.

SOLUTION The second terms of the binomial factors must be such that their product is -6 and their sum is -1. Hence they must have opposite signs, and the one with greater absolute value must be negative in order to give its sign to their sum. The required factors of -6 are -3 and 2, and

$$x^2 - x - 6 = (x - 3)(x + 2)$$

as can be verified by multiplying these factors.

A knowledge of the types of factoring which we have just considered enables us to cope with more complex problems dealing with fractions.

EXAMPLE 5 Express the product $\frac{2a}{a-2} \cdot \frac{a^2 - 4}{6a^2}$ as a single fraction, and express the result in lowest terms.

SOLUTION Now

$$\frac{2a}{a-2} \cdot \frac{a^2 - 4}{6a^2} = \frac{2a}{a-2} \cdot \frac{(a-2)(a+2)}{2a \cdot 3a} = \frac{a+2}{3a}$$

EXAMPLE 6 Express the quotient $\frac{x^2 + 5x + 6}{a} \div \frac{x+3}{a^2}$ as a single fraction in lowest terms.

SOLUTION We write

$$\begin{aligned} \frac{x^2 + 5x + 6}{a} \div \frac{x+3}{a^2} &= \frac{x^2 + 5x + 6}{a} \cdot \frac{a^2}{x+3} \\ &= \frac{(x+2)(x+3)}{a} \cdot \frac{a \cdot a}{x+3} \\ &= a(x+2) \end{aligned}$$

EXERCISES

Make use of the distributive property to factor the following expressions:

1 $14x^2 + 42x$
2 $4 - 8x - 16y$
3 $x^2 - xy + xz$
4 $36xy - 6xy^2$
5 $10x^2y + 15xy - 5y$
6 $c^2x - cx^2 + cx$
7 $2x^3 + 4x^2 - 6x$
8 $\frac{1}{4}am + \frac{1}{4}bm - \frac{1}{4}cm$

Find the following products:

9 $-4xy(x - a)$

10 $b(x + b + 1)$

11 $-ab(a - b)$

12 $a(a - b - c)$

13 $-x(-x + xy)$

14 $\frac{4}{3}a(\frac{1}{2}a - \frac{1}{4})$

15 $5n(3n - 2n^2 + 1)$

16 $\frac{1}{2}ab(-b - a)$

Find the indicated squares:

17 $(x + 2)^2$

18 $(y - 3)^2$

19 $(2a - 3b)^2$

20 $(2x + xy)^2$

21 $(ab - 1)^2$

22 $\left(\frac{x}{2} + 3\right)^2$

23 $(2a + \frac{1}{2})^2$

24 $(x - \frac{2}{3})^2$

Factor the following trinomials into equal binomial factors. Check results by multiplying the factors.

25 $x^2 - 14x + 49$

26 $x^2 + 10x + 25$

27 $y^2 - \frac{2}{3}y + \frac{1}{9}$

28 $x^2 - 16x + 64$

29 $4b^2 + 1 + 4b$

30 $m^2 - 4mn + 4n^2$

31 $9x^2 - 42x + 49$

32 $36 + 12bc + b^2c^2$

Using the identities (3.30) and (3.37), factor the following expressions:

33 $x^2 - 100$

34 $9x^2 - 1$

35 $x^2 - 64$

36 $19x^2 - 19y^2$

37 $16a^2 - 9x^2y^2$

38 $\frac{a^2}{4} - y^2$

39 $\frac{x^2}{36} - \frac{y^2}{25}$

40 $m^2n - n$

Factor the following trinomials. Verify your results by substituting the value -2 for the letter symbol.

41 $x^2 - 6x - 40$

42 $y^2 + 2y - 24$

43 $a^2 - 7a + 10$

44 $y^2 - 8y + 7$

45 $x^2 + 26x + 48$

46 $a^2 - 7a + 12$

47 $x^2 + \frac{3}{2}x + \frac{1}{2}$

48 $y^2 - \frac{4}{3}y + \frac{1}{3}$

Simplify the following fractions, and state the values of the letter symbols which must be excluded:

49 $\frac{3x^2 - 3y^2}{x^2 - 2xy + y^2}$

50 $\frac{a^2 - 3a + 2}{a^2 + 2a - 8}$

Perform the indicated operations, and write results as a single fraction in lowest terms. Indicate the values of the letter symbols which must be excluded.

51 $\dfrac{100 - a^2}{x + y} \cdot \dfrac{y + x}{10 + a}$

52 $\dfrac{x^2 - y^2}{a^2 - b^2} \cdot \dfrac{a + b}{x - y}$

53 $\dfrac{a^2 - a - 6}{3x - 27} \cdot \dfrac{3}{a - 3}$

54 $\dfrac{3x - 3y}{a^2 - b^2} \div \dfrac{x - y}{a + b}$

55 $\dfrac{9 - 6x + x^2}{x - 1} \cdot \dfrac{1 - x}{x - 3}$

56 $\dfrac{x^2 - 16}{25 - y^2} \div \dfrac{x + 4}{5 + y}$

57 $\dfrac{4}{x - 2} - \dfrac{5}{4 - x^2}$

58 $\dfrac{5}{1 + 2x} - \dfrac{2x}{1 - 2x} - \dfrac{4(1 - 3x)}{1 - 4x^2}$

59 $\dfrac{1}{x + 1} + \dfrac{3}{(x + 1)^2} - \dfrac{2}{x^2 - 1}$

60 $\dfrac{1}{x^2 - 2x + 1} - \dfrac{1}{x^2 + 2x + 1}$

3.8
MULTIPLICATION AND DIVISION INVOLVING POSITIVE INTEGRAL EXPONENTS

The growth of a mathematical science is more than a steady accumulation of knowledge in regard to basic theory; the development of a workable symbolism is necessary to the continuation of the growth. In each stage of mathematics we must learn many conventions, signs, and symbols and practice using them before we can explore and exploit the next stage.

nTH POWER

Recall from Sec. 1.8 the definition of a^n, the *nth power of a,* when n is a positive integer; namely,

$$a^n = a \cdot a \cdot a \cdots a \qquad n \textit{ factors } a \tag{3.39}$$

BASE

EXPONENT

where the expression "n factors a" means that a is used as a factor n times. In a^n, a is the **base** and n is the **exponent.** We sometimes read a^n as "a to the n" or "a to the nth power."

Exponents give us a shorthand way of writing the product of any number of equal factors. For example, 8^{12} represents the product of 12 factors, each factor being 8, and it is surely shorter than writing $8 \cdot 8 \cdot 8 \cdot 8 \cdot 8 \cdot 8 \cdot 8 \cdot 8 \cdot 8 \cdot 8 \cdot 8 \cdot 8$.

The following are some illustrations of the use of the definition (3.39):

$(-2)^3 = (-2) \cdot (-2) \cdot (-2) = -8$
$2^3 = 2 \cdot 2 \cdot 2 = 8$
$(\frac{2}{3})^2 = \frac{2}{3} \cdot \frac{2}{3} = \frac{4}{9}$
$(3 \cdot 3)(x \cdot x)(y \cdot y \cdot y) = 9x^2y^3$
$3a^3 = 3(a \cdot a \cdot a)$
$(3a)^3 = 3a \cdot 3a \cdot 3a = 27a^3$

Note that in the product

$$a^3 \cdot a^2 = (a \cdot a \cdot a)(a \cdot a) = a \cdot a \cdot a \cdot a \cdot a = a^5$$

the exponent in the result is the sum of the exponents in the factors.

We shall show that this is true in general.

For any two positive integers m and n, we have

$$a^m = a \cdot a \cdots a \qquad m \textit{ factors } a$$

and

$$a^n = a \cdot a \cdots a \qquad n \textit{ factors } a$$

Therefore,

$$\begin{aligned} a^m \cdot a^n &= (m \text{ factors } a) \cdot (n \text{ factors } a) \\ &= a \cdot a \cdots a \qquad m + n \textit{ factors } a \end{aligned}$$

or

$$a^m \cdot a^n = a^{m+n} \tag{3.40}$$

Notice that in (3.40) both powers in the product on the left side have the same base a. The following are some illustrations of the use of the result (3.40):

$$x^2 \cdot x^3 = x^5 \qquad 2^4 \cdot 2^6 = 2^{10}$$
$$(x^2y^3)(x^4y) = (x^2 \cdot x^4)(y^3 \cdot y) = x^6y^4$$

Observe that

$$(a^2)^3 = a^2 \cdot a^2 \cdot a^2 = (a \cdot a)(a \cdot a)(a \cdot a) = a \cdot a \cdot a \cdot a \cdot a \cdot a = a^6$$

For any two positive integers m and n we have

$$\begin{aligned} (a^m)^n &= a^m \cdot a^m \cdots a^m \qquad && n \textit{ factors } a^m \\ &= a^{m+m+\cdots+m} && n \textit{ terms in the exponent} \end{aligned}$$

$$(a^m)^n = a^{mn} \tag{3.41}$$

The following are some examples of the use of (3.41):

$$(x^2)^3 = x^6 \qquad (2^4)^6 = 2^{24} \qquad (a^5)^4 = a^{20}$$

Sometimes we have the product of two numbers raised to a power, as

$$(3b)^2 = 3b \cdot 3b = (3 \cdot 3)(b \cdot b) = 3^2b^2 = 9b^2$$
$$(ab)^3 = ab \cdot ab \cdot ab = (a \cdot a \cdot a)(b \cdot b \cdot b) = a^3b^3$$

For any positive integer m,

$$\begin{aligned} (ab)^m &= ab \cdot ab \cdots ab \qquad && m \textit{ factors } ab \\ &= (a \cdot a \cdots a)(b \cdot b \cdots b) && m \textit{ factors } a \\ & && m \textit{ factors } b \end{aligned}$$

or

$$(ab)^m = a^mb^m \tag{3.42}$$

To illustrate,

$$(3x)^4 = 3^4x^4 = 81x^4 \qquad (-3a^2)^3 = (-3)^3(a^2)^3 = -27a^6$$

We must be careful to raise all factors of the base to the power of the exponent. Also, we must be careful not to raise to the power of the exponent a number not affected by the exponent. For example,

$$2x(3y)^2 = 2x(3)^2(y)^2 = 18xy^2 \qquad -(4x)^2 = -16x^2$$

We may want to raise a quotient $\frac{a}{b}$ $(b \neq 0)$ to a power, as

$$\left(\frac{4}{5}\right)^2 = \frac{4}{5}\cdot\frac{4}{5} = \frac{4^2}{5^2} = \frac{16}{25} \qquad \left(\frac{a}{6}\right)^2 = \frac{a}{6}\cdot\frac{a}{6} = \frac{a^2}{6^2} = \frac{a^2}{36}$$

For any positive integer m, we have

$$\left(\frac{a}{b}\right)^m = \frac{a}{b}\cdot\frac{a}{b}\cdots\frac{a}{b} \qquad m \text{ factors } \frac{a}{b}$$

$$= \frac{a\cdot a\cdots a}{b\cdot b\cdots b} \qquad \begin{matrix} m \text{ factors } a \\ m \text{ factors } b \end{matrix}$$

or

$$\left(\frac{a}{b}\right)^m = \frac{a^m}{b^m} \tag{3.43}$$

The following are some examples of the use of (3.43):

$$\left(\frac{2x}{3}\right)^3 = \frac{(2x)^3}{3^3} = \frac{8x^3}{27} \qquad \frac{4^4}{(12)^4} = \left(\frac{4}{12}\right)^4 = \left(\frac{1}{3}\right)^4 = \frac{1}{81}$$

We have proved the following laws of operation with positive integral exponents m and n:

(I) $a^m \cdot a^n = a^{m+n}$

(II) $(a^m)^n = a^{mn}$

(III) $(ab)^m = a^m b^m$

(IV) $\left(\frac{a}{b}\right)^m = \frac{a^m}{b^m} \qquad b \neq 0$

Study the following illustrations of the use of exponents in division:

$$\frac{2^5}{2^3} = \frac{2\cdot2\cdot2\cdot2\cdot2}{2\cdot2\cdot2} = 2^2 \qquad \frac{a^5}{a^3} = \frac{a\cdot a\cdot a\cdot a\cdot a}{a\cdot a\cdot a} = a^2$$

Notice that the exponent in the numerator is larger than the exponent in the denominator. Also, the exponent in the quotient is equal to the exponent in the numerator minus the exponent in the denominator, and this is true in general, as we shall now show.

Let m and n be positive integers with $m > n$. Then

$$\frac{a^m}{a^n} = \frac{a \cdot a \cdots a}{a \cdot a \cdots a} \qquad \begin{array}{l} m \textit{ factors } a \textit{ in the numerator} \\ n \textit{ factors } a \textit{ in the denominator} \end{array}$$

In the fraction on the right the numerator and denominator may both be divided by n factors a, with $m - n$ factors a left in the numerator and 1 in the denominator. That is,

$$\frac{a^m}{a^n} = a^{m-n} \qquad m > n \tag{3.44}$$

To illustrate,

$$\frac{a^{27}}{a^{24}} = a^{27-24} = a^3 \qquad \frac{3^6 \cdot 3^4}{3^8} = \frac{3^{10}}{3^8} = 3^2 = 9$$

On the other hand, observe that

$$\frac{2^3}{2^5} = \frac{2 \cdot 2 \cdot 2}{2 \cdot 2 \cdot 2 \cdot 2 \cdot 2} = \frac{1}{2^2} \qquad \frac{a^3}{a^5} = \frac{a \cdot a \cdot a}{a \cdot a \cdot a \cdot a \cdot a} = \frac{1}{a^2}$$

Note that in each of these cases the exponent in the denominator is larger than the exponent in the numerator, and the exponent which appears in the denominator of the result is equal to the exponent in the denominator minus the exponent in the numerator.

Let m and n be positive integers with $m < n$. Then

$$\frac{a^m}{a^n} = \frac{a \cdot a \cdots a}{a \cdot a \cdots a} \qquad \begin{array}{l} m \textit{ factors } a \textit{ in the numerator} \\ n \textit{ factors } a \textit{ in the denominator} \end{array}$$

In the fraction on the right, the numerator and denominator may both be divided by m factors a, leaving 1 in the numerator and $n - m$ factors a in the denominator. We then have

$$\frac{a^m}{a^n} = \frac{1}{a^{n-m}} \qquad m < n \tag{3.45}$$

Some examples are

$$\frac{a^{11}}{a^{14}} = \frac{1}{a^{14-11}} = \frac{1}{a^3} \qquad \frac{2^7}{2^{11}} = \frac{1}{2^{11-7}} = \frac{1}{2^4} = \frac{1}{16}$$

$$\left(\frac{3^4}{3^6}\right)^3 = \left(\frac{1}{3^{6-4}}\right)^3 = \left(\frac{1}{3^2}\right)^3 = \frac{1^3}{(3^2)^3} = \frac{1}{3^6} = \frac{1}{729}$$

We can now add two more laws of operations with positive integral exponents to those listed on page 138. They are:

(V) $\dfrac{a^m}{a^n} = a^{m-n} \qquad m > n;\ a \neq 0$

(VI) $\dfrac{a^m}{a^n} = \dfrac{1}{a^{n-m}} \qquad m < n;\ a \neq 0$

Sometimes after we perform the indicated operations by use of one or more of the laws of exponents, we can simplify the resulting fraction by dividing the numerator and denominator by any common factors. To illustrate,

$$\left(\frac{2a^2}{x^2}\right)^3 \cdot \left(\frac{x}{a}\right)^5 = \frac{2^3 \cdot a^6}{x^6} \cdot \frac{x^5}{a^5} = \frac{8a}{x}$$

EXERCISES

In Exercises 1 to 8, find the value of the given expression.

1 $(-2)^4$ 2 3^3 3 $(-3)^3$
4 $(-\frac{2}{5})^2$ 5 $2 + 3^2$ 6 $2^3 + 5^3$
7 $(3 \cdot 4)^5$ 8 $(0.2)^2$

In Exercises 9 to 14, complete the equality by putting the appropriate base within the parentheses.

9 $-32 = (\ \ )^5$ 10 $-\frac{27}{8} = (\ \ )^3$
11 $0.008 = (\ \ )^3$ 12 $-64 = (\ \ )^3$
13 $\frac{625}{81} = (\ \ )^4$ 14 $0.0001 = (\ \ )^4$

Perform the indicated operations, and simplify when possible. Assume that all exponents are positive integers.

15 $a \cdot a^5$ 16 $x^2 \cdot x^m$ 17 $5x^4 \cdot x^3$
18 $b^3 \cdot b^{10}$ 19 $(a^2)^5$ 20 $(x^2)^m$
21 $5(x^4)^3$ 22 $(b^3)^{10}$ 23 $(5x^4)^3$
24 $(2a)^2$ 25 $5(3b)^2$ 26 $4(3a^3)^2$
27 $\left(\frac{a}{3}\right)^4$ 28 $\left(\frac{a^2}{b}\right)^3$ 29 $\left(\frac{a}{3b}\right)^2$
30 $2x\left(\frac{a}{b}\right)^3$ 31 $5y(y^2)^3$ 32 $\left(\frac{2a}{3b}\right)^3$
33 $\left(\frac{a^2}{2b}\right)^3$ 34 $(a \cdot b^2)^m$ 35 $(3^2)^3$
36 $(2 \cdot 5)^3$ 37 $(2^3 \cdot x^2)^3$ 38 $\left(\frac{2^3}{x^2}\right)^3$
39 $\left(\frac{1}{2a^2}\right)^4$ 40 $a^2 \cdot a^3 \cdot a^4$ 41 $\frac{(6a)^3}{(2a)^3}$
42 $\frac{(28b^2)^2}{(14b)^2}$ 43 $\frac{(7x^3)^4}{(7x)^4}$ 44 $\frac{5(x^4)^2}{(5x)^2}$
45 $\frac{3(a^3)^4}{(2a^2)^4}$

In Exercises 46 to 57, find the value of the given expression.

46 $(2\frac{1}{4})^3$ 47 $(4\frac{2}{5})^2$ 48 $(1.5)^2$
49 $(0.12)^3$ 50 $(1\frac{1}{5})^4$ 51 $(5\frac{2}{3})^3$

52 $(3.14)^2$	53 $(5\frac{1}{7})^1$	54 $(\frac{2}{3})^2 \cdot (\frac{2}{5})^3$
55 $6 \cdot (1\frac{1}{2})^3$	56 $10^3 \cdot (0.2)^2$	57 $10^3 \cdot 10^2$

In Exercises 58 to 63, express the given number as a power of 2.

58 8	59 32	60 4^3
61 $(16)^2$	62 $(32)^3$	63 $(64)^2$

In Exercises 64 to 68, express the given number as a power of 3.

64 27	65 243	66 9^2
67 $(27)^4$	68 $(81)^5$	

Perform the indicated operations, and simplify when possible. Assume that literal exponents are positive integers.

69 $\dfrac{a}{a^3}$	70 $\dfrac{a^7}{a^2}$	71 $\dfrac{5a^4}{a^2}$
72 $\dfrac{a^3}{7a^2}$	73 $\dfrac{4^8}{4^{10}}$	74 $\dfrac{2^3 + 2^2}{2^5}$
75 $\left(\dfrac{a^6}{a^{12}}\right)^3$	76 $\left(\dfrac{a^3}{a}\right)^m$	77 $\dfrac{a^{2m+1}}{a^{m+1}}$
78 $\left(\dfrac{5^{41}}{5^{39}}\right)^2$	79 $a^3\left(\dfrac{b}{a}\right)^2$	80 $2y^2\left(\dfrac{x}{y^2}\right)^4$
81 $x\left(\dfrac{y}{x^2}\right)^m$	82 $\dfrac{(z^{4m})^2}{z^{6m}}$	83 $\dfrac{x^2y^9}{x^3y^7}$
84 $\left(\dfrac{x^2}{3y^4}\right)^2 \cdot \left(\dfrac{6y}{x}\right)^2$	85 $\dfrac{a^{m+n}}{a^{n+1}}$	86 $\dfrac{3y^{2m}}{(4y^m)^2}$
87 $\left(\dfrac{x}{2y}\right)^2 \cdot \left(\dfrac{4y^2}{3x}\right)^3$	88 $\dfrac{(x^4y)(xy^4)}{(xy)^5}$	

In Exercises 89 to 94, find the value of the given expression.

89 $\dfrac{6^{10}}{6^8}$	90 $\dfrac{4^5}{2^4 \cdot 2^7}$	91 $\dfrac{10^7 \cdot 10^5}{10^{10}}$
92 $\dfrac{10^4 \cdot 10^5}{10^3 \cdot 10^6}$	93 $\dfrac{2^3 \cdot 3^4}{2^4 \cdot 3^2}$	94 $\dfrac{10^{13}}{10^{13}}$

3.9 NEGATIVE INTEGRAL EXPONENTS AND ZERO EXPONENTS

So far in our operations with exponents we have required that the exponents be positive integers. We shall now extend our definitions to give meanings to symbols such as 7^0 and 5^{-2}. Our definitions will be selected in such a way that our laws for positive integral exponents will apply when the exponents may be negative integers or zero. For it would be inconvenient to have $a^m \cdot a^n$ equal to a^{m+n} when m and n are positive integers but equal to something else when m and n are negative integers. This selection of definitions so that our new symbols will conform to previously established laws is not a new idea; for, as you will recall, we followed the same procedure in Chap. 2 when we defined new symbols and new operations so that they would be consistent with previously established definitions and laws.

Let us consider what meaning to give to a^{-n} where $a \neq 0$ and n is a positive integer. By law VI in Sec. 3.8,

$$\frac{a^3}{a^5} = \frac{1}{a^{5-3}} = \frac{1}{a^2}$$

If law V of Sec. 3.8 were to hold for $m < n$, we should have

$$\frac{a^3}{a^5} = a^{3-5} = a^{-2}$$

In order to give a^{-2} a meaning consistent with these results, we should set $a^{-2} = \frac{1}{a^2}$. Thus we are led to *define* a^{-n} to be the reciprocal of a^n; that is, for all $a \neq 0$,

a^{-n}

$$a^{-n} = \frac{1}{a^n} \tag{3.46}$$

From (3.46) it follows that

$$a^n = \frac{1}{a^{-n}} \tag{3.47}$$

The following are some examples of the use of the definition (3.46):

$$3^{-2} = \frac{1}{3^2} = \frac{1}{9} \qquad (3c)^{-4} = \frac{1}{(3c)^4} = \frac{1}{3^4 \cdot c^4} = \frac{1}{81c^4}$$

$$3c^{-4} = \frac{3}{c^4} \qquad \left(\frac{2}{3}\right)^{-3} = \frac{1}{\left(\frac{2}{3}\right)^3} = \frac{3^3}{2^3} = \left(\frac{3}{2}\right)^3 = \frac{27}{8}$$

The student should note carefully the distinction between $(3c)^{-4}$ and $3c^{-4}$.

Observe that by use of the definition (3.46) we have

$$a^m \cdot a^{-n} = a^m \cdot \frac{1}{a^n} = \frac{a^m}{a^n} = a^{m-n} \qquad a \neq 0$$

and when $m \neq n$, we see that law I of Sec. 3.8 can be applied to negative integers as well as positive integers. A similar check of laws II, III, and IV of Sec. 3.8 will show that they also can be applied to negative integral exponents. Further, laws V and VI of Sec. 3.8

will now hold for any integers m and n $(m \neq n)$, and in fact laws V and VI become the same law, since

$$a^{m-n} = \frac{1}{a^{n-m}} \qquad m \neq n$$

Now we wish to give a suitable meaning to zero as an exponent. Let us assume that $a \neq 0$ and that n is a positive integer. If the laws for integral exponents are to hold for the zero exponent, we have

$$a^n \cdot a^0 = a^{n+0} = a^n \qquad \textit{by law I}$$

$$\frac{a^n}{a^0} = a^{n-0} = a^n \qquad \textit{by law V}$$

and

$$1 = \frac{a^n}{a^n} = a^{n-n} = a^0 \qquad \textit{by law V}$$

In order to give a^0 a meaning consistent with these results, we *define*

$$a^0 = 1 \tag{3.48}$$

for all $a \neq 0$. To illustrate,

$$2^0 = 1 \qquad (-3)^0 = 1 \qquad (19)^0 = 1 \qquad (-6\tfrac{7}{8})^0 = 1$$

Note that an expression may contain a zero exponent and yet not have the value 1. Some examples are the following:

$$-3^0 = -(3^0) = -(1) = -1$$

$$2^{-3} - 5^0 = \frac{1}{2^3} - 1 = \frac{1}{8} - 1 = -\frac{7}{8}$$

Suppose that we wish to write $\frac{3ab^0}{c^2}$ without a denominator, that is, express it as a fraction whose denominator is 1. We see that if the denominator c^2 of $\frac{3ab^0}{c^2}$ is multiplied by c^{-2}, the result is $c^2 \cdot c^{-2} = 1$. But if the denominator of $\frac{3ab^0}{c^2}$ is multiplied by c^{-2}, the numerator must also be multiplied by c^{-2} or the value of our expression will be changed. Hence we write

$$\frac{3ab^0}{c^2} = \frac{3ac^{-2}}{c^2c^{-2}} = \frac{3ac^{-2}}{c^0} = 3ac^{-2}$$

If we wish to express $\frac{3x^0y^{-2}}{z^{-3}}$ without negative exponents, we can write

$$\frac{3x^0y^{-2}}{z^{-3}} = \frac{3 \cdot 1 \cdot y^{-2}}{z^{-3}} \cdot \frac{y^2z^3}{y^2z^3} = \frac{3y^{-2}y^2z^3}{z^{-3}y^2z^3} = \frac{3z^3}{y^2}$$

We can obtain this result in another way by writing

$$\frac{3x^0y^{-2}}{z^{-3}} = \frac{3 \cdot 1 \cdot \frac{1}{y^2}}{\frac{1}{z^3}} = \frac{3z^3}{y^2}$$

As is customary, we use $\dfrac{\frac{a}{b}}{\frac{c}{d}}$ to mean $\frac{a}{b} \div \frac{c}{d}$. Hence $\dfrac{\frac{a}{b}}{\frac{c}{d}} = \frac{a}{b} \div \frac{c}{d} = \frac{a}{b} \cdot \frac{d}{c}$.

EXERCISES

Find the value of the following:

1 5^{-2}
2 4^{-3}
3 $\frac{1}{3^{-2}}$
4 7^{-3}
5 $7^0 + 2^{-1}$
6 $(5x^0)^2$
7 $(5x^2)^0$
8 $(b^{-1})^0$
9 $5^2 \cdot 3^{-3}$
10 $5^2 + 3^{-3}$
11 $(5 + 5^{-1})^{-1}$
12 $(a + 2b)^0$
13 $3^{-1}(4c)^0$
14 $\frac{1}{2^{-3} \cdot 5^{-3}}$
15 $\left(\frac{2}{3}\right)^{-4}$
16 $\frac{(273x)^0}{2^{-5}}$

Express the following without negative or zero exponents. Express fractions in lowest terms.

17 a^{-2}
18 $\frac{1}{b^{-3}}$
19 $\frac{a^{-2}}{b^{-3}}$
20 $\frac{1}{a^{-2}b^{-3}}$
21 $(2x)^{-2}$
22 $(-2x)^{-2}$
23 $(2x^{-1})^2$
24 $\frac{1}{(2x)^{-2}}$
25 $\frac{x^{-2}}{(-2y^0)^{-2}}$
26 $a^{-1} + b^{-1}$
27 $\frac{a^{-2} - b^{-2}}{a^{-1} - b^{-1}}$
28 $5a^0b^{-2}c^{-3}$

Write the following without a denominator, using negative exponents if necessary:

29 $\frac{1}{x^4}$
30 $\frac{1}{x^2y^2}$
31 $\frac{3a}{c^2d^3}$
32 $\frac{4a^{-1}}{d^4}$
33 $\frac{1}{ab^{-2}c^0}$
34 $\frac{3x}{4y^0}$

35 $\dfrac{4^{-1}a}{3^{-1}b}$ 36 $\dfrac{1}{16}a^{-2}$ 37 $\dfrac{1}{a^{-1}}$

38 $\dfrac{1}{a^{-1} + b^{-2}}$ 39 $\dfrac{9^0 y^{-1}}{a^2}$ 40 $\dfrac{a^2}{9^0 y^{-1}}$

In Exercises 41 to 46, find the value of the given expression.

41 $\left(\dfrac{2}{3}\right)^{-2}$ 42 $10^4 \div 10^{-2}$ 43 $10^4 \div 10^2$

44 $\left(\dfrac{1}{2}\right)^3 \div \left(\dfrac{1}{2}\right)^2$ 45 $(2\frac{3}{4})^0 \cdot (2\frac{3}{4})^{-1}$ 46 $(3\frac{1}{2})^{-2}$

47 Complete the following table:

WHEN $x =$	3	2	1	0	-1	-2	-3
THEN $1^x =$							
$2^x =$		4		1		$\frac{1}{4}$	
$3^x =$							
$5^x =$							
$10^x =$							
$(\frac{1}{2})^x =$						4	
$(\frac{3}{4})^x =$							

48 Complete the following table:

WHEN $x =$	3	2	1	0	-1	-2	-3
THEN $(-1)^x =$							
$(-2)^x =$	-8			1			$-\frac{1}{8}$
$(-3)^x =$							
$(-5)^x =$							
$(-10)^x =$							
$(-\frac{1}{2})^x =$							-8
$(-\frac{3}{4})^x =$							

In Exercises 49 to 54, express the given number as a power of 2.

49 128 50 4^{-3} 51 2

52 $\dfrac{1}{32}$ 53 $\dfrac{1}{8}$ 54 $\dfrac{1}{8^2}$

In Exercises 55 to 60, express the given number as a power of 3.

55 $\frac{1}{27}$ 56 $(243)^{-1}$ 57 $(27)^{-4}$
58 9^{-4} 59 3 60 $\frac{1}{3}$

In Exercises 61 to 66, perform the indicated operation, and express the result as a single power.

61 $10^5 \cdot 10^{-3}$ 62 $10^5 \div 10^{-3}$
63 $10^{-5} \cdot 10^{-3}$ 64 $10^{-5} \div 10^{-3}$
65 $a^5 \cdot a^{-3} \cdot a^4$ 66 $\dfrac{b^6 \cdot b^{-4}}{b^{-3} \cdot b^5}$

In Exercises 67 to 72, express the given number as a power of 10.

67 1,000 68 10,000 69 100,000
70 0.1 71 0.01 72 0.001

In Exercises 73 to 78, find the exponent x.

73 $2^x = 32$ 74 $2^x = \frac{1}{64}$ 75 $(\frac{1}{2})^x = 16$
76 $5^x = 0.04$ 77 $9^x = 1$ 78 $(\frac{4}{3})^x = \frac{16}{9}$

Express the following without zero or negative exponents. Reduce fractions to lowest terms.

79 $\dfrac{a^2 - ab^{-1}}{b - a^{-1}}$ 80 $ab^{-1} + ba^{-1}$
81 $(x + y)^{-1}(x^{-1} + y^{-1})$ 82 $(a^{-2} - b^{-2})(a + b)^{-1}$
83 $(a^{-1} + b^{-1})^{-1}$ 84 $\dfrac{(2b)^{-1} - 2^{-1}b}{b^{-1} + 1^{-1}}$

3.10 RADICALS AND FRACTIONAL EXPONENTS

Before defining fractional exponents we shall consider radicals briefly. If a is a number and n is any positive integer such that $a^n = b$, then a is called an **nth root**, or a *root of order n,* of b. Thus 3 is a root of order 2 of 9, since $3^2 = 9$; -5 is a root of order 3 of -125, since $(-5)^3 = -125$; and -2 is a fourth root of 16, since $(-2)^4 = 16$. Usually we speak of a root of order 2 as a *square root* and a root of order 3 as a *cube root.*

nth ROOT

In this section we shall confine our attention to *rational nth roots.* Thus when we speak of the fourth roots of 16, we shall be speaking of 2 and -2, for they are the only rational fourth roots of 16. Nonrational nth roots will be considered in Chap. 5.

If a positive number b has rational nth roots, it will have two distinct rational nth roots when n is even and only one rational nth root when n is odd. Thus 25 has the square roots 5 and -5, 81 has the fourth roots 3 and -3, 64 has the sixth roots 2 and -2,

$\frac{25}{9}$ has the square roots $\frac{5}{3}$ and $-\frac{5}{3}$, and 64 has the cube root 4. We shall not consider the case when b is negative and n is even. However, when b is negative and n is odd, b may have a rational nth root, in which case it will be negative. Thus -2 is a cube root of -8 since $(-2)^3 = -8$. In case b has rational nth roots, we define the **principal nth root** of b to be the positive nth root of b when b is positive and to be the negative nth root of b when b is negative and n is odd. To illustrate, 2 is the principal square root of 4, -2 is the principal fifth root of -32, and 6 is the principal square root of 36.

PRINCIPAL nth ROOT

RADICAL

The **radical** $\sqrt[n]{b}$ shall stand for the principal nth root of b; that is,

$$\sqrt[n]{b} = a \qquad \text{if} \qquad a^n = b \tag{3.49}$$

RADICAL SIGN

RADICAND

INDEX

and if *a has the same sign as b*. The symbol $\sqrt{\ }$ is called the **radical sign,** the number b is called the **radicand,** and the positive integer n is called the **index,** or **order,** of the radical; the symbol $\sqrt[n]{b}$ is read "the nth root of b." In a square root the index 2 is usually omitted from the symbol. Thus we customarily write $\sqrt{25}$ for $\sqrt[2]{25}$.

In $\sqrt[3]{-125}$, -125 is the radicand, and 3 is the index. Further, -5 is the principal cube root of -125; that is,

$$\sqrt[3]{-125} = -5 \qquad \text{since} \qquad (-5)^3 = -125$$

Similarly,

$$\sqrt[4]{16} = 2 \qquad \text{since} \qquad 2^4 = 16$$

and

$$\sqrt[5]{-243} = -3 \qquad \text{since} \qquad (-3)^5 = -243$$

For all n and b we note that the equality

$$b = (\sqrt[n]{b}\,)^n \tag{3.50}$$

is a consequence of the definition (3.49) of the radical $\sqrt[n]{b}$. To illustrate,

$$16 = (\sqrt[4]{16}\,)^4 \qquad \text{and} \qquad -243 = (\sqrt[5]{-243}\,)^5$$

If you construct a table showing the powers up through the eighth of the integers between -6 and 6, it will aid you in recognizing many principal nth roots. For example, from the following such table for powers of 3,

IF $n =$	1	2	3	4	5	6	7	8
THEN $3^n =$	3	9	27	81	243	729	2,187	6,561

we see that $\sqrt[6]{729} = 3$, $\sqrt[7]{2{,}187} = 3$, $\sqrt[8]{6{,}561} = 3$.

The following are some additional illustrations of the evaluations of expressions containing radicals and powers:

$$\sqrt[3]{8a^6} = 2a^2 \qquad \text{since} \qquad (2a^2)^3 = 8a^6$$

and

$$\sqrt[5]{-1} = -1 \qquad \text{since} \qquad (-1)^5 = -1$$

Also

$$(\sqrt[5]{-32}\,)^3 = (-2)^3 = -8 \qquad \sqrt[3]{\frac{x^3}{8}} = \frac{x}{2}$$

$$\left(\frac{1}{\sqrt[4]{81}}\right)^4 = \frac{1^4}{(\sqrt[4]{81}\,)^4} = \frac{1}{81} \qquad \sqrt[4]{\frac{16}{81}} - \sqrt{\frac{1}{4}} = \frac{2}{3} - \frac{1}{2} = \frac{1}{6}$$

Note carefully that the principal nth root of a radicand must have the same sign as the radicand. To illustrate, $\sqrt{a^2}$ must be positive (unless a is 0), since a^2 is positive. Therefore, if $a > 0$, $\sqrt{a^2} = a$; if $a < 0$, then $\sqrt{a^2} = -a$, since $-a$ is positive when $a < 0$. These results can be written as

$$\sqrt{a^2} = |a|$$

Thus,

$$\sqrt{16b^2} = 4|b| \qquad \sqrt[4]{16x^4} = 2|x|$$

The definitions which we gave in Sec. 3.9 enabled us to extend our laws for operations with positive integral exponents to negative integral and zero exponents. We now consider the problem of giving meaning to the symbol $a^{1/n}$, where n is any positive integer, in such a way that fractional exponents will obey the previously developed laws for operations with integral exponents.

What meaning shall we attach to $a^{1/2}$? If law II of Sec. 3.8 is to hold, we shall have

$$(a^{1/2})^2 = a^{2(1/2)} = a$$

But $(\sqrt{a}\,)^2 = a$. It appears that we should define $a^{1/2}$ to mean $\sqrt{a}$.

Similarly

$$(a^{1/n})^n = a^{n(1/n)} = a$$

But, by (3.50), $(\sqrt[n]{a}\,)^n = a$. For this reason we *define* $a^{1/n}$ to mean $\sqrt[n]{a}$; that is,

$$a^{1/n} = \sqrt[n]{a} \tag{3.51}$$

Observe that $a^{1/n}$ has meaning only when a is a number which has a principal nth root; so when n is even, a is nonnegative.

What meaning shall we attach to $8^{2/3}$? If law II of Sec. 3.8 is to hold, we shall have

$$(8^{1/3})^2 = 8^{2(1/3)} = 8^{2/3}$$

But, by (3.51), $(8^{1/3})^2 = (\sqrt[3]{8})^2$. It seems that we should define $8^{2/3}$ to mean the same thing as $(\sqrt[3]{8})^2$.

Similarly,

$$(a^{1/n})^m = a^{m(1/n)} = a^{m/n}$$

But, by (3.51), $(a^{1/n})^m = (\sqrt[n]{a})^m$. To give $a^{m/n}$ a meaning consistent with these results, we *define* $a^{m/n}$ by the equality

$a^{m/n}$

$$a^{m/n} = (\sqrt[n]{a})^m \tag{3.52}$$

The definitions (3.51) of $a^{1/n}$ and (3.52) of $a^{m/n}$ were made so that law II holds. It can be shown that all the laws of exponents previously given are satisfied by fractional exponents defined by (3.51) and (3.52).

By law II of Sec. 3.8

$$(a^{1/n})^m = (a^m)^{1/n}$$

provided a is positive when n is even

When this condition is satisfied, we can write

$$a^{m/n} = (\sqrt[n]{a})^m = \sqrt[n]{a^m} \tag{3.53}$$

This equality says, in words, that $a^{m/n}$ may be interpreted to mean either the mth power of the nth root a or the nth root of the mth power of a, where a is positive when n is even. Thus

$$(25)^{3/2} = [(25)^{1/2}]^3 = 5^3 = 125$$

and

$$(25)^{3/2} = [(25)^3]^{1/2} = (15{,}625)^{1/2} = 125$$

Clearly the former evaluation is the easier. Other illustrations are

$$8^{4/3} = (8^{1/3})^4 = 2^4 = 16 \qquad 8^{2/3} = (8^{1/3})^2 = 2^2 = 4$$

$$(4z^4)^{1/2} = 4^{1/2} \cdot (z^4)^{1/2} = 2z^{4/2} = 2z^2$$

$$(-27)^{2/3} = [(-27)^{1/3}]^2 = (-3)^2 = 9$$

In the same way that we defined $a^{-n} = \dfrac{1}{a^n}$, where n is an integer, we now *define*

$a^{-m/n}$

$$a^{-m/n} = \frac{1}{a^{m/n}} \tag{3.54}$$

where m and n are any integers ($n \neq 0$). To illustrate,

$$(32)^{-4/5} = \frac{1}{(32)^{4/5}} = \frac{1}{[(32)^{1/5}]^4} = \frac{1}{2^4} = \frac{1}{16}$$

$$\frac{-(32z^{10})^{1/5}}{8^{-2/3}} = \frac{-(32)^{1/5}(z^{10})^{1/5}}{\frac{1}{8^{2/3}}} = \frac{-2z^2}{\frac{1}{2^2}} = -8z^2$$

EXERCISES

Give the principal square root of each of the following:

1 49

2 121

3 $4a^6$

4 $\frac{x^4}{36}$

Give the principal cube root of each of the following:

5 $-x^9$

6 -27

7 $1{,}000$

8 $-a^3b^6$

Give the roots as indicated in the following:

9 $\sqrt[6]{64}$

10 $\sqrt[7]{128}$

11 $\sqrt[3]{729}$

12 $-\sqrt[3]{-216}$

13 $\sqrt{225}$

14 $\sqrt[5]{x^{10}}$

15 $\sqrt{81x^4y^6}$

16 $\sqrt[4]{81a^4b^8}$ $(a > 0)$

Give the value of the following expressions in simplest form:

17 $\sqrt[5]{-32} + \sqrt[4]{256}$

18 $\sqrt[3]{27a^6} + \sqrt{36a^4}$

19 $\sqrt[3]{216} - \sqrt[7]{-1}$

20 $\sqrt{81a^4} - \sqrt{16a^2}$ $(a > 0)$

Find the values of the following expressions:

21 $(\sqrt{4})^2$

22 $(\sqrt[3]{-8})^3$

23 $\left(\frac{1}{\sqrt[3]{-27}}\right)^3$

24 $(\sqrt[3]{-216})^2$

25 $\left(\frac{1}{\sqrt[3]{125}}\right)^4$

26 $\left(\frac{\sqrt{16}}{\sqrt[3]{-27}}\right)^2$

27 $\sqrt[3]{-216}\,\sqrt[3]{-64}$

28 $\frac{\sqrt[5]{-32}\,\sqrt[4]{16}}{\sqrt[3]{-64}}$

29 $\left(\frac{\sqrt[5]{243}}{\sqrt[6]{64}}\right)^4$

30 $\left(\frac{1}{\sqrt[7]{128}}\right)^0$

31 $\left(\frac{\sqrt[3]{216}}{-7}\right)^{-2}$

32 Find the edge e of a square whose area A is 64 square inches (recall that $A = e^2$; so $e = \sqrt{A}$).

33 Find the edge of a square whose area is: 100 square inches; 121 square feet; 81 square yards.

34 Find the radius r of a circle whose area A is 49π square inches. Recall that $A = \pi r^2$; so $r^2 = \frac{A}{\pi}$, and $r = \sqrt{\frac{A}{\pi}}$.

35 Find the radius of a circle whose area is: 25π square feet; 9π square feet; 169π square yards.

36 The areas of two circles are 225π square inches and 9π square inches. What is the ratio of the length of the radius of the larger circle to the length of the radius of the smaller circle?

37 Find the edge e of a cube whose volume V is 125 cubic inches (recall that $V = e^3$; so $e = \sqrt[3]{V}$).

38 Find the edge of a cube whose volume is: 64 cubic inches; 216 cubic feet; 729 cubic yards.

Evaluate:

39 $(36)^{1/2}$ 40 $(49)^{1/2}$ 41 $4^{-1/2}$

42 $9^{-1/2}$ 43 $\frac{1}{(25)^{-1/2}}$ 44 $\frac{9^{1/2}}{4^{-1/2}}$

45 $\frac{9^{-1/2}}{4^{1/2}}$ 46 $\frac{1}{4^{1/2}9^{1/2}}$ 47 $(121)^{-1/2}$

48 $\frac{(100)^{-1/2}}{(81)^{-1/2}}$ 49 $(125)^{1/3}$ 50 $(27)^{-1/3}$

51 $(16)^{1/4}$ 52 $(-27)^{-1/3}$ 53 $\frac{1}{8^{-1/3}}$

54 $\frac{9^{1/2}}{(27)^{1/3}}$ 55 $(32)^{2/5}$ 56 $(64)^{2/3}$

57 $(-32)^{3/5}$ 58 $(-32)^{4/5}$ 59 $(-8)^{4/3}$

60 $(-8)^{5/3}$ 61 $(-8)^{-2/3}$ 62 $\frac{7}{(-32)^{-3/5}}$

63 $(121)^{3/2}$ 64 $(-216)^{4/3}$ 65 $4^{3/2}$

66 $(16)^{5/4}$ 67 $(16^{-1/2})^3$ 68 $(-8^{-2/3})^{-1}$

69 $(-32^{-4/5})^0$ 70 $(32^0)^{-3}$ 71 $(-32^0)^{3/5}$

72 $\frac{(36)^{-1/2}}{(25)^{3/2}}$ 73 $\frac{40}{(100)^{-3/2}}$ 74 $\frac{(8^{2/3})^{1/2}}{(32)^{4/5}}$

Simplify:

75 $(4x^2)^{-1/2}$ 76 $(9x^4)^{3/2}$

77 $-5(x^3y^6)^{1/3}$ 78 $7(z^3w^6)^{-2/3}$

3.11 DECIMAL REPRESENTATION OF RATIONAL NUMBERS

In Sec. 1.11 we explained how the place-value principle enables us to represent any integer by the use of the base-ten numeral system. The place value given to a digit depends on the place the digit occupies relative to the units; for example,

$$6{,}257 = 6(1{,}000) + 2(100) + 5(10) + 7(1)$$

or

$$6{,}257 = 6(10)^3 + 2(10)^2 + 5(10)^1 + 7(10)^0 \tag{3.55}$$

DECIMAL POINT

Let us place a period or dot (.), called the **decimal point,** at the right of the units place and extend the principle of place value to the right of the decimal point.

What value should a digit have if it appears immediately to the right of the decimal point? Let us consider the numeral 6,257.4, and write

$$6{,}257.4 = 6(10)^3 + 2(10)^2 + 5(10)^1 + 7(10)^0 + 4(10)^{-1} \tag{3.56}$$

comparable to (3.55). Then the digit 4 in 6,257.4 represents $4(10)^{-1}$ or $\frac{4}{10}$ or 4 *tenths*. The first place to the right of the decimal point is the *tenths place.*

What value should a digit have if it appears immediately to the right of the tenths place? What place value should the digit 3 have in the numeral 6,257.43? Following the pattern suggested by (3.55) and (3.56), it is natural to write

$$6{,}257.43 = 6(10)^3 + 2(10)^2 + 5(10)^1 + 7(10)^0 + 4(10)^{-1} + 3(10)^{-2} \tag{3.57}$$

With this agreement, the digit 3 represents $3(10)^{-2}$ or $\frac{3}{100}$ or 3 *hundredths*. The first place to the right of the tenths place is the *hundredths place.*

If c is a digit and if c appears in the third place to the right of the decimal point, the digit represents $c(10)^{-3}$ or $\dfrac{c}{1{,}000}$ or c *thousandths*. In general, if c appears in the nth place to the right of the decimal point, it represents $c(10)^{-n}$ or $\dfrac{c}{10^n}$.

If k and n are nonnegative integers and if each of $a_1, a_2, a_3, \ldots, a_k$ and each of $b_1, b_2, b_3, \ldots, b_n$ is one of the ten digits 0, 1, 2, 3, 4, 5, 6, 7, 8, 9, then

$$\begin{aligned} a_k \cdots a_3a_2a_1 \,.\, b_1b_2b_3 \cdots b_n &= a_k(10)^{k-1} + \cdots + a_3(10)^2 + a_2(10)^1 + a_1(10)^0 \\ &\quad + b_1(10)^{-1} + b_2(10)^{-2} + \cdots + b_n(10)^{-n} \end{aligned} \tag{3.58}$$

A numeral in the form

$$a_k \cdots a_3a_2a_1 \,.\, b_1b_2b_3 \cdots b_n \qquad \text{or} \qquad -(a_k \cdots a_3a_2a_1 \,.\, b_1b_2b_3 \cdots b_n)$$

DECIMAL is called a (finite) **decimal.** The places occupied by digits to the right
DECIMAL PLACES of the decimal point are called **decimal places.** Thus 8.3, −12.05, 0.623 are decimals with one decimal place, two decimal places, and three decimal places respectively.

Digits to the right of the decimal point in a numeral indicate the number of tenths, hundredths, thousandths, ten-thousandths, and so forth. The numeral 81,325.47819, shown in the illustration, is read "eighty-one thousand, three hundred twenty-five, *and* forty-seven thousand, eight hundred nineteen hundred-thousandths." When a decimal is read, the word "and" is said only at the decimal point.

Ten-thousands	Thousands	Hundreds	Tens	Units	Decimal point	Tenths	Hundredths	Thousandths	Ten-thousandths	Hundred-thousandths
8	1	3	2	5	.	4	7	8	1	9

Some other illustrations are

12.05 "twelve and five hundredths"
257.105 "two hundred fifty-seven and one hundred five thousandths"

According to the extended principle of place value, symbolized by (3.58), the place value of a digit is *one-tenth* the place value which that digit would have if it appeared in the next place *to the left.* Stated another way, the place value of a digit is *ten times* the place value it would have if it appeared in the next place *to the right.* From this fact it follows that when we multiply a decimal by 10, the decimal point in the result is located one place to the right of its original position; when we divide a decimal by 10, the decimal point in the result is located one place to the left of its original position. To illustrate,

$$(3.7)(10) = 37 \qquad (42.0)(10) = 420$$

$$\frac{3.7}{10} = 0.37 \qquad \frac{0.023}{10} = 0.0023$$

In general, when a decimal is multiplied by a positive integral power of 10, the decimal point in the result is located to the right of its original position by as many places as there are zeros in the multiplier. When a decimal is divided by a positive integral power of 10, the decimal point in the result is located to the left of its original position by as many places as there are zeros in the divisor. For example,

$$(26.4)(10)^2 = 2{,}640 \qquad (0.6243)(1{,}000) = 624.3$$

$$\frac{26.4}{10^2} = 0.264 \qquad \frac{0.6243}{1{,}000} = 0.0006243$$

The results in the preceding paragraph, together with equality (3.9) in Sec. 3.2, can be used to write any (finite) decimal as a fraction representing a rational number. To illustrate,

$$4.25 = \frac{4.25}{1} = \frac{(4.25)(100)}{1(100)} = \frac{425}{100}$$

$$0.1026 = \frac{0.1026}{1} = \frac{(0.1026)(10{,}000)}{1(10{,}000)} = \frac{1{,}026}{10{,}000}$$

Thus 4.25 represents the rational number $\frac{425}{100}$, or, in lowest terms, $\frac{17}{4}$; and 0.1026 represents the rational number $\frac{1{,}026}{10{,}000}$ or $\frac{513}{5{,}000}$. The fact that any (finite) decimal represents a rational number can also be seen by observing that equality (3.58) says that any (finite) decimal represents the sum of rational numbers; and, since the set of rational numbers is closed under addition, the sum in (3.58) represents a rational number.

A given rational number can be expressed in decimal form by making use of the well-known long-division algorithm shown in the following illustrations.

$$\begin{array}{r}
4.816 \\
125\overline{)602.000} \\
\underline{500} \\
102\,0 \\
\underline{100\,0} \\
2\,00 \\
\underline{1\,25} \\
750 \\
\underline{750} \\
0
\end{array}
\qquad
\begin{array}{r}
2.6363\cdots \\
11\overline{)29.0000} \\
\underline{22} \\
7\,0 \\
\underline{6\,6} \\
40 \\
\underline{33} \\
70 \\
\underline{66} \\
40 \\
\underline{33} \\
7
\end{array}$$

These computations show that the rational number $\frac{602}{125}$ is represented by the (finite) decimal 4.816 and that the rational number $\frac{29}{11}$ is represented by the decimal 2.6363 · · · , where the dots · · · indicate that the decimal does not "stop" or "terminate"; indeed the block of digits 63 goes on repeating without end. Such a decimal as $2.6363 \cdots$ is called a *repeating infinite decimal.* We have now seen that, although a finite decimal always represents a rational number, a rational number may not always be represented by a finite decimal. However, it is true that any rational number will be represented by either a finite decimal or a repeating decimal. This may be seen by considering the long-division process used for expressing a rational number in decimal form. The divisor in the division process is a positive integer, and after each division the remainder is a nonnegative integer that is *less than the divisor.* Hence there are only a finite number of possible remainders, and at some stage the remainder must either be zero or be the same as some previous remainder. In the first case (a zero remainder) the division terminates and the decimal is a finite (or terminating) decimal; in the second case (a repeated remainder) the decimal will start to repeat.

Since a finite decimal, such as 4.816, can be considered a repeating infinite decimal with the digit zero repeating, $4.816 = 4.816000 \cdots$, we have the following result: *Every rational number can be represented by a repeating infinite decimal.*

It is natural to ask whether the *converse* of this last statement is true. The converse of the statement in italics is: *Every repeating infinite decimal represents a rational number.* This is indeed true, but we shall not give a demonstration of its truth here.

If we *assume* that a repeating infinite decimal does represent a rational number, and if we *assume* that the same rules of arithmetic apply to repeating infinite decimals as apply to finite decimals, then we can determine the fractional representation of the rational number. To illustrate, suppose that

$$2.0\overline{13}\,\overline{13} \cdots$$

where the block of digits 13 continues repeating, does represent a rational number. Denote this rational number by N, and assume the usual rules of arithmetic apply. Then

$$N = 2.0\overline{13}\,\overline{13} \cdots \qquad 100N = 201.3\overline{13}\,\overline{13} \cdots$$

and

$$\begin{aligned} 100N &= 201.3\overline{13}\,\overline{13} \cdots \\ N &= 2.0\overline{13}\,\overline{13} \cdots \qquad \textit{subtract} \\ \hline 99N &= 199.3 \end{aligned}$$

So

$$N = \frac{199.3}{99} = \frac{1{,}993}{990}$$

The student should note that in using the method of this paragraph to find N, two very *strong assumptions* have been made, and we have *not* demonstrated the truth of these assumptions.

If the decimal representation of a rational number is *not* a finite decimal, we *round off* the result to the nearest tenth, hundredth, thousandth, etc., as required. For example, from our previous calculations we see that $\frac{29}{11} = 2.6$ when rounded off to one decimal place, or to the nearest tenth; $\frac{29}{11} = 2.64$ when rounded off to two decimal places, or to the nearest hundredth; $\frac{29}{11} = 2.636$ when rounded off to three decimal places, or to the nearest thousandth.

There are no principles used in the addition and subtraction of decimals that have not already been discussed in connection with integers and fractions. Since only like things can be added or subtracted, it follows that tenths must be added to *tenths, hundredths to hundredths,* and so forth. In addition or subtraction of decimals, keep the decimal points in a vertical line, and use place value as we have explained.

EXAMPLE 1 Add 57.35, 8.83, and 0.045.

SOLUTION

$$\begin{array}{r} 57.35 \\ 8.83 \\ 0.045 \\ \hline 66.225 \end{array}$$

Carrying in adding decimals is similar to carrying in adding integers, and so is borrowing.

EXAMPLE 2 Subtract 24.89 from 36.784.

SOLUTION

$$\begin{array}{r} 36.784 \\ 24.890 \\ \hline 11.894 \end{array}$$

Multiplication of decimals is based on the multiplication of fractions. Consider the following examples:

$$(0.6)(0.4) = \frac{6}{10} \cdot \frac{4}{10} = \frac{24}{100} = 0.24$$

$$(0.08)(0.7) = \frac{8}{100} \cdot \frac{7}{10} = \frac{56}{1{,}000} = 0.056$$

$$(2.76)(1.4) = \frac{276}{100} \cdot \frac{14}{10} = \frac{3{,}864}{1{,}000} = 3.864$$

In general, *the number of decimal places in the product of two decimals is the sum of the number of the decimal places in the factors multiplied.*

EXAMPLE 3 Multiply 26.14 by 0.522.

SOLUTION

$$\begin{array}{r} 26.14 \\ 0.522 \\ \hline 5228 \\ 5228 \\ 13070 \\ \hline 13.64508 \end{array}$$

Note that it is unnecessary to place units under units, tenths under tenths, and so forth, as in addition. In this example we place the right digit of the multiplier under the right digit of the multiplicand, for we can multiply hundredths by thousandths.

Using equality (3.9), we can write the quotient of two (finite) decimals as the quotient of two integers. For example,

$$10.24 \div 3.2 = \frac{10.24}{3.2} = \frac{(10.24)(100)}{(3.2)(100)} = \frac{1{,}024}{320}$$

Then the long-division algorithm can be used to write the decimal representation of the result.

EXERCISES

Using (3.56), (3.57), and (3.58) as guides, write each of the following numbers as a sum involving powers of 10:

1 27.34
2 357.632
3 0.006
4 1.7524

Write the word form of each of the following numbers:

5 31.25
6 415.673
7 0.005
8 2.6342

Write each of the following word forms as a decimal.

9 Seventy-eight and seven hundredths.
10 Eight hundred fifty-six and twenty-three thousandths.
11 Three hundred eight hundred-thousandths.
12 Three hundred, and eight hundred-thousandths.

In Exercises 13 to 16, express the given fraction as a decimal.

13 $\frac{4}{5}$
14 $\frac{7}{4}$
15 $\frac{7}{8}$
16 $\frac{13}{20}$

In Exercises 17 to 20, express the given fraction or mixed number as a decimal to the nearest thousandth.

17 $\frac{1}{3}$
18 $\frac{5}{6}$
19 $\frac{4}{11}$
20 $4\frac{2}{9}$

In Exercises 21 to 24, express the given decimal as a proper fraction.

21 0.25
22 0.035
23 0.475
24 0.675

In Exercises 25 to 28, find the rational number that is represented by the given repeating infinite decimal.

25 $4.1\overline{36}\,\overline{36}\cdots$
26 $0.37\overline{251}\,\overline{251}\cdots$
27 $1.\overline{24}\,\overline{24}\cdots$
28 $2.2\overline{34}\,\overline{34}\cdots$

Add:

29
417.23
548.98

30
0.305
1.009
0.027

31
3,256.789
721.35

32
9,216.24
27.93
7,429.62

Subtract:

33
8.82
5.43

34
86.00
5.66

35
9.8723
0.8888

36
10.0000
0.0009

37 Multiply 3,254.54 by 25.182.
38 Multiply 0.0034 by 7.7.
39 Multiply 0.0101 by 0.02.
40 Multiply 321.2 by 0.133.
41 Divide 12,548.25 by 438.75.
42 Divide 6.671 by 0.7.
43 Divide 0.14 by 0.007.
44 Divide 0.3591 by 0.3.
45 If coal costs $19.50 per ton, what is the cost of $3\frac{1}{2}$ tons?
46 At $125 per acre, what is the cost to the nearest dime of three lots which contain $2\frac{1}{3}$, $1\frac{1}{2}$, and $1\frac{1}{6}$ acres?
47 Find the cost of the following: 15 pounds of sugar at $0.14 per pound; 4 dozen oranges at $0.27 per dozen; 2 pounds of coffee at $1.04 per pound. What is the total cost of all these items?
48 If $26.22 is divided equally among 23 persons, how much does each person get?

In Exercises 49 to 52, use 3.14 for the value of π, and round off the answer to two decimal places.

49 The length of the radius of a circle is 15 inches. What is the circumference of the circle? What is the area of the circle?
50 A can, which is in the shape of a right circular cylinder, is 2 inches in radius and 4.2 inches tall. What is its volume? What is its lateral area?
51 The length of the radius of a sphere is 3 inches. Find the volume and the surface area of the sphere.
52 The diameter of a steel ball is 4 inches. Find, to two decimal places, the weight of the ball if the steel weighs 486 pounds per cubic foot.

3.12 THE FIELD PROPERTIES

A primary purpose of this chapter has been to define and discuss the set Ra of rational numbers. By our agreement to identify the rational number $\frac{a}{1}$ with the integer a, we have made the set I of integers a subset of the set Ra of rational numbers. Thus we have "enlarged" the set I of integers to produce a new set of numbers. This new set possesses all of the properties possessed by the set of integers and some additional ones. Let us review the properties of the set Ra of rational numbers.

If $a = \frac{p}{q}$ and $b = \frac{r}{s}$ are rational numbers, we have defined what it means to say that a and b are equal ($a = b$). We agreed that

$$\frac{p}{q} = \frac{r}{s} \quad \text{if and only if} \quad ps = qr$$

Since p, q, r, and s are integers, ps and qr are also integers. Using the properties of equality of integers, we could prove that equality in the set of rational numbers has the same properties E_1, E_2, and E_3 given on page 89 for integers.

Addition ($+$) and multiplication ($\cdot$) have been defined on the set Ra of rational numbers in such a way that the set Ra is *closed* under each of these operations. Further, these operations have the uniqueness, commutative, associative, and distributive properties.

Therefore, the set Ra constitutes a set of numbers in accordance with the definition in Sec. 1.13; that is,

for every a, b, $c \in Ra$,
E_1: $a = a$
E_2: if $a = b$, then $b = a$
E_3: if $a = b$ and $b = c$, then $a = c$

P_1: $a + b \in Ra$ and $a + b$ is unique*
$a \cdot b \in Ra$ and $a \cdot b$ is unique*
P_2: $a + b = b + a$
$a \cdot b = b \cdot a$
P_3: $a + (b + c) = (a + b) + c$
$a \cdot (b \cdot c) = (a \cdot b) \cdot c$
P_4: $a(b + c) = ab + ac$

These are the seven properties possessed by any set of numbers; in particular, we have seen that the set I of integers, the set N_0 of nonnegative integers, and the set N of positive integers all have these properties.

In Sec. 3.4, page 107, we saw that there is a *unit element,* that is, an *identity under multiplication,* in the set of rational numbers. In Sec. 3.3 (page 102) we also showed that there is a *zero element,* that is, an *identity under addition,* in the set of rational numbers. Thus, the set Ra has the following two properties:

P_5: there exists a member $1 \in Ra$ with the property that for any $a \in Ra$, $a \cdot 1 = 1 \cdot a = a$
P_6: there exists a member $0 \in Ra$ with the property that for any $a \in Ra$, $a + 0 = 0 + a = a$

In the discussion of subtraction in the set Ra of rational numbers in Sec. 3.5 we saw that every rational number possesses a *negative.* In other words, corresponding to any given rational number a there is a rational number, denoted by $-a$, with the property that

$$a + (-a) = 0$$

The rational number $-a$ is called *the*† *inverse of a under addition.* So, we have shown that every rational number has an inverse under addition.

In Sec. 3.6 we defined the concept of the reciprocal of a rational number, and we showed that the reciprocal of the nonzero rational number $\frac{p}{q}$ is $\frac{q}{p}$. Thus, the reciprocal of a given rational number exists and is a rational number *provided the given rational number is not the zero element.* From the definition of a reciprocal it is apparent that the product of a number and its reciprocal is 1 (the unit element). This is also seen from an examination of the product

$$\frac{p}{q} \cdot \frac{q}{p} = \frac{pq}{qp} = \frac{1}{1} = 1$$

* Recall that $a + b$ is unique and $a \cdot b$ is unique provided that if $a = c$ and $b = d$, then $a + b = c + d$ and $a \cdot b = c \cdot d$.

† See the footnote on page 90.

Because of this property, the reciprocal of a number is called *the** **inverse** of that number **under multiplication.** The fact that every rational number except the zero rational has an inverse under multiplication is a feature that distinguishes the set Ra of rational numbers from the sets I, N_0, and N.

INVERSE UNDER MULTIPLICATION

The results summarized in the two preceding paragraphs enable us to state two more properties of the set Ra of rational numbers.

P_7: if 0 is the identity under addition, and if a is any member of Ra, then there exists a number $-a \in Ra$ with the property that

$$a + (-a) = (-a) + a = 0$$

P_8: if 1 is the identity under multiplication, and if a is any member of Ra, except the identity under addition, then there exists a number $a' \in Ra$ with the property that

$$a \cdot a' = a' \cdot a = 1$$

We have seen [equation (3.27)] that the subtraction of the rational number q from the rational number p is the same as the addition of $-q$ to p. Therefore, the requirement that the property P_7 be true for a set of numbers is equivalent to requiring that subtraction always be possible. Similarly, we have shown [equation (3.28)] that the division of the rational number p by the rational number q ($q \neq 0$) is the same as multiplying p by the reciprocal of q. Therefore, the requirement that property P_8 be true for a set of numbers is equivalent to requiring that division always be possible except when the divisor is the zero element. Hence, because of P_7 and P_8 we can say

for any $a, b \in Ra$,
$a - b \in Ra$
$a \div b \in Ra$ *provided* $b \neq 0$

NUMBER SYSTEM

Recall, from Sec. 1.13, that a **number system** is a set S together with a concept of equality and two operations, denoted by $=$, $+$, $\cdot$, respectively, which possess the seven properties E_1 to E_3 and P_1 to P_4. If, in addition, a number system also has the properties P_5 to P_8, the system is called a **field.** Since the rational number system does possess all 11 of the properties E_1 to E_3 and P_1 to P_8, it is a field, and we often call the rational number system the *field of rational numbers.* The integer number system, that is, the set I of integers together with the equality, addition, and multi-

FIELD

* It can be proved that if, in a set of numbers, a number has an inverse under multiplication, then there is only one such inverse.

plication defined in Chap. 2, satisfies E_1 to E_3 and P_1 to P_7. However, it does not satisfy P_8, and hence the integer number system is *not* a field. Similarly, the positive integer number system is not a field since it does not satisfy P_6, P_7, or P_8.

Let us consider the following sets, where a, b, and c are rational numbers, and let us try to determine the members of each set.

$$S_1 = \{x \in Ra \mid x + b = 0\}$$

$$S_2 = \{x \in Ra \mid ax = c\}$$

$$S_3 = \{x \in Ra \mid ax + b = 0\}$$

For S_1 we are seeking a rational number which can be put in place of x in the sentence $x + b = 0$ to produce a true statement. From the definition of the negative of a rational number, it is apparent that $-b$ is the number we wish. Every rational number has a negative which is also a rational number, so S_1 is not empty and

$$S_1 = \{x \in Ra \mid x + b = 0\} = \{-b\}$$

In S_2 we are seeking a rational number which can be put in place of x in the sentence $ax = c$ to produce a true statement. If we were to replace x by $\frac{1}{a}$, the reciprocal of a, then ax would have the value 1; now $1 \cdot c = c$, and these facts suggest that we replace x by $\frac{1}{a} \cdot c = \frac{c}{a}$. Indeed, we see that $a \cdot \frac{c}{a} = \frac{ac}{a} = \frac{c}{1} = c$, and hence, if we replace x by $\frac{c}{a}$ in the sentence $ax = c$, we produce a true statement. We note that $\frac{c}{a}$, the quotient of two rational numbers, is a rational number provided $a \neq 0$. So, with the requirement that $a \neq 0$, S_2 is not empty and

$$S_2 = \{x \in Ra \mid ax = c \text{ and } a \neq 0\} = \left\{\frac{c}{a}\right\}$$

In S_3 we are seeking a rational number which can be put in place of x in the sentence $ax + b = 0$ to produce a true statement. From a consideration of our result for S_1, we see that if we replace x by a number so that ax becomes $-b$, then we shall have found the member of S_3. This, together with our result for S_2, suggests that we replace x by $\frac{-b}{a}$ in the sentence $ax + b = 0$. Making this replacement, we have

$$a\left(\frac{-b}{a}\right) + b = 0$$

or

$$\frac{a(-b)}{a} + b = 0$$

or

$$(-b) + b = 0$$

which is a true statement. We note that $-b$ is a rational number and that $\frac{-b}{a} = -\frac{b}{a}$ is a rational number provided $a \neq 0$. So, with the requirement that $a \neq 0$, S_3 is not empty and

$$S_3 = \{x \in Ra \mid ax + b = 0 \text{ and } a \neq 0\} = \left\{-\frac{b}{a}\right\}$$

The equation

$$ax + b = 0 \qquad a \neq 0 \tag{3.59}$$

GENERAL FIRST-DEGREE EQUATION

is called the **general first-degree equation.** Therefore *we have shown that, in the set of rational numbers, the solution set of the general first-degree equation is not empty.*

In Sec. 3.5 we defined an order relation in the set of rational numbers by saying that if a and b are rational numbers and if there exists a *positive* rational number c with the property that

$$a - b = c$$

then a is *greater than* b, for which we write $a > b$, and also b is *less than* a, for which we write $b < a$. An alternative (and equivalent) definition that is usually somewhat easier to apply is the following.

GREATER THAN

LESS THAN

Let $a = \frac{p}{q}$ and $b = \frac{r}{s}$ be two rational numbers written so that q and s are *both positive integers.* We say that $\frac{p}{q}$ is **greater than** $\frac{r}{s}$, and write $\frac{p}{q} > \frac{r}{s}$, if and only if $ps > qr$; in this case we also say that $\frac{r}{s}$ is **less than** $\frac{p}{q}$, and write $\frac{r}{s} < \frac{p}{q}$. That is, for rational numbers $\frac{p}{q}$ and $\frac{r}{s}$, *written with q and s as positive integers,*

$$\begin{aligned} \frac{p}{q} > \frac{r}{s} \qquad &\text{if and only if} \qquad ps > qr \\ \frac{r}{s} < \frac{p}{q} \qquad &\text{if and only if} \qquad rq < sp \end{aligned} \tag{3.60}$$

To illustrate,

$$\frac{2}{7} < \frac{3}{8} \qquad \text{because} \qquad 2 \cdot 8 < 7 \cdot 3$$

$$\frac{-4}{9} > \frac{-9}{18} \qquad \text{because} \qquad (-4) \cdot 18 > 9 \cdot (-9)$$

Definitions (3.60) give the meaning of "greater than" and "less than," in the set of rational numbers, in terms of "greater than" and "less than" in the set of integers. Using (3.60) and the properties of order in the set of integers given in Sec. 2.6, it can be shown that the order relation ($>$) in the set Ra as given by (3.60) has the following properties:

For a, b, $c \in Ra$,
O_1: either $a > b$, or $b > a$, or $a = b$
O_2: if $a > b$ and $b > c$, then $a > c$
O_3: if $a > b$, then $a + c > b + c$
O_4: if $a > b$, then $ac > bc$ provided $c > 0$
$ac < bc$ *provided* $c < 0$

Note that if $\frac{p}{q}$ is a *positive* rational number, then

$$\frac{p}{q} > \frac{0}{1} \qquad \text{because} \qquad p \cdot 1 > q \cdot 0$$

Therefore, since $\frac{0}{1} = 0$, to say that $\frac{p}{q} > 0$ is equivalent to saying that $\frac{p}{q}$ is positive. Similarly, $\frac{p}{q} < 0$ is equivalent to saying $\frac{p}{q}$ is a *negative* rational number.

If a field also has properties O_1, O_2, O_3, and O_4, it is called an **ordered field.** Since the rational number system has these properties, it is an ordered field. Note that the integer number system, the nonnegative integer number system, and the positive integer number system are ordered, but they are *not* fields.

ORDERED FIELD

As we shall see in Chap. 4, the general first-degree inequality

$$ax + b > 0 \qquad a \neq 0$$

always has a nonempty solution set in the set of rational numbers.

Although the rational number system suffices to give nonempty solution sets for the general first-degree equation $ax + b = 0$ and the general first-degree inequality $ax + b > 0$, it is *not* true that the set

$$\{x \in Ra \mid ax^2 + bx + c = 0\}$$

is nonempty for all choices of a, b, $c \in Ra$, with $a \neq 0$. This deficiency will be discussed further in Chap. 5 where we shall be especially concerned with finding solution sets for such equations as $x^2 - 2 = 0$ and $x^2 + 1 = 0$.

EXERCISES

1 Is there an identity element under addition in the set N of positive integers? If there is, identify it. If there is not, tell why.

2 Is there an identity element under multiplication in the set N of positive integers? If there is, identify it. If there is not, tell why.

3 Answer question 1 for the set N_0 of nonnegative integers.

4 Answer question 2 for the set N_0 of nonnegative integers.

5 Do any numbers in the set N of positive integers have inverses under addition? If there are any, identify them and their inverses.

6 Do any numbers in the set N of positive integers have inverses under multiplication? If there are any, identify them and their inverses.

7 Answer question 5 for the set N_0 of nonnegative integers.

8 Answer question 6 for the set N_0 of nonnegative integers.

9 Answer question 6 for the set I of integers.

10 Which is the greater, $\frac{4}{9}$ or $\frac{17}{37}$?

11 Which is the greater, $\dfrac{-1}{3}$ or $\dfrac{-9}{31}$?

12 Which is the greater, -3 or $-\frac{5}{2}$?

13 Which of the following are true?

(a) $-\dfrac{1}{7} > -\dfrac{6}{35}$

(b) $\dfrac{4}{-3} < \dfrac{-43}{33}$

(c) $-6 > \dfrac{-29}{5}$

(d) $\dfrac{2}{9} < \dfrac{19}{82}$

4

First-degree Equations and Inequalities

4.1 INTRODUCTION

IDENTITY

CONDITIONAL SENTENCE

SOLUTION

Recall that in Sec. 2.1 we introduced the concept of an *open sentence* S_x, containing a variable x whose universe is a set A. If the sentence S_x is *true* for *every* member of the universe A of the variable x, the sentence is called an **identity.** If the sentence is *false* for *at least one* member of the universe of the variable, the sentence is called a **conditional sentence.** The *solution set*—or simply the **solution**—(in the universe A) of the sentence S_x is the set of all members of the universal set A that satisfy the sentence S_x. The solution set, in the universe A, of the open sentence S_x is denoted by

$$\{x \in A \mid S_x\} \tag{4.1}$$

In Secs. 2.1 and 2.5 we discussed several examples of sets of the form (4.1) where the universal set A was the set I of integers or a subset of I. Now, since we have the set Ra of rational numbers to work with, the universe A of the variable can be the set Ra or a subset of Ra.

As we pointed out in Sec. 2.5, the most common types of open sentences encountered in elementary mathematics are equations, inequalities, and compound sentences whose components are equations or inequalities. Much of elementary algebra is concerned with determining as much as possible about the members of sets of the form (4.1) where S_x is one of the types mentioned in the preceding sentence. In the examples in Chap. 2 we determined the numbers that were members of $\{x \in A \mid S_x\}$ by intuitive means or by substitution, in turn, of the members of A in place of x in the sentence S_x to determine which members of A made S_x a true statement. While these basic methods will always be used at some stage in determining the solution of a sentence, it may not be possible to apply them directly to a given sentence.

To illustrate, if we are asked to determine the solution of $x = 7$ in the set Ra of rational numbers, that is, if we are asked to determine the members of $\{x \in Ra \mid x = 7\}$, we "see" that the rational number 7 is the only member of the universal set that can be put in place of x in $x = 7$ to produce a true statement. Therefore, the solution of the equation $x = 7$ in the set Ra is $\{7\}$. On the other hand, if we are asked to determine the solution of

$$x + \frac{5}{3} = \frac{1}{3}x + 2$$

in the set Ra, it is not so easy to "see" what rational number, or numbers, we should put in place of x in $x + \frac{5}{3} = \frac{1}{3}x + 2$ to produce a true statement. Moreover, since the number of rational numbers is unlimited, it is not possible to substitute each member of the universe in turn in place of x in the sentence. However, using methods that will be discussed in this chapter, we can show that

$$\{x \in Ra \mid x + \tfrac{5}{3} = \tfrac{1}{3}x + 2\} = \{x \in Ra \mid 2x = 1\}$$

that is, the solution of $x + \frac{5}{3} = \frac{1}{3}x + 2$ is the *same* as the solution of $2x = 1$. Now, we can "see" that the rational number $\frac{1}{2}$ is the only number in the universal set that can be put in place of x in $2x = 1$ to produce a true statement, and we have

$$\{x \in Ra \mid x + \tfrac{5}{3} = \tfrac{1}{3}x + 2\} = \{x \in Ra \mid 2x = 1\} = \{\tfrac{1}{2}\}$$

So the solution of $x + \frac{5}{3} = \frac{1}{3}x + 2$ in the set* Ra is $\{\frac{1}{2}\}$.

In this chapter we shall discuss methods that can be used with certain types of sentences (such as in the above example) to enable us to determine the members of $\{x \in A \mid S_x\}$. As the student studies this chapter he should refer to the definitions and examples in Secs. 2.1 and 2.5.

4.2 SOLVING EQUATIONS OF THE FIRST DEGREE

Recall from Sec. 1.7 that an *algebraic expression* is obtained by combining letter symbols and numbers by the use of the four fundamental operations of algebra (addition, multiplication, subtraction, and division). An **algebraic equation** is a statement that one algebraic expression is equal to another algebraic expression where the symbol $=$ is used to stand for the words "is equal to." The two expressions are called sides of the equation. In the equation $a = b$ we speak of a as the left side and b as the right side. In this section we shall consider open sentences that are in the form of algebraic equations with one variable and in which the first power is the highest power of the variable that appears; such a sentence is called an *equation of the first degree.* Some examples are

ALGEBRAIC EQUATION

$$x = 7 \qquad x + \tfrac{5}{3} = \tfrac{1}{3}x + 2 \qquad 4x = 12 \qquad x - 3 = 5$$

$$6x - 21 + \tfrac{3}{2}x = 17 - 4(x + 1) \qquad \tfrac{1}{2}x = 6$$

To solve an equation of the first degree means to find the members of the universe of the variable that satisfy the equation; or, more briefly, to solve an equation means to find the solution set of the equation. Each member of the solution set of an equation is called

*Note that if we had been asked for the solution of $x + \frac{5}{3} = \frac{1}{3}x + 2$ in the set I of integers, we should have found the solution to be $\varnothing$, the empty set.

ROOT OF AN EQUATION

a **root** of the equation. Simple equations, such as $x = 7$, $4x = 12$, $x - 3 = 5$, and $\frac{1}{2}x = 6$, can be solved immediately by examining the equation, recalling what is meant by a solution, and making use of the addition and multiplication facts previously committed to memory. To illustrate, if the universe of the variable is the set of rational numbers,

For $x = 7$, $\{7\}$ is the solution and 7 is the root.

For $4x = 12$, $\{3\}$ is the solution and 3 is the root.

For $x - 3 = 5$, $\{8\}$ is the solution and 8 is the root.

For $\frac{1}{2}x = 6$, $\{12\}$ is the solution and 12 is the root.

However, most equations cannot be solved so easily. Intuitive reactions may be sufficient to contend with simple situations, where it is possible to keep in mind the things involved and their relationships. But for more complicated problems it may be helpful to have a formal procedure to aid in finding solutions of equations.

The most common procedure used as an aid in finding the solution of an equation of a complicated nature is based on the concept of *equivalent equations*. Two equations are said to be **equivalent** if they have the same solution set. For example, if the universe of the variable is Ra, the equations $x = 7$ and $2x = 14$ are equivalent, because

EQUIVALENT EQUATIONS

$$\{x \in Ra \mid x = 7\} = \{x \in Ra \mid 2x = 14\}$$

If each of two equations is equivalent to a third equation, the two equations are equivalent to each other. For, if A is the solution set of the first equation, B is the solution set of the second equation, and C is the solution set of the third equation, and if $A = C$ and $B = C$, then $A = B$.

In order to find the solution of a given first-degree equation in one variable, we replace the given equation by an equivalent equation simple enough to make its solution set intuitively apparent. As we shall see, it will be possible to find an equation of the form $x = k$ that is equivalent to the given equation; and since the solution of $x = k$ is $\{k\}$, the solution of the given equation is also $\{k\}$.

To use the method just described, we need to know what can be done to a given equation to produce an equation that is equivalent to the given equation. There are two things that can be done to an equation that will *always* produce an equation equivalent to the given equation. We describe the procedures for doing these things below and also give a symbolic form of them, using the symbols a, b, and c to represent *algebraic expressions*. The validity of these procedures can be established by use of the field properties (Sec. 3.12).

1 *The same quantity can be added to or subtracted from each side of an equation to produce a new equation that is equivalent to the original equation.* Symbolically,

$$a = b \qquad \text{if and only if} \qquad a + c = b + c \tag{4.2}$$

To illustrate, the equations $2x - 3 = 6 - x$ and $3x = 9$ are equivalent, since the second equation is obtained by adding $3 + x$ to each side of the first equation. Also, the equations $3x + 6 = 2x - 2$ and $x = -8$ are equivalent, since the second equation is obtained by subtracting $2x + 6$ from each side of the first equation.

2 *Each side of an equation can be multiplied by or divided by the same* **nonzero** *quantity* to produce a new equation that is equivalent to the original equation.* Symbolically,

$$\text{for } c \neq 0, \quad a = b \qquad \text{if and only if} \qquad ac = bc \tag{4.3}$$

To illustrate, the equations $2x = 12$ and $6x = 36$ are equivalent, since the second equation is obtained by multiplying each side of the first equation by 3. Also the equations $5x = 35$ and $x = 7$ are equivalent, since the second equation is obtained by dividing each side of the first equation by 5.

The fact that the operations symbolized by (4.2) and (4.3) always produce equivalent equations is a consequence of the fact that the operations of addition and multiplication (and therefore subtraction and division) are unique in any number system (property P_1 in Secs. 1.13, 2.6, and 3.12).

The statements (4.2) and (4.3) constitute a basis for solving any equation of the first degree with one variable. We shall see how these statements can be used to find an equation of the form $x = k$ that is equivalent to a given first-degree equation with one variable.

EXAMPLE 1 Find the solution (in the universe Ra) of

$$x - 1\tfrac{1}{5} = 2\tfrac{3}{4} \tag{4.4}$$

SOLUTION We see that if $1\frac{1}{5}$ is added to the left side of the equation, that side becomes simply x. Therefore, we use the operation symbolized by (4.2) and add $1\frac{1}{5}$ to each side of (4.4) to obtain the equation $x = 2\frac{3}{4} + 1\frac{1}{5}$, which is equivalent to equation (4.4). Now $2\frac{3}{4} = \frac{11}{4}$ and $1\frac{1}{5} = \frac{6}{5}$, so

$$2\tfrac{3}{4} + 1\tfrac{1}{5} = \frac{11}{4} + \frac{6}{5} = \frac{55 + 24}{20} = \frac{79}{20}$$

* It is to be understood that this quantity does not contain the variable. In cases in which the multiplier or divisor does contain the variable, the new equation may not be equivalent to the original equation. Such cases are discussed in Sec. 4.7.

and the given equation (4.4) is equivalent to

$x = \frac{79}{20}$

Clearly $\{\frac{79}{20}\}$ is the solution of this equation, and hence it is also the solution of (4.4).

We are claiming that when $\frac{79}{20}$ is put in place of x in $x - 1\frac{1}{5} = 2\frac{3}{4}$ we obtain a true statement. We can check this claim by showing that $\frac{79}{20} - 1\frac{1}{5} = 2\frac{3}{4}$ is a true statement.

EXAMPLE 2 Find the members of $\{x \in Ra \mid \frac{2}{7}x = \frac{3}{5}\}$. That is, find the solution of $\frac{2}{7}x = \frac{3}{5}$.

SOLUTION We see that if we multiply $\frac{2}{7}x$ by $\frac{7}{2}$ (the reciprocal of $\frac{2}{7}$) we obtain simply x. So, using (4.3), we find that $x = \frac{7}{2} \cdot \frac{3}{5}$ is equivalent to the given equation, or

$\{x \in Ra \mid \frac{2}{7}x = \frac{3}{5}\} = \{x \in Ra \mid x = \frac{7}{2} \cdot \frac{3}{5}\}$

Since $\frac{7}{2} \cdot \frac{3}{5} = \frac{21}{10}$, we have

$\{x \in Ra \mid \frac{2}{7}x = \frac{3}{5}\} = \{x \in Ra \mid x = \frac{21}{10}\} = \{\frac{21}{10}\}$

The check consists in noting that $\frac{2}{7} \cdot \frac{21}{10} = \frac{3}{5}$ is a true statement.

In finding the solution of an equation, the student should always keep in mind the definition of "the solution of an equation" and, if possible, find the solution by inspection. No matter how the solution is found, it should be checked by substitution in the given equation as we have done in Examples 1 and 2.

EXAMPLE 3 Find the solution (in the universe I) of $x + 6 = 10$.

SOLUTION By inspection we see that $\{4\}$ is the solution of the equation, since 4 is the only integer that can be put in place of x in $x + 6 = 10$ to produce a true statement. The check consists in noting that $4 + 6 = 10$ is a true statement.

Of course we could have used (4.2) and by subtracting 6 from each side of the given equation have found that $x = 4$ is equivalent to the equation $x + 6 = 10$. The first method is preferable.

EXAMPLE 4 Find the members of $\{x \in I \mid 4x = \frac{7}{2}\}$.

SOLUTION Since the universe of the variable x is the set of integers, we are looking for the set of all integers that can be put in place of x in $4x = \frac{7}{2}$ to produce a true statement. We know that the set of integers is closed under multiplication; hence whenever x is replaced by an integer, the quantity $4x$ is an integer. Therefore $\{x \in I \mid 4x = \frac{7}{2}\} = \varnothing$. Also, we could have divided both sides of $4x = \frac{7}{2}$ by 4, and using (4.3), have found that

$$\{x \in I \mid 4x = \tfrac{7}{2}\} = \{x \in I \mid x = \tfrac{7}{2} \div 4\}$$
$$= \{x \in I \mid x = \tfrac{7}{8}\}$$

So

$$\{x \in I \mid 4x = \tfrac{7}{2}\} = \varnothing$$

The equations which we have solved so far have required at most only one operation; that is, we have simply added, subtracted, multiplied, or divided in order to produce an equation of the form $x = k$. We now learn how to solve equations for which more than one operation is required to obtain an equivalent equation of the form $x = k$.

Solving an equation has been likened to untying a knot. In this respect, solving a simple equation with a single operation is like untying a single knot of a special sort. How would you untie a composite knot made up of a sequence of special small knots? You would untie the big knot by first untying the little knot which was made last, then untying the next to the last small knot, and so on. In the same manner, to solve an equation which has been constructed by using several operations, it is frequently desirable to look for an operation which could have been performed last, then to undo this operation by performing the inverse operation on both sides of the equation, and after that, to look for another operation which can be undone, and so on.

In Examples 5 to 10 the universe of the variable is Ra.

EXAMPLE 5 Solve $4x + 5 = 21$.

SOLUTION We observe that 5 has been added to the left side. So we perform the inverse of this operation; that is, we subtract 5 from each side of the equation to obtain the equation

$$4x = 16 \tag{4.5}$$

By inspection it can be seen that $\{4\}$ is the solution of the equation, since 4 is the only rational number that can be put in place of x in $4x = 16$ to produce a true statement.

An alternative procedure would be to note that the left side of (4.5) has been multiplied by 4. So we perform the inverse operation; that is, we divide each side by 4 and get the equation

$$x = 4 \tag{4.6}$$

Since we have performed successively the operations symbolized by (4.2) and (4.3), we know that each of equations (4.5) and (4.6) is equivalent to the given equation. Clearly $\{4\}$ is the solution of $x = 4$; so we know that $\{4\}$ is the solution of the given equation. The check consists in noting that $4(4) + 5 = 21$, or $16 + 5 = 21$, is a true statement.

EXAMPLE 6 Solve $3x + 2x - 8 = 5x - 3x + 7$.

SOLUTION Combining like terms on the left side and like terms on the right side, we get

$5x - 8 = 2x + 7$

We subtract $2x$ from each side and obtain

$3x - 8 = 7$

Next, we add 8 to each side and get

$3x = 15$

Finally, dividing each side by 3, we get

$x = 5$

All four of these equations are equivalent, and since $\{5\}$ is the solution of the last equation, $\{5\}$ is also the solution of the given equation. The check consists in noting that $5 \cdot 5 - 8 = 2 \cdot 5 + 7$ is a true statement.

Parentheses frequently appear in equations, and sometimes when they do not occur in an equation as it is given, they are introduced to avoid confusion. Computations and algebraic manipulations involving parentheses are carried out with the use of the distributive property. For example, consider the expression $6x - a + (3x - 4a)$; here $+(3x - 4a)$ can be written as $+1(3x - 4a)$, which by the distributive property, becomes $+3x - 4a$. So the original expression can be written as $6x - a + 3x - 4a$ or $9x - 5a$. As another example, consider the expression $7a - (5a + 2x + 4)$; here $-(5a + 2x - 4)$ can be written as $-1(5a + 2x - 4)$, which by the distributive property, becomes $-5a - 2x + 4$. Hence, $7a - (5a + 2x - 4) = 7a - 5a - 2x + 4 = 2a - 2x + 4$.

As illustrated by these examples, parentheses preceded by a plus sign can be interpreted as $+1$ times the expression in the parentheses, and parentheses preceded by a minus sign can be interpreted as -1 times the expression in the parentheses. Then the distributive property can be employed, and we have the following operational rules:

1 Parentheses preceded by a plus sign may be removed without changing the sign of any term.

2 Parentheses preceded by a minus sign may be removed by changing the sign of each term within the parentheses.

Similarly, if parentheses are preceded by a multiplication factor other than $+1$ or -1, the distributive property leads to a third rule.

3 Parentheses preceded by a multiplication factor may be removed by multiplying each term within the parentheses by that factor.

To illustrate this third rule,

$$5x + 2ax - 4a(x - 3a + 2) = 5x + 2ax - 4ax + 12a^2 - 8a$$
$$= 5x - 2ax + 12a^2 - 8a$$

$$2y + 5 + 3(2y - 7) = 2y + 5 + 6y - 21 = 8y - 16$$

EXAMPLE 7 Solve $5(x - 2) - (x - 3) = 4 + (2 - x)$.

SOLUTION First rewrite the equation by performing the indicated operations. Then proceed as usual. We get

$$5x - 10 - x + 3 = 4 + 2 - x \quad \text{or} \quad 4x - 7 = 6 - x \quad \text{or} \quad 5x = 13$$

So the solution is $\{\frac{13}{5}\}$. You should check this.

Recall that in the term $3x$ we say 3 is the coefficient of x; in $\frac{2}{3}y$, $\frac{2}{3}$ is the coefficient of y; and in ax, a is the coefficient of x. An equation with fractional coefficients may be changed into an equivalent equation with integers for coefficients by multiplying both sides of the given equation by the least common denominator of the fractional coefficients. This process is called clearing the equation of fractions. It is usually desirable to enclose each numerator in parentheses before clearing an equation of fractions.

EXAMPLE 8 Solve $\frac{x + 6}{2} - \frac{3x + 36}{4} = 4$.

SOLUTION The least common denominator is 4. Multiplying both sides of the given equation by 4, we obtain

$$2(x + 6) - (3x + 36) = 16$$
$$2x + 12 - 3x - 36 = 16$$
$$-x - 24 = 16 \quad \text{or} \quad -x = 40 \quad \text{or} \quad x = -40$$

So the solution is $\{-40\}$. The check is left to the student.

EXAMPLE 9 Solve $\frac{16x - 13}{6} = \frac{3x + 5}{2} - \frac{4 - x}{3}$.

SOLUTION The least common denominator of the fractions which appear in the equation is 6. Multiplying both sides by 6, we get

$$16x - 13 = 3(3x + 5) - 2(4 - x)$$
$$16x - 13 = 9x + 15 - 8 + 2x$$
$$5x = 20 \quad \text{or} \quad x = 4$$

So the solution is $\{4\}$. You should check this.

EXAMPLE 10 Solve $(x - 3)(x - 4) = x^2 - 2$.

SOLUTION Removing parentheses, we get

$$x^2 - 7x + 12 = x^2 - 2$$

Since the term x^2 occurs on both sides, by subtracting x^2 from both sides we get

$-7x + 12 = -2$ or $-7x = -14$

Since the coefficient of x is -7, we divide both sides by -7 and obtain

$\frac{-7x}{-7} = \frac{-14}{-7}$ or $\frac{-7}{-7}x = 2$ or $x = 2$

This last equation is equivalent to the given equation, and hence the required solution is $\{2\}$. The student should check this.

Using the operations symbolized by the statements (4.2) and (4.3), we can start with any first-degree equation with one variable and produce an equation of the form

$$ax + b = 0 \qquad a \neq 0 \tag{4.7}$$

that is equivalent to the given equation. For this reason equation (4.7) is called the *general first-degree equation* (in one variable). We have seen in Sec. 3.12 that *in the field of rational numbers* the solution set of (4.7) is not empty. Indeed, the desire to have a set of numbers in which the solution of the general first-degree equation is not empty was one of the motives that led us to the construction of the set of rational numbers.

By using (4.2) we see that equation (4.7) is equivalent to

$$ax = -b$$

By (4.3) we see this equation is equivalent to

$$x = -\frac{b}{a}$$

Clearly there is only one rational number that can be put in place of x in $x = -\frac{b}{a}$ to produce a true statement, namely, $-\frac{b}{a}$. Therefore, *in the field of rational numbers, the solution set of the general first-degree equation contains exactly one member.*

EXERCISES

In Exercises 1 to 8, simplify the given expression.

1 $x + 8 - 8$

2 $5y \div 5$

3 $s - 17 + 17$

4 $\frac{3}{4}a \div \frac{3}{4}$

5 $m + \frac{5}{2} - \frac{5}{2}$

6 $(\frac{1}{4}x)(4)$

7 $x - 4.5 + 4.5$

8 $1.1x \div 1.1$

Tell what mathematical operation to perform, and with what number, on the following expressions in order to obtain x as a result.

9 $8x$
10 $x - 7$
11 $x + 9$
12 $\frac{3}{4}x$
13 $\frac{x}{3}$
14 $\frac{x}{2}$
15 $\frac{1}{4}x$
16 $2.2x$
17 $x - 52$
18 $0.8x$
19 $x - \frac{5}{3}$
20 $\frac{4}{5}x$

Tell what to do to the following expressions to obtain x as a result.

21 $\frac{1}{2}x - 3$
22 $2x + 7$
23 $5x - 8$
24 $\frac{1}{2}x + 5$

In Exercises 25 to 28, an equation and several numbers are given. Determine which of the numbers are members of the solution set of the equation.

25 $4x - 17 = 3$; $4, 5, 0, 17$
26 $x^2 - 5x + 6 = 0$; $-2, -1, 0, 1, 2, 3$
27 $8x^3 - 27 = 0$; $3, 2, \frac{1}{2}, \frac{3}{2}, -\frac{3}{2}$
28 $2x^2 = 2x(x - 3) + 6$; $2, 1, 0, 3$

Find the solution of each of the following equations in the universe Ra. If you cannot find the solution by inspection, tell what operation (addition, subtraction, multiplication, or division) you must perform on both sides of the equation, and with what number, in order to produce an equivalent equation of the form $x = k$.

29 $3x = 9$
30 $x - 7 = 15$
31 $4x = 18$
32 $\frac{1}{2}x = 3$
33 $\frac{x}{6} = 1$
34 $\frac{x}{9} = 0.7$
35 $10x = 2$
36 $\frac{1}{2} + x = 11\frac{1}{2}$
37 $m - 1.5 = 4.5$
38 $x + 1\frac{1}{3} = 3\frac{1}{4}$
39 $x + 0.7 = 5.1$
40 $1.1x = 0.66$
41 $(4\frac{1}{2})x = 9$
42 $4x = 0$
43 $12 = 6x$
44 $1.7 = \frac{x}{3}$
45 $7x = -3$
46 $-9x = 5$
47 $(-5\frac{2}{3})x = 34$
48 $\frac{x}{-3} = -2$
49 $11 = \frac{x}{-2}$
50 $\frac{3}{2}x = -6$
51 $1.44x = 288$
52 $(2 + \frac{1}{4})x = -13$
53 $x + 5.26 = 0.13$
54 $x + 10 = 10$

Solve the following equations. The universe of each variable is Ra. List the steps you perform in finding the solution, as indicated for Exercise 55. Check the results.

55 $2x + 1 = 5$

Step 1. Subtract 1 from each side.
Step 2. Divide each side by 2.

We obtain $x = 2$; so the solution is $\{2\}$.

56 $-8x - 1 = 7$

57 $\frac{1}{2}y + 9 = 4$

58 $3x + 2.5 = 8.5$

59 $\frac{z}{4} - 2 = 7$

60 $\frac{3}{4}m + 16 = 31$

61 $5t + 4 = 17$

62 $8x - \frac{7}{2} = 7$

63 $3 + \frac{y}{4} = 8$

64 $-\frac{1}{3}x - \frac{1}{2} = 5\frac{1}{2}$

65 $2.1x - 0.34 = 0.08$

In Exercises 66 to 94, solve the given equation in the universe Ra, and check the solutions.

66 $\frac{2}{3}x + \frac{7}{3} = \frac{9}{2}$

67 $3x + 7 = x + 19$

68 $x - 3 + 5x = 2x + 5$

69 $3y + 1 - y = y + 4$

70 $x - 1 + 2x - 3 = x$

71 $\frac{1}{4}x = 10 - x$

72 $y - 3(y + 4) = 2$

73 $3(x - 2) = 2(x - 6)$

74 $y^2 + 3 - (y + 1)(y + 2) = 0$

75 $3(x + 1) + x^2 = x^2 + 12$

76 $(2x - 1)(x - 7) - 2(x^2 - 3) = 25$

77 $\frac{x - 3}{4} - \frac{2x - 1}{5} = 5$

78 $\frac{x}{5} + \frac{x + 5}{2} = 6$

79 $\frac{6x + 5}{4} - \frac{3x}{2} = x$

80 $0.3x + 0.055 = -0.033 - 0.5x$

81 $\frac{3x - 11}{9} + 2 = 0$

82 $\frac{x + 1}{2} - \frac{2x}{5} = \frac{x - 1}{2}$

83 $\frac{3x - 2}{5} = \frac{x + 1}{2} - \frac{3 - 7x}{10}$

84 $\frac{2x}{4} + \frac{3(x - 5)}{6} = 7$

85 $\frac{5(y - 3)}{4} = \frac{y}{5} - 2$

86 $\frac{2(x - 0.12)}{6} - \frac{3(x + 2.4)}{4} = 12(1.2)$

87 $8 - 4(2x - 3) = 8(x - 1)$

88 $\frac{1}{2}x - 9 = \frac{3}{4}x - 5$

89 $\frac{1}{3}(x + \frac{3}{2}) + \frac{1}{5}(x + 2) = x + 1\frac{1}{4}$

90 $\frac{2z - 5}{4} + \frac{4z - 1}{3} = \frac{2z - 3}{2}$

91 $4x + (x - 2)(x - 3) - 3 = x^2 + 5$

92 $\frac{1 - 3x}{4} = \frac{1 - x}{2} - \frac{1 - 2x}{3}$

93 $0.8(4y - 5.5) - 2.1(2y - 4) = 0.25(3y - 5)$

94 $\frac{9(x - 2)}{4} - \frac{7(x - 1)}{3} = 6x + 1$

95 For what value of u is the expression $\frac{3(u - 2)}{4}$ equal to 10? To 0? To -6?

96 For what value of z is the expression $\frac{3(z+1)}{4}$ equal to 12? To 0? To -7?

97 Centigrade temperatures, C, can be converted into Fahrenheit temperatures, F, by use of the formula $F = \frac{9}{5}C + 32$. Find the Fahrenheit temperature which corresponds to: 20°C; 10°C; 0°C; −10°C; −25°C.

4.3 LITERAL EQUATIONS

LITERAL EQUATION

An equation may have other letters in it besides the variable (or variables). Such an equation is called a **literal equation.** Some illustrations are

$$x + b = a \qquad p = 2l + 2w \qquad d = rt$$

The members of the solution set of a literal equation will not be specific arithmetic numerals, as was the case in the solutions in our previous problems, but instead the members of the solution set contain letter symbols. Literal equations of the first degree in one variable are solved by the methods discussed in the preceding section, the only difference being that the solution may involve letter symbols. *It is necessary that we know which of the letters in a literal equation is to be considered the variable.* In this section we shall explicitly identify the variable for each equation. In practice the variable is usually implied by the context. If, in an equation, say, $ax + 4b = c + d$, we wish to indicate that x is the variable and that we are to find the solution of the equation, we shall say,

"*Solve* $ax + 4b = c + d$ *for* x"

EXAMPLE 1 Solve $ax - 2bc = d$ for x.

SOLUTION Consider each letter, except x, which appears in the given equation as a constant. Adding $2bc$ to both sides of the given equation, we get

$$ax = 2bc + d$$

Now, assuming $a \neq 0$, we divide each side of this equation by a and obtain the equation

$$x = \frac{2bc + d}{a}$$

which is equivalent to the given equation. Therefore the solution is $\left\{\frac{2bc + d}{a}\right\}$.

In this example and elsewhere we assume that whenever we perform a division, the divisor is not zero.

EXAMPLE 2 Solve $ax - bx = a^2 - b^2$ for x.

SOLUTION First we indicate the combination of like terms by writing the given equation in the form

$$(a - b)x = a^2 - b^2$$

Dividing both sides by $a - b$, we have

$$x = \frac{a^2 - b^2}{a - b} = \frac{(a - b)(a + b)}{a - b} = a + b \qquad a - b \neq 0$$

Therefore $\{a + b\}$ is the solution of the given equation.

Many formulas of algebra, geometry, physics, chemistry, and business mathematics are literal equations which contain several letters. It is frequently important to be able to find an equation that is equivalent to the given formula and which has a particular letter standing alone on the left side of the equation. When this has been done, we say that we have solved the given formula (or equation) for that particular letter in terms of the other letters.

EXAMPLE 3 Recall that the area A of a rectangle is given by the formula $A = lw$, where l is the length and w the width of the rectangle. Solve this formula for w in terms of A and l.

SOLUTION Since

$$lw = A$$

we divide both sides by l and obtain

$$w = \frac{A}{l}$$

We note that if $l \neq 0$, the equation $w = \frac{A}{l}$ is equivalent to the equation $A = lw$.

EXAMPLE 4 The formula $\frac{1}{r} = \frac{1}{s} + \frac{1}{t}$ occurs in physics. Solve it for s in terms of r and t.

SOLUTION We multiply both sides by rst and get

$$st = rt + rs \qquad \text{or} \qquad ts - rs = rt \qquad \text{or} \qquad (t - r)s = rt$$

Dividing both sides of the last equation by $t - r$, it being understood that $t \neq r$, we obtain

$$s = \frac{rt}{t - r}$$

which is equivalent to the given formula.

EXERCISES

Solve the following equations for x:

1 $3ax + 5a = 7a$

2 $2(3x + 2a) - 3(x + 3a) = 4a$

3 $a(2x + 5) - b(3x - 2) = 0$

4 $\dfrac{x - 8a}{3} = 2(3a - 2x)$

5 $\dfrac{4a - x}{3} = \dfrac{x - 4a}{2}$

6 $\dfrac{1}{a} - \dfrac{1}{x} = \dfrac{1}{x} - \dfrac{1}{b}$

7 $mnx - a = anx - m$

8 $6ax - b^2 + 2ab = a^2 + 6bx$

Solve the following equations or formulas for the letter indicated:

9 $ax + b + cx = d$, for x

10 $ax + bx = c$, for x

11 $A = \frac{1}{2}bh$, for b

12 $A = P(1 + rt)$, for t

13 $I = Prt$, for r

14 $I = Prt$, for t

15 $I = Prt$, for P

16 $d = rt$, for r

17 $d = rt$, for t

18 $8z - 2n = 5n$, for n

19 $v = k + gt$, for t

20 $v = k + gt$, for k

21 $s = \frac{1}{2}at^2$, for a

22 $F = \dfrac{kmM}{d^2}$, for m

23 $F = \dfrac{kmM}{d^2}$, for M

24 $P = 2L + 2W$, for L

25 $V = LWH$, for W

26 $\mathrm{F} = \frac{9}{5}\mathrm{C} + 32$, for C

27 $f = ma$, for m

28 $S = \dfrac{a - rl}{1 - r}$, for a

29 $A = \frac{1}{2}(b + c)h$, for b

30 $A = \frac{1}{3}(x + y + z)$, for x

31 $l = a + (n - 1)d$, for a

32 $l = a + (n - 1)d$, for n

33 $l = a + (n - 1)d$, for d

34 $\dfrac{1}{r} = \dfrac{1}{s} + \dfrac{1}{t}$, for t

Solve the following equations for y:

35 $\dfrac{2y}{b} = \dfrac{b}{2} - 3$

36 $(a + b^2)y = a^2 + b$

37 $(y - a)^2 - (y + b)^2 = 2ab$

38 $2by + a^2 = 4b^2 + ay$

39 $\dfrac{y}{ab} - \dfrac{y}{bc} = \dfrac{1}{ac}$

40 $5by - 7a = 2ay + 4b$

Solve the following equations or formulas for the letter indicated:

41 $s = Vt + 16t^2$, for V

42 $V = \frac{1}{3}\pi r^2 h$, for h

43 $C = 2\pi r$, for r

44 $A + B + C = 180°$, for B

45 $I = \frac{E}{R}$, for E

46 $VP = bT$, for P

47 $S = \frac{a}{l - r}$, for a

48 $I = \frac{nE}{R + nr}$, for E

49 $\frac{1}{C} = \frac{1}{r} + \frac{1}{s} + \frac{1}{t}$, for r

50 $S = \frac{n}{2}(a + l)$, for n

51 $E = mc^2$, for m

52 $S = \pi r^2 + \pi rL$, for L

4.4 STATED PROBLEMS

STATED PROBLEM

A written or verbal statement which expresses some condition or conditions of equality concerning one or more unknown quantities is called a **stated problem.** In this section and the next we consider stated problems which are concerned with one unknown quantity. In order to solve such a stated problem by means of algebra, you must first translate the condition of equality into an equation. This process, called *algebraic translation,* consists of two steps:

1 Deciding what is unknown, and representing it by some appropriate letter, say x

2 Restating the condition of equality of the problem in the form of an equation containing the variable x (in doing this an appropriate figure may be helpful)

Solving stated problems is a significant test of your skill and understanding of basic algebra. For when you can do this, you will know that:

1 You can read carefully and interpret what you have read.

2 You can translate what you have read into an algebraic equation or equations.

3 You can solve the equation or equations.

You will gain confidence and assurance by working stated problems. By doing them you will be motivated to learn well the algebraic techniques we have studied in the preceding sections. Give stated problems your careful attention.

Stated problems are so varied that no general instructions beyond the comments we have made in the initial paragraph of this section can be formulated. We shall consider a few selected types of such problems, with some suggestions and one or more examples for each type.

UNIFORM-MOTION PROBLEMS

Many interesting and useful problems are solved by use of the formula

$$d = rt \tag{4.8}$$

Here d stands for the number of units of distance, r the number of units of distance traveled in one unit of time, and t the number of units of time. For brevity we call d the *distance*, r the *rate*, and t the *time*. Formula (4.8) holds only if the *rate is constant;* by this we mean that the moving object passes over equal distances in any two equal intervals of time. Such motion is referred to as *uniform motion*, or motion with a *uniform rate*. If the distance is in *miles* and the time in *hours*, then the rate is in *miles per hour*. The unit of rate is always the unit of distance per unit of time. Equation (4.8) is a formula for the distance when the rate and the time are known. If we solve it for r, we get

$$r = \frac{d}{t} \tag{4.9}$$

as a formula for the uniform rate in terms of the distance and the time. Similarly, we may solve (4.8) for t and obtain

$$t = \frac{d}{r} \tag{4.10}$$

as a formula for the time in terms of the distance and the rate. You should not tax your memory with memorizing all three of the formulas (4.8), (4.9), and (4.10). Remember (4.8), and be able to obtain (4.9) or (4.10) from it.

All problems on motion in this book will be concerned with uniform motion (on a straight line). So when we speak of the "rate" in a problem, it is to be understood that we mean "uniform rate." The following is an example of such a problem.

A car travels 180 miles in 6 hours. Its rate r is given by

$$r = \frac{180}{6} = 30 \text{ miles per hour}$$

Equivalently, 180 miles $= 180 \cdot 5{,}280$ feet, 6 hours $= 6 \cdot 60 \cdot 60$ seconds, and

$$r = \frac{180 \cdot 5{,}280}{6 \cdot 60 \cdot 60} = 44 \text{ feet per second}$$

EXAMPLE 1 Two airplanes A and B start at the same time from points that are 1,425 miles apart and fly toward each other. A's rate is 225 miles per hour, and B's rate is 250 miles per hour. How long will it take the two airplanes to meet?

SOLUTION Clearly the unknown in this problem is the time required for the two airplanes to meet. So we let

x = time in hours for the two airplanes to meet

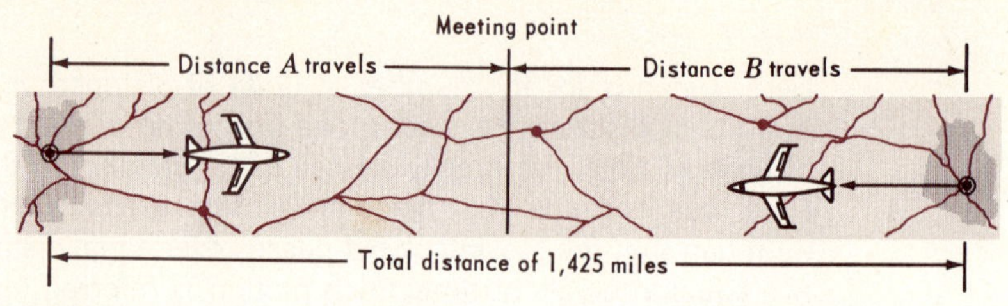

FIGURE 4.1

The verbal statement expresses an equality concerning distance, since (see Fig. 4.1)

$$\text{(distance } A \text{ travels)} + \text{(distance } B \text{ travels)} = 1{,}425 \text{ miles} \tag{1}$$

For airplane A, rate $= 225$, time $= x$, distance traveled $= 225x$.
For airplane B, rate $= 250$, time $= x$, distance traveled $= 250x$.
Substituting these expressions in (1), we have the desired equation,

$$225x + 250x = 1{,}425 \tag{2}$$

Whereas the variable in this equation is the time x, the equation is *set up*, or *balanced*, in terms of distance, for it expresses an equality about distance. In this problem the universe of the variable x is the set of positive rational numbers. So we are seeking the members of

$$\{x \in Ra \mid 225x + 250x = 1{,}425 \text{ and } x > 0\}$$

Proceeding with the solution of (2), we get

$$475x = 1{,}425 \qquad x = \frac{1{,}425}{475} = 3$$

Therefore

$$\{x \in Ra \mid 225x + 250x = 1{,}425 \text{ and } x > 0\}$$
$$= \{x \in Ra \mid x = 3 \text{ and } x > 0\} = \{3\}$$

and the two airplanes meet each other 3 hours after they start.

LEVER PROBLEMS

A lever is a rigid bar with one point of support F called the *fulcrum*. A familiar instance of a lever is a seesaw, or teeterboard. If a weight W is applied to the lever at the distance X from the fulcrum F, X is called the *lever arm* of W and the product WX is called the *moment* of W with respect to the fulcrum. In physics it is shown that for a simple lever to be in equilibrium, or balance, the product

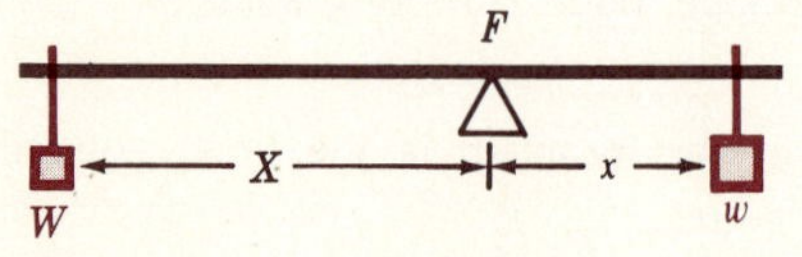

FIGURE 4.2

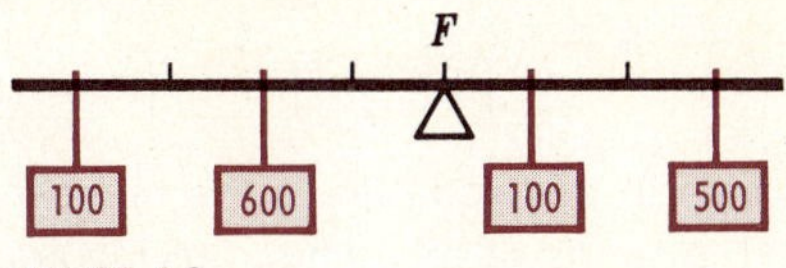

FIGURE 4.3

of the weight on one side of the fulcrum and the length of its lever arm must be equal to the product of the weight on the other side of the fulcrum and the length of its lever arm. Thus for the lever pictured in Fig. 4.2,

$$WX = wx \tag{4.11}$$

More generally, if more than two weights are applied to a lever, for the lever to be in equilibrium the sum of the moments of the weights on one side of the fulcrum must be equal to the sum of the moments of the weights on the other side of the fulcrum.

If weights of 100 and 600 pounds are placed at a distance of 4 feet and 2 feet, respectively, to the left of the fulcrum (Fig. 4.3) and weights of 100 and 500 pounds are placed at a distance of 1 foot and 3 feet, respectively, to the right of the fulcrum, the lever is in a position of equilibrium. For we have

$$100 \cdot 4 + 600 \cdot 2 = 100 \cdot 1 + 500 \cdot 3$$
$$400 + 1{,}200 = 100 + 1{,}500$$

In all problems on levers in this book, it will be assumed that the weight of the lever is negligible.

EXAMPLE 2 How many pounds of force must be exerted at one end of a lever 9 feet long to lift a 350-pound rock if the fulcrum is 2 feet from the rock?

SOLUTION The unknown in this problem is the number of pounds of force which must be exerted; let x stand for this quantity. The lever described in the stated problem is represented in Fig. 4.4. The stated problem implies that

$$\begin{pmatrix}\text{moment of force} \\ \text{exerted downward} \\ \text{to left of fulcrum}\end{pmatrix} = \begin{pmatrix}\text{moment of force} \\ \text{exerted downward} \\ \text{to right of fulcrum}\end{pmatrix} \tag{1}$$

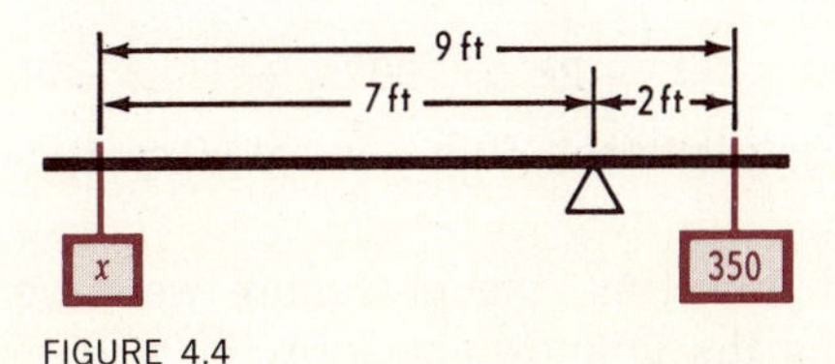

FIGURE 4.4

Therefore the desired equation is

$$7x = 2 \cdot 350 \tag{2}$$

Note that while the variable in this equation is the number of pounds x, the equation is balanced in terms of moments. Solving equation (2), we get $7x = 700$, or $x = 100$. So a force of 100 pounds must be exerted in the manner described to lift the 350-pound rock.

MIXTURE PROBLEMS

Some problems involve putting certain substances of known values or strengths into a mixture of required value or strength. Mixture problems which we work in this section involve the mixture of commodities with specified prices. A basic consideration in setting up the equation for a stated problem of this type is that when two substances are mixed,

$$\begin{pmatrix}\text{value of first}\\ \text{substance used}\end{pmatrix} + \begin{pmatrix}\text{value of second}\\ \text{substance used}\end{pmatrix} = \begin{pmatrix}\text{value of}\\ \text{mixture}\end{pmatrix} \tag{4.12}$$

Some mixture problems involve percentages, and we shall study these problems in the next section.

EXAMPLE 3 A grocer wishes to make up a 75-pound mixture to sell for 48 cents per pound from two grades of nuts which sell for 40 cents and 60 cents per pound, respectively. How many pounds of each grade of nuts should be used?

SOLUTION Let $x =$ the number of pounds of the 40-cent nuts used; then $75 - x =$ the number of pounds of 60-cent nuts used. Now x pounds of nuts at 40 cents per pound are worth $40x$ cents. Also $75 - x$ pounds of nuts at 60 cents per pound are worth $60(75 - x)$ cents. The value of the mixture, which weighs 75 pounds and is valued at 48 cents per pound, is $48 \cdot 75 = 3{,}600$ cents. From the statement of the problem we have the equality

$$\begin{pmatrix}\text{value of 40-cent}\\ \text{nuts used}\end{pmatrix} + \begin{pmatrix}\text{value of 60-cent}\\ \text{nuts used}\end{pmatrix} = \begin{pmatrix}\text{value of}\\ \text{mixture}\end{pmatrix} \tag{1}$$

If these values are expressed in cents, we have, corresponding to the equality (1), the equation in x,

$$40x + 60(75 - x) = 3{,}600 \tag{2}$$

Dividing both sides by 10, we get

$$4x + 6(75 - x) = 360$$

$$4x + 450 - 6x = 360 \qquad \text{or} \qquad -2x = -90 \qquad \text{or} \qquad x = 45$$

So 45 pounds of 40-cent nuts and 30 pounds of 60-cent nuts should be used.

In solving stated problems, the procedure we have suggested requires representing the unknown quantity by a letter symbol (x

in the above examples). This letter symbol then becomes the *variable* in an equation that states the condition of equality in the problem. It is very important that we have clearly in mind the *universe* of the variable in each particular problem. In each of the three preceding examples the universe of the variable is the set of positive rational numbers. In each of the two examples to follow we shall see that the universe will be some set other than the set of positive rational numbers.

DIGIT AND NUMBER PROBLEMS

Recall that any two-digit number may be written as

10 (tens digit) + (units digit)

or

$$10t + u$$

where u = the units digit and t = the tens digit. Note that the universe of each of the variables t and u is

$$\{0, 1, 2, 3, 4, 5, 6, 7, 8, 9\}$$

EXAMPLE 4 In a two-digit number, the units digit is 3 less than the tens digit. The number has a value 26 times that of the units digit. Find the number; that is, determine the two digits.

SOLUTION Let x be the tens digit; then $x - 3$ is the units digit. The number is $10x + (x - 3)$. Since we are told the number has a value 26 times that of the units digit, the desired equation is

$$10x + (x - 3) = 26(x - 3)$$

Note that the universe of the variable x is $\{0, 1, 2, 3, 4, 5, 6, 7, 8, 9\}$. Removing the parentheses and solving for x, we obtain

$$10x + x - 3 = 26x - 78 \qquad \text{or} \qquad 11x - 3 = 26x - 78$$

Therefore we want to find the member of

$$\{x \in A \mid 11x - 3 = 26x - 78\} \quad \text{where} \quad A = \{0, 1, 2, 3, 4, 5, 6, 7, 8, 9\}$$

Now $11x - 3 = 26x - 78$ is equivalent to

$$-15x = -75 \qquad \text{and to} \qquad x = 5$$

So the tens digit is 5 and the units digit is 2, and the number is $10(5) + 2 = 52$. If we had found the original equation to be equivalent to an equation of the form $x = k$ where k was some number other than a member of $\{0, 1, 2, 3, 4, 5, 6, 7, 8, 9\}$, then the solution set of the original equation would have been empty and there would have been no integer with the required properties.

EXAMPLE 5 Find three consecutive integers whose sum is 48.

SOLUTION Let $x =$ the smallest of the three integers. Then $x + 1$ and $x + 2$ are the other two integers. We have

$$x + (x + 1) + (x + 2) = 48 \qquad \text{or} \qquad 3x + 3 = 48$$

or $3x = 45$. The universe of the variable in this equation is the set I of integers. Now

$$\{x \in I \mid 3x = 45\} = \{15\}$$

so the smallest integer is 15 and the other two integers are 16 and 17. If the solution set of the equation in the *set I of integers* had been the empty set, this would have indicated that there were no three consecutive integers with the prescribed property.

EXERCISES

1 Two boys leave a town at 6:00 A.M. driving in opposite directions. If one boy travels at an average speed of 54 miles per hour and the other travels at an average speed of 60 miles per hour, at what time will they be 513 miles apart?

2 Two motorists start toward each other at 4:30 P.M. from towns which are 225 miles apart. If their respective average speeds are 30 miles per hour and 45 miles per hour, at what time will they meet?

3 Two cars start toward each other from points 200 miles apart. If one car travels 2 miles per hour faster than the other, and if it takes 4 hours for the cars to meet, what is the average speed of each car?

4 Two cars starting at the same point and traveling in opposite directions were 240 miles apart at the end of 4 hours. If one traveled 6 miles per hour faster than the other, find the rate of each car.

5 A man traveled a distance of 265 miles. He drove at an average speed of 40 miles per hour during the first part of the trip and at 35 miles per hour during the remaining part. If he made the trip in 7 hours, how long did he travel at 40 miles per hour?

6 A boy leaves town and starts walking toward home at an average speed of 4 miles per hour. When he is part of the way home, he begins to run at an average speed of 8 miles per hour. He reaches home 1 hour after he left town. How long did he walk, and how long did he run, if his home is 5 miles from town?

7 A, B, and C weigh 100, 120, and 200 pounds, respectively. A and C sit on opposite ends of a $13\frac{1}{2}$-foot seesaw. B sits 2 feet from A. Where must the fulcrum be placed in order that A and B may balance C?

8 A lever is 9 feet long, and the fulcrum is 2 feet from the end under a weight to be lifted. How heavy a weight can a man lift if he exerts a force of 140 pounds on the other end?

9 How many pounds of nuts worth 50 cents per pound should be mixed with another kind of nuts worth 30 cents per pound in order to make a mixture of 120 pounds of nuts worth 42 cents per pound?

10 Ten pounds of a mixture of two types of candy cost $11.20. Find how many pounds of each type of candy were used if the first kind cost $1 per pound and the second kind cost $1.20 per pound.

11 In a two-digit number, the tens digit is 2 more than the units digit. The number itself is 16 times the units digit. Find the number.

12 One-third of a certain number is 6 less than one-half of the number. What is the number?

13 The sum of the digits of a two-digit number is 11. If the digits are reversed, the resulting number is 27 more than the original number. Find the digits and the number.

14 The sum of the digits of a two-digit number is 15. If the digits are reversed, the number thus formed is 27 less than the original number. Find the digits and the two numbers.

15 In a certain two-digit number, the units digit is 3 less than the tens digit. The sum of the digits is 7. Find the two digits. What is the number?

16 One-half of a number plus one-third of the same number equals $5\frac{5}{6}$. Find the number.

17 Two cars start from the same point, one traveling east and the other west. The first one travels three-fourths as fast as the second. In 6 hours they are 378 miles apart. Find the rate of each.

18 Using a bar 6 feet long, where would a 160-pound man have to place the fulcrum to lift a 600-pound motor?

19 At 9 A.M. an airplane traveling at the rate of 350 miles per hour is 65 miles behind an airplane traveling at 300 miles per hour in the same direction. When will the first airplane overtake the second?

20 A and B start at the same time from two towns 288 miles apart and travel toward each other. A travels 8 miles per hour faster than B, and they meet at the end of 6 hours. Find the rate of each.

21 A and B together weigh 360 pounds. They balance on a seesaw when A is 9 feet and B is 7 feet from the fulcrum. Find the weight of each.

22 A grocer mixes one kind of coffee which sells for 60 cents per pound with another kind which sells for 75 cents per pound. How many pounds of each must he take to make a mixture of 80 pounds to sell for 66 cents per pound?

23 The difference of the squares of two consecutive positive integers is 41. What are the integers?

24 A contractor employs 100 men, with a daily payroll of $1,040. The unskilled laborers earn $8 per day, and the skilled workmen earn $14 per day. How many of each are employed?

25 The sum of three numbers is -8. The second number is $\frac{1}{3}$ of the first, and the third number is 15 less than the first. Find the numbers.

26 Find two numbers such that their sum is 43 and $\frac{5}{3}$ of the smaller exceeds $\frac{2}{5}$ of the larger by 20.

27 Three years ago a father was five times as old as his daughter. Three years from now he will be only three times as old as his daughter. Find the daughter's age now.

28 Thirty coins consisting of nickels and dimes have a value of \$2.25. How many dimes are there?

29 Are there three consecutive integers whose sum is 87? Try to find them.

30 Are there three consecutive integers whose sum is 86? Try to find them.

31 Are there three consecutive even integers whose sum is 78? Try to find them.

32 Are there three consecutive even integers whose sum is 76? Try to find them.

33 A 60-pound weight is on a lever 8 feet to the left of the fulcrum, and a 45-pound weight is on the opposite side of the fulcrum 6 feet from the fulcrum. Where should a weight of 20 pounds be placed to give equilibrium?

4.5 PERCENTAGE—SIMPLE INTEREST

Percentage is widely used in modern business. It is used by merchants in expressing markups on goods, by banks in expressing interest, by governments in computing taxes, by newspapers in reporting population gains, by highway patrols in reporting accident rates, and by insurance companies in computing rates. It affects our daily lives to such an extent that it is imperative that we understand clearly what it is and how it is used. We cannot afford to have vague ideas about its meaning.

"Percentage" is the generalized noun form of "percent," which is an anglicized form of the Latin *per centum*. Sometimes the terms percent and percentage are used interchangeably. Thus we may ask, "18 is what percent of 24?" rather than, "18 is what percentage of 24?"

The familiar symbol for percent is %. It may be used in place of the term percent in calculations or in tabular or display matter. The symbol was probably derived from one of the abbreviations used in business arithmetics written by the Italians in the fifteenth century.

PERCENT

The term **percent** means "hundredths." To illustrate, 63 percent, or 63%, means 63 hundredths. Recall that we write 63 hundredths as 0.63, or $\frac{63}{100}$. Similarly,

$$17.5\% = \frac{17.5}{100} = 0.175 \qquad 3\% = \frac{3}{100} = 0.03$$

If a mixture is 63 percent water by volume, then $\frac{63}{100}$ of its volume is water. That is, in 100 gallons of the mixture there are 63 gallons of water, or in 100 quarts of the mixture there are 63 quarts of water. If a bank pays interest on savings deposits at the rate of 4 percent per year, it pays $4 interest per year on each $100 in a savings deposit, or it pays 4 cents interest per year on each 100 cents in a savings deposit. The whole of something is 100 percent of it. Thus, if a man loses all of his money, he loses 100 percent of it.

We have noted that a percent may be expressed as a decimal. In a comparable way a decimal may be changed to a percent. To illustrate,

$$0.275 = \frac{27.5}{100} = 27.5\% \qquad 3.21 = \frac{321}{100} = 321\%$$

The procedure for changing a decimal to a percent and a percent to a decimal may be stated as follows:

To change a decimal to a percent, multiply the decimal by 100, *and annex the percent sign.*

To change a percent to a decimal, divide the percent by 100, *and omit the percent sign.*

BASE RATE PERCENTAGE

When we take a percent of a certain number, that number is called the **base;** the percent we take is called the **rate;** and the result obtained is called the **percentage.** If we let B represent the base, R the rate, and P the percentage, then the relationship among these three quantities is expressed by the formula

$$P = RB$$

The rate R is usually expressed as a decimal number.

EXAMPLE 1 Find 22.8 percent of 6.25.

SOLUTION We see that $R = 0.228$ and $B = 6.25$. Substituting these values in the formula, we obtain $P = (0.228)(6.25) = 1.425$.

EXAMPLE 2 What percent of 12.5 is 7.75?

SOLUTION Since $P = 7.75$ and $B = 12.5$, we have

$$7.75 = R(12.5) \qquad \text{and} \qquad R = \frac{7.75}{12.5} = 0.62 = 62\%$$

EXAMPLE 3 Find a number such that 24 percent of it is 4.07.

SOLUTION Let B be the number. Then

$$4.07 = 0.24B$$

So

$$B = \frac{4.07}{0.24} = 16.96 \qquad \text{to two decimal places}$$

EXAMPLE 4 A salesman earns a commission of 20 percent on each of his sales. What does he earn on a \$55 sale?

SOLUTION Let P be the salesman's commission. Then $P = 0.20(\$55)$. So $P = \$11$.

EXAMPLE 5 How many pounds of cream containing 30 percent butterfat should be added to 750 pounds of milk containing 3.5 percent butterfat to give milk containing 4 percent butterfat?

SOLUTION The unknown in this problem is the number of pounds of cream which is to be added; let x be this number. The condition of equality here is

$$\begin{pmatrix}\text{number of pounds}\\ \text{of butterfat in the}\\ \text{given milk}\end{pmatrix} + \begin{pmatrix}\text{number of}\\ \text{pounds of butter-}\\ \text{fat in the cream}\end{pmatrix} = \begin{pmatrix}\text{number of}\\ \text{pounds of butter-}\\ \text{fat in the mixture}\end{pmatrix}$$

and the corresponding equation is

$0.035(750) + 0.30x = 0.04(750 + x)$ or
$35(750) + 300x = 40(750 + x)$
$7(750) + 60x = 8(750 + x)$ or $5{,}250 + 60x = 6{,}000 + 8x$

Hence

$52x = 750$ and $x = 14.42$ to two decimal places

So 14.42 pounds (approximately) of cream should be added.

Note that in setting up the equation in the last example, we recognized that the weight of the mixture was the sum of the weight of the given milk and weight of the cream which was added.

In each of the five preceding examples, it is understood that the universe of the variable is the set of rational numbers.

Sometimes, for comparison, it is desirable to use percents greater than 100. If a boy had five dogs last year but now has six dogs, he now has 20 percent more dogs than he had last year and he now has 120 percent of the number of dogs that he had last year. If a man had 50 acres of beans last year and has 70 acres of beans this year, he has an increase of 40 percent over his last year's acreage and his present bean acreage is 140 percent of his last year's acreage. If a man's income was \$4,000 in 1960 and \$10,000 in 1966, his income increased 150 percent from 1960 to 1966 and his 1966 income was 250 percent of his 1960 income; that is, his 1966 income was 2.5 times his 1960 income, and the difference in the two incomes was 1.5 times the 1960 income.

Problems in simple interest are like problems in percentage, with the additional factor of time. Let us recall the terms generally used in simple interest problems and specify corresponding letter symbols used to represent them.

INTEREST

Interest I is the rent paid by a borrower for the use of money. The interest in a situation depends upon three things: the amount of money borrowed, the length of time for which the borrower has the use of the money, and the interest rate agreed upon by the borrower and the lender.

PRINCIPAL

The **principal** P is the sum of money loaned.

INTEREST RATE

The **interest rate** r is the percent of the principal that is to be paid at the end of a stated period of time for the use of the money. The number of periods of time is referred to simply as *time* and is usually expressed in years. An interest rate of 4% means that \$4 will be paid for the use of each \$100 of the principal for a period of 1 year.

AMOUNT

The **amount** A is the sum of the principal and the interest earned during the period of the loan.

Suppose that a man borrows \$100 subject to the agreement between him and the lender that he will repay the money at the end of 5 years, together with interest at the rate of 4% per year on the principal of \$100. What is the interest? What is the amount paid to the lender at the end of 5 years?

The interest for 1 year is 100(0.04) = \$4; for 5 years the interest is $5 \cdot 4 = \$20$. So we can write

$$I = 100(0.04)5 = \$20$$

The amount paid to the lender at the end of the period of 5 years is

$$A = \$100 + \$20 = \$120$$

SIMPLE INTEREST

If interest is computed on the original principal only (as in the illustration above), it is called **simple interest.** From the definition of simple interest, the interest for one period is Pr, and the interest I for t periods is given by

$$I = Prt \tag{4.13}$$

The amount A, which is to be paid by the borrower to the lender at the end of time t, is given by

$$A = P + I = P + Prt$$

or

$$A = P(1 + rt) \tag{4.14}$$

Observe that formula (4.13) contains four different letter symbols and it may be solved for any one of these symbols in terms of the others. The same statement applies to formula (4.14). Moreover, formulas (4.13) and (4.14) together involve five letter symbols. If the values of any three of these symbols are known, the values of the other two may be found by using equalities (4.13) and (4.14).

We shall follow the usual practice of assuming that a given interest rate is an annual rate.

In computing interest for a fraction of a year, each month is usually considered to be $\frac{1}{12}$ of a year. Sometimes a day is considered $\frac{1}{365}$ of a year (or $\frac{1}{366}$ of a leap year), and sometimes a day is assumed to be $\frac{1}{360}$ of a year. Simple interest computed in the first way is called *exact simple interest,* and in the second way *ordinary simple interest.* In this book we shall consider each month $\frac{1}{12}$ of a year and each day $\frac{1}{360}$ of a year.

EXAMPLE 6 What is the simple interest on \$5,000 for 9 months at 4%? What is the amount?

SOLUTION Here $P = \$5{,}000$, $r = 0.04$, and $t = \frac{3}{4}$, since 9 months $= \frac{3}{4}$ of a year. Substituting these values in formula (4.13), we get for the interest

$$I = Prt = 5{,}000(0.04)\tfrac{3}{4} = \$150$$

The amount is given by

$$A = P + I = \$5{,}000 + \$150 = \$5{,}150$$

EXAMPLE 7 What principal will amount to \$960 in 4 years at 5% simple interest?

SOLUTION Here $A = \$960$, $r = 0.05$, and $t = 4$. We know values for all the letter symbols in formula (4.14) except P. Let us solve (4.14) for P; we do this by dividing both sides by $1 + rt$, obtaining

$$P = \frac{A}{1 + rt}$$

Substituting the given values in this formula, we get for the principal

$$P = \frac{960}{1 + (0.05)4} = \frac{960}{1.2} = \$800$$

EXAMPLE 8 Find the time required for \$2,400 to yield \$360 in simple interest at 6%.

SOLUTION In this case $P = \$2{,}400$, $r = 0.06$, and $I = \$360$. We know values for all the letter symbols in formula (4.13) except t. We solve (4.13) for t by dividing both sides by Pr, and we have

$$t = \frac{I}{Pr}$$

Substituting in this formula the values designated above, we get

$$t = \frac{360}{2{,}400(0.06)} = \frac{360}{144} = 2\tfrac{1}{2} \text{ years}$$

Formula (4.14) tells us that

$$A = P(1 + rt)$$

Solving this for P, we obtain

$$P = \frac{A}{1 + rt} \tag{4.15}$$

We may consider P and A equivalent values of the same sum of money at two different times. That is, P, the value at the beginning of the period of t years, is equivalent to A at the end of the period, and conversely. This relation is shown in the following line diagram:

$$P \xleftrightarrow{t \text{ years at simple interest rate } r} A$$

PRESENT VALUE

The principal P is also called the **present value** of the amount A. It is the sum which must be invested now at the simple interest rate r to yield an amount A in t years. To describe this situation, we sometimes say P will *accumulate* to A in t years at the simple interest rate r, and call A the *accumulated value* of P.

EXAMPLE 9 Find the present value of \$5,000 due in $1\frac{1}{2}$ years at 5% simple interest. Draw a line diagram for this problem.

SOLUTION The line diagram is

$$P \xleftrightarrow{1\frac{1}{2} \text{ years at 5\% simple interest}} \$5{,}000$$

Here $A = \$5{,}000$, $r = 0.05$, $t = \frac{3}{2}$. Substituting these values in formula (4.15), we get

$$P = \frac{5{,}000}{1 + \frac{3}{2}(0.05)} = \frac{5{,}000}{1.075} = \$4{,}651.16$$

TRUE DISCOUNT

The difference between A and P, $A - P$, is called the **true discount** on A. Since $A = P + I$, or $I = A - P$, the true discount on A is the same as the simple interest on P for the same period of time. We denote the true discount on A by D_t; so

$$D_t = A - P$$

In dealing with a sum of money, it is essential to know the time connected with that sum. The amount \$1,000 a year hence is not the same as \$1,000 today, under usual business conditions. The phrase "Money is worth a certain percent" means that money can be invested at that rate.

EXAMPLE 10 What is the present value of $1,000 due in 1 year if money is worth 5%? What is the true discount on $1,000 for 1 year at 5% simple interest?

SOLUTION Here $A = \$1{,}000$, $t = 1$, and $r = 0.05$. Substituting these values in formula (4.15), we obtain (to the nearest cent)

$$P = \frac{1{,}000}{1 + 1(0.05)} = \frac{1{,}000}{1.05} = \$952.38$$

Further, the true discount D_t is given by

$$D_t = A - P = \$47.62$$

This means that, if money is worth 5%, then $952.38 today is as good as $1,000 one year from now. Observe that the simple interest on $952.38 for 1 year at 5% is given by $I = 952.38(1)(0.05) = \$47.62$ and that $\$952.38 + \$47.62 = \$1{,}000$. The true discount on $1,000 is the simple interest on $952.38.

EXERCISES

In any of these exercises (and similar ones which may appear later), if the answer does not come out to be a finite decimal, it should be given to two decimal places.

1 Find 32 percent of 25.
2 What is 19 percent of 500?
3 What is 42 percent of 25?
4 Find 96 percent of 64.
5 24 is what percent of 64?
6 128 is what percent of 512?
7 Two hundred thirty-one and twelve hundredths is what percent of three hundred twenty-one?
8 Seventy-five is what percent of eighty?
9 Find a number such that 21 percent of it is 42.
10 Twenty-two and two tenths is 80 percent of what number?
11 Eighteen and five tenths is 74 percent of what number?
12 Find a number such that 87 percent of it is 165.3.
13 If 30 percent of a 1,700-pound load of peaches were spoiled, how many pounds of peaches were spoiled?
14 If 74 percent of the total weight of a hog can be made into edible products, how many pounds of food can be obtained from a 231-pound hog?
15 If fish loses 28 percent of its weight in cleaning, how many pounds of raw fish are needed to obtain 144 pounds of cleaned fish? How many pounds of cleaned fish can be obtained from 250 pounds of raw fish? How many pounds of waste are in 10 tons of fish?
16 A concrete mixture is 1 part cement, 2 parts sand, and 2 parts gravel by weight. What percent of the mixture is cement? Sand? Gravel? How many

pounds of cement are in 3 tons of the mixture? How many pounds of the mixture can be made with 14,000 pounds of gravel?

17 A 3,500-pound automobile contains 70 pounds of chromium, 105 pounds of lead, and 427 pounds of rubber. What percent of its total weight is chromium? Lead? Rubber?

18 How many pounds of lead should be added to 100 pounds of an alloy containing 40 percent lead in order to produce an alloy which is 50 percent lead?

19 An alloy is 23 percent copper and 31 percent tin. How many pounds of copper and tin are in 150 pounds of the alloy? How many pounds of the alloy can be made from 50 pounds of tin?

20 Green gold is 60 percent gold, 35 percent silver, and 5 percent copper. How much silver is in 90 ounces of green gold? How many ounces each of gold, silver, and copper are in 2,000 ounces of green gold?

21 Eighteen-carat gold is 75 percent gold and 25 percent a mixture of copper and silver. How many ounces of copper are in 160 ounces of 18-carat gold which contains 10 ounces of silver? What percent of the 160 ounces is silver?

22 An automobile radiator contains 20 quarts of a solution which is 20 percent alcohol and 80 percent water. How much of the solution must be drained out and replaced by pure alcohol to give a solution which is 40 percent alcohol?

23 A solution is 18 percent alcohol. How many gallons of water must be added to 400 gallons of the solution so that it will be only a 10 percent solution of alcohol?

24 How many pounds of water must be evaporated from a ton of brine that is 6 percent salt to obtain a brine that is 10 percent salt?

25 If $\frac{1}{5}$ of an alloy is replaced by pure copper, the result is an alloy which is 50 percent copper. What percent of the original alloy was copper?

26 On a spelling test a student missed 3 out of the first 15 words, and thereafter he missed 4 out of every 15 words. How many words were on the test if he spelled correctly 75 percent of the words?

27 Monel metal is 27 percent copper, 68 percent nickel, and 5 percent other metals. How many ounces of copper and how many ounces of nickel are needed to produce 300 ounces of monel metal? How many ounces of pure copper would have to be added to 300 ounces of monel metal to produce an alloy which is 50 percent copper?

28 Fifty gallons of milk containing 4 percent butterfat is mixed with twenty gallons of milk containing 5 percent butterfat. How many gallons of cream containing 20 percent butterfat must be added to obtain a cereal cream containing 10 percent butterfat?

29 Three hundred pounds of an alloy that is 40 percent tin is to be produced by fusing an alloy that is 36 percent tin with an alloy that is 41 percent tin. How many pounds of each alloy should be used?

30 Forty-nine gallons of a 20 percent spray solution is to be formed by

mixing an 18 percent spray solution with a 25 percent spray solution. How many gallons of each solution should be used?

31 Two alloys contain 60 percent silver and 75 percent silver, respectively. In what proportions must they be mixed in order to obtain an alloy which is 65 percent silver?

32 Two numbers are such that the larger minus twice the smaller is equal to 540, and 5 percent of the larger is equal to 20 percent of the smaller. Find the numbers.

33 How much water must be added to 2 quarts of a solution which is 90 percent alcohol in order to make a new solution which is 50 percent alcohol?

34 How many pounds of water must be added to 75 pounds of brine solution which contains 20 percent salt in order to make a new solution which contains 6 percent salt?

35 How many pounds of cream which contains 35 percent butterfat and how many pounds of milk which contains $3\frac{1}{2}$ percent butterfat should be mixed in order to make 600 pounds of milk which contains 4 percent butterfat?

36 One solution of glycerin and water contains 50 percent glycerin, and a second solution contains 15 percent glycerin. How many gallons of each solution should be mixed to make 100 gallons that is 45 percent glycerin?

37 A solution is 55 percent alcohol. How much water should be added to 18 quarts of the solution so that it will be only 15 percent alcohol?

38 How many pounds of water must be evaporated from 100 pounds of brine that is 8 percent salt in order that the remaining part will be 10 percent salt?

39 An automobile radiator contains 21 quarts of a solution which is 20 percent alcohol and 80 percent water. How much of the solution must be drained off and replaced by pure alcohol in order to make a solution which is 35 percent alcohol?

40 Find $\frac{1}{2}$ of 1 percent of 2,000.

41 Find 0.7 percent of 2,180.

42 Find 0.005 percent of 12.

43 Find 0.025 percent of 0.64.

44 Find 325 percent of 60.

45 Find 140 percent of 0.035.

46 Find 632 percent of 25.

47 Find 120 percent of 45.

48 12 is what percent of 3,000?

49 7 is what percent of 1,400?

50 0.03 is what percent of 1,200?

51 4.5 is what percent of 500?

52 210 is what percent of 60?

53 0.078 is what percent of 0.0012?

54 Duralumin is $\frac{1}{2}$ of 1 percent magnesium. How many pounds of duralumin can be made with 40 pounds of magnesium?
55 What is the income of a man who finds that 0.15 percent of his income is $9.60?
56 One ton of a commercial feed contains 5 pounds of salt. What percent of the feed is salt?
57 A town had a population of 5,200 in 1940 and a population of 13,000 in 1960. The increase in the population is what percent of the 1940 population? The 1960 population is what percent of the 1940 population?
58 A man invested $12,800 and lost $160. What percent of his investment did he lose?
59 A mixture contains 0.4 percent of a liquid dye. How much dye must be added to 10 quarts of the mixture to produce a mixture which contains 4 percent dye?
60 Wrought iron is 99.8 percent pure iron. How many pounds of impurities are in 22.78 tons of wrought iron?
61 A kind of steel contains $\frac{3}{4}$ of 1 percent of carbon. How many tons of this steel can be made with 810 pounds of carbon?

Find the simple interest and amount of a principal:

62 Of $1,800 for $2\frac{1}{2}$ years at 4%.
63 Of $2,400 for 8 months ($\frac{2}{3}$ of 1 year) at 6%.
64 Of $4,000 for 6 months at 5%.
65 Of $5,784 for 4 years and 3 months at 4%.
66 Of $4,200 for 4 months at $4\frac{1}{2}$%.
67 Of $300 for 2 years at $3\frac{1}{4}$%.

68 A student borrows $200 for 9 months at 6% simple interest. How much will he owe the lender at the end of the period?
69 A philanthropist wishes to endow a scholarship fund at a university to provide for $1,050 per year. If the university earns $3\frac{1}{2}$% on its investments, how much should the philanthropist give the university?
70 In what time will $1,250 earn $375 at 4% simple interest?
71 What is the rate of simple interest when $2,500 earns $87.50 in 6 months?
72 How long must $1,000 be kept at 5% simple interest to become the amount of $1,250?
73 How long will it take for a principal P invested at 4% simple interest to double itself? *Hint:* In the formula $A = P(1 + rt)$, put $A = 2P$, $r = 0.04$, and solve for t.
74 How long will it take for a principal invested at 6% simple interest to double itself?
75 How long will it take for a principal invested at 8% simple interest to double itself?

76 At what rate of simple interest will \$3,600 amount to \$3,780 in 1 year and 3 months?

77 Derive a formula to express the time t in terms of the principal P, the amount A, and the rate r of simple interest.

78 Derive a formula to express the rate of interest r in terms of the principal P, the simple interest I, and the time t.

79 Derive a formula to express the rate of simple interest r in terms of the principal P, the amount A, and the time t.

80 A business invested \$30,000 for 1 year at simple interest, with part invested at 6% and the remainder at 3%. How much money was invested at each rate of interest if the total return was the same as if all the money had been invested at 4%?

81 Find the accumulated value of \$2,000 in 2 years at 5% simple interest. What is the interest on the \$2,000?

82 What is the present value of \$2,000 due in 2 years if money is worth 5%? What is the true discount on the \$2,000?

83 What is the present value of \$6,000 due in 8 months if money is worth 6%? What is the true discount?

84 What is the present value, at 3% simple interest, of \$2,400 due in 10 months? What is the true discount?

85 Find the present value of \$1,000 due in 1 year at the rate of 4% simple interest; 5% simple interest; 6% simple interest.

86 Find the present value of \$1,000 at 6% simple interest due in 1 year; in 2 years; in 3 years.

87 Using the definition of D_t given by $D_t = A - P$, show that $D_t = \dfrac{Art}{1 + rt}$.

88 Using the formula in Exercise 87, show that D_t on A is equal to I on P.

89 Mr. Jones can sell his house now for \$20,000 or for \$20,100 two months from now. If money is worth 4% at simple interest, which is the better offer for him on the basis of present value, and how much better is it?

90 If \$10 is paid at the end of a period of 45 days for the use of \$100 over that period, what is the rate of simple interest paid?

91 What principal is it necessary to invest now at $3\frac{1}{2}$% simple interest in order to obtain \$10,665 at the end of 5 years?

4.6 BANK DISCOUNT

Notes are used frequently in business. A *promissory note,* or simply a **note,** is a written promise made by one person to another and signed by the *maker,* in which the maker agrees to pay a certain sum of money at a fixed or determinable future date. The date on

NOTE

$ 1000.00 Atlanta, Ga. July 5, 19 67

One hundred twenty days AFTER DATE I PROMISE TO PAY TO

THE ORDER OF James R. Rich

-------- One thousand and 00/100 --------- DOLLARS

AT The Textile National Bank

VALUE RECEIVED.

NO. 19 DUE Nov. 2, 1967 Richard E Poor

FIGURE 4.5

DATE OF MATURITY

MATURITY VALUE

which the money is due is called the **date of maturity,** and the sum of money due is called the **maturity value.**

A note may or may not bear interest. In Fig. 4.5 the note is non-interest-bearing. This note is a promise to pay Mr. Rich $1,000 on Nov. 2, 1967. When the time is given indirectly as the time between dates, we follow the custom of counting the last day, but not the first. The date on which the note of Fig. 4.5 becomes due may be checked as follows:

Number of days remaining in July	26
Number of days in August	31
Number of days in September	30
Number of days in October	31
Number of days in November to be counted	2
Total number of days	120

The $1,000 is called the *face* of the note, and the 120 days is called the *period* of the note. The maturity value of this note is $1,000, since the note does not bear interest.

The note in Fig. 4.6 is interest-bearing. The interest on $2,000

$ 2000.00 San Francisco, Calif., Oct. 10 19 67

Thirty days AFTER DATE I PROMISE TO PAY TO

THE ORDER OF

------- Two thousand and 00/100 ---------- DOLLARS

AT The City National Bank

VALUE RECEIVED WITH INTEREST AT 6% PER ANNUM.

NO. 42 DUE Nov. 9, 1967 Richard E Poor

FIGURE 4.6

for 30 days at 6% is $10, and this must be added to the face, $2,000, to find the maturity value. So the maturity value of the note in Fig. 4.6 is $2,010.

BANK DISCOUNT

Bank discount is simple interest computed on the maturity value of the note, but it is deducted in advance. If, for example, $1,000 is borrowed at a bank for 1 year at 6%, the $60 charge is deducted at the time the loan is made and the borrower receives $940. The bank discount is $60. The sum of money which the borrower receives on the day of the loan is called the **proceeds.** In the example just given, the proceeds are $940, although the borrower is said to have received a $1,000 loan.

PROCEEDS

Denote the bank discount by D_b, the maturity value by A, the period of discount in years by t, and the annual rate of discount by d. Then from the definition of bank discount, it follows that

$$D_b = Adt \tag{4.16}$$

If P_b denotes the proceeds, which is the difference between the maturity value and the bank discount, then

$$P_b = A - D_b = A - Adt$$

or

$$P_b = A(1 - dt) \tag{4.17}$$

EXAMPLE 1 A man goes to a bank and requests a loan of $960 for a period of 8 months. The bank's discount rate is 6% per year. What is the bank discount? What are the proceeds of this loan?

SOLUTION Here $A = \$960$, $d = 0.06$, and $t = \frac{2}{3}$. Substituting these values in formula (4.16), we get for the bank discount

$$D_b = Adt = 960(0.06)\tfrac{2}{3} = \$38.40$$

For the proceeds we have

$$P_b = A - D_b = \$960.00 - \$38.40 = \$921.60$$

When a note is a negotiable instrument, there may be three parties involved instead of only two. One person may borrow from another and the second person take the note to a bank before the due date to obtain cash. Consider the following example:

EXAMPLE 2 A note of $1,500 due in 6 months at 6% simple interest is discounted at a bank after 2 months of the 6 months has passed. Find the bank discount and the proceeds if the rate of discount used by the bank is 7%.

SOLUTION First we find the maturity value of the note. Substituting the values $P = 1{,}500$, $r = 0.06$, $t = \frac{1}{2}$ in the formula $A = P(1 + rt)$, we get for the maturity value

$$A = 1{,}500[1 + (0.06)(\tfrac{1}{2})] = 1{,}500(1.03) = \$1{,}545$$

We now proceed as in Example 1. On substituting the values $A = 1{,}545$, $d = 0.07$, $t = \frac{1}{3}$ in formula (4.16), we obtain for the bank discount

$$D_b = Adt = 1{,}545(0.07)(\tfrac{1}{3}) = \$36.05$$

The proceeds are given by

$$P_b = A - D_b = \$1{,}545 - \$36.05 = \$1{,}508.95$$

The following question is frequently asked: What is the rate of interest earned by the bank in performing a transaction at a given discount rate? In Example 1 above, A was permitted the use of \$921.60 for 8 months under the condition that he pay the bank \$960 at the end of the period. To determine the interest rate corresponding to the given discount rate, we must find that rate of interest at which \$921.60 will amount to \$960 in $\frac{2}{3}$ of a year. Substituting the values $P = 921.60$, $A = 960$, $t = \frac{2}{3}$ in the formula $A = P + Prt$, we get

$$960 = 921.60 + (921.60)(\tfrac{2}{3})r$$

Solving for r, we get

$$r = \frac{38.40}{614.40} = \frac{1}{16} = 0.0625 = 6.25\%$$

In this transaction the bank discount rate is 6%, but the bank actually received simple interest at the rate of 6.25%.

We now use similar reasoning to derive a formula for simple interest rate r in terms of the bank discount rate d. In general, the simple interest rate r and the bank discount rate d are said to be *corresponding* for a given value of the time t when each leads to the same accumulated value A of P_b in t years. If A is the accumulated value of P_b for t years at simple interest rate r (as shown in the following line diagram),

$$P_b \xleftrightarrow{t \text{ years at simple interest rate } r} A$$

then by (4.14) we have

$$A = P_b(1 + rt) \tag{4.18}$$

Further, if the amount A is discounted at the rate d for t years to give the proceeds P_b, then, by (4.17),

$$A = \frac{P_b}{1 - dt} \tag{4.19}$$

Equating the values of A given by formulas (4.18) and (4.19), we get

$$P_b(1 + rt) = \frac{P_b}{1 - dt}$$

It follows that

$$1 + rt = \frac{1}{1 - dt}$$

or

$$rt = \frac{1}{1 - dt} - 1 = \frac{1 - 1 + dt}{1 - dt} = \frac{dt}{1 - dt}$$

Therefore, we have

$$r = \frac{d}{1 - dt} \qquad (4.20)$$

as a general formula for the interest rate r which corresponds to the discount rate d.

It is left as an exercise for the student to solve formula (4.20) for d and obtain the formula

$$d = \frac{r}{1 + rt} \qquad (4.21)$$

EXAMPLE 3 What simple interest rate corresponds to a discount rate of 8% if the period of discount is 3 months?

SOLUTION Here $d = 0.08$, and $t = \frac{1}{4}$. Substituting in formula (4.20), we have

$$r = \frac{0.08}{1 - (0.08)(\frac{1}{4})} = \frac{4}{49} = 0.0816 = 8.16\% \qquad \textit{approximately}$$

EXAMPLE 4 A note is discounted at a bank. If the note has 90 days to run, what should be the discount rate in order for the bank to earn 6% simple interest on the transaction?

SOLUTION Substituting $r = 0.06$ and $t = \frac{1}{4}$ in formula (4.21), we obtain

$$d = \frac{0.06}{1 + (0.06)(\frac{1}{4})} = \frac{12}{203} = 0.0591 = 5.91\% \qquad \textit{approximately}$$

EXERCISES

1 A non-interest-bearing note due in 90 days was discounted by a bank at a discount rate of 4%. The proceeds of the note were \$2,970. Find its maturity value.

2 Sam Jones goes to a bank and requests a loan of $1,000 for a period of 6 months. The bank's discount rate is 7% per year. What is the bank discount?

3 A note for $2,000 bears interest at 4% and is due in 60 days. It is discounted 25 days before it is due by a bank at a discount rate of 6%. Find the proceeds.

4 A note of $400 due in 6 months with interest at 5% is discounted by a bank 3 months before the note was due so as to yield proceeds of $401.80. Find the discount rate.

Using the method of Example 2 in this section, find the bank discount and the proceeds for the notes in Exercises 5 to 16.

	FACE VALUE OF NOTE	PERIOD OF NOTE	INTEREST RATE, %	PERIOD OF DISCOUNT	DISCOUNT RATE, %
5	$2,000	60 days	4	25 days	6
6	1,900	120 days	4	60 days	$4\frac{1}{2}$
7	1,000	60 days	3	25 days	4
8	5,500	60 days	None	58 days	4
9	500	9 months	None	7 months	5
10	500	9 months	6	7 months	5
11	375	6 months	None	97 days	6
12	375	6 months	5	97 days	6
13	850	8 months	5	6 months	6
14	4,250	4 months	4	3 months	5
15	800	1 year	4	1 year	4
16	800	1 year	None	6 months	8

17 A bank wants to earn interest at the rate of 8%. At what discount rate expressed as a percent to hundredths should the bank discount a note for a period of 90 days?

18 A student received $188 cash as the proceeds of a note for 1 year, at the end of which period he agrees to pay the bank $200. Find the bank's discount rate. What is the corresponding simple interest rate?

19 In formula (4.20), let $d = 0.06$, and compute the value of r for $t = 1$ month; 2 months; 3 months; 4 months; 6 months; 1 year. In each case express r as a percent to hundredths.

20 In formula (4.21), let $r = 0.06$, and compute the value of d for $t = 1$ month; 2 months; 3 months; 4 months; 6 months; 1 year. In each case express d as a percent to hundredths.

4.7 FRACTIONAL EQUATIONS

FRACTIONAL EQUATION

An equation which has its variable in the denominator of one or more fractions appearing in the equation is called a **fractional equation.** We shall consider the type of fractional equation with the property that the equation obtained by multiplying each side by the least common denominator of the fractions is an equation of first degree with one variable.

When each side of an equation is multiplied by an expression containing the variable, the new equation *may not* be equivalent to the original equation. This is because the solution set of the new equation may contain numbers that are not members of the solution set of the original equation, or the solution set of the new equation may fail to contain some numbers that are members of the solution set of the original equation. Therefore the members of the solution set of the new equation (called the *derived equation*) should be tested to see whether or not they do satisfy the original equation. Examples 1 and 2 illustrate the procedure to be used in such cases.

EXAMPLE 1 Solve the equation

$$\frac{36-4x}{x^2-9} - \frac{2+3x}{3-x} = \frac{3x-2}{x+3} \tag{1}$$

in the field of rational numbers.

SOLUTION We first rewrite the equation, changing the sign of the second fraction on the left side and the sign of the denominator of that fraction. We get

$$\frac{36-4x}{x^2-9} + \frac{3x+2}{x-3} = \frac{3x-2}{x+3} \tag{2}$$

Since $x^2 - 9 = (x-3)(x+3)$, it is apparent that $x^2 - 9$ is the least common denominator of the fractions appearing in equation (2). Multiplying each side of equation (2) by $x^2 - 9$, we obtain

$$36 - 4x + (x+3)(3x+2) = (x-3)(3x-2)$$

which, upon expanding the products and combining like terms, becomes

$$42 + 7x + 3x^2 = 3x^2 - 11x + 6 \tag{3}$$

Equation (3) is the *derived equation,* and by the use of the operations described in Sec. 4.2 we find that it is equivalent to

$$18x = -36 \qquad \text{and to} \qquad x = -2$$

Therefore, $\{-2\}$ is the solution of the derived equation (3).

TESTING OF POSSIBLE SOLUTION The number -2 is the member of the solution set of equation (3), but it may not satisfy the original equation (1). Substituting -2 in place of x in equation (1), we ask:

Is $\dfrac{36+8}{4-9} - \dfrac{2-6}{3+2} = \dfrac{-6-2}{-2+3}$ a true statement?

Does $\dfrac{44}{-5} - \dfrac{-4}{5} = \dfrac{-8}{1}$?

Does $-\frac{44}{5} + \frac{4}{5} = -8$?

The answer is *yes*. So $\{-2\}$ is the solution of equation (1).

EXAMPLE 2 Solve the equation

$$\frac{2}{x-2} - \frac{1}{x-1} = \frac{1}{x^2 - 3x + 2} \qquad (1)$$

in the field of rational numbers.

SOLUTION Observe that $x^2 - 3x + 2 = (x-2)(x-1)$ is the least common denominator of the fractions in equation (1). Multiplying each side of equation (1) by $(x-2)(x-1)$, we obtain

$$2(x-1) - (x-2) = 1$$

which, upon expanding the products and combining like terms, becomes

$$x = 1 \qquad (2)$$

Equation (2) is the derived equation, and its solution is $\{1\}$. We must now inquire whether the number 1 satisfies the original equation (1).

TESTING OF POSSIBLE SOLUTION Putting 1 in place of x in equation (1), we are led to ask:

Is $\frac{2}{1-2} - \frac{1}{1-1} = \frac{1}{1-3+2}$ a true statement?

It is impossible for this to be so, for we are asked to consider the nonpermissible division by zero in the second fraction on the left side and in the fraction on the right side. Therefore the statement cannot be true, and 1 is *not* a member of the solution set of equation (1). Indeed, the solution of equation (1) is the empty set.

Some stated problems lead to fractional equations, especially those called *work problems* and *rate problems*. To illustrate the treatment of such problems, we give an example of each kind.

EXAMPLE 3 A can do a piece of work in 6 days, and B can do the same piece of work in 3 days. How long will it take A and B working together to do the work?

SOLUTION Since A can do the work in 6 days, in 1 day he will do $\frac{1}{6}$ of it. Similarly, since B can do the work in 3 days, in 1 day he will do $\frac{1}{3}$ of it.

Let $x =$ the number of days required by A and B working together to do the work. Then $\frac{1}{x} =$ the part of the work they can do together in 1 day.

The basic equality for this problem is

$$\begin{pmatrix}\text{part of work}\\ A \text{ can do in}\\ \text{1 day}\end{pmatrix} + \begin{pmatrix}\text{part of work}\\ B \text{ can do in}\\ \text{1 day}\end{pmatrix} = \begin{pmatrix}\text{part of work}\\ A \text{ and } B \text{ together}\\ \text{can do in 1 day}\end{pmatrix} \qquad (1)$$

In terms of x the equation corresponding to the equality (1) is $\frac{1}{6} + \frac{1}{3} = \frac{1}{x}$.
The derived equation is $x + 2x = 6$, or

$$3x = 6$$

and the solution of this equation is $\{2\}$. Observe that 2 satisfies the original equation. An alternative way of obtaining the solution is to note that

$$\frac{1}{x} = \frac{1}{6} + \frac{1}{3} = \frac{1}{2}$$

It follows that the solution is $\{2\}$. Therefore, both men working together can do the work in 2 days.

EXAMPLE 4 On a river whose current flows at the rate of 3 miles per hour, a motorboat takes as long to travel 12 miles downstream as to travel 8 miles upstream. At what rate could the boat travel in still water?

SOLUTION Let x miles per hour = rate of the boat in still water. Then

$(x + 3)$ miles per hour = rate of boat downstream

and

$(x - 3)$ miles per hour = rate of boat upstream

The underlying equality of this stated problem is

$$\begin{pmatrix}\text{time required to}\\ \text{travel 12 miles}\\ \text{downstream}\end{pmatrix} = \begin{pmatrix}\text{time required to}\\ \text{travel 8 miles}\\ \text{upstream}\end{pmatrix} \tag{1}$$

From the basic formula for uniform motion, $d = rt$, we get

$$t = \frac{d}{r} \tag{2}$$

Using (2) and the expressions we obtained above for the rate downstream and for the rate upstream, we have

$$\begin{pmatrix}\text{time required to}\\ \text{travel 12 miles}\\ \text{downstream}\end{pmatrix} = \frac{12}{x+3} \qquad \begin{pmatrix}\text{time required to}\\ \text{travel 8 miles}\\ \text{upstream}\end{pmatrix} = \frac{8}{x-3}$$

Hence the equation for the problem is

$$\frac{12}{x+3} = \frac{8}{x-3}$$

Multiplying both sides of this equation by $(x + 3)(x - 3)$ to clear it of fractions, we get

$$12(x - 3) = 8(x + 3) \qquad \text{or} \qquad 12x - 36 = 8x + 24$$

This derived equation is equivalent to

$$4x = 60 \qquad \text{and to} \qquad x = 15$$

Note that 15 satisfies the original equation. Hence the boat travels 15 miles per hour in still water.

EXERCISES

Solve the following equations. The universe of the variable in each equation is the set of all rational numbers for which no denominator is zero. Remember to test the solution of the derived equation to see whether or not it is a solution of the given equation.

1 $\dfrac{4}{3x} - \dfrac{3}{4x} = \dfrac{7}{16}$

2 $\dfrac{8}{x-4} = 3$

3 $\dfrac{3}{x} - 1 = \dfrac{3}{2}$

4 $\dfrac{x-1}{x+3} = \dfrac{x-2}{x+1}$

5 $\dfrac{4}{x-1} = \dfrac{3}{x-2}$

6 $\dfrac{3}{2x} + \dfrac{7}{16} = \dfrac{4}{3x}$

7 $\dfrac{x+3}{x} = \dfrac{x+9}{x+4}$

8 $\dfrac{2x+3}{5x-1} = \dfrac{2x-7}{5x+6}$

9 $\dfrac{4}{x^2-4} = \dfrac{1}{x-2} + \dfrac{1}{x+2}$

10 $\dfrac{x+7}{x+5} - \dfrac{12}{x-5} = 1$

11 $\dfrac{z}{z+2} - \dfrac{z}{z-2} = \dfrac{z+20}{z^2-4}$

12 $\dfrac{4}{7x} + \dfrac{3}{x+7} = \dfrac{4}{x^2+7x}$

13 A can mow a lawn in 20 minutes, and B can do it in 30 minutes. How long does it take the two to mow the lawn together if they use two lawn mowers?

14 A can do a piece of work alone in 45 minutes. With the help of B he can do it in 20 minutes. How long will it take B alone to do the work?

15 One faucet can fill a tub in 28 minutes, and another faucet can fill it in 21 minutes. How long will it take to fill one-half of the tub when both faucets are open?

16 A tank can be filled by one pipe in 2 hours and by another in 3 hours and can be emptied by a third pipe in 6 hours. How long will it take to fill the tank when all three pipes are open?

17 What number must be added to both the numerator and the denominator of the fraction $\frac{3}{5}$ to obtain the fraction $\frac{5}{9}$?

18 In a given fraction the denominator exceeds the numerator by 20. If the numerator and the denominator are decreased by 1, the value of the fraction is $\frac{1}{2}$. Find the given fraction.

19 A working alone can do a job in a days, and B working alone can do the same job in b days. How long will it take A and B working together to do the job?

20 A boat can travel 38 miles downstream in the same time that it takes to go 26 miles upstream. If its rate in still water is 16 miles an hour, what is the rate of the current?

In Exercises 21 to 28, solve the given equation for x, y, or z, whichever appears. Remember to test the solution of the derived equation to see whether or not it is a solution of the given equation.

21 $\dfrac{3}{y-2} - \dfrac{5}{y+2} = \dfrac{2}{y^2-4}$

22 $\dfrac{y-1}{2y+4} - \dfrac{y+3}{2y-4} = \dfrac{2}{y^2-4}$

23 $\dfrac{z+a}{z-a}=\dfrac{b+a}{b-a}$

24 $\dfrac{2x+1}{2}+\dfrac{x^2}{8-2x}=\dfrac{8}{4-x}$

25 $\dfrac{z}{4-z}+3=\dfrac{5+4z}{2z}$

26 $\dfrac{14-y^2}{y^2-7y+12}=\dfrac{y+2}{y-4}-\dfrac{2y+3}{y-3}$

27 $\dfrac{x-2}{x-3}+\dfrac{x}{x-2}=\dfrac{2x^2}{x^2-5x+6}$

28 $\dfrac{3}{x+1}-\dfrac{1}{x-2}=\dfrac{2}{x+2}$

29 What number should be added to the numerator and the denominator of the fraction $\frac{4}{7}$ to obtain the reciprocal of $\frac{4}{7}$?

30 What are the two numbers whose sum is 70 and whose quotient is $\frac{3}{2}$?

31 The inlet pipe to a tank can fill the tank in 45 minutes, and the outlet pipe can empty it in 30 minutes. If both pipes are opened when the tank is full, how long will it be before the tank is emptied?

32 A boat can travel 40 miles downstream in the same time it takes to go 30 miles upstream. If its rate in still water is 20 miles per hour, what is the rate of the current?

33 The sum of two numbers is 37. If the larger is divided by the smaller, the quotient is 3 with remainder 5. Find the numbers.

34 A tank can be filled by one pipe in 3 hours and by another in 4 hours and can be emptied by a third pipe in 12 hours. If all three pipes are opened when the tank is empty, how long will it take to fill the tank?

35 A and B can run around a circular mile track in 6 minutes and 10 minutes, respectively. If they start at the same time from the same place, in how many minutes will they pass each other if they run around the track in the same direction; if they run around the track in opposite directions?

36 A can mow 2 lawns in 35 minutes, and B can mow 3 lawns in 70 minutes. If they work together, using 2 lawn mowers, how long will it take them to mow 5 lawns?

4.8
INEQUALITIES OF THE FIRST DEGREE

For each of the sets N, N_0, I, and Ra that we have considered, we have established an "ordering" of the members in the set by defining what it means to say that one number is greater than another number. For each of the sets of numbers we have considered, the definition of "greater than" can be stated as follows: If a and b are numbers and if there exists a *positive* number c with the property that $a-b=c$, or equivalently, $b+c=a$, we say that a is **greater than** b and b is **less than** a, and write $a>b$ and $b<a$. To illustrate,

GREATER THAN

LESS THAN

$6>2$ because $6-2=4$

$-6<-2$ because $(-2)-(-6)=-2+6=4$

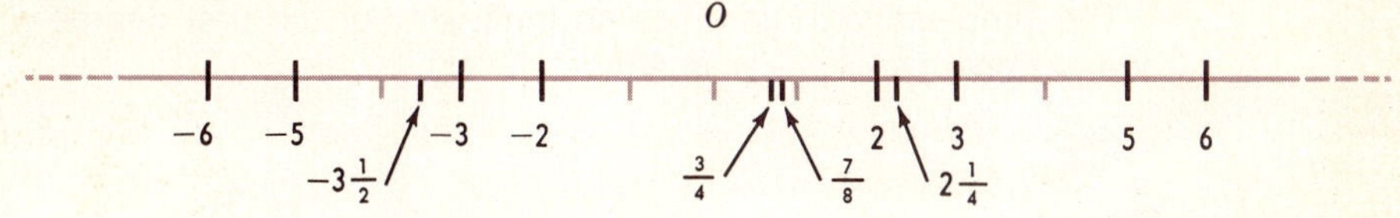

FIGURE 4.7

$3 > -5$	because	$3 - (-5) = 3 + 5 = 8$
$-3 < 5$	because	$5 - (-3) = 5 + 3 = 8$
$\frac{3}{4} < \frac{7}{8}$	because	$\frac{7}{8} - \frac{3}{4} = \frac{7}{8} - \frac{6}{8} = \frac{1}{8}$
$-3\frac{1}{2} < 2\frac{1}{4}$	because	$2\frac{1}{4} - (-3\frac{1}{2}) = \frac{9}{4} + \frac{7}{2} = \frac{23}{4}$

As we have seen, every number that we have considered can be represented by a point on the number scale; see Figs. 3.6, 2.1, and 1.13. If a is greater than b, that is, if $a > b$, then, on the number scale, the point corresponding to a lies to the *right* of the point corresponding to b. The points corresponding to the numbers in the above examples are shown in Fig. 4.7.

ALGEBRAIC INEQUALITY

An **algebraic inequality** is a statement that one algebraic expression is greater than* (or less than) another algebraic expression, where the symbol $>$ (or $<$) is used to stand for the words "is greater than" (or "is less than"). The two expressions are called sides of the inequality. In the inequality $a > b$ we speak of a as the left side and b as the right side. In this section we shall consider open sentences containing one variable that are in the form of algebraic inequalities or are compound sentences whose components are inequalities and in which the only nonzero power of the variable that appears is the first power. Such a sentence is called an *inequality of the first degree*. Some examples are

$x > 7$ $\quad x + \frac{5}{3} < \frac{1}{3}x + 2$ $\quad 4x > 12$
$x - 3 < 5$ $\quad 6x - 21 + \frac{3}{2}x > 17 - 4(x + 1)$
$x > 5$ and $x < 16$ $\quad x > 2$ or $x < -5$

To *solve an inequality* of the first degree means to find the members of the universe of the variable that satisfy the inequality; or more briefly, to solve an inequality means to find the solution set of the inequality. To illustrate, *if the universe of the variable is the set N_0 of nonnegative integers,*

The solution of $x < 5$ is $\{0, 1, 2, 3, 4\}$

The solution of $x - 3 < 3$ is $\{0, 1, 2, 3, 4, 5\}$

The solution of $6x < 20$ is $\{0, 1, 2, 3\}$

* Also a statement, using the symbol $\geqslant$ (or $\leqslant$), that one algebraic expression is greater than or equal to (or less than or equal to) another algebraic expression is called an algebraic inequality.

The simplest form in which an inequality of the first degree with one variable can appear is either

$$x > a \tag{4.22}$$

or

$$x < b \tag{4.23}$$

The solution of $x > a$ in the universe U is the set S of all members of U that are *greater than a;* that is, the solution of (4.22) is the set

$$S = \{x \in U \mid x > a\} \tag{4.24}$$

Similarly, the solution of $x < b$ in the universe U is the set T of all members of U that are *less than b;* that is, the solution of (4.23) is the set

$$T = \{x \in U \mid x < b\} \tag{4.25}$$

To illustrate, the solution of $x > 9$ in the universe of rational numbers is

$$\{x \in Ra \mid x > 9\}$$

or "the set of all rational numbers greater than 9." The solution of $x < 6$ in the universe of positive integers is "the set of all positive integers less than 6," or

$$\{x \in N \mid x < 6\} = \{1, 2, 3, 4, 5\}$$

However, most inequalities cannot be solved so easily, and it is helpful to have a formal procedure to aid in finding solutions of inequalities. The most common procedure used is similar to the one used in solving equations and is based on the concept of *equivalent inequalities.* Two inequalities are said to be **equivalent** if they have the same solution set. For example, if the universe of the variable is the set N of positive integers, the inequalities $x < 7$ and $2x < 14$ are equivalent, because

EQUIVALENT INEQUALITIES

$$\{x \in N \mid x < 7\} = \{x \in N \mid 2x < 14\} = \{1, 2, 3, 4, 5, 6\}$$

Clearly, as is the case for equations, if each of two inequalities is equivalent to a third inequality, the two inequalities are equivalent to each other.

In order to find the solution of a given first-degree inequality in one variable, we replace the given inequality by an equivalent inequality of a simple enough form that its solution set is intuitively apparent. To use this procedure we need to know what operations can be performed on a given inequality to produce an inequality that is equivalent to the given inequality. We state below three operations that *always* produce an inequality that is equivalent to the inequality with which we start. We also give a symbolic form of the operations, using the symbols a, b, and c to represent *algebraic expressions*. The fact that these operations always produce equivalent inequalities is a consequence of the properties O_1, O_2, O_3, and O_4 of the order relation given in Secs. 2.6 and 3.12.

1 *The same quantity can be added to or subtracted from each side of an inequality to produce a new inequality that is equivalent to the original inequality.* Symbolically,

$$a > b \quad \text{if and only if} \quad a + c > b + c \tag{4.26}$$

To illustrate, the inequalities $2x - 3 > 6 - x$ and $3x > 9$ are equivalent, since the second inequality is obtained by adding $3 + x$ to each side of the first inequality. Also, the inequalities $3x + 6 < 2x - 2$ and $x < -8$ are equivalent, since $x < -8$ is obtained when $2x + 6$ is subtracted from each side of $3x + 6 < 2x - 2$.

2 *Each side of an inequality can be multiplied by or divided by the same* **positive** *quantity* to produce a new inequality that is equivalent to the original inequality.* Symbolically,

$$\textit{for } c > 0, \quad a > b \quad \text{if and only if} \quad ac > bc \tag{4.27}$$

To illustrate, the inequalities $2x > 12$ and $6x > 36$ are equivalent, since the second inequality is obtained when both sides of the first inequality are multiplied by 3. Also, the inequalities $5x < 35$ and $x < 7$ are equivalent, since $x < 7$ is obtained when each side of $5x < 35$ is divided by 5.

3 *If each side of an inequality is multiplied by or divided by the same* **negative** *quantity* and if the sense of the inequality is reversed, the new inequality is equivalent to the original inequality.* Symbolically,

$$\textit{for } c < 0, \quad a > b \quad \text{if and only if} \quad ac < bc \tag{4.28}$$

To illustrate, the inequalities $2x > -9$ and $-2x < 9$ are equivalent, since the second inequality is obtained by multiplying each side of $2x > -9$ by -1 and reversing the sense of the inequality. Also $-3x < 9$ and $x > -3$ are equivalent, since the second inequality is obtained by dividing each side of the first inequality by -3 and reversing the sense of the inequality.

* It is to be understood that this quantity does not contain the variable. In cases in which the multiplier or divisor does contain the variable, the new inequality will not, in general, be equivalent to the original inequality.

We should be especially careful in using the operations symbolized by (4.27) and (4.28). The operation symbolized by (4.28) states that when each side of an inequality is multiplied by or divided by a *negative* quantity, the *sense* of the inequality *must be reversed* if the new inequality is to be equivalent to the original inequality.

The statements (4.26) to (4.28) constitute a basis for solving inequalities of the first degree with one variable. We shall see how these statements can be used to find an inequality of the form $x > a$ or $x < b$ that is equivalent to a given first-degree inequality with one variable.

EXAMPLE 1 Find the solution of

$$x - 5 > 14 \tag{4.29}$$

if the universe of the variable is the set N of positive integers.

SOLUTION We see that if 5 is added to the left side of (4.29), that side becomes simply x. Therefore, we use the operation symbolized by (4.26) and add 5 to each side of (4.29) to obtain the inequality $x > 19$, which is equivalent to the inequality (4.29). Since the solution of $x > 19$, in the set N of positive integers, is the set of all positive integers greater than 19, we have that the solution of $x - 5 > 14$ in the set of positive integers is

$$\{x \in N \mid x > 19\}$$

EXAMPLE 2 If the universe of the variable is the set Ra of rational numbers, find the solution of

$$\tfrac{2}{3}x < \tfrac{1}{4} \tag{4.30}$$

SOLUTION We see that if we multiply $\frac{2}{3}x$ by $\frac{3}{2}$ (the reciprocal of $\frac{2}{3}$), we obtain simply x. So, since $\frac{3}{2}$ is positive, we use (4.27) to find that $x < \frac{3}{2} \cdot \frac{1}{4}$ is equivalent to (4.30), or

$$\{x \in Ra \mid \tfrac{2}{3}x < \tfrac{1}{4}\} = \{x \in Ra \mid x < \tfrac{3}{2} \cdot \tfrac{1}{4}\}$$

Since

$$\frac{3}{2} \cdot \frac{1}{4} = \frac{3}{8}$$

we have

$$\{x \in Ra \mid \tfrac{2}{3}x < \tfrac{1}{4}\} = \{x \in Ra \mid x < \tfrac{3}{8}\}$$

and we have found the solution of (4.30) to be the set of rational numbers less than the rational number $\frac{3}{8}$.

EXAMPLE 3 If the universe of the variable is the set N_0 of nonnegative integers, solve the inequality

$$x + 9 - 9x > 2x + 2 - 7x + 1 \tag{4.31}$$

SOLUTION Combining like terms, we get

$$-8x + 9 > -5x + 3$$

We add $5x$ to each side and obtain

$$-3x + 9 > 3$$

Next, we subtract 9 from each side and get

$$-3x > -6$$

Since we have used only the operation symbolized by (4.26) to arrive at this inequality, we know that $-3x > -6$ is equivalent to the inequality (4.31). Dividing each side of $-3x > -6$ by -3 and reversing the sense of the inequality (since $-3 < 0$), we find by use of (4.28) that

$$x < 2$$

is an inequality equivalent to $-3x > -6$ and hence equivalent to (4.31). Therefore, the solution of

$$x + 9 - 9x > 2x + 2 - 7x + 1$$

in the set N_0 of nonnegative integers is the collection of nonnegative integers less than 2, or

$$\{x \in N_0 \mid x < 2\} = \{0, 1\}$$

From an examination of the preceding examples we note that although for first-degree *equations* we can always list or tabulate the members of the solution set, we cannot in general do so for the solution set of first-degree *inequalities*. Quite often the solution set of an inequality is an infinite* set. Therefore, in general it is not possible to check the solution of an inequality, as we do for the solution of an equation, by substitution of the members of the solution set in place of the variable in the given open sentence. However, a partial check of the solution can be made by selecting two or three members of the solution set and verifying that the given inequality is true for these numbers.

Frequently it is helpful to represent the solution of an equation or inequality graphically by the use of the number scale for rational numbers. To illustrate, the solution set $\{x \in N \mid x > 19\}$ of inequality (4.29) in Example 1 can be represented as shown in Fig. 4.8, where it is understood that the graph extends indefinitely

FINITE SET
INFINITE SET

* If the number of members in a set can be expressed as a positive integer, or if a set is the empty set, it is called a *finite set;* otherwise it is called an *infinite set.*

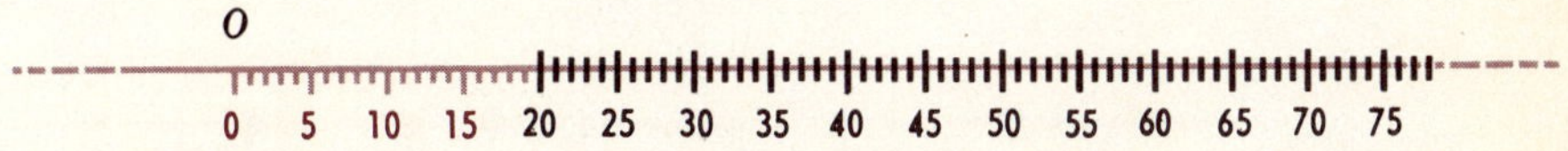

0

0 5 10 15 20 25 30 35 40 45 50 55 60 65 70 75

FIGURE 4.8

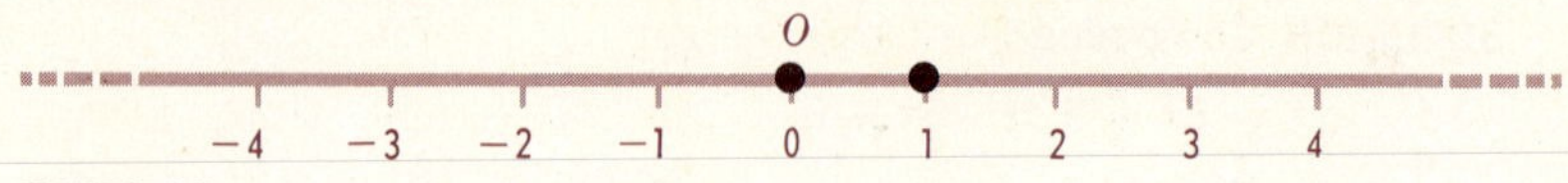

FIGURE 4.9

to the right. The solution set $\{x \in N_0 \mid x < 2\} = \{0, 1\}$ of the inequality (4.31) in Example 3 can be represented as shown in Fig. 4.9.

The representation of $\{x \in Ra \mid x < \frac{3}{8}\}$, the solution set of inequality (4.30) in Example 2, on the number scale requires some additional comment. This set is composed of all the rational numbers less than $\frac{3}{8}$. We recall that to each rational number there corresponds a point on the number scale and that between any two distinct rational numbers there is always another rational number. Hence (see Fig. 4.10), if two points A and B on the number scale represent rational numbers, there is always a third point C between A and B that represents a rational number. Then there is a point D between C and B that represents a rational number, and a point E between A and C that represents a rational number, and so on. Thus, if we were to attempt to indicate all the points on the number scale between A and B that represent rational numbers, very soon the dots representing these points would run together and it would *appear* that *every* point between A and B represents a rational number. For example, if we were asked to represent graphically all the rational numbers between 1 and 3 on the number scale, the representation would appear as in Fig. 4.11.

As we shall see in the next chapter, there are points on the number scale that do *not* correspond to rational numbers. However, the points corresponding to rational numbers are so "close together" that when we attempt to represent these points by printed dots or dots made by a pencil, we obtain a graph such as that in Fig. 4.11. That is, when we are asked to graph the set of all rational numbers between 1 and 3, we are unable to distinguish between the set of points corresponding to these rational numbers and the set of all points on the line between the points 1 and 3.

EXAMPLE 4 If the universe of the variable is Ra, find the solution of the sentence

$$3x > -6 \quad \text{and} \quad 1 - 3x > x - 11 \tag{4.32}$$

and represent the solution graphically.

FIGURE 4.10

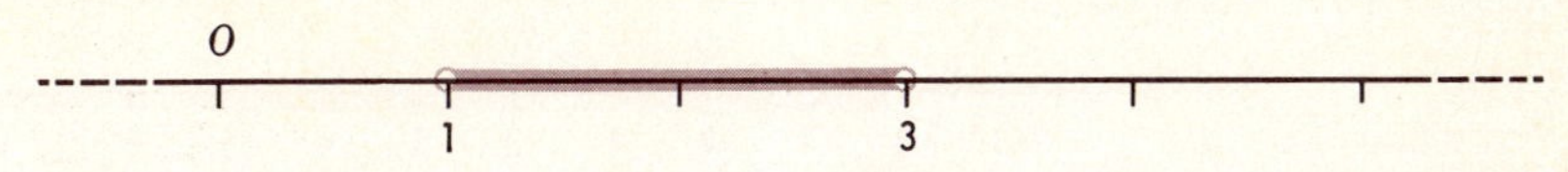

FIGURE 4.11

SOLUTION Note that the sentence (4.32) is in the form of a *conjunction* of two inequalities, and recall from Sec. 2.5 that the solution of a sentence of the form "P_x and Q_x" is the *intersection* of the solution of P_x and the solution of Q_x. Therefore we first find the solution of $3x > -6$ and the solution of $1 - 3x > x - 11$. We see that $3x > -6$ is equivalent to $x > -2$; so the solution of $3x > -6$ is

$$\{x \in Ra \mid x > -2\} \tag{4.33}$$

Also we see that $1 - 3x > x - 11$ is equivalent to $12 > 4x$, which in turn is equivalent to $3 > x$ or $x < 3$. Hence the solution of $1 - 3x > x - 11$ is

$$\{x \in Ra \mid x < 3\} \tag{4.34}$$

Now, by the use of statement (2.42) in Sec. 2.5, we have that the solution of the given sentence in (4.32) is the intersection of the sets given by (4.33) and (4.34):

$$\{x \in Ra \mid -2 < x\} \cap \{x \in Ra \mid x < 3\} \tag{4.35}$$

or

$$\{x \in Ra \mid -2 < x \text{ and } x < 3\}$$

As we pointed out in Sec. 2.5, it is customary to write the compound sentence "$-2 < x$ and $x < 3$" as $-2 < x < 3$. Using this notation, we write the solution of (4.32) as

$$\{x \in Ra \mid -2 < x < 3\} \tag{4.36}$$

recognizing that this designates the same set as (4.35).

To represent (4.36) graphically, we must represent all the rational numbers between -2 and 3, not including either -2 or 3. As we pointed out in the paragraph preceding Example 4, the points representing rational numbers are so close together that the graphical representation of (4.36) will appear as all the points between the points corresponding to -2 and 3. This is shown in Fig. 4.12, where we have used small circles at -2 and 3 to indicate that these points are not included in the solution. Note that the solution shown graphically in Fig. 4.12 is the intersection of the sets indicated in Fig. 4.13(*a*) and (*b*).

EXAMPLE 5 Find the solution of the sentence

$$4x - 9 < -5 \qquad \text{or} \qquad x + 18 > 33 - 2x \tag{4.37}$$

where the universe of the variable is Ra. Represent the solution graphically.

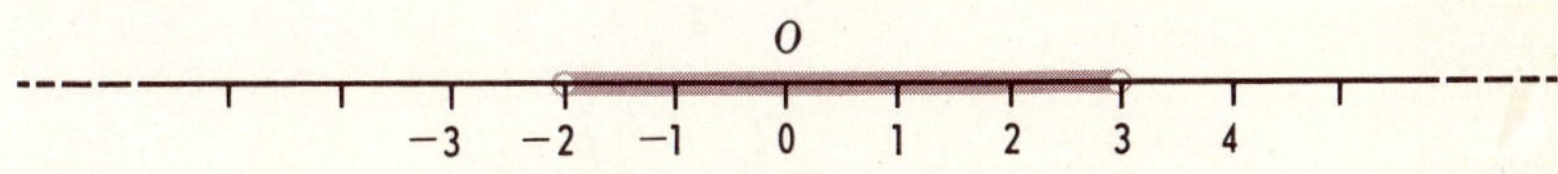

FIGURE 4.12

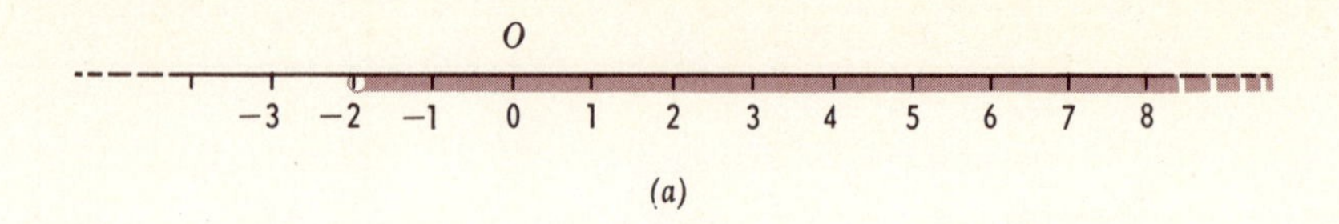

(a)

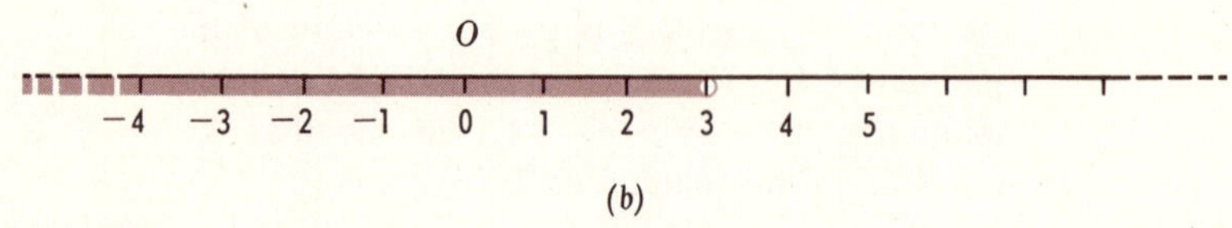

(b)

FIGURE 4.13

SOLUTION The sentence (4.37) is in the form of a disjunction of the two inequalities; so we recall from Sec. 2.5 that the solution of a sentence of the form "P_x or Q_x" is the *union* of the solution of P_x and the solution of Q_x. To find the solution of $4x - 9 < -5$, we note that this inequality is equivalent to $4x < 4$ and hence to $x < 1$. So the solution of $4x - 9 < -5$ is

$$\{x \in Ra \mid x < 1\} \tag{4.38}$$

To find the solution of $x + 18 > 33 - 2x$, we see that this inequality is equivalent to $3x > 15$ and hence to $x > 5$. Therefore the solution of $x + 18 > 33 - 2x$ is

$$\{x \in Ra \mid x > 5\} \tag{4.39}$$

By the use of statement (2.43) in Sec. 2.5, we have that the solution of the given sentence (4.37) is the union of the sets given by (4.38) and (4.39).

$$\{x \in Ra \mid x < 1\} \cup \{x \in Ra \mid x > 5\} \tag{4.40}$$

A graphical representation of the solution (4.40) is given in Fig. 4.14.

EXERCISES

In each part of Exercises 1 and 2, place the appropriate symbol $>$ or $<$ between the given numbers.

1 3, 8; −1, 3; −2, −7; 1.414, 1.413
2 4, 9; −2, 4; −4, −8; −2.691, −2.690

In Exercises 3 to 6, think of x as plotted on the number scale, and express

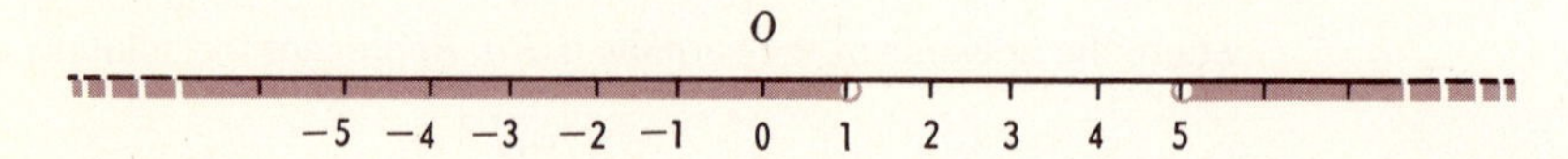

FIGURE 4.14

the given statement by means of an appropriate inequality of the form $x > a$ or $x < b$.

3 x lies to the right of 4
4 x lies to the right of -3
5 x lies to the left of 5
6 x lies to the left of -2

In Exercises 7 to 18, solve the given inequality if the universe of the variable is the set Ra of rational numbers. For each exercise, select four members of your solution set, and verify that the given inequality is true for each of these numbers.

7 $5x - 25 > 0$
8 $6x + 37 > 0$
9 $15 - 6x < 0$
10 $4 + 3x < 10$
11 $5x - 5 > -9 + 3x$
12 $2x - 2 > \frac{8}{3}x - 6$
13 $2x - 9 < 5x - 3$
14 $x - 7 < -4x - 1$
15 $3x - \frac{1}{5} < \frac{5}{3}x + 3$
16 $\frac{1}{4}(x + 10) > \frac{2}{3}x + 5$
17 $2x + 4 < \frac{1}{2}x - 4$
18 $\frac{1}{2}(3x + 7) < \frac{7}{3}x - 4$

19 Represent the solution of Exercise 7 graphically on a number scale.
20 Represent the solution of Exercise 15 graphically on a number scale.
21 Represent graphically the *intersection* of the solution of Exercise 10 and the solution of Exercise 13.
22 Represent graphically the *union* of the solution of Exercise 7 and the solution of Exercise 17.

In each of Exercises 23 to 26, tabulate the given set.

23 $\{x \in I \mid x^2 < 9\}$
24 $\{x \in I \mid 4x^2 < 25\}$
25 $\{x \in N \mid x^2 = 16\}$
26 $\{x \in N_0 \mid x^2 < 16\}$

In Exercises 27 to 32, find the solution of the given sentence for the specified universe of the variable. Represent each solution graphically on the number scale.

27 $2x < 10$ and $6 - x < 3$ universe is I
28 $6x + 1 > 9 - 2x$ and $x > 9 - 2x$ universe is Ra
29 $2x - 1 > 7 - 2x$ or $2x < -4$ universe is Ra
30 $2(x - 4) - x < 0$ or $2x > 20$ universe is I
31 $2x - 1 = 4 + x$ and $2x < x - 4$ universe is Ra
32 $6x = 12 + 2x$ or $x > 4$ universe is Ra

33 We have made repeated use of the fact that for a, b, c rational numbers, *if* $c > 0$ *and if* $a > b$, *then* $ac > bc$. This statement can be proved as follows. By hypothesis, $a > b$. This means that $a - b = p$, where p is a *positive* rational number. We note that $ac - bc = (a - b)c$, and so we have $ac - bc = pc$, where p is a positive number. Since c is positive by hypothesis,

pc is a positive number; that is, $ac - bc$ is a positive number. But this says that $ac > bc$. Use a similar argument to prove that *if* $c < 0$ *and if* $a > b$, *then* $ac < bc$.

4.9
EQUATIONS OF THE FIRST DEGREE WITH TWO VARIABLES

An equation of the form

$$ax + by = c \tag{4.41}$$

where a, b, and c are constants and a and b are not both zero, is called an *equation of the first degree* in the variables x and y. A pair of numbers (x_1, y_1) is said to *satisfy* the equation (4.41) if when x is replaced by x_1 and y is replaced by y_1, equation (4.41) becomes a true statement. The symbol x_1 is read "x sub one," and y_1 is read "y sub one." We call (x_1, y_1) an *ordered pair of numbers.*

ORDERED PAIR

We have an **ordered pair** of numbers when we have two numbers, one of which is designated as the first and the other is designated as the second. Note that the ordered pair (3, 4) is different from the ordered pair (4, 3).

In this section we shall assume that every letter symbol represents a rational number and that the universe of each variable is the set Ra.

SOLUTION OF AN EQUATION WITH TWO VARIABLES

The **solution,** or solution set, of an equation of the first degree with two variables is the set of all ordered pairs of numbers that satisfy the equation. The solution of such an equation is a set with an unlimited number of members (an infinite set). To illustrate, the ordered pair (4, 2) satisfies the equation

$$x - y = 2 \tag{4.42}$$

because upon replacing x by 4 and y by 2 we obtain $4 - 2 = 2$, which is a true statement. So (4, 2) belongs to the solution set of equation (4.42). But there are many other ordered pairs that belong to the solution of (4.42); for example, $(0, -2)$, $(7, 5)$, and $(-9, -11)$ also satisfy the equation. In fact, an ordered pair that satisfies (4.42) may be obtained by assigning a value to either variable and then solving the resulting first-degree equation (which contains only one variable). For instance, if we replace y by 13 in (4.42), we get

$$x - 13 = 2 \tag{4.43}$$

We observe that $\{15\}$ is the solution of (4.43), and so the ordered pair (15, 13) satisfies equation (4.42), as may be verified by substitution.

The determination of a large number of the ordered pairs in the solution of (4.42) is facilitated by first solving this equation for one of the variables in terms of the other. To illustrate, adding y to each side of (4.42), we find that

$$x = y + 2 \tag{4.44}$$

is equivalent to (4.42). Assigning to y any value we please from the universe of the variables, we can determine quickly from (4.44) a value of x such that these *corresponding* values of x and y constitute an ordered pair that satisfies (4.44) and hence that satisfies (4.42). Thus, in (4.44) if we assign to y the value 100, we obtain the equation $x = 100 + 2$; therefore (102, 100) satisfies (4.42).

We found above that the ordered pair (4, 2) satisfies equation (4.42); that is, (4, 2) is a member of the solution set of (4.42). Is (2, 4) a member of the solution of this equation? Substituting 2 in place of x and 4 in place of y in (4.42), we get the statement $2 - 4 = 2$, which is false. Therefore (2, 4) does not satisfy equation (4.42) and is not a member of the solution set of that equation.

The solution set of the equation (4.42) is represented by the symbols

$$\{(x, y) \mid x - y = 2\}$$

This is read "the set of ordered pairs (x, y) for which $x - y = 2$ is true."

Now let us consider the pair of equations

$$\begin{aligned} x - y &= 2 \\ 2x + 3y &= 9 \end{aligned} \tag{4.45}$$

SOLUTION OF A PAIR OF EQUATIONS

An ordered pair of numbers that satisfies *both* equations is said to *satisfy the pair of equations.* The **solution** of such a pair of equations is the set of ordered pairs that satisfy the pair of equations. Thus the solution of the pair of equations (4.45) is symbolized by

$$\{(x, y) \mid x - y = 2 \text{ and } 2x + 3y = 9\}$$

which is the same set as

$$\{(x, y) \mid x - y = 2\} \cap \{(x, y) \mid 2x + 3y = 9\}$$

There are several methods that can be used as aids in finding the solution of a pair of equations of the first degree with two variables. We shall illustrate one of these methods, usually called the *method of elimination by addition and subtraction,* in the following examples.

EXAMPLE 1 Find the solution of the pair of equations

$$x + y = 8 \tag{1}$$

$$x - y = 2 \tag{2}$$

SOLUTION Adding the left side of (2) to the left side of (1) and the right side of (2) to the right side of (1), we obtain the equation

$2x = 10 \quad \text{or} \quad x = 5$

Subtracting the left side of (2) from the left side of (1) and the right side of (2) from the right side of (1), we obtain the equation

$2y = 6 \quad \text{or} \quad y = 3$

The pair of equations

$x = 5$
$y = 3$

is equivalent to the given pair of equations; that is,

$\{(x, y) \mid x + y = 8 \text{ and } x - y = 2\} = \{(x, y) \mid x = 5 \text{ and } y = 3\}$

It is clear that $\{(x, y) \mid x = 5 \text{ and } y = 3\} = \{(5, 3)\}$, and so the solution of the given pair of equations is $\{(5, 3)\}$.

EXAMPLE 2 Solve the pair of equations

$2x + 3y = 7$ (1)
$3x - 2y = 4$ (2)

SOLUTION Here neither the absolute values of the coefficients of x nor the absolute values of the coefficients of y in the two equations are equal, and simply adding or subtracting corresponding sides will not produce an equation that contains only one variable. However, we may proceed as follows.

Multiply each side of equation (1) by 2 to get

$4x + 6y = 14$

and multiply each side of equation (2) by 3 to get

$9x - 6y = 12$

The pair

$4x + 6y = 14$ (3)
$9x - 6y = 12$ (4)

is equivalent to the given pair (1), (2). Now if we add the corresponding sides of (3) and (4), we obtain

$13x = 26 \quad \text{or} \quad x = 2$

Multiply each side of (1) by 3 to get

$6x + 9y = 21$

and multiply each side of (2) by 2 to get

$6x - 4y = 8$

The pair

$6x + 9y = 21$ (5)
$6x - 4y = 8$ (6)

is equivalent to the given pair (1), (2). Now if we subtract equation (6) from equation (5), we obtain

$13y = 13 \qquad \text{or} \qquad y = 1$

We now have found the pair of equations

$x = 2$
$y = 1$

which is equivalent to the given pair of equations (1), (2). Thus

$$\{(x, y) \mid 2x + 3y = 7 \text{ and } 3x - 2y = 4\} = \{(x, y) \mid x = 2 \text{ and } y = 1\} = \{(2, 1)\}$$

is the solution of the given pair of equations.

EXAMPLE 3 Find the solution of the pair of equations

$$2x - 3y = 6 \qquad (1)$$
$$-4x + 6y = -12 \qquad (2)$$

SOLUTION If we multiply each side of (1) by 2, we get $4x - 6y = 12$, and the pair

$$4x - 6y = 12 \qquad (3)$$
$$-4x + 6y = -12 \qquad (4)$$

is equivalent to the given pair of equations (1), (2). Now if we add the corresponding sides of equations (3) and (4), we get

$0 + 0 = 0 \qquad \text{or} \qquad 0 = 0$

The pair

$$4x - 6y = 12 \qquad (5)$$
$$0 = 0 \qquad (6)$$

is equivalent to the pair (3), (4) and hence to the original pair of equations (1), (2).

It is apparent that every ordered pair of rational numbers will satisfy equation (6), since $0 = 0$ is a true statement. Therefore

$$\{(x, y) \mid 2x - 3y = 6 \text{ and } -4x + 6y = -12\}$$
$$= \{(x, y) \mid 4x - 6y = 12 \text{ and } 0 = 0\}$$
$$= \{(x, y) \mid 4x - 6y = 12\} \cap \{(x, y) \mid 0 = 0\}$$
$$= \{(x, y) \mid 4x - 6y = 12\} = \{(x, y) \mid 2x - 3y = 6\}$$

That is, the solution of the given pair of equations (1), (2) is the infinite set of ordered pairs of numbers that satisfy $2x - 3y = 6$. Note that $2x - 3y = 6$ is equivalent to $-4x + 6y = -12$ (multiply each side of the first equation by -2), so that the solution is also the set of ordered pairs that satisfy $-4x + 6y = -12$.

Our procedure as illustrated in Examples 1, 2, and 3 is to endeavor to obtain a pair of equations of the form

$x = k$
$y = m$

which is equivalent to the given pair of equations. If this can be done, as it was in Examples 1 and 2, then the solution of the given pair of equations is

$$\{(x, y) \mid x = k \text{ and } y = m\} = \{(k, m)\}$$

In Example 3 we have a situation in which it is not possible to find a pair of equations of the form $x = k$, $y = m$ that is equivalent to the given pair of equations. In this example the given equations are equivalent to each other, hence the solution set is an infinite set of ordered pairs.

We have given examples in which the solution set of a pair of first-degree equations with two variables contains only one ordered pair of numbers. We have also given an example showing that the solution set of a pair of equations may be an infinite set. Another possibility is illustrated in the following example.

EXAMPLE 4 Find the solution of the pair of equations

$$2x - 5y = 1 \tag{1}$$
$$6x - 15y = 9 \tag{2}$$

SOLUTION Multiplying each side of equation (1) by 3, we obtain $6x - 15y = 3$, and the pair of equations

$$6x - 15y = 3 \tag{3}$$
$$6x - 15y = 9 \tag{4}$$

is equivalent to the given pair of equations. Subtracting equation (3) from equation (4), we get

$$0 = 6$$

The pair

$$6x - 15y = 9$$
$$0 = 6$$

is equivalent to the original pair of equations, that is,

$$\begin{aligned} \{(x, y) \mid 2x - 5y = 1 \text{ and } 6x - 15y = 9\} \\ &= \{(x, y) \mid 6x - 15y = 9 \text{ and } 0 = 6\} \\ &= \{(x, y) \mid 6x - 15y = 9\} \cap \{(x, y) \mid 0 = 6\} \end{aligned}$$

Since $0 = 6$ is a false statement, we see that

$$\{(x, y) \mid 0 = 6\} = \varnothing$$

and therefore the solution of the given pair of equations is

$$\{(x, y) \mid 6x - 15y = 9\} \cap \varnothing = \varnothing$$

There are no ordered pairs of numbers that satisfy the given pair of equations.

A graphical interpretation of the solution of a pair of equations of the first degree with two variables is given in Sec. 7.5. This interpretation will shed light on the case in which the solution of such a pair of equations is the empty set.

From the examples of this section we conclude that the solution set of a given pair of equations of the first degree with two variables may contain one ordered pair, an unlimited number of ordered pairs, or no ordered pairs.

Observe that in the process illustrated in the preceding examples the numbers used to multiply each side of the given equations are chosen so that, in the resulting equivalent pair of equations, the coefficients of one of the variables either are equal or are the negatives of each other.

Many stated problems can be solved by making use of a pair of equations with two variables.

EXAMPLE 5 The sum of two numbers is 32, and their difference is 4. Find the numbers.

SOLUTION Let x represent one number, and let y represent the other number. Then

$$x + y = 32 \qquad \text{and} \qquad x - y = 4$$

Adding corresponding sides of these two equations we get $2x = 36$ or $x = 18$; subtracting corresponding sides of the original equations we get $2y = 28$ or $y = 14$. The numbers are 18 and 14.

EXAMPLE 6 A boy has 25 coins consisting of nickels and dimes. If the value of his money is \$2.10, find the number of nickels and the number of dimes that the boy has.

SOLUTION The stated problem has two equalities in it, namely,

$$\begin{pmatrix}\text{number}\\ \text{of nickels}\end{pmatrix} + \begin{pmatrix}\text{number}\\ \text{of dimes}\end{pmatrix} = \begin{pmatrix}\text{total number}\\ \text{of coins}\end{pmatrix}$$

$$\begin{pmatrix}\text{value of}\\ \text{nickels}\end{pmatrix} + \begin{pmatrix}\text{value of}\\ \text{dimes}\end{pmatrix} = \begin{pmatrix}\text{total value}\\ \text{of coins}\end{pmatrix}$$

Let x represent the number of nickels and y represent the number of dimes. Then we have the pair of equations

$$x + y = 25$$
$$0.05x + 0.10y = 2.10$$

which is equivalent to the pair

$$x + y = 25$$
$$5x + 10y = 210$$

and to the pair

$$x + y = 25$$
$$x + 2y = 42$$

We find that $\{(8, 17)\}$ is the solution of the last pair of equations, so there are 8 nickels and 17 dimes.

EXERCISES

Solve the following pairs of equations:

1 $4x + 7y = 23$
$x - 7y = 21$

2 $5m - 3n = 12$
$2m - 3n = 3$

3 $4x + 3y = 3$
$10x - 6y = 6$

4 $4x + 7y = 2$
$3x + 5y = 1$

5 $r + s = 24$
$r - s = 37$

6 $15x + 7y = 37$
$9x - 10y = 8$

7 $2.4x + 2y = 11.1$
$x + y = 5$

8 $5x + 10y = 51$
$3x - 15y = 28\frac{1}{2}$

9 $x - y = 3$
$2x + 3y = 16$

10 $y = 2x$
$3x + 2y = 21$

11 $2x + 3y = 7$
$3x + y = 7$

12 $3x + 2y = 0$
$2x + 5y = -11$

13 $2u + 3v = -3$
$3u - 4v = 38$

14 $x = \frac{3}{4}y$
$x - y = 2$

15 $0.01x - 0.02y = 0$
$x - 10y = 8$

16 $\frac{x}{3} = 2y - 14$
$3x - 4y = 0$

17 $x - \frac{1}{2}y = 1$
$\frac{1}{2}x - \frac{3}{4}y = -\frac{5}{2}$

18 $0.04x + 0.02y = 5$
$0.5(x - 2) - 0.4y = 29$

19 $x = 6$
$\frac{1}{2}x - y = 0$

20 $\frac{x}{5} + \frac{y}{2} = 4$
$\frac{x-1}{3} = \frac{y+2}{2}$

Solve the following pairs of equations, where x and y are the variables:

21 $mx + ny = a$
$mx - ny = b$

22 $ax + by = e$
$cx + dy = f$

23 $mx - ny = m^2 + n^2$
$x - y = 2n$

24 $ax - y = a$
$x + ay = 1$

25 The sum of two numbers is 26, and their difference is 8. Find the numbers.

26 The sum of two numbers is 15. Three times the first number added to four times the second number is 55. Find the numbers.

27 Find two numbers whose sum is a and whose difference is b.

28 The cost of 6 bats and 9 balls is \$22.05, and the cost of 8 bats and 6 balls is \$23.10. Find the cost of 1 bat and the cost of 1 ball.

29 A canoeist can row downstream 4 miles in $\frac{1}{2}$ hour and can row upstream 2 miles in $\frac{2}{3}$ hour. What is his rate in still water, and what is the rate of the stream?

30 The sum of the digits of a certain two-digit number is 9. If the digits

were interchanged, the new number would be 9 less than twice the original number. Find the number.

31 An airplane traveled 450 miles in 3 hours with the wind. The return trip for the same distance but against the wind required 5 hours. Find the rate of the airplane in still air and the rate of the wind.

32 A man can row 12 miles downstream in 2 hours, and he can row 9 miles upstream in 3 hours. Find the rate at which he can row in still water and the rate of the current.

33 The length of a rectangle exceeds the width by 7 inches. The perimeter is 44 inches. Find the dimensions of the rectangle.

34 Find two numbers such that the first plus three times the second is 5, and two times the first plus the second is 5.

35 A man is now four times as old as his son. Six years from now the father will be three times as old as the son. Find their present ages.

36 A grocer mixes one kind of coffee, which sells for 85 cents per pound, with another kind, which sells for $1 per pound. How many pounds of each should be used to make 100 pounds of coffee to sell for 95 cents per pound?

37 A motorboat operating at full power during the entire trip goes 12 miles downstream in 2 hours and returns in 3 hours. What would be the speed of the boat in still water, and what is the rate of the current?

5

Further Extensions of the Number System

5.1 THE NEED FOR EXTENDING THE RATIONAL NUMBER SYSTEM

As we have seen in previous chapters (in Sec. 4.2, for example), if a, $b \in Ra$ and $a \neq 0$, then

$$\{x \in Ra \mid ax + b = 0\} = \left\{-\frac{b}{a}\right\}$$

In other words, *in the set of rational numbers,* the solution set of the general first-degree equation $ax + b = 0$, $a \neq 0$, is *never* empty. A similar statement is not true for the general second-degree equation $ax^2 + bx + c = 0$. It may happen that for a, b, $c \in Ra$,

$$\{x \in Ra \mid ax^2 + bx + c = 0\} = \varnothing$$

A simple example of this is given by the equation $x^2 - 2 = 0$. It can be shown that

$$\{x \in Ra \mid x^2 - 2 = 0\} = \varnothing \tag{5.1}$$

This means that there is no rational number whose square is 2. Recall from Sec. 3.10 that if the symbol $\sqrt{k}$ represents a number, it is the number c with the property that $c^2 = k$. So the statement (5.1) is equivalent to the statement that

$$\sqrt{2} \text{ is not a rational number} \tag{5.2}$$

[The truth of statements (5.1) and (5.2) is established in Appendix B.]

We have pointed out that, on a number scale, between any two distinct points that represent rational numbers there is always a third point that represents a rational number. Thus the points representing rational numbers are arbitrarily close together. This property is described by saying that the rational numbers are *dense* on the number line. However, making use of the Pythagorean theorem (discussed in Sec. 5.8) and the fact that $\sqrt{2}$ is not a rational number, we can show that there are points on the number scale that do not represent rational numbers. Let us construct a square whose sides are one unit in length on a number scale, as shown in Fig. 5.1. The diagonal of the square is the hypotenuse of an isosceles right triangle whose legs have length 1. By the Pythagorean theorem the square of the length c of the hypotenuse is 2 and $c = \sqrt{2}$. Drawing a circle with center at O and with radius c we determine a point B on the number line whose distance from the origin is $\sqrt{2}$. B is therefore a point which does not correspond to a rational number.

By an extension of the method used in Appendix B it can be shown that $\sqrt{3}$, $\sqrt{5}$, and many other numbers are not rational numbers,

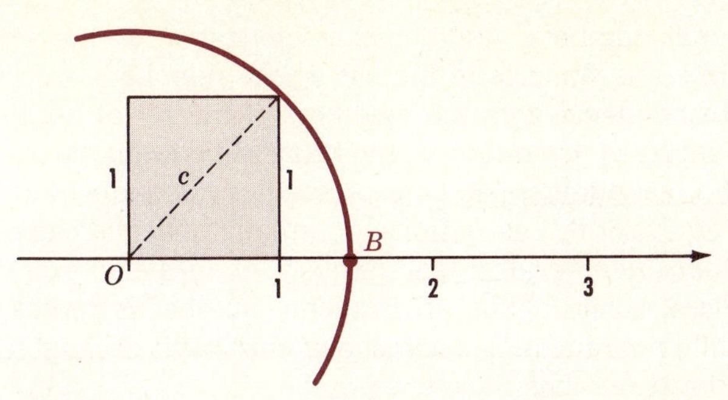

FIGURE 5.1

and by constructions similar to the one used in the preceding paragraph we can show that there are points on the number scale representing these nonrational numbers.

Thus, if we wish to have a set of numbers with the properties that each second-degree equation will have a nonempty solution set and that each point on a line will correspond to a number, the rational number system is inadequate. It is essentially the second of these properties—the desire to have a number that corresponds to each point on a line—that leads to the development of the set of numbers that is called the set of real numbers.

5.2 REAL NUMBERS

We recall that the number scale for rational numbers was obtained by considering a line and setting up a correspondence that associated a point on the line with each rational number. If we agree that we wish to have a number corresponding to each point on the line, the set of numbers that is necessary to accomplish this correspondence is the set of real numbers. In other words, we can characterize the set of **real numbers** as the set whose members can be put in a one-to-one correspondence with the points on a line. The set of real numbers will be denoted by the symbol Re.

SET OF REAL NUMBERS

Another way of thinking about the set of real numbers is by considering the set of all decimals. In Sec. 3.11 we saw that each rational number is represented by a decimal that is either a finite decimal or an infinite repeating decimal. Conversely, it is true that each decimal that is either finite or infinite repeating represents a rational number. So an infinite nonrepeating decimal does not represent a rational number. If we wish to have every decimal (finite, infinite repeating, and infinite nonrepeating) correspond to a number we must extend our set of numbers beyond the set of

Re AS THE SET OF DECIMALS

rational numbers. Another characterization of the set Re of real numbers is that Re is the set whose members are in one-to-one correspondence with the members of the set of decimals. That is, the set Re of real numbers can be identified with the set of decimals.

IRRATIONAL NUMBER

If a real number can be expressed as the quotient of two integers or, equivalently, as a finite or infinite repeating decimal, it is a *rational number.* If a real number *cannot* be so expressed it is an **irrational number;** thus an irrational number is represented by an infinite nonrepeating decimal and each such decimal represents an irrational number.

We have mentioned in Sec. 5.1 three irrational numbers, namely $\sqrt{2}$, $\sqrt{3}$, and $\sqrt{5}$. There are many more irrational numbers. For example, $\sqrt{7}$ and $\sqrt{13}$ are irrational, as is $6.101001000100001 \cdots$, where the decimal is infinite with the rule of formation clearly indicated. Further, if a is a positive integer that is not the square of an integer, then $\sqrt{a}$ and $-\sqrt{a}$ are irrational.

We can never write a finite or a repeating infinite decimal that is equal to a given irrational number. Frequently, however, we are able to find a decimal approximation to an irrational number. For example, if a is a positive integer that is not the square of an integer, we can find a decimal approximation of $\sqrt{a}$ to as many decimal places as we wish. To illustrate, let us find a decimal approximation of $\sqrt{2}$.

Observe that

$$(1.4)^2 = 1.96 \qquad \text{and} \qquad (1.5)^2 = 2.25$$

Hence $\sqrt{2}$ lies between 1.4 and 1.5. Likewise,

$$(1.41)^2 = 1.9881 \qquad \text{and} \qquad (1.42)^2 = 2.0124$$

So $\sqrt{2}$ is between 1.41 and 1.42. Moreover,

$$(1.414)^2 = 1.999396 \qquad \text{and} \qquad (1.415)^2 = 2.002225$$

Therefore, $\sqrt{2}$ lies between 1.414 and 1.415. Continuing in this manner, we can by a trial-and-error process approximate $\sqrt{2}$ to any number of decimal places we wish. From the results that we have just stated, we can say that $\sqrt{2}$ is approximately 1.41, but we cannot state definitely what the digit is in the third decimal place; the above results indicate that most probably it is 4, since $(1.414)^2$ is closer to 2 than $(1.415)^2$.

We say that $\sqrt{2} = 1.4$ *correct to one decimal place* since 1.4 is the closest approximation to $\sqrt{2}$ that can be written by using only one decimal place; that is, $\sqrt{2}$ is closer to 1.4 than it is to 1.3 or to 1.5. Similarly, we say that $\sqrt{2} = 1.41$ *correct to two decimal places* since $\sqrt{2}$ is closer to 1.41 than it is to 1.40 or to 1.42. Further, $\sqrt{2} = 1.414$ *correct to three decimal places.*

The calculation of decimal approximations of square roots is a tedious process. In Sec. 5.7 we give two methods of calculating such approximations. The square roots of many numbers have been calculated and listed in tables for our reference. The table on this page is sufficient for the needs of this course. It gives to three decimal places the square roots of the integers from 1 to 100.

There are many other irrational numbers besides those determined by square roots. For example, $\sqrt[3]{2}$, $\sqrt[3]{4}$, $\sqrt[3]{5}$ are irrational. However, $\sqrt[3]{8} = 2$ and $\sqrt[3]{27} = 3$ are rational. The ratio of the circumference of a circle to its diameter, usually designated by π, is also an irrational number, and there are countless other examples.

Approximate square roots of numbers from 1 to 100

N	$\sqrt{N}$	N	$\sqrt{N}$	N	$\sqrt{N}$	N	$\sqrt{N}$
1	1.000	26	5.099	51	7.141	76	8.718
2	1.414	27	5.196	52	7.211	77	8.775
3	1.732	28	5.292	53	7.280	78	8.832
4	2.000	29	5.385	54	7.348	79	8.888
5	2.236	30	5.477	55	7.416	80	8.944
6	2.449	31	5.568	56	7.483	81	9.000
7	2.646	32	5.657	57	7.550	82	9.055
8	2.828	33	5.745	58	7.616	83	9.110
9	3.000	34	5.831	59	7.681	84	9.165
10	3.162	35	5.916	60	7.746	85	9.220
11	3.317	36	6.000	61	7.810	86	9.274
12	3.464	37	6.083	62	7.874	87	9.327
13	3.606	38	6.164	63	7.937	88	9.381
14	3.742	39	6.245	64	8.000	89	9.434
15	3.873	40	6.325	65	8.062	90	9.487
16	4.000	41	6.403	66	8.124	91	9.539
17	4.123	42	6.481	67	8.185	92	9.592
18	4.243	43	6.557	68	8.246	93	9.644
19	4.359	44	6.633	69	8.307	94	9.695
20	4.472	45	6.708	70	8.367	95	9.747
21	4.583	46	6.782	71	8.426	96	9.798
22	4.690	47	6.856	72	8.485	97	9.849
23	4.796	48	6.928	73	8.544	98	9.899
24	4.899	49	7.000	74	8.602	99	9.950
25	5.000	50	7.071	75	8.660	100	10.000

5.3
PROPERTIES OF THE REAL NUMBER SYSTEM

The student should realize that before the designation "number" can properly be given to the members of Re, there must be a concept of equality and two operations on the set that have the properties, designated by E_1, E_2, E_3, P_1, P_2, P_3, and P_4 in Sec. 1.13, which characterize a number system.

In considering equality and the operations of addition and multiplication (as well as subtraction and division) on the set Re of real numbers, we can of course use the definitions previously given *if* only rational numbers are involved. When irrational numbers are involved it is helpful to think of the set of real numbers as being identified with the set of decimals.

In addition to the properties E_1, E_2, E_3, P_1, P_2, P_3, and P_4 mentioned above, the set Re of real numbers also shares with the set Ra of rational numbers the properties designated by P_5, P_6, P_7, P_8 in Sec. 3.12. That is, *the real number system is a field.*

POSITIVE REAL NUMBER

NEGATIVE REAL NUMBER

A real number a is said to be **positive** if it is represented on the number scale by a point to the *right* of the origin. A real number is said to be **negative** if it is represented on the number scale by a point to the *left* of the origin.

If a and b are real numbers, and if there is a *positive* real number c with the property that

$$a - b = c$$

or equivalently,

$$a = b + c$$

GREATER THAN

we say that a is **greater than** b and write

$$a > b$$

LESS THAN

In this situation we also say that b is **less than** a and write

$$b < a$$

If a is positive, then $a - 0 = a$ is positive; therefore we write $a > 0$ to indicate that a is *positive*. Similarly we write $a < 0$ to indicate that a is *negative*. This definition of "greater than" produces an "ordering" of the set Re, and this "order" has the properties designated by O_1, O_2, O_3, and O_4 in Sec. 3.12.

The results stated in the preceding three paragraphs mean that the real number system is an ordered field, just as the rational number system is an ordered field.

The one characteristic that distinguishes the real number system from the rational number system is the property that every point on the number scale corresponds to a real number and conversely. In other words, *a one-to-one correspondence can be established between the members of the set of real numbers and the points on a line.* The rational number system does not have this property. This property is sometimes called the **completeness property,** or the continuity property. An ordered field that possesses this completeness property is called a **complete ordered field.**

COMPLETENESS PROPERTY

COMPLETE ORDERED FIELD

We list here the properties that characterize the real number system.

For every $a, b, c \in Re$,

E_1: $a = a$

E_2: if $a = b$, then $b = a$

E_3: if $a = b$ and $b = c$, then $a = c$

P_1: $a + b \in Re$, and $a + b$ is unique
$a \cdot b \in Re$, and $a \cdot b$ is unique

P_2: $a + b = b + a$
$a \cdot b = b \cdot a$

P_3: $a + (b + c) = (a + b) + c$
$a \cdot (b \cdot c) = (a \cdot b) \cdot c$

P_4: $a(b + c) = ab + ac$

P_5: There exists a member $1 \in Re$, called the *unit* element, with the property that for any $a \in Re$, $a \cdot 1 = 1 \cdot a = a$.

P_6: There exists a member $0 \in Re$, called the *zero* element, with the property that for any $a \in Re$, $a + 0 = 0 + a = a$.

P_7: If 0 is the zero element described in P_6, and if a is any member of Re, then there exists a number $-a \in Re$ with the property that $a + (-a) = (-a) + a = 0$.

P_8: If 1 is the unit element described in P_5, and if a is any member of Re, except the zero element 0, then there exists a number $a' \in Re$ with the property that $a \cdot a' = a' \cdot a = 1$.

O_1: either $a > b$, or $b > a$, or $a = b$

O_2: if $a > b$ and $b > c$, then $a > c$

O_3: if $a > b$, then $a + c > b + c$

O_4: if $a > b$, then $ac > bc$ *provided* $c > 0$
$ac < bc$ *provided* $c < 0$

CP: A one-to-one correspondence can be established between the members of Re and the points on a line.

All of the other properties and rules of computation of the set Re of real numbers are consequences of the 16 properties listed above that characterize a complete ordered field. We are familiar with many of these properties and rules from our previous experience, and we probably use them automatically without realizing that they are in fact "theorems," that is, statements that can be proved by the use of the properties of a complete ordered field. Some of the more important and frequently used theorems of the real number system are listed below.

In these statements the letter symbols represent members of the set Re of real numbers.

$a = b$ if and only if $a + c = b + c$ (5.3)
$a \cdot b = 0$ if and only if $a = 0$ or $b = 0$ (5.4)
$(-1)a = -a$ (5.5)
$-(-a) = a$ (5.6)
$(-a) \cdot b = -(a \cdot b)$ (5.7)
$(-a) \cdot (-b) = a \cdot b$ (5.8)
$a \cdot a = a^2 \geqslant 0$ (5.9)
For $c \neq 0$, $a = b$ if and only if $ac = bc$ (5.10)
If $a < b$, then there is a *rational* number r such that $a < r < b$. (5.11)
If p and q are rational numbers and $p < q$, then there is a *real* number a such that $p < a < q$. (5.12)
If a is positive, then there is a positive real number k such that $k^2 = a$. (5.13)
If a is a positive real number and n is a positive integer, then there is a positive real number k with the property that $k^n = a$. (5.14)
If a is a negative real number and n is an *odd* positive integer, then there is a negative real number k with the property that $k^n = a$. (5.15)

5.4 RADICALS

As we mentioned in Sec. 5.2, if a is an integer, then $\sqrt{a}$ is not rational *unless* a is the square of an integer. Also the number denoted by the radical $\sqrt[n]{a}$ and called the principal nth root of a (Sec. 3.10) is not rational *unless* a is the nth power of a rational number. We are told by properties (5.14) and (5.15) in Sec. 5.3 that if a is a real number and n is a positive integer, then $\sqrt[n]{a}$ is a real number, subject only to the provision that n be odd when a is negative.

The most commonly encountered irrational numbers are those which can be expressed in terms of nth roots of integers. Very few irrational numbers except those which can be expressed in terms of radicals are likely to be encountered by the average person, even though it has been established that there are an infinite number of irrational numbers that are not expressible in terms of radicals. The number, denoted by π, that expresses the quotient of the length of the circumference of a given circle divided by the length of the diameter of that circle, is an irrational number that cannot be expressed in terms of radicals. The number denoted by e, which is encountered in the study of logarithms and calculus, is another irrational number that cannot be expressed in terms of radicals. The numbers π and e are examples of a class of irrational numbers called *transcendental numbers.*

Because radicals occur so often in the consideration of irrational numbers, it is well to obtain some experience in working with radicals.

Recall from Sec. 3.10 that the radical $\sqrt[n]{a}$ is called the principal nth root of a and is defined by

$$\sqrt[n]{a} = b \tag{5.16}$$

if $b^n = a$, and if b *has the same sign as* a

RADICAL SIGN
RADICAND
ORDER
INDEX

The symbol $\sqrt{\ }$ is the **radical sign,** the number a is the **radicand,** and the number n is the **order,** or *index,* of the radical. Laws of operations with radicals follow from laws of operations with exponents, since

$$\sqrt[n]{a} = a^{1/n} \tag{5.17}$$

The commonly used laws of radicals are stated below.

$$\sqrt[n]{ab} = \sqrt[n]{a}\sqrt[n]{b} \tag{5.18}$$

This equality follows from the fact that the law of exponents $(ab)^p = a^p b^p$ holds for $p = \dfrac{1}{n}$.

Also,

$$\sqrt[n]{\frac{a}{b}} = \frac{\sqrt[n]{a}}{\sqrt[n]{b}} \tag{5.19}$$

since the law of exponents $\left(\dfrac{a}{b}\right)^p = \dfrac{a^p}{b^p}$ holds for $p = \dfrac{1}{n}$.

Further,

$$\sqrt[n]{a^m} = (\sqrt[n]{a})^m = a^{m/n} \tag{5.20}$$

as we discussed in Sec. 3.10.

Finally,

$$\sqrt[m]{\sqrt[n]{a}} = \sqrt[mn]{a} \tag{5.21}$$

since the law of exponents $(a^p)^r = a^{pr}$ holds for $p = \dfrac{1}{n}$ and $r = \dfrac{1}{m}$.

The laws (5.18) to (5.21) are frequently used to change the form of a radical to make it more convenient to use. There is no one form that will be the best for all problems. The most convenient form of the radical will depend on the particular problem. We shall study three methods of changing the form of a radical.

1 REMOVAL OF FACTORS FROM THE RADICAND

In a radical of order n, the radicand should be examined for factors which are perfect nth powers. Such factors can be removed from the radicand by using (5.18). Consider the following illustrations:

$$\sqrt{32} = \sqrt{16 \cdot 2} = \sqrt{16}\sqrt{2} = 4\sqrt{2}$$
$$\sqrt[3]{-81} = \sqrt[3]{(-27)(3)} = \sqrt[3]{-27}\sqrt[3]{3} = -3\sqrt[3]{3}$$
$$\sqrt[3]{(x+y)^5} = \sqrt[3]{(x+y)^3(x+y)^2} = \sqrt[3]{(x+y)^3}\sqrt[3]{(x+y)^2}$$
$$= (x+y)\sqrt[3]{(x+y)^2}$$
$$\sqrt{27a^6b^3} = \sqrt{9 \cdot 3 \cdot a^6 \cdot b^2 \cdot b} = \sqrt{9a^6b^2}\sqrt{3b} = 3a^3b\sqrt{3b}$$

2 CHANGING THE ORDER OF A RADICAL

In a radical of order n, if the radicand is a perfect mth power, and if m is a factor of n, equality (5.20) can be used to reduce the order of the radical. For example,

$$\sqrt[4]{25} = \sqrt[4]{5^2} = 5^{2/4} = 5^{1/2} = \sqrt{5}$$
$$\sqrt[6]{27x^3} = \sqrt[6]{(3x)^3} = (3x)^{3/6} = (3x)^{1/2} = \sqrt{3x}$$
$$\sqrt[9]{-27} = \sqrt[9]{(-3)^3} = (-3)^{3/9} = (-3)^{1/3} = \sqrt[3]{-3}$$

The order of a radical can always be increased by using equality (5.21) as shown by the following examples:

$$\sqrt{3} = \sqrt{\sqrt{3^2}} = \sqrt[4]{9}$$
$$\sqrt[3]{4x^2} = \sqrt[3]{\sqrt{(4x^2)^2}} = \sqrt[6]{16x^4}$$
$$\sqrt[4]{3a} = \sqrt[4]{\sqrt[3]{(3a)^3}} = \sqrt[12]{27a^3}$$

3 PUTTING A MULTIPLIER UNDER THE RADICAL SIGN

An expression of the form $b\sqrt[n]{a}$ can be written in the form $\sqrt[n]{c}$ by using in reverse the procedure discussed in method 1. To illustrate,

$$2\sqrt{3} = \sqrt{4}\sqrt{3} = \sqrt{4 \cdot 3} = \sqrt{12}$$
$$3\sqrt[3]{4} = \sqrt[3]{3^3}\sqrt[3]{4} = \sqrt[3]{27 \cdot 4} = \sqrt[3]{108}$$
$$(x+y)\sqrt[3]{(x+y)^2} = \sqrt[3]{(x+y)^3}\sqrt[3]{(x+y)^2} = \sqrt[3]{(x+y)^3(x+y)^2}$$
$$= \sqrt[3]{(x+y)^5}$$

Recall that only like quantities can be combined by addition or subtraction. This means that two or more expressions involving radicals cannot be combined by the operations of addition and subtraction unless the radicals are all of the same order with the same radicand or unless the form of the radicals can be changed so that this is true. To illustrate,

$$4\sqrt{3} - 8\sqrt{3} = -4\sqrt{3} \qquad \sqrt[3]{28} + 5\sqrt[3]{28} - 2\sqrt[3]{28} = 4\sqrt[3]{28}$$

EXAMPLE 1 Add $\sqrt[3]{48}$ and $3\sqrt[3]{6}$.

SOLUTION The order of the two radicals is the same, but the radicands are different. We may try to change either radicand to the same value as the other. We observe that $\sqrt[3]{48} = \sqrt[3]{8 \cdot 6} = \sqrt[3]{8}\,\sqrt[3]{6} = 2\sqrt[3]{6}$, which provides the desired change. Therefore

$$\sqrt[3]{48} + 3\sqrt[3]{6} = 2\sqrt[3]{6} + 3\sqrt[3]{6} = 5\sqrt[3]{6}$$

The same result would have been obtained by writing

$$\sqrt[3]{6} = \tfrac{1}{2}\sqrt[3]{8}\,\sqrt[3]{6} = \tfrac{1}{2}\sqrt[3]{48}$$

so that

$$\sqrt[3]{48} + 3\sqrt[3]{6} = \sqrt[3]{48} + \tfrac{3}{2}\sqrt[3]{48} = \tfrac{5}{2}\sqrt[3]{48}$$

The reader should verify that $5\sqrt[3]{6} = \tfrac{5}{2}\sqrt[3]{48}$.

Multiplication and division of radicals are accomplished by use of (5.18) and (5.19) and the rule that $a^m \cdot a^n = a^{m+n}$. To use (5.18) and (5.19), the order of the radicals must be the same. Consider the following examples:

$$\sqrt{5}\,\sqrt{6} = \sqrt{30} \qquad \frac{\sqrt[3]{15x^2y}}{\sqrt[3]{5xy^2}} = \sqrt[3]{\frac{15x^2y}{5xy^2}} = \sqrt[3]{\frac{3x}{y}}$$

$$\sqrt{5}\,\sqrt[4]{25} = \sqrt{5}\,\sqrt[4]{5^2} = \sqrt{5}\,\sqrt{5} = 5$$

EXAMPLE 2 Perform the division $\dfrac{\sqrt[3]{25}}{\sqrt{5}}$.

SOLUTION Since the orders of the radicals are not the same, the forms of the radicals must be changed in order to perform the division. We can change two radicals of different orders into two radicals of the same order by taking the new order to be a multiple of both of the original orders. In this example 6 is the lowest common multiple of the orders 2 and 3 of the given radicals. So we write

$$\frac{\sqrt[3]{25}}{\sqrt{5}} = \frac{\sqrt[6]{(25)^2}}{\sqrt[6]{5^3}} = \sqrt[6]{\frac{(25)^2}{5^3}} = \sqrt[6]{\frac{(5^2)^2}{5^3}} = \sqrt[6]{\frac{5^4}{5^3}} = \sqrt[6]{5}$$

EXAMPLE 3 Find the product of $3 + 4\sqrt{5}$ and $3 - 4\sqrt{5}$.

SOLUTION Using the rule for finding the product of the sum and difference of two numbers, we have

$$(3 + 4\sqrt{5})(3 - 4\sqrt{5}) = 3^2 - (4\sqrt{5})^2 = 9 - 16 \cdot 5 = -71$$

When a fraction contains radicals, it is frequently useful to change the form of the fraction so that either the numerator or the denominator is free of radicals. When the form of a fraction is changed so that no radicals appear in the numerator, we say that we have *rationalized the numerator*. If the form of a fraction is changed so that no radicals appear in the denominator, we have *rationalized the denominator*.

RATIONALIZE THE NUMERATOR OR THE DENOMINATOR

Rationalization of either the numerator or the denominator usually results in a simpler appearance of the fraction. Whether one rationalizes the numerator or the denominator depends on the use to be made of the fraction. If a decimal approximation of the fraction is needed or if computations involving addition or subtraction are to be made, then it is usually helpful to rationalize the denominator. For example, if we wish an approximation of $\frac{2}{\sqrt{3}}$ correct to two decimal places, we could use the table of square roots to write $\frac{2}{\sqrt{3}} = \frac{2}{1.732}$ and then proceed by long division to obtain an approximation. However, the arithmetic is much simpler if we write

$$\frac{2}{\sqrt{3}} = \frac{2\sqrt{3}}{\sqrt{3}\sqrt{3}} = \frac{2(1.732)}{3} = \frac{3.464}{3} = 1.15 \quad \text{to two decimal}$$

places.

Similarly, if we wish to compute the value of

$$\frac{2+\sqrt{2}}{1-\sqrt{2}} + \frac{6}{\sqrt{3}+1}$$

correct to two decimal places, we should probably wish to rationalize the denominators of the two fractions before performing the addition. On the other hand, there are occasions when radicals are used in more advanced mathematics in which it is useful to rationalize the numerator.

When the radicals are of order 2, rationalizing either the numerator or the denominator may be done by using the identity

$$(a+b)(a-b) = a^2 - b^2$$

EXAMPLE 4 Write the fraction $\frac{2+\sqrt{2}}{1-\sqrt{2}}$ with no radicals in the denominator.

SOLUTION If $1-\sqrt{2}$ is multiplied by $1+\sqrt{2}$, the result will contain no radicals, since $(1-\sqrt{2})(1+\sqrt{2}) = 1^2 - (\sqrt{2})^2 = 1 - 2 = -1$. So we can write

$$\frac{2+\sqrt{2}}{1-\sqrt{2}} = \frac{2+\sqrt{2}}{1-\sqrt{2}} \cdot \frac{1+\sqrt{2}}{1+\sqrt{2}} = \frac{2+3\sqrt{2}+2}{-1} = -4-3\sqrt{2}$$

EXAMPLE 5 Write $\dfrac{\sqrt{5} + 2\sqrt{3}}{3\sqrt{5} - 6\sqrt{3}}$ without radicals in the numerator.

SOLUTION To rationalize the numerator, we multiply $\sqrt{5} + 2\sqrt{3}$ by $\sqrt{5} - 2\sqrt{3}$. Hence we have

$$\frac{\sqrt{5} + 2\sqrt{3}}{3\sqrt{5} - 6\sqrt{3}} = \frac{\sqrt{5} + 2\sqrt{3}}{3\sqrt{5} - 6\sqrt{3}} \cdot \frac{\sqrt{5} - 2\sqrt{3}}{\sqrt{5} - 2\sqrt{3}} = \frac{(\sqrt{5})^2 - (2\sqrt{3})^2}{3 \cdot 5 - 12\sqrt{15} + 12 \cdot 3}$$

$$= \frac{5 - 4 \cdot 3}{15 + 36 - 12\sqrt{15}} = \frac{-7}{51 - 12\sqrt{15}}$$

EXERCISES

In Exercises 1 to 15, change the form of the radical by removing all possible factors from the radicand.

1 $\sqrt{500}$

2 $\sqrt[3]{-40}$

3 $\sqrt{72a^3b^5}$

4 $\sqrt[3]{(x^2 - 3y^2)^6}$

5 $\sqrt{x^4 + 2x^6y^2}$

6 $\sqrt[3]{108(x + y)^7}$

7 $\sqrt[3]{64x^8 - 192x^6y^2}$

8 $\sqrt{3x^2 - 12x + 12}$

9 $\sqrt{75x^3}$

10 $\sqrt{x^{-4} - 3x^{-2}y^2}$

11 $\sqrt{45a^{-4}}$

12 $\sqrt[3]{-(x^2 + y^2)^4}$

13 $\sqrt[3]{243}$

14 $\sqrt[3]{(a + b)(a^3 - 3a^2b + 3ab^2 - b^3)}$

15 $\sqrt[5]{\dfrac{96a^6}{y^8}}$

In Exercises 16 to 23, reduce the radical to the lowest possible order, and remove all possible factors from the radicand.

16 $\sqrt[4]{\frac{4}{9}}$

17 $\sqrt[12]{16x^4y^8}$

18 $\sqrt[6]{a^2b^2c^4}$

19 $\sqrt[6]{1{,}000}$

20 $\sqrt[9]{-27}$

21 $\sqrt[4]{0.64}$

22 $\sqrt[6]{0.027x^3}$

23 $\sqrt[8]{\dfrac{x^2 + 6xy + 9y^2}{81}}$

In Exercises 24 to 29, change the form of the radical by writing it with the indicated order.

24 $\sqrt{5}$ with order 8

25 $\sqrt[3]{x + y}$ with order 6

26 $\sqrt[6]{2a - b}$ with order 18

27 $\sqrt[4]{\dfrac{a - 2}{4}}$ with order 8

28 $\sqrt[3]{-3}$ with order 9

29 $\sqrt{10}$ with order 6

In each of Exercises 30 to 35, write all the expressions as radicals of the same order.

30 $\sqrt{2}$, $\sqrt[3]{3x^2}$, $\sqrt[6]{4}$

31 $\sqrt{3ax}$, $\sqrt[5]{6a^3x^2}$

32 $\sqrt[3]{4a}$, $\sqrt{5ax}$, $\sqrt[4]{a^3x^2}$

33 $\sqrt{2mx}$, $\sqrt[7]{4m^5x^2}$

34 $\sqrt[3]{\dfrac{2x^2}{3y}}$, $\sqrt[4]{\dfrac{4x^3}{y^2}}$

35 $\sqrt{\dfrac{2x}{3y}}$, $\sqrt[4]{\dfrac{x^2}{2y^3}}$, $\sqrt[8]{\dfrac{9x^5}{7y^2}}$

In Exercises 36 to 41, rewrite the expressions by putting the multiplier under the radical sign.

36 $5\sqrt[3]{2}$

37 $(x + y)\sqrt{x - y}$

38 $\dfrac{3a}{b}\sqrt{2}$

39 $5x\sqrt[3]{x}$

40 $2y\sqrt[4]{x - y}$

41 $(2x + 1)\sqrt{26}$

In Exercises 42 to 45, write the given expression as a fraction times a radical which contains no fraction.

42 $\sqrt{\dfrac{2}{5}}$

43 $\sqrt[3]{\dfrac{3}{100}}$

44 $\sqrt{3 - \dfrac{4}{9x}}$

45 $\sqrt{\dfrac{x}{y} + \dfrac{y}{x}}$

In Exercises 46 to 63, perform the indicated operation when possible, and write the result with the radicals reduced to the lowest possible order and with all possible factors removed from the radicands.

46 $\sqrt{2} - 3\sqrt{2}$

47 $\sqrt{12} + \sqrt{27}$

48 $3\sqrt{98} - 4\sqrt{18}$

49 $\sqrt[3]{5} - \sqrt[3]{40} + 2\sqrt[6]{25}$

50 $\sqrt{2} \cdot \sqrt{3}$

51 $\sqrt{8} \cdot \sqrt{6}$

52 $\sqrt{2}(6\sqrt{2} + \sqrt{72} + \sqrt{18})$

53 $(4\sqrt{3} - 3\sqrt{2})(4\sqrt{3} + 3\sqrt{2})$

54 $\sqrt{4x^2 - y^2} \div \sqrt{2x + y}$

55 $\sqrt[3]{x^2 - 4} \div \sqrt{x^2 - 5x + 6}$

56 $(5\sqrt{6} + 15\sqrt{18} - 5\sqrt{32}) \div 5\sqrt{2}$

57 $(3\sqrt{3} - \sqrt{5})(2\sqrt{3} + 3\sqrt{5})$

58 $4\sqrt{16} - 2\sqrt{8}$

59 $\sqrt{a - x} \cdot \sqrt{a + x}$

60 $\sqrt{3}(\sqrt{18} - \sqrt{6})$

61 $\sqrt[3]{25} \div \sqrt{5}$

62 $\sqrt[6]{4a^2} - a\sqrt[9]{8a^3} + 2\sqrt[3]{2a}$

63 $\sqrt[5]{64} - 3\sqrt[5]{2}$

In Exercises 64 to 73, rationalize the denominator.

64 $\dfrac{4}{2 - \sqrt{7}}$

65 $\dfrac{2}{\sqrt{a} + \sqrt{b}}$

66 $\dfrac{8}{5 + \sqrt{3}}$

67 $\dfrac{3}{\sqrt{7} - \sqrt{2}}$

68 $\dfrac{\sqrt{a} + \sqrt{b}}{a\sqrt{a} + b\sqrt{b}}$

69 $\dfrac{\sqrt{2} - \sqrt{5}}{\sqrt{3} - \sqrt{2}}$

70 $\dfrac{\sqrt{xy}}{\sqrt{x} - \sqrt{y}}$

71 $\dfrac{\sqrt[3]{36}}{\sqrt{3}}$

72 $\dfrac{a + \sqrt{b}}{2a - 3\sqrt{b}}$

73 $\dfrac{x + y - \sqrt{x - y}}{x + y + \sqrt{x - y}}$

In Exercises 74 to 83, rationalize the numerator.

74 $\dfrac{a - 2\sqrt{b}}{a + 2\sqrt{b}}$

75 $\dfrac{\sqrt{2} - \sqrt{5}}{\sqrt{3} - \sqrt{2}}$

76 $\dfrac{\sqrt{a} - \sqrt{b}}{a\sqrt{a} + b\sqrt{b}}$

77 $\dfrac{3 + \sqrt{2}}{2 - \sqrt{3}}$

78 $\dfrac{\sqrt[3]{x} + 1}{\sqrt[3]{x}}$

79 $\dfrac{3\sqrt{6} - \sqrt{18}}{4\sqrt{2}}$

80 $\dfrac{\sqrt{x + a} - \sqrt{x}}{a}$

81 $\dfrac{2\sqrt{3} + 4\sqrt{2}}{\sqrt{3} - 2\sqrt{2}}$

82 $\dfrac{\sqrt{x^2 - 4} + \sqrt{x}}{\sqrt{x + 2}}$

83 $\dfrac{\sqrt{x + 1} + 1}{\sqrt{x}}$

In Exercises 84 to 90, perform the indicated operations when possible, and write the results with the radicals reduced to the lowest possible order and with all possible factors removed from each radicand. If radicals appear in the denominator, rationalize the denominator.

84 $\dfrac{2\sqrt{3} + 4\sqrt{2}}{\sqrt{3} - 2\sqrt{2}} + \dfrac{\sqrt{3} + 2\sqrt{5}}{4\sqrt{3} - \sqrt{5}}$

85 $\dfrac{3\sqrt{2}}{2 - 3\sqrt{2}} \cdot \dfrac{\sqrt{2} - 1}{\sqrt{2}}$

86 $\dfrac{2\sqrt{a}}{\sqrt{a} - \sqrt{b}} - \dfrac{3\sqrt{b}}{2(\sqrt{a} - \sqrt{b})}$

87 $\sqrt[3]{x^4 - y^4} \div \sqrt{x^2 - y^2}$

88 $\dfrac{2\sqrt{5}}{\sqrt{7} + 2\sqrt{6}} \cdot \dfrac{3\sqrt{2} - 4\sqrt{6}}{\sqrt{2}}$

89 $\sqrt[3]{5} - \sqrt[3]{40} + 3\sqrt[6]{25}$

90 $\dfrac{2\sqrt{3} + 4\sqrt{2}}{\sqrt{3} - 2\sqrt{2}} + \dfrac{4\sqrt{2} - 5\sqrt{3}}{\sqrt{3} + 2\sqrt{2}}$

5.5 COMPLEX NUMBERS

With the extension of the rational number system to produce the real number system, we have a set of numbers with the property that the solution set of

$$x^n - a = 0$$

is nonempty *whenever* n is a positive integer and a is a *positive* real number; that is,

$$\text{for } n \in N \text{ and } a > 0, \quad \{x \in Re \mid x^n - a = 0\} \neq \varnothing \tag{5.22}$$

In fact

$$\{x \in Re \mid x^n - a = 0\} = \{\sqrt[n]{a}\}$$

However, the set Re of real numbers will not provide a nonempty solution set for all equations of the form $x^n - a = 0$ when a is *negative.* To illustrate, let us consider the equation

$$x^2 + 1 = 0 \tag{5.23}$$

If there is a real number c that satisfies (5.23), then $c^2 + 1 = 0$ or $c^2 = -1$. But property (5.9) of the real number system, listed in Sec. 5.3, says that the square of any real number is nonnegative; therefore there cannot be a real number c for which $c^2 = -1$. So we have

$$\{x \in Re \mid x^2 + 1 = 0\} = \varnothing$$

In general,

for $a < 0$ *and* n *an even positive integer*

$$\{x \in Re \mid x^n - a = 0\} = \varnothing$$

Thus, even though the real number system does provide nonempty solution sets for many more equations than does the rational number system, there are still many equations for which the real number system fails to provide nonempty solution sets.

To remove this deficiency and to provide a nonempty solution of any polynomial equation* of any degree, the complex number system is introduced. The set of complex numbers will in particular contain numbers whose squares are negative.

To produce the set C of complex numbers, we introduce a number, symbolized by i, with the property that its square is -1; that is, i is the number defined by

$$i^2 = -1 \tag{5.24}$$

Heretofore we have not had a meaning for the symbol $\sqrt{-a}$ where a is a positive number. We are now in a position to give a definition of this symbol. It seems reasonable that if we are to give a meaning to $\sqrt{-a}$, we should want the symbol to represent a

*Polynomial equations are discussed in the next chapter.

number whose square is $-a$. Such a definition is

for $a > 0$,
$$\sqrt{-a} = \sqrt{a} \cdot i \tag{5.25}$$

since assuming that multiplication with the number i is commutative and associative, we can write

$$(\sqrt{a} \cdot i)(\sqrt{a} \cdot i) = (\sqrt{a}\sqrt{a})(i \cdot i) = a(-1) = -a$$

Definition (5.25) gives a unique meaning to the square root of a negative number. We shall agree to make use of (5.25) and henceforth write the square root of a negative number as the product of a real number and the number i. To illustrate,

$$\sqrt{-25} = \sqrt{25}\, i = 5i \qquad -\sqrt{-16} = -\sqrt{16}\, i = -4i$$
$$\sqrt{-81b^2} = \sqrt{81b^2}\, i = 9bi \qquad b > 0$$

From (5.24) we may construct the following table of powers of i:

$$i^3 = i^2 \cdot i = (-1)i = -i$$
$$i^4 = i^2 \cdot i^2 = (-1)(-1) = 1$$
$$i^5 = i^4 \cdot i = 1 \cdot i = i$$
$$i^6 = i^5 \cdot i = i \cdot i = -1$$
$$i^7 = i^6 \cdot i = (-1)i = -i$$

From these equalities it should be clear that any positive or negative integral power of i is equal to one of the four numbers 1, -1, i, $-i$. To illustrate,

$$i^{16} = (i^4)^4 = 1^4 = 1 \qquad i^{17} = i^{16} \cdot i = 1 \cdot i = i$$
$$i^{-3} = \frac{1}{i^3} = \frac{1}{-i} = -\frac{1}{i} = -\frac{1}{i} \cdot \frac{i}{i} = \frac{-i}{i^2} = \frac{-i}{-1} = i$$

COMPLEX NUMBER

REAL PART

IMAGINARY PART

IMAGINARY NUMBER

If a and b are real numbers, any expression of the form $a + bi$ is called a *complex number*. Either a or b, or both a and b, may be zero. The number a is called the *real part* and the number b is called the *imaginary part* of the complex number $a + bi$. If $a = 0$, then $a + bi = bi$ is called an *imaginary number*. If $b = 0$, then $a + bi = a + 0i = a$ is a *real number*.

In speaking of $a + bi$ as a number, we have assumed that the meaning of equality of two of these symbols can be defined and that the operations of addition, subtraction, multiplication, and division can be defined for these symbols in such a way that the commutative, associative, and distributive properties which we have studied for real numbers will hold for complex numbers as well. This is indeed true, and we now state suitable definitions for equality and the four fundamental arithmetic operations.

EQUALITY OF COMPLEX NUMBERS

Two complex numbers are **equal** *if and only if their real parts are equal and their imaginary parts are equal.* That is,

$a + bi = c + di$ if and only if $a = c$ and $b = d$

In particular, if $a + bi = 0$, then $a = 0$ and $b = 0$.

EXAMPLE State what values x and y must have in order that

$$x + 3xi + 2yi + 4y = 14 + 12i$$

SOLUTION Rewrite the given expression as

$$(x + 4y) + (3x + 2y)i = 14 + 12i$$

Equating the real parts of the left- and right-hand sides of this expression, we get

$$x + 4y = 14 \tag{1}$$

Equating the imaginary parts of the left- and right-hand sides, we get

$$3x + 2y = 12 \tag{2}$$

Solving equations (1) and (2) for x and y, we find that $x = 2$ and $y = 3$.

SUM OF TWO COMPLEX NUMBERS

The **sum** *of two complex numbers* is the complex number whose real part is the sum of the real parts of the addends and whose imaginary part is the sum of the imaginary parts of the addends. That is,

$$(a + bi) + (c + di) = (a + c) + (b + d)i \tag{5.26}$$

For example,

$$(2 + 3i) + (5 + 7i) = (2 + 5) + (3 + 7)i = 7 + 10i$$

DIFFERENCE OF TWO COMPLEX NUMBERS

The **difference** *of two complex numbers* is the complex number whose real part is the difference of the real parts of the minuend and subtrahend and whose imaginary part is the difference of the imaginary parts of the minuend and subtrahend. Thus

$$(a + bi) - (c + di) = (a - c) + (b - d)i \tag{5.27}$$

To illustrate,

$$(2 + 3i) - (5 + 7i) = (2 - 5) + (3 - 7)i = -3 - 4i$$

CONJUGATE COMPLEX NUMBERS

If two complex numbers differ only in the sign of their imaginary part, they are called **conjugate** *complex numbers.* That is,

$a + bi$ and $a - bi$

are conjugates. Since

$(a + bi) + (a - bi) = 2a$ and $(a + bi) - (a - bi) = 2bi$

we see that the sum of two conjugate complex numbers is real and the difference of two conjugate complex numbers is imaginary.

PRODUCT OF TWO COMPLEX NUMBERS

The **product** *of two complex numbers* $a + bi$ and $c + di$ is the complex number which is obtained by multiplying them in accordance with the rules for multiplying real numbers and replacing i^2 by -1 in the result. This gives

$$(a + bi)(c + di) = ac + adi + bci + bdi^2$$

Replacing i^2 by -1 and collecting the real and imaginary parts, we have

$$(a + bi)(c + di) = (ac - bd) + (ad + bc)i \qquad (5.28)$$

To illustrate,

$$(2 + 3i)(5 + 7i) = 10 + 14i + 15i + 21i^2 = -11 + 29i$$

It is useful to note that the product of two conjugate complex numbers is a positive real number, since

$$(a + bi)(a - bi) = a^2 - b^2i^2 = a^2 + b^2$$

For example,

$$(3 + 4i)(3 - 4i) = 9 - 16i^2 = 9 + 16 = 25$$

and

$$(2 + 7i)(2 - 7i) = 2^2 + 7^2 = 4 + 49 = 53$$

QUOTIENT OF TWO COMPLEX NUMBERS

The **quotient** *of two complex numbers* $(a + bi) \div (c + di)$ is the complex number $x + yi$ such that $(c + di)(x + yi) = a + bi$.

Division of one complex number by another can be accomplished by use of the property that the product of two conjugate complex numbers is a real number. To divide $a + bi$ by $c + di$, express the quotient as a fraction $\frac{a + bi}{c + di}$, and multiply both numerator and denominator by the conjugate $c - di$ of the denominator. Thus

$$\frac{a + bi}{c + di} = \frac{(a + bi)(c - di)}{(c + di)(c - di)} = \frac{(ac + bd) + (bc - ad)i}{c^2 + d^2}$$

Therefore

$$\frac{a + bi}{c + di} = \frac{ac + bd}{c^2 + d^2} + \frac{bc - ad}{c^2 + d^2}\, i \qquad (5.29)$$

For example,

$$\frac{3 + i}{3 + 4i} = \frac{(3 + i)(3 - 4i)}{(3 + 4i)(3 - 4i)} = \frac{9 - 9i - 4i^2}{9 - 16i^2} = \frac{13 - 9i}{25} = \frac{13}{25} - \frac{9}{25}\, i$$

In working with expressions which contain complex numbers, these numbers should be written in the form $a + bi$ before any of the fundamental arithmetic operations are performed. For example,

$$\begin{aligned}(5 - \sqrt{-2})(4 - 3\sqrt{-2}) &= (5 - \sqrt{2}\,i)(4 - 3\sqrt{2}\,i) \\ &= 20 - 4\sqrt{2}\,i - 15\sqrt{2}\,i + 3(2)i^2 \\ &= 20 - 19\sqrt{2}\,i - 6 = 14 - 19\sqrt{2}\,i\end{aligned}$$

It can be seen that the complex number system is a field. In other words, if C denotes the set of complex numbers, then

for *every* $\alpha = a + bi$, $\beta = c + di$, $\gamma = e + gi$ belonging to C,

E_1: $\alpha = \alpha$

E_2: if $\alpha = \beta$, then $\beta = \alpha$

E_3: if $\alpha = \beta$ and $\beta = \gamma$, then $\alpha = \gamma$

P_1: $\alpha + \beta \in C$, and $\alpha + \beta$ is unique
$\alpha \cdot \beta \in C$, and $\alpha \cdot \beta$ is unique

P_2: $\alpha + \beta = \beta + \alpha$
$\alpha \cdot \beta = \beta \cdot \alpha$

P_3: $\alpha + (\beta + \gamma) = (\alpha + \beta) + \gamma$
$\alpha \cdot (\beta \cdot \gamma) = (\alpha \cdot \beta) \cdot \gamma$

P_4: $\alpha(\beta + \gamma) = \alpha\beta + \alpha\gamma$

P_5: There exists a complex number ϵ, called the unit element, with the property that for any $\alpha \in C$, $\alpha \cdot \epsilon = \epsilon \cdot \alpha = \alpha$.

P_6: There exists a complex number δ, called the zero element, with the property that for any $\alpha \in C$, $\alpha + \delta = \delta + \alpha = \alpha$.

P_7: If δ is the zero element described in P_6, and if α is any member of C, then there exists a number $-\alpha \in C$ with the property that $\alpha + (-\alpha) = -\alpha + \alpha = \delta$.

P_8: If ϵ is the unit element described in P_5, and if α is any member of C except the zero element δ, then there exists a number $\alpha' \in C$ with the property that $\alpha \cdot \alpha' = \alpha' \cdot \alpha = \epsilon$.

Since equality, addition, and multiplication of complex numbers are defined in terms of equality, addition, and multiplication of real numbers, properties E_1 to E_3 and P_1 to P_4 for complex numbers can be established by using the fact that the set Re is a set of numbers. To illustrate, let us establish the fact that for every $\alpha = a + bi$ and $\beta = c + di$ in the set C,

$$\alpha + \beta = \beta + \alpha$$

By definition (5.26) we know that

$$\alpha + \beta = (a + bi) + (c + di) = (a + c) + (b + d)i$$

and that

$$\beta + \alpha = (c + di) + (a + bi) = (c + a) + (d + b)i$$

The definition of equality of complex numbers tells us that

$$(a + c) + (b + d)i = (c + a) + (d + b)i$$

if and only if

$$a + c = c + a \qquad \text{and} \qquad b + d = d + b$$

Now a, b, c, and d are members of Re, and we know that addition in Re is commutative (property P_2 on page 235). Therefore we know that $a + c = c + a$ and $b + d = d + b$, and we have thus proved that $\alpha + \beta = \beta + \alpha$. The remainder of the properties E_1 to E_3 and P_1 to P_4 can be established in a similar way.

To show that P_5 holds, we simply note that the complex number $1 = 1 + 0i$ has the property required of the unit element. Let $\alpha = a + bi$ be any complex number; then by use of (5.28) we have

$$\begin{aligned}(a + bi)(1 + 0i) &= (a \cdot 1 - b \cdot 0) + (a \cdot 0 + b \cdot 1)i \\ &= a + bi\end{aligned}$$

UNIT ELEMENT IN C

Thus $1 + 0i$ is the unit element in the set C of complex numbers.

To show that P_6 holds, we note that the complex number $0 = 0 + 0i$ has the property required of the zero element. If $\alpha = a + bi$ is any complex number, then by (5.26) we have

$$(a + bi) + (0 + 0i) = (a + 0) + (b + 0)i = a + bi$$

ZERO ELEMENT IN C

Hence, $0 + 0i$ is the zero element in the set C of complex numbers.

To show that P_7 holds, we note that for any $\alpha = a + bi$ in the set C,

$$\begin{aligned}(a + bi) + [-a + (-b)i] &= [a + (-a)] + [b + (-b)]i \\ &= 0 + 0i\end{aligned}$$

so,

$$-\alpha = -a - bi$$

To show that P_8 holds, let $\alpha = a + bi$ be any complex number *except the zero element;* that is, a and b can be any real numbers with the condition that either $a \neq 0$ or $b \neq 0$. We wish to find a complex number $x + yi$ with the property that

$$(a + bi)(x + yi) = 1 + 0i$$

From (5.28) we have

$$(a + bi)(x + yi) = (ax - by) + (ay + bx)i$$

If $(ax - by) + (ay + bx)i$ is to equal the unit element $1 + 0i$, we must have

$$\begin{aligned}ax - by &= 1 \\ ay + bx &= 0\end{aligned} \qquad (5.30)$$

Equations (5.30) form a pair of equations in which the variables are x and y. By using the methods of Sec. 4.9, the pair (5.30) can be solved, and we find that

$$x = \frac{a}{a^2 + b^2} \qquad y = \frac{-b}{a^2 + b^2} \qquad \text{(Note that } a^2 + b^2 \neq 0.)$$

Therefore,

$$(a + bi)\left(\frac{a}{a^2 + b^2} - \frac{b}{a^2 + b^2}i\right) = 1 + 0i$$

and property P_8 is established.

Although the complex number system is a field, it is *not* an *ordered* field. It is not possible to give a definition of "greater than" in the set of complex numbers that will satisfy the properties O_1, O_2, O_3, O_4 given in Sec. 5.3. Since the statement that "α is positive" is equivalent to "α is greater than 0," we cannot speak of "positive" or "negative" complex numbers.

EXERCISES

Express the following in terms of i:

1 $\sqrt{-36}$

2 $-\sqrt{-49}$

3 $\sqrt{-\frac{4}{9}}$

4 $\sqrt{-13b^2}$ $\quad b > 0$

5 $\sqrt{-\frac{3}{4}}$

6 $\sqrt{-9b}$

7 $\sqrt{-72}$

8 $\sqrt{-128a^2b^2}$ $\quad a > 0,\ b > 0$

In Exercises 9 to 12, find the values which x and y must have for the given equality to hold.

9 $2 - 5i = x + yi$

10 $2x - 3i = -yi - 4$

11 $x - 2y + xi - yi = 2 + 5i$

12 $x + 2xi - 7y - 6yi = -15 - 10i$

Find the simplest form (in terms of i or a real number) of each of the following expressions:

13 i^{14}

14 $-i^{12}$

15 $-i^{16}$

16 $\dfrac{1}{i^5}$

17 $(-i^5)^3$

18 $\dfrac{1}{-i^9}$

19 i^{20}

20 $\dfrac{1}{i^{10}}$

Give the conjugate of each of the following complex numbers:

21 $3 - 8i$

22 $-7 - 5i$

23 $\frac{2}{3} + \frac{4}{5}i$

24 $1 - \sqrt{2}\,i$

In Exercises 25 to 36, perform the indicated operation, and express the result in the form $a + bi$, where a and b are real numbers.

25 $\sqrt{-9} + \sqrt{-25} + \sqrt{64}$
26 $(1 + 4i) + (3 + 2i)$
27 $(4 - 3i) - (5 - 6i)$
28 $(2 - 3i)(5 + 4i)$
29 $(3 - \sqrt{-3})(5 + 2\sqrt{-3})$
30 $(3 + 2\sqrt{2}\,i)(-4 + 3\sqrt{2}\,i)$
31 $(1 - i^2)(1 + i^2)$
32 $(5 + \sqrt{-4})(5 - \sqrt{-4})$
33 $(3 + 4i) \div (5 - i)$
34 $(1 + \sqrt{2} + \sqrt{3}\,i)(-1 + \sqrt{2} - \sqrt{3}\,i)$
35 $(2 + 3\sqrt{-2}) \div (1 + \sqrt{-2})$
36 $\dfrac{6}{4 + \sqrt{-3}}$

37 Given $x = -3 + 2i$, find the value of $x^2 + 6x + 13$.
38 Is $3 - i$ a root of the equation $y^2 - 6y + 10 = 0$?

Show that if $\alpha = a + bi$, $\beta = c + di$, and $\gamma = e + gi$ are any complex numbers, then:

39 $\alpha \cdot \beta = \beta \cdot \alpha$
40 $\alpha + (\beta + \gamma) = (\alpha + \beta) + \gamma$
41 $\alpha(\beta + \gamma) = \alpha\beta + \alpha\gamma$

5.6 FRACTIONS

In Chap. 3 we defined a fraction as a symbol of the form $\frac{a}{b}$, where a and b are integers and $b \neq 0$. However, in Secs. 5.3 and 5.4 we have used the word "fraction" with a broader meaning. Let us agree that a **fraction** is any symbol of the form $\frac{a}{b}$, where a and b are any numbers whatever with $b \neq 0$; and further, the fraction $\frac{a}{b}$ is to be identified with the quotient $a \div b$. The properties and operations which we used in Chap. 3 for fractions considered as quotients of integers carry over in an obvious manner to "fractions" in the sense just defined.

FRACTION

To illustrate, the fraction whose numerator is $\sqrt{6}$ and whose denominator is $\sqrt{3}$ may be written

$$\frac{\sqrt{6}}{\sqrt{3}} = \frac{\sqrt{2 \cdot 3}}{\sqrt{3}} = \frac{\sqrt{2}\,\sqrt{3}}{\sqrt{3}} = \sqrt{2}$$

Similarly, the fraction whose numerator is $\sqrt{6} + \sqrt{2}$ and whose denominator is $\sqrt{2}$ may be written

$$\frac{\sqrt{6} + \sqrt{2}}{\sqrt{2}} = \frac{\sqrt{2}\,\sqrt{3} + \sqrt{2}}{\sqrt{2}} = \sqrt{3} + 1$$

Also, recall that for the quotient $\frac{x}{y}$ which results when $\frac{a}{b}$ is divided by $\frac{c}{d}$, we write

$$\frac{x}{y} = \frac{a}{b} \div \frac{c}{d} \tag{5.31}$$

Sometimes we write this quotient as

$$\frac{x}{y} = \frac{\frac{a}{b}}{\frac{c}{d}} \tag{5.32}$$

COMPLEX FRACTION

and call $\frac{x}{y}$ when written in this form a complex* fraction. A **complex fraction** is a fraction which has one or more fractions in the numerator, or in the denominator, or in both. For example, $\dfrac{\frac{x}{y} + \frac{y}{x}}{\frac{1}{y} - \frac{1}{x}}$ is a complex fraction.

TO SIMPLIFY A COMPLEX FRACTION

To simplify a complex fraction means to express the complex fraction as a single fraction in simplest form. In order to simplify a complex fraction we may first express the numerator of the complex fraction as a single fraction $\frac{a}{b}$ and express the denominator of the complex fraction as a single fraction $\frac{c}{d}$ so that the complex fraction is put in the form $\dfrac{\frac{a}{b}}{\frac{c}{d}}$. To simplify this complex fraction we notice that if we multiply the denominator $\frac{c}{d}$ by its reciprocal $\frac{d}{c}$ the product of course is 1 and we may proceed as follows:

$$\frac{\frac{a}{b}}{\frac{c}{d}} = \frac{\frac{a}{b} \cdot \frac{d}{c}}{\frac{c}{d} \cdot \frac{d}{c}} = \frac{\frac{a}{b} \cdot \frac{d}{c}}{1} = \frac{a}{b} \cdot \frac{d}{c} = \frac{ad}{bc}$$

that is,

$$\frac{\frac{a}{b}}{\frac{c}{d}} = \frac{a}{b} \cdot \frac{d}{c} = \frac{ad}{bc} \tag{5.33}$$

*The word "complex" as used here does *not* refer to a complex number as used in Sec. 5.5.

This of course is consistent with the fact that $\dfrac{\dfrac{a}{b}}{\dfrac{c}{d}} = \dfrac{a}{b} \div \dfrac{c}{d}$.

The line which separates the numerator $\dfrac{a}{b}$ and the denominator $\dfrac{c}{d}$ of the complex fraction on the right-hand side of (5.32) is called the *main division line.* To avoid confusion we draw it longer than the other lines of division.

Note that $\dfrac{3}{\dfrac{4}{5}} = \dfrac{3}{1} \div \dfrac{4}{5} = \dfrac{3}{1} \cdot \dfrac{5}{4} = \dfrac{15}{4}$ but $\dfrac{\dfrac{3}{4}}{5} = \dfrac{3}{4} \div \dfrac{5}{1} = \dfrac{3}{4} \cdot \dfrac{1}{5} = \dfrac{3}{20}$. The symbol $\begin{matrix} 3 \\ \overline{4} \\ \overline{5} \end{matrix}$ is meaningless.

EXAMPLE 1 Simplify $\dfrac{\dfrac{x}{y} + \dfrac{y}{x}}{\dfrac{1}{y} - \dfrac{1}{x}}$

SOLUTION We first combine the fractions in the numerator of the complex fraction into a single fraction and also combine the fractions in the denominator of the complex fraction into a single fraction, and then use (5.33). We have

$$\frac{\dfrac{x}{y} + \dfrac{y}{x}}{\dfrac{1}{y} - \dfrac{1}{x}} = \frac{\dfrac{x^2 + y^2}{xy}}{\dfrac{x - y}{xy}} = \frac{x^2 + y^2}{xy} \cdot \frac{xy}{x - y} = \frac{x^2 + y^2}{x - y}$$

EXAMPLE 2 Simplify $\dfrac{1 - \dfrac{b^2}{a^2}}{1 - \dfrac{b}{a}}$

SOLUTION Now

$$\frac{1 - \dfrac{b^2}{a^2}}{1 - \dfrac{b}{a}} = \frac{\dfrac{a^2 - b^2}{a^2}}{\dfrac{a - b}{a}} = \frac{a^2 - b^2}{a^2} \cdot \frac{a}{a - b} = \frac{(a - b)(a + b)}{a^2} \cdot \frac{a}{a - b} = \frac{a + b}{a}$$

EXAMPLE 3 Express the fraction $\dfrac{ab^{-1} - a^{-1}b}{a^{-1} - b^{-1}}$ without negative exponents, and simplify.

SOLUTION We have

$$\frac{ab^{-1} - a^{-1}b}{a^{-1} - b^{-1}} = \frac{a \cdot \dfrac{1}{b} - \dfrac{1}{a} \cdot b}{\dfrac{1}{a} - \dfrac{1}{b}} = \frac{\dfrac{a}{b} - \dfrac{b}{a}}{\dfrac{1}{a} - \dfrac{1}{b}} = \frac{\dfrac{a^2 - b^2}{ab}}{\dfrac{b-a}{ab}}$$

$$= \frac{(a-b)(a+b)}{ab} \cdot \frac{ab}{b-a} = -(a+b)$$

EXERCISES

In Exercises 1 to 6, express the given fraction as a fraction whose denominator contains no radicals.

1 $\dfrac{\sqrt{12}}{\sqrt{3}}$ 2 $\dfrac{\sqrt{30}}{\sqrt{5}}$ 3 $\dfrac{\sqrt{14} + \sqrt{2}}{\sqrt{2}}$

4 $\dfrac{\sqrt{16x^2y^6}}{\sqrt{4y^4}}$ 5 $\dfrac{\sqrt{5} + \sqrt{15}}{\sqrt{5}}$ 6 $\dfrac{\sqrt{8} - \sqrt{18}}{\sqrt{2}}$

In Exercises 7 to 18, simplify the given complex fraction.

7 $\dfrac{\dfrac{2}{3} - \dfrac{1}{4}}{\dfrac{3}{5} - \dfrac{7}{3}}$ 8 $\dfrac{3 - \dfrac{2}{xy}}{\dfrac{a}{x} - \dfrac{3b}{y}}$

9 $\dfrac{\dfrac{a-b}{x+y}}{\dfrac{a^2-b^2}{x^2-y^2}}$ 10 $\dfrac{\dfrac{a}{b}}{c}$

11 $\dfrac{a}{\dfrac{b}{c}}$ 12 $\dfrac{a - \dfrac{b^2}{a}}{a^2 - \dfrac{b^4}{a^2}}$

Hint for Exercise 12: Note that $a^4 - b^4 = (a^2 - b^2)(a^2 + b^2)$.

13 $\dfrac{x^2 - x + 1 - \dfrac{1}{x+1}}{1 - \dfrac{1}{x+1}}$ 14 $\dfrac{2 - \dfrac{1}{b}}{2 + \dfrac{1}{b} - \dfrac{1}{b^2}}$

15 $\dfrac{\dfrac{x^2}{y^2} - 1}{\dfrac{x^2}{y^2} - \dfrac{2x}{y} + 1}$ 16 $\dfrac{1 - \dfrac{1}{1+x}}{1 + \dfrac{1}{x-1}}$

17 $\dfrac{\dfrac{x-y}{x+y} - \dfrac{x+y}{x-y}}{\dfrac{x-y}{x+y} + \dfrac{x+y}{x-y}}$ 18 $\dfrac{\dfrac{6x-15}{x^2-4x+4}}{\dfrac{10-4x}{4-x^2}}$

In Exercises 19 to 24, express the given fraction without negative exponents, and simplify.

19 $\dfrac{a^{-1} + b^{-1}}{a^{-2} + b^{-2}}$ 20 $\dfrac{x^0 + 4y^{-1}}{x^{-1} + 3y^{-2}}$ 21 $\dfrac{1}{a^{-2} + b^{-2}}$

22 $\dfrac{a^{-1} + b^{-1}}{(a + b)^{-1}}$ 23 $\dfrac{a^{-1}b^{-1}}{\dfrac{1}{a^{-1}} + \dfrac{1}{b^{-2}}}$ 24 $\dfrac{\dfrac{1}{(x - y)^{-1}}}{x^{-2} - y^{-2}}$

5.7 EXTRACTION OF SQUARE ROOTS

To compute the (usually approximate) square root of a number which does not appear in a table of square roots, or in case no table is available, a method of estimation and refinement may be employed. This method is particularly suited to use with a desk calculator or a digital computer, and it is illustrated in Examples 1, 2, and 3.

EXAMPLE 1 Find $\sqrt{729}$.

SOLUTION We notice that $20^2 = 400$ and $30^2 = 900$, so $\sqrt{729}$ must lie between 20 and 30. Since 729 is considerably closer to 900 than it is to 400, $\sqrt{729}$ will be closer to 30 than to 20.

As a first estimate of $\sqrt{729}$ let us select 26, and then proceed to refine this estimate. We know that if 26 were equal to $\sqrt{729}$, then $729 \div 26$ would be equal to 26. Therefore, let us carry out the division,

```
     28
26)729
    52
   209
   208
     1
```

Since the quotient when 729 is divided by 26 is 28, we know that $\sqrt{729}$ lies between 26 and 28, and since the remainder in this division is so near zero $\sqrt{729}$ should lie about half way between 26 and 28. So, we take for our next estimate the number 27 and, as before, divide 729 by this estimate.

```
     27
27)729
    54
   189
   189
     0
```

This division shows that $\sqrt{729} = 27$.

In Example 1, 729 was a perfect square and $\sqrt{729}$ could be expressed exactly. The method of estimation and refinement by division can be used to compute to any desired accuracy the square root of a number which is not a perfect square. This use is illustrated in the following two examples.

EXAMPLE 2 Compute $\sqrt{69.05}$ correct to two decimal places.

SOLUTION Since $8^2 = 64$ and $9^2 = 81$, $\sqrt{69.05}$ must lie between 8 and 9. As a first estimate let us choose 8, and then divide 69.05 by 8.

```
    8.63
8)69.05
  64
   5 0
   4 8
     25
     24
      1
```

We see that $\sqrt{69.05}$ lies between 8 and 8.63. For our second estimate we can choose any number between 8 and 8.63; let us choose $\frac{1}{2}(8 + 8.63) = 8.31$, the two-decimal-place approximation of the arithmetic mean (or average) of 8 and 8.63. We now carry out the division $(69.05) \div (8.31)$.

```
           8.309
8.31)69.05 000
     66 48
      2 57 0
      2 49 3
          7 700
          7 479
            221
```

Therefore $\sqrt{69.05}$ lies between 8.31 and 8.309, and since the two decimal place approximation of any number between 8.31 and 8.309 is 8.31, we have found that

$$\sqrt{69.05} = 8.31 \qquad \text{correct to two decimal places}$$

EXAMPLE 3 Find $A = \sqrt{534.29}$ correct to four decimal places.

SOLUTION Note that $20^2 = 400$ and $30^2 = 900$. Select 25 as a first estimate of A.

```
      21.37
25)534.29
   50
    34
    25
     92
     75
     179
     175
       4
```

Select $\frac{1}{2}(25 + 21.37) = 23.18$ as a second estimate of A.

```
           23.0496
23.18)534.29 0000
      463 6
       70 69
       69 54
        1 15 00
          92 72
          22 280
          20 862
           1 4180
           1 3908
              272
```

Select $\frac{1}{2}(23.18 + 23.050) = 23.115$ as a third estimate.

```
               23.1144
23.115)534.290 0000
        462 30
         71 990
         69 345
          2 645 0
          2 311 5
            333 50
            231 15
            102 350
             92 460
              9 8900
              9 2460
                6440
```

Select $\frac{1}{2}(23.115 + 23.1144) = 23.1147$ as a fourth estimate.

```
                23.11472
23.1147)534.2900 00000
        462 294
         71 9960
         69 3441
          2 6519 0
          2 3114 7
            3404 30
            2311 47
            1092 830
             924 588
             168 2420
             161 8029
               6 43910
               4 62294
               1 81616
```

As a result of this division we see that

$\sqrt{534.29} = 23.1147$ correct to four decimal places

Students who have some knowledge of the use of digital computers will notice that the method illustrated in the above examples is especially adaptable for use on such computers. The method also has the advantage of requiring only an understanding of the meaning of "square root" and the operation of division.

There is a scheme for directly computing square roots that has been in use since about the fourth century. This scheme, or "algorithm," is considerably more complicated than the estimation and refinement, or "successive approximation," method illustrated above. As a matter of historical interest we give an example of the use of the direct computational algorithm in Example 4.

EXAMPLE 4 Find $\sqrt{1{,}743.262}$ correct to three decimal places.

SOLUTION

```
                         4  1. 7  5  2
                       √17 43.26 20 00
                        16
        80 + 1 = 81)     1 43
                           81
      820 + 7 = 827)      62 26
                          57 89
    8340 + 5 = 8345)       4 37 20
                           4 17 25
  83500 + 2 = 83502)        19 95 00
                            16 70 04
                             3 24 96
```

Group the digits in the radicand in pairs, beginning at the decimal point and moving both to the left and to the right, annexing as many zeros as needed. Find the largest perfect square in the pair on the extreme left. The largest perfect square in 17 is 16, and its square root is 4. Place the 4 above the pair 17 as the first digit in the result. Subtract the square of 4 from 17, and bring down the next pair, 43. Double 4, the part of the square root already found, and annex a zero. This gives 80, which is used as a trial divisor of 143; 80 will go into 143 one time. Add 1 to the trial divisor to obtain 81 as a divisor, and place 1 over the group 43 as the second digit in the result. Multiply the divisor 81 by 1, write the result under the dividend 143, subtract, and bring down the next pair, 26; this gives 6,226. Now double 41, the part of the square root already found, and annex a zero. This gives 820, which is used as a trial divisor of 6,226; 820 will go into 6,226 about seven times. Add 7 to the trial divisor to obtain 827 as a divisor, and place 7 over the group 26 as the third digit (the first decimal place) in the result. Multiply the divisor 827 by 7, write the result under the dividend 6,226, subtract, and bring down the next pair, 20. Continue the same process as shown. Since the remainder after the last division is less than one-half of the divisor, $\sqrt{1{,}743.262} = 41.752$ correct to three decimal places.

EXERCISES

Find the square root of each of the following numbers. If a given number is not a perfect square, find its square root correct to three decimal places.

1	3,939	2	273,529	3	2
4	552.25	5	7.809	6	3,472.5
7	943	8	5	9	47,524
10	2.1904	11	72.25	12	15,765

5.8 THE THEOREM OF PYTHAGORAS

In a right triangle the side opposite the right angle is called the **hypotenuse.** The theorem of Pythagoras states that *in any right triangle the square of the length of the hypotenuse is equal to the sum of the squares of the lengths of the two sides.* If c is the length of the hypotenuse of the right triangle ABC (see Fig. 5.2) and if a and b are the lengths of the other two sides, then the theorem simply states that

HYPOTENUSE

$$c^2 = a^2 + b^2 \tag{5.34}$$

This well-known theorem is one of the most useful in all mathematics, and it has been proved in scores of ways. The proof that

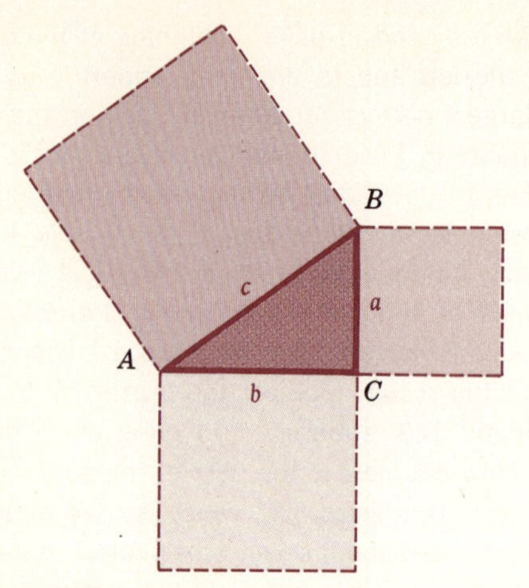

FIGURE 5.2

we present here is based on the concept of similar triangles (which are discussed in Sec. 9.1).

Let ABC be a triangle with the right angle at C. Construct a perpendicular CD from C to the side AB. Then the triangles ADC and ABC in Fig. 5.3 are similar, since angle CAD in triangle ADC is the angle CAB in triangle ABC; angle ADC in triangle ADC is equal to angle ACB in triangle ABC because the angles are both right angles; and angle ACD in triangle ADC is equal to angle ABC in triangle ABC because they are the third angles in triangles whose other two angles are respectively equal. Since the triangles are similar, the ratios of the lengths of corresponding sides are equal, and we have

$$\frac{AB}{AC} = \frac{AC}{AD} \tag{5.35}$$

Therefore,

$$\overline{AC}^2 = AB \cdot AD \tag{5.36}$$

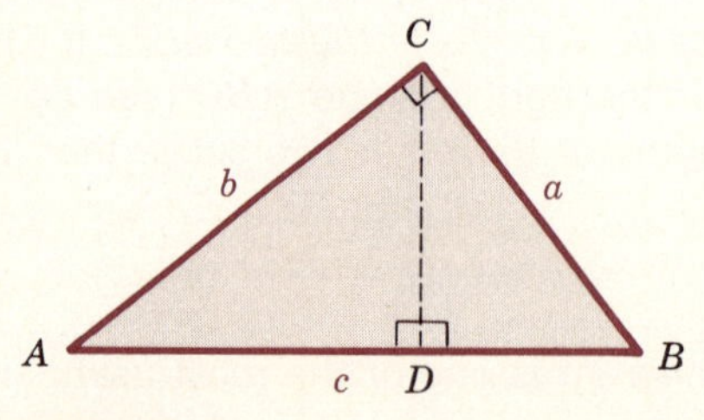

FIGURE 5.3

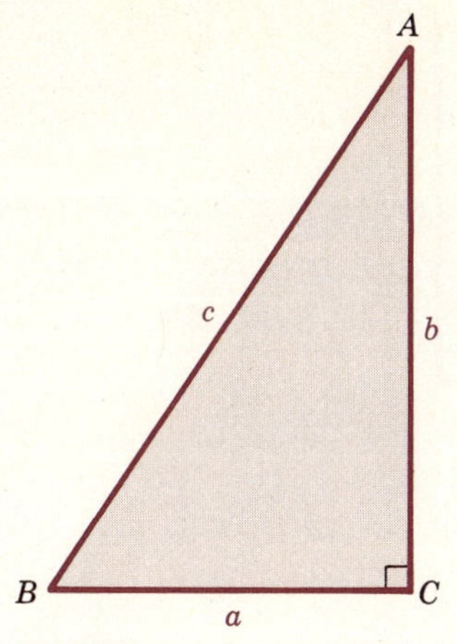

FIGURE 5.4

Likewise, we can show that triangle ABC is similar to triangle CBD and hence

$$\frac{AB}{BC} = \frac{BC}{BD} \tag{5.37}$$

or

$$\overline{BC}^2 = AB \cdot BD \tag{5.38}$$

Adding the corresponding sides of equations (5.36) and (5.38), we obtain

$$\overline{AC}^2 + \overline{BC}^2 = AB \cdot AD + AB \cdot BD = AB(AD + BD) = \overline{AB}^2$$

Using our earlier notation, we may express this result for the triangle of Fig. 5.4 in the form

$$c^2 = a^2 + b^2$$

which completes the proof.

The *converse* of the theorem of Pythagoras, which is true, states that *if* $c^2 = a^2 + b^2$, *then angle C is a right angle.*

EXAMPLE 1 Suppose that in a right triangle one side is 4 feet long and the hypotenuse is 5 feet long. Find the length of the other side.

SOLUTION For such a problem you should always draw the triangle, labeling and numbering the sides according to what is given in the problem in relation to the general triangle in Fig. 5.4.

In the present example we are given that $c = 5$ and one side, say a, is equal to 4. The triangle for this example is shown in Fig. 5.5. Substituting

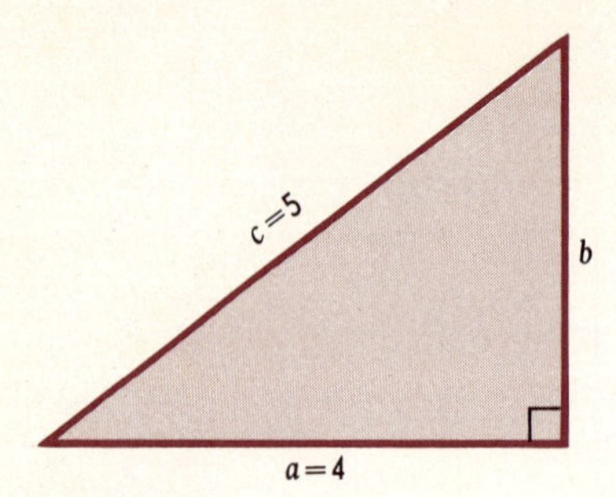

FIGURE 5.5

$a = 4$ and $c = 5$ in the formula $a^2 + b^2 = c^2$, we get

$4^2 + b^2 = 5^2$
$16 + b^2 = 25$
$b^2 = 25 - 16$
$b^2 = 9$

Hence $b = 3$.

EXAMPLE 2 If the sides of a right triangle are 4 feet and 6 feet, respectively, find the length of the hypotenuse in radical form and also in approximate decimal form to three decimal places.

SOLUTION Since the sides are given, we let $a = 4$ and $b = 6$, and draw the triangle of Fig. 5.6. Substituting these values in the formula $c^2 = a^2 + b^2$, we get

$c^2 = 4^2 + 6^2 = 16 + 36 = 52$

So $c = \sqrt{52} = 7.211$, approximately, by use of the table of approximate square roots.

EXERCISES

In Exercises 1 to 10, you will need to find the square root of a number in order to find the required answer. Determine this square root by the

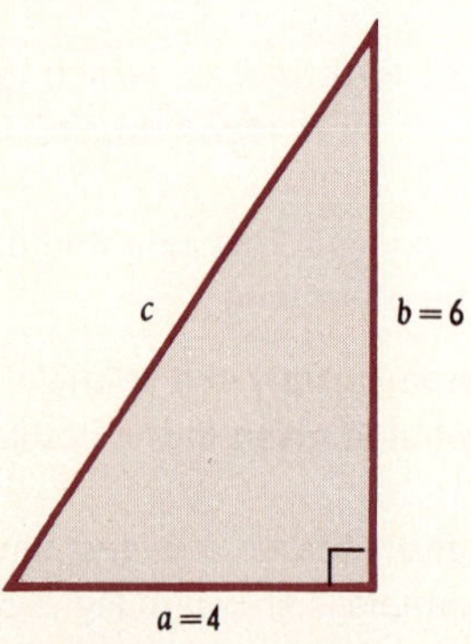

FIGURE 5.6

approximation and refinement method of the preceding section, and give the answer correct to one decimal place.

1 The hypotenuse of a right triangle is 23 feet in length, and one side is 18 feet long. Find the length of the other side.
2 A baseball diamond is a square whose side is 90 feet. Find the distance from the first base to the third base.
3 One side of a right triangle is 20 feet long, and the other side is 40 feet long. How long is the hypotenuse?
4 A guy wire 100 feet long is attached to a vertical tower 85 feet above the ground. How far from the base of the tower may the other end be fastened?
5 A rope is needed to reach from the top of a tent pole to a peg in the ground. The pole is 17.5 feet high, and the peg is 15.3 feet from the foot of the pole. How long a rope is needed?
6 A box (rectangular parallelepiped) has the inside dimensions 6 feet, 8 feet, and 9 feet. What is the length of the longest steel rod that can be placed inside the box?
7 An automobile travels 25 miles east and then 20 miles north. What is the shortest distance from its stopping point to its starting point?
8 A log is 24 inches in diameter. Find the length of the side of the largest square beam which can be cut from it.
9 In a right triangle one side is 8 feet in length and the hypotenuse is 14 feet in length. Use the theorem of Pythagoras to find the length of the third side.
10 The base of a rectangle is twice the length of its altitude, and the rectangle is inscribed in a circle with diameter 30 inches in length. Find the length of the sides of the rectangle.
11 Show that the area A of an equilateral triangle is given by $A = \dfrac{s^2\sqrt{3}}{4}$, where s is the length of the side of the triangle. *Hint:* Let h denote the altitude of the triangle (Fig. 5.7). Use the theorem of Pythagoras and the property of an equilateral triangle that the altitude bisects the side to which it is drawn and is perpendicular to this side.

Use the formula in Exercise 11 as an aid in working Exercises 12 to 15.

12 What is the area of an equilateral triangle whose side is 4 feet long? 6 yards long? 8 inches long?
13 What is the length of a side of an equilateral triangle whose area is $9\sqrt{3}$ square feet? $4\sqrt{3}$ square yards? $25\sqrt{3}$ square inches?
14 Find to three decimal places the length of a side of an equilateral triangle whose area is 9 square feet; 4 square yards; 25 square inches.
15 Find to two decimal places the length of the side of the equilateral triangle for which the number of square inches in the area is twice the number of inches in the perimeter.

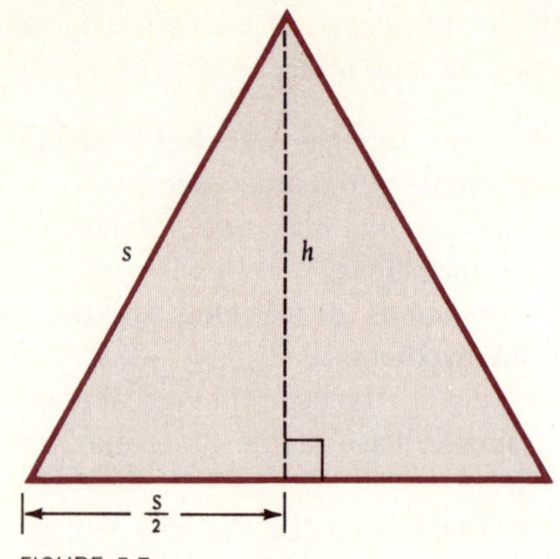

FIGURE 5.7

16 The hypotenuse of a right triangle is 16 yards long, and one side is 11 yards long. Find the length of the other side.

17 What is the length of the diagonal of a square whose side is 10 inches long? Whose side is a inches long?

18 How long must a ladder be to reach from a point on the ground which is 12 feet from the foot of a wall of a building to a point on the wall 18 feet above the ground?

19 A ladder 30 feet long which is resting on the ground just touches the lower edge of a window when the foot of the ladder is 9 feet from the wall. How high is the window above the ground?

20 The diagonal of a square is 2 feet longer than its side. Find the length of the side.

21 Show that for any integer m, the three integers $2m$, $m^2 - 1$, and $m^2 + 1$ are the lengths of the sides and the hypotenuse, respectively, of a right triangle.

22 A rectangular lot is 100 feet long and 50 feet wide. What is the length of its diagonal?

23 Which of the following sets of numbers may be the length of the sides and of the hypotenuse of a right triangle: 6, 8, 10; 4, 5, 6; 1, 2, 3; or 5, 12, 13?

24 The area of a square is 289 square inches. Find the length of a diagonal of the square.

25 A ladder 20 feet long just reaches a window when the foot of the ladder is 4 feet from the foot of the wall. How high above the ground is the window?

26 From a point A, a boat sails 10 miles due east and then 16 miles due north to reach a point B. How far is B from A?

6

Polynomials

6.1 OPERATIONS WITH POLYNOMIALS

POLYNOMIAL IN x

A *polynomial in x* is an algebraic sum in which each term is either a constant or the product of a constant and a positive integral power of the variable x. To illustrate, $3x^5 - 6x^4 - 7x + 2$ is a polynomial in x, and $6a^3 - 4a^2 + 6$ is a polynomial in a.

Usually the universe of the variable is the set Re of real numbers and will be so understood unless clearly indicated to the contrary (sometimes the universe will be the set C of complex numbers).

DEGREE OF A POLYNOMIAL

The **degree** *of a polynomial* is the highest power of the variable (with a nonzero coefficient) which appears in the polynomial.

GENERAL POLYNOMIAL OF DEGREE n IN x

The **general polynomial of degree** n **in** x is an expression of the form

$$a_0x^n + a_1x^{n-1} + a_2x^{n-2} + \cdots + a_{n-1}x + a_n$$

STANDARD FORM

where $a_0, a_1, a_2, \ldots, a_{n-1}, a_n$ are constants with $a_0 \neq 0$ and n is a positive integer. We call the above expression the **standard form** of a polynomial.

The following expressions are some illustrations of polynomials in standard form: $17x^4 + 5x^3 - 6x + 15$ is a polynomial of degree 4 in x; $22y - 10$ is a polynomial of degree 1 in y; 15 is a polynomial of degree 0 in any variable, since, for any variable x, $15 = 15(x)^0$.

The expression $(2x^3 - 4) + 5x(3x^3 - 2x + 6)$ is not a polynomial, but it can be written in the form of a polynomial by use of the distributive, commutative, and associative properties. We write

$$(2x^3 - 4) + 5x(3x^3 - 2x + 6) = 2x^3 - 4 + 15x^4 - 10x^2 + 30x$$

which written in standard form becomes

$$15x^4 + 2x^3 - 10x^2 + 30x - 4$$

So the original expression can be written as a polynomial of degree 4 in x.

Addition or subtraction of polynomials is accomplished by using the commutative, associative, and distributive properties.

EXAMPLE 1 Add $3x^2 + 2x - 5$ and $6x^3 + 2x^2 - 6x + 10$. Write the result in standard form, and state its degree.

SOLUTION In problems of this type it is convenient to arrange the work so that like powers of x appear in a column. This is done below for the given addends, and their sum is found to be $6x^3 + 5x^2 - 4x + 5$, which is of degree 3.

$$\begin{array}{r} 3x^2 + 2x - 5 \\ 6x^3 + 2x^2 - 6x + 10 \\ \hline 6x^3 + 5x^2 - 4x + 5 \end{array}$$

EXAMPLE 2 Subtract $9x^4 + 2x^2 - 3$ from $x^5 - 2x^4 + 3x^3 - 6x + 5$.

SOLUTION We may arrange the work as shown below:

$$\begin{array}{r} x^5 - 2x^4 + 3x^3 \phantom{{}- 2x^2} - 6x + 5 \\ 9x^4 \phantom{{}+ 3x^3} + 2x^2 \phantom{{}- 6x} - 3 \\ \hline x^5 - 11x^4 + 3x^3 - 2x^2 - 6x + 8 \end{array}$$

Multiplication of polynomials is accomplished by recalling the distributive property (property P_4 in Sec. 5.3). Successive applications of the distributive property will show that to find the product of two polynomials, we multiply one of them successively by each term of the other and add the results.

EXAMPLE 3 Multiply $2x^3 - 6x^2 + 3x - 2$ by $x + 4$.

SOLUTION The work may be arranged as follows:

$$\begin{array}{rl} 2x^3 - 6x^2 + 3x - 2 & \\ x + 4 & \\ \hline 2x^4 - 6x^3 + 3x^2 - 2x \phantom{{}- 8} & \textit{multiplication of the first polynomial by } x \\ 8x^3 - 24x^2 + 12x - 8 & \textit{multiplication of the first polynomial by } 4 \\ \hline 2x^4 + 2x^3 - 21x^2 + 10x - 8 & \textit{adding} \end{array}$$

The product is $2x^4 + 2x^3 - 21x^2 + 10x - 8$.

EXAMPLE 4 Find the product $(2x^2 - 6x + 2)(4x^3 - 3x^2 + 5x - 7)$.

SOLUTION Using the polynomial with the smallest number of terms as the multiplier, we may arrange the work as shown below:

$$\begin{array}{r} 4x^3 - 3x^2 + 5x - 7 \phantom{{}+ 52x - 14} \\ 2x^2 - 6x + 2 \phantom{{}+ 52x - 14} \\ \hline 8x^5 - 6x^4 + 10x^3 - 14x^2 \phantom{{}+ 52x - 14} \\ - 24x^4 + 18x^3 - 30x^2 + 42x \phantom{{}- 14} \\ 8x^3 - 6x^2 + 10x - 14 \\ \hline 8x^5 - 30x^4 + 36x^3 - 50x^2 + 52x - 14 \end{array}$$

To divide one polynomial by another, both polynomials are arranged either in decreasing or in increasing powers of the variable. The first term of the quotient is obtained by dividing the first term in the dividend by the first term of the divisor. The divisor is multiplied by the first term in the quotient, and this product is subtracted from the dividend. The result so obtained is used as a new dividend, and the process is continued in a manner which is illustrated in the following examples.

EXAMPLE 5 Divide $x^3 - 4x^2 + 6x - 3$ by $x - 1$.

SOLUTION The work is usually arranged as shown below:

$$\begin{array}{rll} & x^2 - 3x + 3 & \\ x - 1 \,\big) & x^3 - 4x^2 + 6x - 3 & \\ & x^3 - x^2 & \textit{subtract} \\ \hline & -3x^2 + 6x - 3 & \textit{new dividend} \\ & -3x^2 + 3x & \textit{subtract} \\ \hline & 3x - 3 & \textit{new dividend} \\ & 3x - 3 & \textit{subtract} \\ \hline & 0 & \end{array}$$

Here the dividend is $x^3 - 4x^2 + 6x - 3$, the divisor is $x - 1$, and the quotient is $x^2 - 3x + 3$. That is,

$$\frac{x^3 - 4x^2 + 6x - 3}{x - 1} = x^2 - 3x + 3 \tag{1}$$

EXAMPLE 6 Divide $4x^4 + 5x^2 - 7x + 3$ by $2x^2 + 3x - 2$.

SOLUTION Arranging the polynomials in decreasing powers of x and proceeding as above, we have the work shown below:

$$\begin{array}{rl} & 2x^2 - 3x + 9 \\ 2x^2 + 3x - 2 \,\big) & 4x^4 \qquad + 5x^2 - 7x + 3 \\ & 4x^4 + 6x^3 - 4x^2 \\ \hline & -6x^3 + 9x^2 - 7x + 3 \\ & -6x^3 - 9x^2 + 6x \\ \hline & 18x^2 - 13x + 3 \\ & 18x^2 + 27x - 18 \\ \hline & -40x + 21 \end{array}$$

Here the quotient is $2x^2 - 3x + 9$ with a remainder $-40x + 21$, and we may write

$$\frac{4x^4 + 5x^2 - 7x + 3}{2x^2 + 3x - 2} = 2x^2 - 3x + 9 + \frac{-40x + 21}{2x^2 + 3x - 2} \tag{1}$$

or

$$4x^4 + 5x^2 - 7x + 3 = (2x^2 - 3x + 9)(2x^2 + 3x - 2) - 40x + 21 \tag{2}$$

The correctness of the result could be checked by performing the indicated operations on the right side of (2).

We find it convenient to use symbols such as $P_n(x)$, $Q_n(x)$, $R_n(x)$, etc., to represent polynomials of degree n. Thus

$$P_n(x) = a_0x^n + a_1x^{n-1} + a_2x^{n-2} + \cdots + a_{n-1}x + a_n$$

is the general polynomial of degree n in x,

$$Q_4(x) = 2x^4 - 5x^3 + 6x - 5$$

is a polynomial of degree 4 in x, and

$$R_3(x) = 2x^3 - 5x^2 + 2x$$

is a polynomial of degree 3 in x.

If b is any member of the universe of the variable, $P_n(b)$ is the value of $P_n(x)$ when x is replaced by b; that is,

$$P_n(b) = a_0b^n + a_1b^{n-1} + a_2b^{n-2} + \cdots + a_{n-1}b + a_n$$

Thus, for $R_3(x) = 2x^3 - 5x^2 + 2x$,

$$R_3(-2) = 2(-2)^3 - 5(-2)^2 + 2(-2) = -40$$

and

$$R_3(2) = 2(2)^3 - 5(2)^2 + 2(2) = 16 - 20 + 4 = 0$$

POLYNOMIAL EQUATION

A **polynomial equation** of degree n is an equation of the form

$$P_n(x) = 0$$

Since $P_n(x) = 0$ is an open sentence, we have from the definition given in Sec. 2.1 that the solution set or simply the *solution of a polynomial equation* $P_n(x) = 0$ in the set Re of real numbers is the set denoted by

$$\{x \in Re \mid P_n(x) = 0\}$$

We have already studied in Chap. 4 the solution of the general polynomial equation of degree 1,

$$ax + b = 0$$

and we found that

$$\{x \in Re \mid ax + b = 0\} = \left\{-\frac{b}{a}\right\}$$

In subsequent sections of this chapter we shall consider solutions of polynomial equations of degree higher than 1.

In any division problem the equality

$$\frac{dividend}{divisor} = quotient + \frac{remainder}{divisor} \tag{6.1}$$

or

$$dividend = (quotient)(divisor) + remainder \tag{6.2}$$

is always true.

Equality (6.1) for the general case is comparable with (1) in Example 5 and (1) in Example 6 for the special cases, and equality (6.2) is comparable with (2) in Example 6.

EXAMPLE 7 Divide $x^2 + px + q$ by $x - a$, where p, q, and a are constants.

SOLUTION The work is arranged as in previous examples and is shown below:

$$\begin{array}{r l}
 & x + (p + a) \\
x - a \,\big) & x^2 + px + q \\
 & x^2 - ax \\
\hline
 & (p + a)x + q \\
 & (p + a)x - a(p + a) \\
\hline
 & \qquad a(p + a) + q
\end{array}$$

The quotient is $x + (p + a)$ with remainder $q + ap + a^2$. This means that

$$\frac{x^2 + px + q}{x - a} = x + (p + a) + \frac{q + ap + a^2}{x - a}$$

or

$$x^2 + px + q = [x + (p + a)](x - a) + q + ap + a^2$$

Division of one polynomial by another may be carried out until the remainder obtained is a polynomial of degree *less* than the degree of the divisor. For example, in the problem $(16x^4 - 6x^3 + 4x) \div (x^2 + 2x - 1)$ the remainder will be of degree 1 or less, while if with the same dividend the divisor is $x + 3$, the remainder will be of degree 0 (that is, constant).

EXERCISES

In Exercises 1 to 6, write the given expression as a polynomial in standard form, and state its degree.

1 $3x^4 - 5x + 7 + 8x^2 - 3x^4$
2 $2(x^3 - 4x) + 3x(x^2 - 2x) + 6$
3 $(5x^3 + 4x^2 - 6x) - (5x^3 - 8x^2 - 2x + 9)$
4 $3x(x^4 - 8x + 9) - 3x^5 + x^3 + 4$
5 $7(x + 10) - 7x + 16$
6 $-(4x^3 - 2x^2 + 3x + 5) + (4x^3 + 7x^2 + 2)$

In the following exercises, perform the indicated operation. If the result is a polynomial, write it in standard form, and state its degree; if the result is not a polynomial, write it in the form (6.1), and state the degree of the quotient.

7 $(5x^6 - 3x^4 - 8x^3) + (-2x^6 + 3x^4 - 1)$
8 $(2x^3 - 8x^2 + 10) - (5x^3 - 7x^2 + 9)$
9 $(x^3 - x^2 + 3) + (3x^4 + 6x^3 + 8x - 4) - (3x^4 + 5x^3 + 8x - 1)$
10 $(7x^4 + 3x^3 + 6a^2x^2 - 4ax) + (2x^4 + 3x^3 - 6ax)$
11 $(2x - 1)(3x + 2)$
12 $(4x - 7)(2x + 5)$

13 $(x^2 + 5x - 4)(2x - 5)$
14 $(x^2 - 2x + 3)(2x^2 + x - 3)$
15 $(-x^2 - ax + 2a^2)(3x^2 - 6ax + 2a^2)$
16 $(4x^2 - x + 2)(-x^2 + 3x + 5)$
17 $(2x^2 - 9x + 3)(3x^2 + 7x - 1)$
18 $(3x^2 - 8x)(-x^2 + 2x - 5)$
19 $(x^3 + 2x^2 - 5) \div (x + 3)$
20 $(2x^4 - 3x^3 + x^2 + 9x - 9) \div (x^2 - 2x + 3)$
21 $(-4x^4 + 22x^2 - 34x) \div (-2x^2 + 2)$
22 $(6x^3 + 5x^2 - 3x - 2) \div (3x - 2)$
23 $(2x^3 + 5x^2 - 33x + 20) \div (2x - 5)$
24 $(4x^4 - 13ax^3 + 12a^2x^2 - 5a^3x + 2a^4) \div (4x^2 - ax + a^2)$
25 $(3x^4 - 7ax^3 - 4a^2x^2 + 5a^3x - a^4) \div (x^2 - 3ax + a^2)$
26 $(3x^5 + 3x^4 - 10x^3 - 10x^2 - 8x - 8) \div (x^2 + 3x + 2)$

6.2 SOLUTIONS OF QUADRATIC EQUATIONS BY FACTORING

A *quadratic equation* is a polynomial equation of degree 2. The **general quadratic equation** in x may be written as

GENERAL QUADRATIC EQUATION

$$ax^2 + bx + c = 0 \qquad a \neq 0 \tag{6.3}$$

where a, b, and c are constants. Thus

$$x^2 - 5x + 6 = 0 \tag{6.4}$$
$$9x^2 + 4x = 0 \tag{6.5}$$

and

$$x^2 - 25 = 0 \tag{6.6}$$

are quadratic equations in x.

COMPLETE QUADRATIC EQUATION

PURE QUADRATIC EQUATION

If a quadratic equation contains the variable to both the second and the first power, it is called a **complete quadratic equation.** If the quadratic equation contains the variable to the second power but *not* to the first power, it is called a **pure quadratic equation.** Equations (6.4) and (6.5) are complete quadratic equations; equation (6.6) is a pure quadratic equation.

To solve the quadratic equation (6.3) means to find the members of the set

$$\{x \in Re \mid ax^2 + bx + c = 0\}$$

Recall from Chap. 4 that the basic idea used in solving an equation of the first degree in one variable, say x, is to find an equation (or sentence) of as simple a form as possible that is equivalent to the given equation. (For a first-degree equation it is always possible to find an equation of the form $x = k$ that is equivalent to the given equation.) This is the basic idea in solving any equation. We shall try to solve a quadratic equation by finding a sentence containing only first-degree equations that is equivalent to the given quadratic equation.

One procedure for trying to find a sentence containing only first-degree equations that is equivalent to a given quadratic equation is the following:

1 First, write the given quadratic equation in the form $ax^2 + bx + c = 0$.
2 Next, factor the second-degree trinomial $ax^2 + bx + c$ into the product of first-degree polynomials (if this is possible).
3 Finally, use the fundamental property [stated in (5.4) in Sec. 5.3]: *The product of two factors is zero if and only if at least one of these factors is zero.* That is,

$$A \cdot B = 0 \quad \text{if and only if} \quad A = 0 \quad \text{or*} \quad B = 0 \tag{6.7}$$

We apply this property to our particular problem to obtain a sentence in the form of a disjunction† of two first-degree equations.

To illustrate, by factoring we find that $x^2 - x - 2 = 0$ can be written as

$$(x - 2)(x + 1) = 0$$

By using (6.7) we see that this equation is equivalent to the sentence

$$x - 2 = 0 \quad \text{or} \quad x + 1 = 0$$

Thus

$$\{x \mid x^2 - x - 2 = 0\} = \{x \mid x - 2 = 0 \text{ or } x + 1 = 0\}$$

and the solution of $x^2 - x - 2 = 0$ is the union of the solutions of $x - 2 = 0$ and $x + 1 = 0$. Since $\{2\}$ is the solution of $x - 2 = 0$ and $\{-1\}$ is the solution of $x + 1 = 0$, we have found that

$$\{x \mid x^2 - x - 2 = 0\} = \{-1, 2\}$$

ROOT OF AN EQUATION

Each member of the solution set of an equation is called a **root** of the equation. Thus to solve a polynomial equation means to find the *roots* of the equation. In the above illustration we found that -1 and 2 are the roots of $x^2 - x - 2 = 0$.

EXAMPLE 1 Find the roots of the equation $(x - 2)(x - 3) = 0$.

* Recall that we have agreed to use the word "or" in the inclusive sense (see Sec. 2.5).

† The student should review Sec. 2.5 at this point.

SOLUTION Using the property (6.7), we see that a number is a root of

$$(x - 2)(x - 3) = 0$$

if and only if it is a root of one of the equations

$$x - 2 = 0 \qquad x - 3 = 0$$

that is, if and only if it is a root of one of the equations

$$x = 2 \qquad x = 3$$

Hence 2 and 3 are the roots of $(x - 2)(x - 3) = 0$. The roots should always be checked. To check that 2 is a root of the given equation, we note that

$$(2 - 2)(2 - 3) = 0$$

is a true statement.

EXAMPLE 2 Solve the equation

$$x^2 - 4x - 5 = 0 \tag{1}$$

and check the solution.

SOLUTION We want to factor $x^2 - 4x - 5$, the left side of equation (1), into binomial factors. The first terms of the binomials are factors of x^2, so they must be x and x. Write

$$x^2 - 4x - 5 = (x \qquad)(x \qquad)$$

The second terms in the parentheses must be such that their product is -5 and their sum is -4. We find (after some trials and errors, perhaps) that -5 and 1 are such numbers. So

$$x^2 - 4x - 5 = (x - 5)(x + 1)$$

Hence the equation

$$x^2 - 4x - 5 = 0$$

is equivalent to the equation

$$(x - 5)(x + 1) = 0 \tag{2}$$

A number is a root of equation (2) if and only if it is a root of at least one of the equations

$$x - 5 = 0 \qquad x + 1 = 0$$

Hence the roots of (2), and consequently of (1), are 5 and -1.

CHECKING 5 $\quad 5^2 - 4 \cdot 5 - 5 = 25 - 20 - 5 = 0.$
CHECKING -1 $\quad (-1)^2 - 4(-1) - 5 = 1 + 4 - 5 = 0.$

Stated problems may lead to quadratic equations. The following example is illustrative of these:

EXAMPLE 3 A lot is 40 feet long and 26 feet wide. How many feet must be added to both its length and its width to increase the area by 432 square feet?

SOLUTION Let x = the number of feet to be added. Then

$$(40 + x)(26 + x) = 40 \cdot 26 + 432$$

$$x^2 + 66x - 432 = 0 \qquad \text{or} \qquad (x - 6)(x + 72) = 0$$

Therefore a number is a root of the first equation if and only if it is a root of one of the equations

$$x = 6 \qquad x = -72$$

The number -72 has to be rejected. If 72 feet is subtracted from both the length and the width, negative numbers are obtained; negative numbers have no interpretation as dimensions of a rectangle. When $x = 6$, the new dimensions are 46 and 32 and the new area is $46 \cdot 32 = 1{,}472$ square feet. The old area was $40 \cdot 26 = 1{,}040$ square feet; $1{,}472 - 1{,}040 = 432$. Hence 6 feet must be added to each dimension.

EXERCISES

Give the roots of the equations in Exercises 1 to 8.

1 $(x - 3)(x - 4) = 0$

2 $(x - 5)(x + 6) = 0$

3 $x(x + 2) = 0$

4 $(x + 4)(x - 7) = 0$

5 $2x(x - 4) = 0$

6 $(x - 1)(2x - 6) = 0$

7 $\frac{2}{3}(x - 1)(3x - 9) = 0$

8 $x(x + 3) = 0$

In Exercises 9 to 24, find the roots of the given equation by factoring, and check your result.

9 $x^2 + 3x + 2 = 0$

10 $y^2 + 6y + 8 = 0$

11 $z^2 + 7z + 12 = 0$

12 $x^2 - 10x + 21 = 0$

13 $w^2 - 9w + 18 = 0$

14 $s^2 + 11s + 24 = 0$

15 $y^2 + 3y - 40 = 0$

16 $y^2 - y - 42 = 0$

17 $p^2 + 12p + 27 = 0$

18 $y^2 - 18y + 72 = 0$

19 $x^2 - 11x + 30 = 0$

20 $x^2 - 14x + 48 = 0$

21 $x^2 - 4x - 32 = 0$

22 $x^2 - x = 72$

23 $y^2 = 3y + 28$

24 $5x = 24 - x^2$

25 Three times the square of a number is 45 more than six times the number. Find the number.

26 Find a number which when increased by 17 is equal to 60 times the reciprocal of the number.

27 The area of a rectangle is 40 square feet. The length is 3 feet more than the width. Find the dimensions.

28 If the product of two consecutive odd numbers is 143, find the numbers.

29 The square of a certain number diminished by the number itself is 12. What is the number?

30 A train travels 150 miles at a uniform rate. If the rate had been 5 miles per hour more, the journey would have taken 1 hour less. Find the

rate of the train. *Hint:* Recall from Sec. 4.4 that for uniform motion $d = rt$, so that $t = \frac{d}{r}$; obtain expressions for t in terms of r for the two situations, and make use of the underlying equality in the problem. Solve the resulting equation for r.

31 Find a number x such that the product of x and a number which is 7 more than x is 44.

32 The area of a rectangle is 221 square feet, and one side is 4 feet longer than the other. Find the dimensions.

In Exercises 33 to 40, write the given quadratic equation in the general form $ax^2 + bx + c = 0$, with $a > 0$, and give the values of a, b, and c.

33 $3x^2 = 10x - 3$

34 $5x = 6x^2 - 4$

35 $9x^2 + 16 = 0$

36 $4 - 2x + 7x^2 = 0$

37 $5x^2 = 4x$

38 $(2x + 3)(x - 2) = 0$

39 $3x(x - 3) = 2(x - 1)$

40 $\frac{1}{x - 2} + \frac{1}{x + 2} = 3$

In Exercises 41 to 44, find the roots of the given equation in terms of a by factoring, and check your result.

41 $x^2 + 6ax - 16a^2 = 0$

42 $x^2 + 9ax + 14a^2 = 0$

43 $x^2 - 3ax - 10a^2 = 0$

44 $x^2 - 7ax + 12a^2 = 0$

45 The product of the first and second of three consecutive integers is 12 more than six times the third integer. Find the integers.

46 The perimeter of a rectangle is 124 feet, and the area is 672 square feet. Find the length and the width of the rectangle.

47 The sum of the positive integers 1, 2, 3, 4, . . . , n is $\frac{n(n + 1)}{2}$. How many consecutive positive integers, starting with 1, must be taken for their sum to be 435?

48 Use the fact given in Exercise 47 to determine how many consecutive positive integers, starting with 4, must be taken for their sum to be 814.

49 Find the positive integer such that its cube is greater than its square by 132 times the number itself.

50 Are there three consecutive even integers such that the product of the first and the third integer is 7 more than 10 times the second integer?

51 An open box is to be made from a square piece of tin by cutting out a square with sides of 3 inches in length from each corner and turning up the sides of the remaining material. Find the area of the original square if the box is to contain 432 cubic inches.

52 How large must a square piece of metal be to form a box by cutting from each corner a square, as in Exercise 51:

(a) If the box is to have a depth of 4 inches and a volume of 1,024 cubic inches?

(b) If the box is to have a depth of 3 feet and a volume of 27 cubic feet?

6.3 FACTORING ANOTHER TYPE OF EXPRESSION

We want to be able to solve quadratic equations of the type

$$ax^2 + bx + c = 0 \qquad a \neq 1$$

by factoring. In order to do this, we need to consider an additional type of factoring of a trinomial. Let us obtain the product of $ax + b$ and $cx + d$. Using the familiar distributive property, we obtain

$$\begin{aligned}(ax + b)(cx + d) &= (ax + b)cx + (ax + b)d \\ &= acx^2 + bcx + adx + bd\end{aligned}$$

or

$$(ax + b)(cx + d) = acx^2 + (bc + ad)x + bd \tag{6.8}$$

In connection with the product (6.8), observe that:

1 The first terms of the factors when multiplied together give the first term of the trinomial.
2 The last terms of the factors when multiplied together give the last term of the trinomial.
3 The middle term of the trinomial is the sum of the product of the two adjacent terms and the product of the two extreme terms of the factors.

EXAMPLE 1 Find the product of $2x - 3$ and $3x + 1$.

SOLUTION Using the procedure outlined above, we get

$$(2x - 3)(3x + 1) = 2x \cdot 3x + (-9x + 2x) + (-3)(1) = 6x^2 - 7x - 3$$

In factoring a general trinomial of the form $ax^2 + bx + c$ with $a > 0$, it is well to note that:

1 If the third term of the given trinomial is *positive,* then the second terms of its factors both have the *same sign* and this sign is the same as that of the middle term of the trinomial.
2 If the third term of the given trinomial is *negative,* then the second terms of its factors have *opposite* signs.

EXAMPLE 2 Factor $15x^2 + 8x - 12$.

SOLUTION Write $(5x - 2)(3x + 6)$ for a first trial. However, since

$$5x \cdot 6 + (-2)(3x) = 30x - 6x = 24x$$

this combination fails to give the correct middle term.

Next try $(5x + 6)(3x - 2)$. Now

$$(5x + 6)(3x - 2) = 15x^2 + (-10 + 18)x - 12 = 15x^2 + 8x - 12$$

So these are the correct factors. Hence

$$15x^2 + 8x - 12 = (5x + 6)(3x - 2)$$

EXAMPLE 3 Solve the equation

$$4x^2 - 7x + 3 = 0 \qquad (1)$$

SOLUTION We inspect the left side, trying to find correct factors for it. We see that:

The first terms of the two binomial factors must be either $4x$ and x, or $2x$ and $2x$.

The last terms of the binomial factors may be either 3 and 1 or -3 and -1.

The middle term is *negative,* and consequently the connecting sign between the two terms in each of the binomial factors is minus. Therefore, the possibilities are

$$(4x - 3)(x - 1) \qquad (4x - 1)(x - 3) \qquad (2x - 1)(2x - 3)$$

The first of these is found to be correct; so

$$4x^2 - 7x + 3 = (4x - 3)(x - 1)$$

Hence the roots of equation (1) are roots of

$$4x - 3 = 0 \qquad \text{and} \qquad x - 1 = 0$$

and these roots are $\frac{3}{4}$, and 1.

CHECKING $\frac{3}{4}$ We have

$$4\left(\frac{3}{4}\right)^2 - 7\left(\frac{3}{4}\right) + 3 = 4\left(\frac{9}{16}\right) - \frac{21}{4} + 3 = \frac{36 - 84 + 48}{16}$$

$$= \frac{84 - 84}{16} = \frac{0}{16} = 0$$

CHECKING 1 We have

$$4 \cdot 1^2 - 7 \cdot 1 + 3 = 4 - 7 + 3 = 7 - 7 = 0$$

EXAMPLE 4 If we disregard air resistance, the formula

$$s = 100t - 16t^2 \qquad (1)$$

gives the distance s feet above the ground of an object at the end of t seconds, when the object is thrown vertically upward into the air with a velocity of 100 feet per second. How long will it take the object to reach a height of 144 feet above the ground?

SOLUTION Substituting 144 for s in the formula (1), we get

$$144 = 100t - 16t^2 \qquad (2)$$

or

$$16t^2 - 100t + 144 = 0$$

Dividing both sides of the latter equation by 4, we have

$4t^2 - 25t + 36 = 0 \qquad \text{or} \qquad (4t - 9)(t - 4) = 0$

The desired values of t are roots of $4t - 9 = 0$ and $t - 4 = 0$, and these roots are $\frac{9}{4}$ and 4. By substitution you can see that both these values check in equation (2). The object will be 144 feet above the ground at the end of $2\frac{1}{4}$ seconds, and again at the end of 4 seconds. When $t = 2\frac{1}{4}$ seconds, the object is rising; when $t = 4$ seconds, the object is falling toward the ground after having been as high as it can go.

EXERCISES

In Exercises 1 to 4, find the given product.

1 $(3x + 7)(6x - 5)$

2 $(5x + 3)(2x + 7)$

3 $(x - 9)(2x + 1)$

4 $(3x - 5)(2x - 7)$

In Exercises 5 to 12, factor the given expression. Check each result by multiplying the factors.

5 $3x^2 + 7x + 2$

6 $2x^2 - 5x - 3$

7 $6x^2 + 7x - 20$

8 $2y^2 + 11y + 15$

9 $2w^2 + 5w - 3$

10 $18 - 5r - 2r^2$

11 $3y^2 - 4yz + z^2$

12 $2x^2 - 11x - 21$

In Exercises 13 to 24, find the roots of the given equation by factoring, and check your result.

13 $3x = 1 + 2x^2$

14 $42 - 13x + x^2 = 0$

15 $2y^2 - 11y = 21$

16 $3y^2 = 5y + 12$

17 $2x^2 + 7x - 4 = 0$

18 $\dfrac{3}{4x^2} + \dfrac{7}{8x} - \dfrac{5}{2} = 0$

19 $\dfrac{x + 3}{2x - 7} = \dfrac{2x - 1}{x - 3}$

20 $\dfrac{6}{x} - \dfrac{5}{x + 2} = 3$

21 $18x^2 + 21x - 4 = 0$

22 $x(x + 3) = 18$

23 $(x + 3)^2 = x^2 + (x - 3)^2$

24 $\dfrac{4x}{3x - 2} + \dfrac{5}{x} = \dfrac{7}{3}$

In Exercises 25 to 28, find the roots of the given equation in terms of a by factoring, and check your result.

25 $2x^2 + ax - 3a^2 = 0$

26 $6a^2x^2 - 7ax - 3 = 0$

27 $x^2 + 3ax + 2a^2 = 0$

28 $\dfrac{x + 2a}{x + 3a} + 1 = \dfrac{36a^2}{(x + 3a)^2}$

29 The sum of a number and its reciprocal is $\frac{29}{10}$. Find the number.

30 If three times the square of a certain number is increased by the number itself, the sum is 10. Find the number.

31 Find the dimensions of a rectangle, whose area is 357 square feet, if its length exceeds its width by 4 feet.

32 The formula $s = -16t^2 + 100t$ gives the distance s, in feet above the ground, of an object at the end of t seconds when the object is thrown vertically upward into the air with a velocity of 100 feet per second (air resistance neglected). Find the number of seconds at the end of which the object will be 100 feet above the ground.

33 The decrease t, in degrees, between the boiling point of water (212 degrees) at sea level and that at an elevation h feet above sea level is given by the equation $t^2 + 517t - h = 0$. Find t when $h = 7{,}980$.

34 A ball is thrown vertically upward from the ground with a velocity of 64 feet per second, with the distance s above the ground at time t given by the formula $s = 64t - 16t^2$. How many seconds after the ball is thrown will it be 48 feet above the ground?

35 The length of a rectangle is $1\frac{1}{2}$ feet more than the width, and the area is 76 square feet. Find the width, the length, and the perimeter of the rectangle.

36 The perimeter of a rectangle is 26 yards, and the area is $\frac{165}{4}$ square yards. Find the length and the width of the rectangle.

37 The length of a rectangle is $1\frac{1}{4}$ feet more than the width, and the area is $\frac{403}{8}$ square feet. Find the width, the length, and the perimeter of the rectangle.

38 Find a negative fraction such that three times the square of the fraction is 2 more than the fraction. Is there a positive number which also satisfies this condition? If so, what is it?

In Exercises 39 to 44, factor the given expression. Check each result by multiplying the factors.

39 $12x^2 - 17x + 6$

40 $15x^2 - 4x - 3$

41 $3y^2 + 11y + 6$

42 $6 + 5x - 4x^2$

43 $4z^2 - 9z + 2$

44 $12 - 5x - 2x^2$

In Exercises 45 to 56, find the roots of the given equation by factoring, and check your result.

45 $5x^2 - 13x - 6 = 0$

46 $12z^2 - 16z = -5$

47 $3x^2 = 2 + 5x$

48 $3y^2 + 22y + 7 = 0$

49 $2(x + 10) = 9x + 3(4 + 5x^2)$

50 $\dfrac{3 + 2y}{y - 1} + \dfrac{8 - 3y}{3 - y} = 9$

51 $4x^2 + 4x = 3$

52 $x^2 + \frac{10}{3}x + 1 = 0$

53 $\dfrac{16}{x^2 - 1} + \dfrac{6}{x + 1} = 5$

54 $3x^2 - 7x + 2 = 0$

55 $\dfrac{y + 4}{y - 4} + \dfrac{y - 4}{y + 4} = \dfrac{10}{3}$

56 $2x^2 + x - 3 = 0$

6.4
ROOTS OF PURE QUADRATIC EQUATIONS

Recall that a pure quadratic equation is one in which the first-degree term is missing. So

$$ax^2 + c = 0 \qquad a \neq 0 \tag{6.9}$$

is the general form of a pure quadratic equation.

We obtain the roots of a pure quadratic equation by first solving it for the second power of the variable, that is, for x^2, if x is the variable. Then the roots of the given equation will be the *two square roots* of the value for x^2 thus obtained.

EXAMPLE 1 Solve $4x^2 - 100 = 0$.

SOLUTION Adding 100 to both sides of the given equation, we get

$$4x^2 = 100$$

Dividing both sides of this equation by 4, we get

$$x^2 = 25 \tag{1}$$

So the roots of the given equation are the roots of the first-degree equations

$$x = +\sqrt{25} = 5 \qquad x = -\sqrt{25} = -5$$

The last two equations are usually written together in a more compact form by use of the double sign $\pm$.* Using this notation, we pass from the equality (1) to

$$x = \pm\sqrt{25} = \pm 5$$

So the roots of the given equation are 5 and -5, as may be verified by substitution.

EXAMPLE 2 If we disregard air resistance, the distance s, in feet, which a body falls from rest in time t, in seconds, is given by the formula $s = 16t^2$. How long does it take for an object to fall 576 feet from rest?

SOLUTION When $s = 576$, we have

$$16t^2 = 576$$

Dividing both sides by 16, we get

$$t^2 = 36$$

So 6 and -6 are roots of the equation. We see that 6 seconds is required for an object to fall 576 feet. The root -6 has no meaning for this problem.

EXAMPLE 3 Solve the equation $4x^2 + 13 = 25$.

*The sentence "$x = \pm 5$" is a compact form of the sentence "$x = 5$ or $x = -5$."

SOLUTION Adding -13 to each side of the given equation, we get the equivalent equation $4x^2 = 12$. This is equivalent to $x^2 = 3$. So $\sqrt{3}$ and $-\sqrt{3}$ are the roots of the given equation.

It is usual to leave answers in radical form, unless decimal approximations to the solutions are specifically requested. In the present case, if approximate decimal answers are requested, we could read from the table of approximate square roots that $\sqrt{3} = 1.732$. So the roots of the given equation, correct to three decimal places, are 1.732 and -1.732.

A decimal approximation of a root should not be treated as being exact in the checking.

CHECKING 1.732 in $4x^2 + 13 = 25$ Is it true that

$$4(1.732)^2 + 13 = 25?$$
$$4(2.999824) + 13 = 25?$$
$$24.999296 = 25?$$

The decimal 24.999296 is not equal to 25; however, when we round it off to four significant figures, we get 25.00.

CHECKING $\sqrt{3}$ in $4x^2 + 13 = 25$ Is it true that

$$4(\sqrt{3})^2 + 13 = 25?$$
$$4 \cdot 3 + 13 = 25?$$
$$12 + 13 = 25?$$

This is obviously true.

Observe that the solution of each of the above examples could also be obtained by factoring.

Solution of equations such as $x^2 + 10 = 0$, whose roots are not real numbers, will be considered in Sec. 6.5.

EXERCISES

In Exercises 1 to 16, find the roots of the given equation, and check.

1 $x^2 = 16$

2 $4x^2 - 49 = 0$

3 $16x^2 - 9 = 0$

4 $x^2 = 1.21$

5 $x^2 = 10.24$

6 $7x^2 + 2 = 30$

7 $y^2 - 84 = -3$

8 $x^2 + 5 = \frac{29}{4}$

9 $\dfrac{x+2}{3} = \dfrac{4}{x-2}$

10 $4x^2 - 27 = x^2$

11 $144 = 121x^2$

12 $\dfrac{5x+2}{8} = \dfrac{4}{5x-2}$

13 $\dfrac{y}{4} = \dfrac{4}{9y}$

14 $\dfrac{5}{x^2+1} = \dfrac{3}{x^2-3}$

15 $\dfrac{x}{5} = \dfrac{5}{4x}$

16 $2y^2 - 90 = 8$

In Exercises 17 to 20, find the roots of the given equation in terms of a, b, and c, and check.

17 $x^2 = 16a^2$

18 $9x^2 - a^2 = 24a^2$

19 $a^2x^2 + 2abx^2 = c^2 - b^2x^2$

20 $a^2x^2 - b^2 - 2bc - c^2 = 0$

21 The sum of the areas of two squares is 250 square inches. If a side of one is three times as long as that of the other, how long is the side of each?

22 The square of a number exceeds 2 by 0.25. Find the number.

23 The side of one square is three times as long as that of another. The difference of their areas is 288 square feet. Find the side of each square.

24 If we disregard air resistance, the distance s, in feet, which a body falls from rest in time t, in seconds, is given by the formula $s = 16t^2$. How long does it take for a body to fall 9 feet? 16 feet? 25 feet? 36 feet? 49 feet?

In Exercises 25 to 32, solve the given equation. Give the roots in both radical form and approximate form to two decimal places. Simplify the radicals.

25 $2x^2 - 1 = 0$

26 $x^2 - 18 = 0$

27 $x^2 = 54$

28 $\dfrac{x^2}{3} - 19 = 0$

29 $\dfrac{x^2}{2} - 1 = \dfrac{x^2}{5}$

30 $\dfrac{x^2 + 2}{3} = \dfrac{5}{6}$

31 $\dfrac{x + 2}{4} = \dfrac{1}{x - 2}$

32 $(x + 10)^2 = 20x + 189$

33 The distance s, in feet, which a body falls from rest in time t, in seconds, is given by the formula $s = 16t^2$, if we disregard air resistance. How long will it take for a body to fall 40 feet? 60 feet? 80 feet? Give your answers in radical form, and also in decimal form, rounded off to one decimal place.

Use the assumptions and formula in Exercise 33 in working Exercises 34 and 35.

34 How many seconds (to one decimal place) will it take a baseball to fall to the street from the top of the Empire State Building, which is 1,250 feet high?

35 How many seconds (to one decimal place) will it take a baseball to fall to the ground from an airplane which is 2 miles above the earth's surface (1 mile = 5,280 feet)?

In Exercises 36 to 41, solve the equation for the positive value of the indicated literal symbol.

36 $A = \pi r^2$, for r

37 $A = 6e^2$, for e

38 $s = 16t^2$, for t

39 $V = \pi r^2 h$, for r

40 $S = 4\pi r^2$, for r

41 $A = \frac{s^2}{4}\sqrt{3}$, for s

In Exercises 42 to 49, find the roots of the given equation.

42 $\frac{2x-1}{2} = \frac{4}{2x+1}$

43 $8y^2 = 60 - 7y^2$

44 $\frac{x-2}{8} = \frac{4}{x+2}$

45 $\frac{3}{x+8} = \frac{x-8}{75}$

46 $\frac{y+3}{1-y} = \frac{4}{y-7}$

47 $\frac{3}{2x} = \frac{x}{5}$

48 $11 - y^2 = 3y^2 + 2$

49 $\frac{3x-1}{3x-8} = \frac{2x+7}{x+2}$

6.5 COMPLETING THE SQUARE—THE QUADRATIC FORMULA

Recall the formula for the square of a binomial,

$$(x + k)^2 = x^2 + 2kx + k^2$$

We say that the trinomial $x^2 + 2kx + k^2$ is a *perfect square,* since it is equal to the square of the quantity $x + k$.

From a given expression

$$x^2 + px$$

which is not a perfect square, one may construct an expression which is a perfect square by adding the square of one-half of the coefficient of x. In $x^2 + px$:

The coefficient of x is p.

One-half of the coefficient of x is $\frac{p}{2}$.

The square of one-half of the coefficient of x is $\frac{p^2}{4}$. Thus

$$x^2 + px + \frac{p^2}{4} = \left(x + \frac{p}{2}\right)^2$$

is a perfect square, namely, the square of $x + \frac{p}{2}$.

To construct a perfect square from

$$x^2 + 12x$$

we add $(\frac{1}{2} \cdot 12)^2 = 6^2 = 36$. We obtain

$$x^2 + 12x + 36 = (x + 6)^2$$

To construct a perfect square from

$$x^2 - 5x$$

we add $[\frac{1}{2}(-5)]^2 = (-\frac{5}{2})^2 = \frac{25}{4}$. We obtain

$$x^2 - 5x + \frac{25}{4} = (x - \frac{5}{2})^2$$

COMPLETING THE SQUARE

This process of constructing a perfect square from an expression of the form $x^2 + px$ is called **completing the square.** Completing the square is used as an aid in solving quadratic equations, particularly those which we cannot readily solve by factoring.

EXAMPLE 1 Solve the equation

$$x^2 - 4x + 2 = 0 \qquad (1)$$

by completing the square.

SOLUTION First add -2 to both sides of the given equation, so that the left side will be of the form $x^2 + px$. We get

$$x^2 - 4x = -2 \qquad (2)$$

To complete the square on the left side of equation (2), we must add the square of one-half of the coefficient of x, namely $(-2)^2 = 4$, to the left side. To obtain an equation equivalent to (2), we must also add 4 to the right side. We then have

$$x^2 - 4x + 4 = 4 - 2$$

or

$$(x - 2)^2 = 2 \qquad (3)$$

From equation (3) we get

$$x - 2 = \pm\sqrt{2}$$

Thus the given equation is equivalent to the sentence

$$x = 2 + \sqrt{2} \qquad \text{or} \qquad x = 2 - \sqrt{2}$$

So $2 + \sqrt{2}$ and $2 - \sqrt{2}$ are the desired roots.

CHECKING $2 - \sqrt{2}$ We see that

$$\begin{aligned}(2 - \sqrt{2})^2 - 4(2 - \sqrt{2}) + 2 &= 4 - 4\sqrt{2} + 2 - 8 + 4\sqrt{2} + 2 \\ &= 8 - 8 = 0\end{aligned}$$

is a true statement.

You should check $2 + \sqrt{2}$.

EXAMPLE 2 Solve the equation

$$3x^2 - 5x - 2 = 0 \qquad (1)$$

SOLUTION It is much simpler to complete the square when the coefficient of x^2 is 1. Therefore the first step is to divide both sides of equation (1) by 3, which gives

$$x^2 - \frac{5}{3}x - \frac{2}{3} = 0 \qquad (2)$$

Adding $\frac{2}{3}$ to both sides of equation (2), we have

$$x^2 - \tfrac{5}{3}x = \tfrac{2}{3} \tag{3}$$

To complete the square on the left side of equation (3), we need to add the square of one-half of the coefficient of x, namely $[\frac{1}{2}(-\frac{5}{3})]^2 = \frac{25}{36}$. Hence we add $\frac{25}{36}$ to both sides of equation (3) and obtain

$$x^2 - \tfrac{5}{3}x + \tfrac{25}{36} = \tfrac{25}{36} + \tfrac{2}{3}$$

or

$$(x - \tfrac{5}{6})^2 = \tfrac{49}{36} \tag{4}$$

Equation (4) is equivalent to

$$x - \tfrac{5}{6} = \pm\tfrac{7}{6}$$

Thus the given equation is equivalent to the sentence

$$x = \tfrac{5}{6} + \tfrac{7}{6} \qquad \text{or} \qquad x = \tfrac{5}{6} - \tfrac{7}{6}$$

The desired roots are 2 and $-\frac{1}{3}$. You should check each of these solutions in equation (1).

From these examples it should be clear that the procedure for solving a quadratic equation by completing the square may be stated as follows:

1 Collect like terms in the given equation, and write the equation so that the terms in x^2 and x are on the left side and the *constant term* (the term which does not contain the variable) is on the right side of the equation.

2 If the coefficient of x^2 is not 1, divide both sides of the equation by this coefficient.

3 After the equation is written in the form $x^2 + px = q$, the next step is to add to both sides of this equation the square of one-half of the coefficient of x, namely $\dfrac{p^2}{4}$.

4 Rewrite the left side as a perfect square and the right side in simplest form.

5 Take the square root of each side, prefixing the $\pm$ sign to the right side.

6 Solve separately the resulting two first-degree equations.

EXAMPLE 3 Solve the *general quadratic equation*

$$ax^2 + bx + c = 0 \qquad a \neq 0 \tag{1}$$

by completing the square.

SOLUTION Dividing each side of (1) by a, we obtain

$$x^2 + \frac{b}{a}x + \frac{c}{a} = 0$$

which is equivalent to

$$x^2 + \frac{b}{a}x = -\frac{c}{a} \tag{2}$$

To complete the square on the left side of equation (2), we add the square of one-half the coefficient of x, namely $\left(\frac{b}{2a}\right)^2$. Adding $\frac{b^2}{4a^2}$ to each side of (2), we obtain

$$x^2 + \frac{b}{a}x + \left(\frac{b}{2a}\right)^2 = \frac{b^2}{4a^2} - \frac{c}{a}$$

So

$$\left(x + \frac{b}{2a}\right)^2 = \frac{b^2 - 4ac}{4a^2} \tag{3}$$

is equivalent to (1). From (3) we get

$$x + \frac{b}{2a} = \pm\sqrt{\frac{b^2 - 4ac}{4a^2}}$$

and simplifying the radical, we can write*

$$x + \frac{b}{2a} = \pm\frac{\sqrt{b^2 - 4ac}}{2a} \tag{4}$$

The sentence (4) is equivalent to the sentence

$$x = -\frac{b}{2a} \pm \frac{\sqrt{b^2 - 4ac}}{2a}$$

or

$$x = \frac{-b \pm \sqrt{b^2 - 4ac}}{2a}$$

Thus the roots of $ax^2 + bx + c = 0$ are

$$\frac{-b + \sqrt{b^2 - 4ac}}{2a} \quad \text{and} \quad \frac{-b - \sqrt{b^2 - 4ac}}{2a}$$

and

$$\{x \mid ax^2 + bx + c = 0\} = \left\{\frac{-b + \sqrt{b^2 - 4ac}}{2a}, \frac{-b - \sqrt{b^2 - 4ac}}{2a}\right\}$$

The sentence

$$x = \frac{-b \pm \sqrt{b^2 - 4ac}}{2a} \tag{6.10}$$

which is a condensed form of the sentence

$$x = \frac{-b + \sqrt{b^2 - 4ac}}{2a} \quad \text{or} \quad x = \frac{-b - \sqrt{b^2 - 4ac}}{2a}$$

* Remember that (4) is a condensed form of the sentence

$$x + \frac{b}{2a} = \frac{\sqrt{b^2 - 4ac}}{2a} \quad \text{or} \quad x + \frac{b}{2a} = \frac{-\sqrt{b^2 - 4ac}}{2a}$$

QUADRATIC FORMULA

is called the **quadratic formula.** By substituting in this formula the values of a, b, and c of a particular quadratic equation, we can determine the roots of that equation.

The expression $b^2 - 4ac$ that occurs under the radical sign in the quadratic formula is of special significance and interest. If we assume that a, b, and c are rational numbers, then the following four statements are true:

1 If $b^2 - 4ac$ is the square of a rational number, the roots of the quadratic equation will be rational numbers. Thus for

$$3x^2 + 7x - 20 = 0$$

we have, using the quadratic formula,

$$x = \frac{-7 \pm \sqrt{49 + 240}}{6} = \frac{-7 \pm \sqrt{289}}{6} = \frac{-7 \pm 17}{6}$$

Here $b^2 - 4ac = 289 = (17)^2$ and the roots are the rational numbers $\frac{5}{3}$ and -4.

2 If $b^2 - 4ac$ is positive but is not the square of a rational number, the roots of the quadratic equation will be irrational numbers. To illustrate, for

$$3x^2 + 5x - 4 = 0$$

use of the quadratic formula gives

$$x = \frac{-5 \pm \sqrt{25 + 48}}{6} = \frac{-5 \pm \sqrt{73}}{6}$$

Here $b^2 - 4ac = 73$, which is positive but not the square of a rational number, and the roots are irrational numbers.

3 If $b^2 - 4ac$ is negative, the roots of the quadratic equation will not be real numbers; they will be complex numbers. For example, using the quadratic formula to find the roots of the equation

$$x^2 + x + 1 = 0$$

we obtain

$$x = \frac{-1 \pm \sqrt{1 - 4}}{2} = \frac{-1 \pm \sqrt{3}\, i}{2}$$

That is, we obtain

$$-\frac{1}{2} + \frac{\sqrt{3}}{2} i \qquad \text{and} \qquad -\frac{1}{2} - \frac{\sqrt{3}}{2} i$$

as roots.

4 If $b^2 - 4ac$ is zero, the quadratic equation has only one root.

For example, using the quadratic formula to find the roots of

$$x^2 + 6x + 9 = 0$$

we obtain

$$x = \frac{-6 \pm \sqrt{36 - 36}}{2} = -3$$

So the root of the given equation is -3. In this connection we note that the left side of the given quadratic equation is a perfect square, namely, $(x + 3)^2$.

EXERCISES

In Exercises 1 to 6, supply the missing term.

1 $(x - 5)^2 = x^2 - 10x + (\ \ )$

2 $(y - \frac{1}{2})^2 = y^2 - y + (\ \ )$

3 $(x + 4)^2 = x^2 + (\ \ ) + 16$

4 $(x - \frac{3}{4})^2 = x^2 - (\ \ ) + \frac{9}{16}$

5 $(u + \frac{2}{5})^2 = u^2 + \frac{4}{5}u + (\ \ )$

6 $(x + \frac{5}{2})^2 = x^2 + (\ \ ) + \frac{25}{4}$

In Exercises 7 to 12, what constant term is needed to make the trinomial a perfect square?

7 $x^2 - 12x + (\ \ )$

8 $x^2 + \frac{2}{5}x + (\ \ )$

9 $x^2 + \frac{1}{3}x + (\ \ )$

10 $y^2 - \frac{2}{7}y + (\ \ )$

11 $x^2 - 8ax + (\ \ )$

12 $x^2 + \frac{3}{4}x + (\ \ )$

In Exercises 13 to 28, solve the given equation by completing the square. Express irrational roots in radical form, and also to two decimal places with the use of the table of approximate square roots.

13 $2x^2 - x - 1 = 0$

14 $6y^2 + y - 1 = 0$

15 $2z^2 + 3z + 1 = 0$

16 $2x^2 + 3x - 8 = 0$

17 $2x^2 - 6x + 1 = 0$

18 $y^2 - 5y - 5 = 0$

19 $5x^2 - 8x + 1 = 0$

20 $x^2 + 6x + 1 = 0$

21 $6x^2 - x - 12 = 0$

22 $5x^2 - 6x - 4 = 0$

23 $3x^2 + 10x + 6 = 0$

24 $2x^2 - 3x - 4 = 0$

25 $2x^2 - 3x - 3 = 0$

26 $9x^2 - 18x + 4 = 0$

27 $2x^2 + 3x - 9 = 0$

28 $9x^2 - 12x - 8 = 0$

29 A number plus its reciprocal is equal to 3. Find the number.

30 A man wants to make a box from a rectangular piece of tin which is 12 inches wide and 18 inches long by cutting equal squares from the four corners and then turning up the sides. Find how large a square must be cut from each of the four corners if the area of the base of the box is to be 91 square inches.

31 After plowing a uniform border inside of a rectangular field 50 rods long by 40 rods wide, a farmer finds that he has plowed 60 percent of the field. Find the width of the border.

32 The formula $s = -16t^2 + 200t$ gives the distance s, in feet above the ground, of an object at the end of t seconds, when the object is thrown vertically upward into the air with a velocity of 200 feet per second. Find when the object will be 500 feet above the ground. Find when the object will hit the ground.

In Exercises 33 to 40, find the roots of the given quadratic equation by use of the quadratic formula.

33 $2x^2 - 5x + 2 = 0$

34 $3x^2 + 4x = 1$

35 $2x^2 - 35x + 75 = 0$

36 $4x = 15 - 3x^2$

37 $x^2 - 2x - 1 = 0$

38 $3x^2 = 4x + 1$

39 $x + 21 = 10x^2$

40 $2x^2 - 32 = 0$

41 A farmer is plowing a rectangular field which is 40 rods wide and 60 rods long. Find how wide a strip he must plow around the field so that one-half of the area of the field will be plowed.

42 A farmer is plowing a rectangular field which is 20 yards wide and 80 yards long. Find how wide a strip he must plow around the field in order to have three-fourths of the field plowed.

43 An airplane travels 1,400 miles in a certain time. If its average speed had been 60 miles per hour greater, the trip would have taken an hour less. Find the speed of the airplane.

44 An airplane travels 1,500 miles in a certain time. If its average speed had been 70 miles per hour greater, the trip would have taken $1\frac{1}{2}$ hours less. Find the speed of the airplane.

In Exercises 45 to 48, find the roots of the given quadratic equation by use of the quadratic formula.

45 $x^2 + 6x + 13 = 0$

46 $x^2 - 6x + 10 = 0$

47 $2x^2 - 3x + 2 = 0$

48 $6 + 6x - 5x^2 = 0$

6.6 PROPERTIES OF THE ROOTS OF QUADRATIC EQUATIONS

Let s_1 and s_2 denote the roots of the general quadratic equation $ax^2 + bx + c = 0$. From the quadratic formula (6.10) we may write

$$s_1 = \frac{-b + \sqrt{b^2 - 4ac}}{2a} \qquad s_2 = \frac{-b - \sqrt{b^2 - 4ac}}{2a}$$

On adding these values of s_1 and s_2, we get

$$\begin{aligned} s_1 + s_2 &= \frac{-b + \sqrt{b^2 - 4ac}}{2a} + \frac{-b - \sqrt{b^2 - 4ac}}{2a} \\ &= \frac{-b - b + \sqrt{b^2 - 4ac} - \sqrt{b^2 - 4ac}}{2a} = \frac{-2b}{2a} = -\frac{b}{a} \end{aligned}$$

On multiplying the values of s_1 and s_2, we obtain

$$s_1 \cdot s_2 = \frac{-b + \sqrt{b^2 - 4ac}}{2a} \cdot \frac{-b - \sqrt{b^2 - 4ac}}{2a} = \frac{b^2 - (b^2 - 4ac)}{4a^2}$$

$$= \frac{4ac}{4a^2} = \frac{c}{a}$$

We have proved the following: *The sum of the roots s_1 and s_2 of the quadratic equation $ax^2 + bx + c = 0$ ($a \neq 0$) is given by*

$$s_1 + s_2 = -\frac{b}{a} \tag{6.11}$$

and the product of the roots is given by

$$s_1 \cdot s_2 = \frac{c}{a} \tag{6.12}$$

EXAMPLE 1 For the equation $x^2 + 6x + 13 = 0$, verify that the roots satisfy the equalities (6.11) and (6.12).

SOLUTION For the given quadratic equation, $a = 1$, $b = 6$, $c = 13$, and thus:
Corresponding to equation (6.11) we have $s_1 + s_2 = -6$
Corresponding to equation (6.12) we have $s_1 \cdot s_2 = 13$
Using the quadratic formula, we find

$$s_1 = \frac{-6 + \sqrt{36 - 52}}{2} = \frac{-6 + \sqrt{16}\, i}{2} = -3 + 2i$$

$$s_2 = \frac{-6 - \sqrt{36 - 52}}{2} = \frac{-6 - \sqrt{16}\, i}{2} = -3 - 2i$$

So $s_1 + s_2 = -3 + 2i - 3 - 2i = -6$, which is the same as the sum $s_1 + s_2$ given above. Also, from the values of s_1 and s_2,

$$s_1 \cdot s_2 = (-3 + 2i)(-3 - 2i) = 9 - 4i^2 = 13$$

and this is the same as the product $s_1 \cdot s_2$ given above.

EXAMPLE 2 Find the sum and the product of the roots of the equation $21x^2 - 9x + 7 = 0$.

SOLUTION Here $a = 21$, $b = -9$, and $c = 7$. Therefore, by use of (6.11) we get

$$s_1 + s_2 = -\frac{b}{a} = -\frac{-9}{21} = \frac{3}{7}$$

By use of (6.12) we obtain

$$s_1 \cdot s_2 = \frac{c}{a} = \frac{7}{21} = \frac{1}{3}$$

If the value of $s_1 + s_2$ which is given by (6.11) and the value of $s_1 \cdot s_2$ which is given by (6.12) are substituted in

$$x^2 + \frac{b}{a}x + \frac{c}{a} = 0$$

it follows that

$$x^2 - (s_1 + s_2)x + s_1 \cdot s_2 = 0 \tag{6.13}$$

The result (6.13) provides a method of forming a quadratic equation with two given roots.

EXAMPLE 3 Form an equation $ax^2 + bx + c = 0$ whose roots are 2 and $-\frac{2}{3}$.

SOLUTION Let $s_1 = 2$, $s_2 = -\frac{2}{3}$. Then

$$s_1 + s_2 = 2 - \frac{2}{3} = \frac{6-2}{3} = \frac{4}{3} \qquad s_1 \cdot s_2 = 2\left(-\frac{2}{3}\right) = -\frac{4}{3}$$

Substituting these values of $s_1 + s_2$ and $s_1 \cdot s_2$ in (6.13), we get for the desired equation

$$x^2 - \tfrac{4}{3}x - \tfrac{4}{3} = 0 \qquad \text{or} \qquad 3x^2 - 4x - 4 = 0$$

Recall from Sec. 6.5 that the expression $b^2 - 4ac$ which appears in the quadratic formula (6.10) has special significance in determining what kind of numbers the roots of the quadratic equation may be. This expression $b^2 - 4ac$ is called the **discriminant** of the quadratic equation $ax^2 + bx + c = 0$. Suppose that $b^2 - 4ac$ has a negative value, say $-k$ $(k > 0)$. Then from (6.10) the roots s_1 and s_2 of the quadratic equation are given by

DISCRIMINANT

$$s_1 = \frac{-b + \sqrt{-k}}{2a} \qquad \text{and} \qquad s_2 = \frac{-b - \sqrt{-k}}{2}$$

or

$$s_1 = \frac{-b + \sqrt{k}\,i}{2a} \qquad \text{and} \qquad s_2 = \frac{-b - \sqrt{k}\,i}{2a}$$

These results prove this statement: *The roots of a quadratic equation* $ax^2 + bx + c = 0$ *are (nonreal) conjugate complex numbers if and only if the discriminant* $b^2 - 4ac$ *has a negative value.* A graphical interpretation of the discriminant will be discussed in Chap. 7.

EXAMPLE 4 (*a*) With the use of the discriminant, determine whether the roots of the equation $x^2 - 14x + 58 = 0$ are real or not real.

(*b*) Find the roots of the given equation, and verify your answer to (*a*).

SOLUTION (*a*) For $x^2 - 14x + 58 = 0$, we have $a = 1$, $b = -14$, $c = 58$. Then $b^2 - 4ac = (-14)^2 - 4 \cdot 1 \cdot 58 = 196 - 232 = -36$, which is seen to be negative. Hence the roots of the given quadratic equation are not real.

(*b*) By use of the quadratic formula, we obtain

$$x = \frac{-(-14) \pm \sqrt{(-14)^2 - 4 \cdot 1 \cdot 58}}{2 \cdot 1} = \frac{14 \pm \sqrt{-36}}{2} = \frac{14 \pm 6i}{2}$$

$$= 7 \pm 3i$$

The roots are $s_1 = 7 + 3i$ and $s_2 = 7 - 3i$, which are conjugate complex numbers.

EXERCISES

In Exercises 1 to 4, find the roots of the given equation by use of the quadratic formula. Check both of the roots in each exercise.

1 $6x^2 + 5x + 1 = 0$

2 $9y^2 - 6y - 2 = 0$

3 $9z^2 - 6z + 2 = 0$

4 $2x^2 - 5x + 4 = 0$

Find the sum $s_1 + s_2$ and the product $s_1 \cdot s_2$ of the roots of the given equation in Exercises 5 to 8 without solving the equations.

5 $2x^2 - 9x - 5 = 0$

6 $4y^2 + 7y + 17 = 0$

7 $3x^2 - 10x = 0$

8 $11x^2 + 4x + 13 = 0$

Write in the form $ax^2 + bx + c = 0$ an equation whose roots are the numbers given in Exercises 9 to 16.

9 $7, -2$

10 $\frac{3}{5}, \frac{-4}{5}$

11 $2 - 3i, 2 + 3i$

12 $2 + \sqrt{3}, 2 - \sqrt{3}$

13 $\sqrt{17}, -\sqrt{17}$

14 $\frac{-3 + \sqrt{3}}{2}, \frac{-3 - \sqrt{3}}{2}$

15 $\frac{-1 + i\sqrt{23}}{4}, \frac{-1 - i\sqrt{23}}{4}$

16 $1 + i\sqrt{2}, 1 - i\sqrt{2}$

17 Write in the form $ax^2 + bx + c = 0$ an equation which has $1 + \sqrt{3}\,i$ for one root.

In Exercises 18 to 20, (*a*) with the use of the discriminant, determine whether the roots of the given equation are real or nonreal numbers; (*b*) find the roots of the given equation, and verify your answer to (*a*).

18 $2y^2 - 7y + 4 = 0$

19 $3x^2 + 2x + 1 = 0$

20 $5x^2 + 7x + 3 = 0$

6.7
THE REMAINDER THEOREM AND THE FACTOR THEOREM

Recall from Sec. 6.1 the equalities

$$\frac{\textit{dividend}}{\textit{divisor}} = \textit{quotient} + \frac{\textit{remainder}}{\textit{divisor}} \tag{6.14}$$

and

$$\textit{dividend} = (\textit{quotient})(\textit{divisor}) + \textit{remainder} \tag{6.15}$$

Suppose that the dividend in equality (6.14) is the polynomial

$$P_n(x) = a_0x^n + a_1x^{n-1} + a_2x^{n-2} + \cdots + a_{n-1}x + a_n$$

and the divisor is the linear factor $x - r$. Then the quotient, which we denote by $Q_{n-1}(x)$, is a polynomial of degree $n - 1$, and the division may always be carried out until a remainder R which is constant (that is, of degree zero) is obtained. To illustrate,

$$\frac{x^3 - 3x^2 + x - 1}{x - 2} = x^2 - x - 1 + \frac{-3}{x - 2} \tag{6.16}$$

as the student should verify. Here the quotient is $x^2 - x - 1$, the remainder is -3, and we may write

$$P_3(x) = x^3 - 3x^2 + x - 1 = (x - 2)(x^2 - x - 1) - 3 \tag{6.17}$$

Note that $P_3(2) = -3$, and hence we see that $P_3(2)$ has the same value as the remainder that was obtained when $P_3(x)$ was divided by $x - 2$. We find $P_3(2)$ from $P_3(x) = x^3 - 3x^2 + x - 1$ by substituting 2 for x; thus

$$P_3(2) = 2^3 - 3 \cdot 2^2 + 2 - 1 = 8 - 12 + 2 - 1 = -3$$

The equality (6.16) is an identity which is true for all values of x except $x = 2$, while (6.17) is true for all values of x.

The fact that $P_3(2)$ has the same value as the remainder when $P_3(x)$ is divided by $x - 2$ is illustrative of the following theorem, which is called the **remainder theorem.**

REMAINDER THEOREM

If a polynomial $P_n(x)$ is divided by $x - r$ until a constant remainder is obtained, then this remainder has the same value as $P_n(r)$. That is, the remainder R is equal to the result which is obtained when r is substituted for x in $P_n(x)$.

PROOF Denote the quotient in this division by $Q_{n-1}(x)$. We have

$$\frac{P_n(x)}{x - r} = Q_{n-1}(x) + \frac{R}{x - r} \tag{6.18}$$

or

$$P_n(x) = (x - r)Q_{n-1}(x) + R \tag{6.19}$$

The equalities (6.18) and (6.19) for the general situation correspond, respectively, to (6.16) and (6.17) for the illustration considered above.

The equality (6.18) is an identity which is true for all values of x except $x = r$, but the equality (6.19) is an identity which holds for all values of x. If in (6.19) we put $x = r$, we obtain

$$P_n(r) = (r - r)Q_{n-1}(r) + R$$

or

$$P_n(r) = R \tag{6.20}$$

which proves the theorem.

EXAMPLE 1 By use of the remainder theorem, find the remainder when $P_3(x) = x^3 - 4x^2 + 7x - 6$ is divided by (a) $x - 1$; (b) $x - 2$; (c) $x - 3$.

SOLUTION (a) Here $r = 1$, and

$$P_3(1) = 1^3 - 4 \cdot 1^2 + 7 \cdot 1 - 6 = 1 - 4 + 7 - 6 = -2$$

By use of the remainder theorem, which is expressed symbolically by (6.20), we have that

$$R = P_3(1) = -2$$

(b) In this case $r = 2$, and

$$P_3(2) = 2^3 - 4 \cdot 2^2 + 7 \cdot 2 - 6 = 8 - 16 + 14 - 6 = 0$$

By (6.20)

$$R = P_3(2) = 0$$

(c) Here $r = 3$, and

$$P_3(3) = 3^3 - 4 \cdot 3^2 + 7 \cdot 3 - 6 = 27 - 36 + 21 - 6 = 6$$

By (6.20)

$$R = P_3(3) = 6$$

Since $P_n(r) = R$, we may write equality (6.19) in the form

$$P_n(x) = (x - r)Q_{n-1}(x) + P_n(r) \tag{6.21}$$

Recall that a root of an equation in the variable x is a value of x which satisfies the equation. If r is a root of the polynomial equation

$$P_n(x) = a_0x^n + a_1x^{n-1} + a_2x^{n-2} + \cdots + a_{n-1}x + a_n = 0$$

then we have

$$P_n(r) = a_0r^n + a_1r^{n-1} + a_2r^{n-2} + \cdots + a_{n-1}r + a_n = 0$$

But when $P_n(r) = 0$, then (6.21) becomes

$$P_n(x) = (x - r)Q_{n-1}(x)$$

FACTOR THEOREM

We have proved the following theorem, which is called the **factor theorem:**

If a polynomial $P_n(x)$ has the value zero when r is substituted for x, then $x - r$ is a factor of $P_n(x)$.

For the polynomial $P_3(x) = x^3 - 4x^2 + 7x - 6$, we found in Example 1(*b*) above that $P_3(2) = 0$. Therefore, by the factor theorem, $x - 2$ is a factor of $P_3(x)$. That such is indeed the case is verified in the work displayed below:

$$
\begin{array}{r|l}
 & x^2 - 2x + 3 \\
x - 2 & x^3 - 4x^2 + 7x - 6 \\
 & \underline{x^3 - 2x^2} \\
 & -2x^2 + 7x \\
 & \underline{-2x^2 + 4x} \\
 & 3x - 6 \\
 & \underline{3x - 6}
\end{array}
$$

If $x - r$ is a factor of the polynomial $P_n(x)$, then

$$P_n(x) = (x - r)Q_{n-1}(x) \tag{6.22}$$

where $Q_{n-1}(x)$ is the quotient obtained when $P_n(x)$ is divided by $x - r$. When $x = r$, we get from (6.22)

$$P_n(r) = (r - r)Q_{n-1}(r) = 0 \cdot Q_{n-1}(r) = 0$$

CONVERSE OF THE FACTOR THEOREM

This proves the following theorem, which is called the **converse of the factor theorem:**

If $x - r$ is a factor of the polynomial $P_n(x)$, then r is a root of the equation $P_n(x) = 0$.

For the polynomial $P_3(x) = x^3 + 4x^2 - 7x - 10$, we find that $P_3(2) = 0$, $P_3(-1) = 0$, and $P_3(-5) = 0$. Therefore, by the factor theorem, $x - 2$, $x + 1$, and $x + 5$ are factors of $P_3(x)$. That is,

$$x^3 + 4x^2 - 7x - 10 = (x - 2)(x + 1)(x + 5)$$

as should be verified by the student by multiplying out the factors on the right.

The principal use which we shall make at this time of the factor theorem is to find linear factors* of a polynomial of the form

$$P_n(x) = x^n + p_1x^{n-1} + p_2x^{n-2} + \cdots + p_{n-1}x + p_n \tag{6.23}$$

in which the coefficient of the highest power of x (that is, the coefficient of x^n) is 1 and all the other coefficients (that is, $p_1, p_2, \ldots, p_{n-1}$, and p_n) are integers. In seeking a factor $x - r$ of $P_n(x)$, we use the fact that *if r is an integer, then it is a factor of the constant term p_n*. For a given numerical value of p_n, we list the integral factors of p_n and use these as trial values for r to see whether or not we can find a value such that $P_n(r) = 0$. If we can find such a value of r, then by the factor theorem, $x - r$ is a factor of $P_n(x)$.

*A linear factor is a factor of the first degree in the variable.

EXAMPLE 2 Express $P_3(x) = x^3 + 2x^2 - 5x - 6$ as the product of linear factors.

SOLUTION We note that the given polynomial $P_3(x)$ is of the form (6.23). The integral factors of the constant term -6 are ± 1, ± 2, ± 3, and ± 6. Now

$$P_3(1) = 1^3 + 2 \cdot 1^2 - 5 \cdot 1 - 6 = 1 + 2 - 5 - 6 = -8$$
$$P_3(2) = 2^3 + 2 \cdot 2^2 - 5 \cdot 2 - 6 = 8 + 8 - 10 - 6 = 0$$

Since $P_3(2) = 0$, it follows from the factor theorem that $x - 2$ is a factor of $P_3(x)$. Dividing the given expression for $P_3(x)$ by $x - 2$, we obtain the quotient $x^2 + 4x + 3$. So we have

$$x^3 + 2x^2 - 5x - 6 = (x - 2)(x^2 + 4x + 3)$$

Recalling the procedure for factoring a trinomial like $x^2 + 4x + 3$, we observe that

$$x^2 + 4x + 3 = (x + 3)(x + 1)$$

Therefore,

$$x^3 + 2x^2 - 5x - 6 = (x - 2)(x + 3)(x + 1)$$

and the given polynomial is expressed as the product of linear factors.

EXAMPLE 3 Express $P_3(x) = x^3 - 5x^2 - 2x + 10$ as the product of linear factors.

SOLUTION Since $P_3(x)$ is of the form (6.23), we list the integral factors of 10; these are ± 1, ± 2, ± 5, and ± 10. We find that

$$P_3(1) = 1^3 - 5 \cdot 1^2 - 2 \cdot 1 + 10 = 1 - 5 - 2 + 10 = 4$$
$$P_3(2) = 2^3 - 5 \cdot 2^2 - 2 \cdot 2 + 10 = 8 - 20 - 4 + 10 = -6$$
$$P_3(5) = 5^3 - 5 \cdot 5^2 - 2 \cdot 5 + 10 = 125 - 125 - 10 + 10 = 0$$

Because $P_3(5) = 0$, the factor theorem tells us that $x - 5$ is a factor of $P_3(x)$. Dividing the given expression for $P_3(x)$ by $x - 5$, as indicated below, we obtain the quotient $x^2 - 2$.

$$\begin{array}{r|l} & x^2 - 2 \\ \hline x - 5 & x^3 - 5x^2 - 2x + 10 \\ & \underline{x^3 - 5x^2} \\ & \quad - 2x + 10 \\ & \quad \underline{-2x + 10} \end{array}$$

Therefore,

$$x^3 - 5x^2 - 2x + 10 = (x - 5)(x^2 - 2)$$

Recalling that

$$a^2 - b^2 = (a - b)(a + b)$$

we see that $x^2 - 2 = (x - \sqrt{2})(x + \sqrt{2})$. Hence

$$x^3 - 5x^2 - 2x + 10 = (x - 5)(x - \sqrt{2})(x + \sqrt{2})$$

EXAMPLE 4 Solve $x^3 - 5x^2 - 2x + 10 = 0$.

SOLUTION Using the result of Example 3, we can write

$$x^3 - 5x^2 - 2x + 10 = (x - 5)(x + \sqrt{2})(x - \sqrt{2})$$

and the equation becomes

$$(x - 5)(x + \sqrt{2})(x - \sqrt{2}) = 0$$

The equation will be satisfied when one or all of the equations $x - 5 = 0$, $x + \sqrt{2} = 0$, $x - \sqrt{2} = 0$ are satisfied. So the three roots of the given equation are 5, $-\sqrt{2}$, and $\sqrt{2}$.

EXERCISES

In Exercises 1 to 4, verify the remainder theorem for the given polynomial $P_n(x)$ by calculating $P_n(r)$ for the indicated value of r, by finding the remainder R when $P_n(x)$ is divided by $x - r$ until a constant remainder is obtained, and by observing that $P_n(r) = R$.

1 $P_3(x) = 2x^3 - 9x^2 - 3x + 1$, $r = 5$
2 $P_5(x) = x^5 - 3x^3 + 5x - 7$, $r = -2$
3 $P_2(x) = ax^2 + bx + c$, $r = k$
4 $P_3(x) = ax^3 + bx^2 + cx + d$, $r = k$

5 Given $P_4(x) = 2x^4 - 11x^3 + 29x^2 - 3x - 117 = (x - 3)(2x^3 - 5x^2 + 14x + 39)$, find $P_4(3)$. Is $x - 3$ a factor of $P_4(x)$? Is 3 a root of $P_4(x) = 0$?
6 Given $P_3(x) = (x - 2)(x - 4)(x - 6)$, is 2 a root of $P_3(x) = 0$? Why? Is 4 a root of $P_3(x) = 0$? Why? Is 5 a root of $P_3(x) = 0$? Why?

By use of the factor theorem, show that:

7 $x - 1$ is a factor of $4x^3 + 3x^2 - 5x - 2$
8 $z + 1$ is a factor of $z^4 - 1$
9 $x + 2$ is a factor of $x^3 - 2x^2 - 5x + 6$
10 $x - 3$ is a factor of $x^4 - 7x^2 - 6x$

In Exercises 11 to 14, use the factor theorem to determine whether the given binomial is a factor of the given polynomial.

11 $x + 3$, $x^4 - 7x^2 + 11x + 19$
12 $y - 4$, $y^4 - 40y^2 + 64$
13 $x - a$, $x^3 - a^3$
14 $x - 3$, $x^3 - x^2 + 2x + 4$

With the use of the factor theorem, express each of the following polynomials as the product of linear factors:

15 $x^3 + 6x^2 + 11x + 6$
16 $z^3 + 4z^2 - 7z - 10$
17 $x^4 + 5x^3 - 10x^2 - 80x - 96$
18 $y^3 - 9y^2 + 26y - 54$

In Exercises 19 to 22, find the linear factors of the left member of the given equation, and state the roots of the equation.

19 $x^3 - x^2 - 4x + 4 = 0$
20 $x^3 - 2x^2 - 5x + 6 = 0$
21 $x^4 - 2x^3 - 7x^2 + 20x - 12 = 0$
22 $x^4 - 13x^3 + 47x^2 - 59x + 24 = 0$

23 The volume of a rectangular box is 105 cubic inches. Find the dimensions of the box if they are three consecutive odd integers.
24 An open rectangular box is to be made from a square piece of tin, 14 inches on a side, by cutting out equal squares from the corners and turning up the sides. How large should these squares be if the volume of the box is to be 192 cubic inches?
25 The volume of a rectangular box is 396 cubic feet. If the height exceeds the width by 5 feet and the length exceeds the height by 2 feet, find the dimensions of the box.
26 How long is the edge of a cube such that after a slice 1 inch thick is cut off from one side of the cube, the volume of the remaining solid is 100 cubic inches?

6.8 METHODS OF FACTORING

We have previously developed the following formulas for factoring:

$$ab + ac + ad = a(b + c + d) \qquad (6.24)$$
$$a^2 + 2ab + b^2 = (a + b)^2 \qquad (6.25)$$
$$a^2 - 2ab + b^2 = (a - b)^2 \qquad (6.26)$$
$$a^2 - b^2 = (a - b)(a + b) \qquad (6.27)$$
$$x^2 + (a + b)x + ab = (x + a)(x + b) \qquad (6.28)$$
$$acx^2 + (bc + ad)x + bd = (ax + b)(cx + d) \qquad (6.29)$$

We now use the factor theorem to develop additional formulas for factoring. Consider the polynomial

$$P_3(x) = x^3 - a^3$$

Now

$$P_3(a) = a^3 - a^3 = 0$$

So by the factor theorem, $x - a$ is a factor of $x^3 - a^3$. To find the other factor, we divide $x^3 - a^3$ by $x - a$, as shown below, and find the quotient to be $x^2 + ax + a^2$.

$$\begin{array}{r l}
 & x^2 + ax + a^2 \\
x - a \,\big) & \overline{x^3 - a^3} \\
 & \underline{x^3 - ax^2} \\
 & \quad ax^2 - a^3 \\
 & \quad \underline{ax^2 - a^2x} \\
 & \qquad a^2x - a^3 \\
 & \qquad \underline{a^2x - a^3}
\end{array}$$

Therefore,

$$x^3 - a^3 = (x - a)(x^2 + ax + a^2) \tag{6.30}$$

The trinomial $x^2 + ax + a^2$ cannot be factored without introducing complex numbers. We shall agree that, unless specifically called for, complex numbers will not be introduced in factors.

As an illustration of the use of (6.30), let us factor $8b^3 - 27a^3$. We have

$$\begin{aligned} 8b^3 - 27a^3 = (2b)^3 - (3a)^3 &= (2b - 3a)[(2b)^2 + (2b)(3a) + (3a)^2] \\ &= (2b - 3a)(4b^2 + 6ba + 9a^2) \end{aligned}$$

For the polynomial

$$P_3(x) = x^3 + a^3$$

note that

$$P_3(-a) = (-a)^3 + a^3 = -a^3 + a^3 = 0$$

Therefore, $x - (-a) = x + a$ is a factor of $x^3 + a^3$. Dividing $x^3 + a^3$ by $x + a$, we find the quotient to be $x^2 - ax + a^2$. So we have

$$x^3 + a^3 = (x + a)(x^2 - ax + a^2) \tag{6.31}$$

The trinomial $x^2 - ax + a^2$ cannot be factored without introducing complex numbers. As an illustration of the use of (6.31), we have

$$8b^3 + 27a^3 = (2b)^3 + (3a)^3 = (2b + 3a)(4b^2 - 6ba + 9a^2)$$

If

$$P_n(x) = x^n - a^n$$

where n is any positive integer, we have

$$P_n(a) = a^n - a^n = 0$$

So $x - a$ is a factor of $x^n - a^n$, and the other factor can be found by dividing $x^n - a^n$ by $x - a$.

EXAMPLE 1 Factor $x^5 - a^5$.

SOLUTION From the result just established we know that $x - a$ is a factor of $x^5 - a^5$. The division of $x^5 - a^5$ by $x - a$ is shown below:

$$\begin{array}{r|l} & x^4 + x^3a + x^2a^2 + xa^3 + a^4 \\ x - a & x^5 - a^5 \\ & x^5 - x^4a \\ \hline & x^4a - a^5 \\ & x^4a - x^3a^2 \\ \hline & x^3a^2 - a^5 \\ & x^3a^2 - x^2a^3 \\ \hline & x^2a^3 - a^5 \\ & x^2a^3 - xa^4 \\ \hline & xa^4 - a^5 \\ & xa^4 - a^5 \\ \hline \end{array}$$

From this division we see

$$x^5 - a^5 = (x - a)(x^4 + x^3a + x^2a^2 + xa^3 + a^4)$$

Consider the polynomial $P_n(x) = x^n + a^n$. We have the following two cases:

CASE 1 n AN ODD INTEGER
Here

$$P_n(-a) = (-a)^n + a^n = (-1)^n \cdot (a^n) + a^n = -a^n + a^n = 0$$

since (-1) raised to an odd power is -1. Hence, *when n is an odd integer, $x - (-a) = x + a$ is a factor of $x^n + a^n$.*

However, observe that, when n is odd,

$$P_n(a) = (a)^n + a^n = a^n + a^n \neq 0$$

and $x - a$ is not a factor of $x^n + a^n$.

CASE 2 n AN EVEN INTEGER
In this case

$$P_n(-a) = (-a)^n + a^n = (-1)^n(a^n) + a^n = a^n + a^n \neq 0$$

since (-1) raised to an even power is 1.

Also, when n is even,

$$P_n(a) = (a)^n + a^n = a^n + a^n \neq 0$$

So neither $x + a$ nor $x - a$ is a factor of $x^n + a^n$ when n is even.

In most situations we are confronted with factoring a binomial or a trinomial. In either case we should first factor out any common factor in accordance with (6.24). To complete the factoring of a trinomial, use (6.25), (6.26), (6.28), or (6.29). To complete the factoring of a binomial, use (6.27), (6.30), or (6.31), or the more general procedures for factoring $x^n - a^n$ and $x^n + a^n$.

If we are given a polynomial to factor, we should first factor out any common factor in accordance with (6.24). Additional factors may be found by using the factor theorem of the preceding section.

The following are some illustrations of factoring:

$$4a^4 - 4a^2b^2 = 4a^2(a^2 - b^2) = 4a^2(a - b)(a + b)$$

$$\begin{aligned}16x^4 - 81 &= (2x)^4 - 3^4 = [(2x)^2 - 3^2][(2x)^2 + 3^2] \\ &= (2x - 3)(2x + 3)(4x^2 + 9)\end{aligned}$$

$$49x^2 + 70xy + 25y^2 = (7x)^2 + 2(7x)(5y) + (5y)^2 = (7x + 5y)^2$$

EXAMPLE 2 Factor $a^6 + b^6$.

SOLUTION We may treat the given expression as the sum of two cubes and write

$$a^6 + b^6 = (a^2)^3 + (b^2)^3$$

Then, using (6.31), we get

$$a^6 + b^6 = (a^2)^3 + (b^2)^3 = (a^2 + b^2)(a^4 - a^2b^2 + b^4)$$

EXAMPLE 3 Factor $6x^2 - x - 15$.

SOLUTION The procedure for factoring a trinomial of this type was discussed in detail in Sec. 6.3. Using that procedure, we find that

$$6x^2 - x - 15 = (2x + 3)(3x - 5)$$

EXAMPLE 4 Factor $10x^5 + 320y^5$.

SOLUTION We have

$$10x^5 + 320y^5 = 10(x^5 + 32y^5) = 10[x^5 + (2y)^5]$$

We know that $x + 2y$ is a factor of this expression. By division of $x^5 + 32y^5$ by $x + 2y$ the other factor is determined as in Example 1, and we find that

$$\begin{aligned}10x^5 + 320y^5 &= 10[x^5 + (2y)^5] \\ &= 10(x + 2y)(x^4 - 2x^3y + 4x^2y^2 - 8xy^3 + 16y^4)\end{aligned}$$

Sometimes it is advantageous to group the terms of a given expression in a particular way in order to facilitate factoring. Consider the following examples:

EXAMPLE 5 Factor $x^2 + 10x + 25 - y^2$.

SOLUTION We write

$$x^2 + 10x + 25 - y^2 = (x^2 + 10x + 25) - y^2 = (x + 5)^2 - y^2$$

which is the difference of two squares. Using (6.27), we have

$(x + 5)^2 - y^2 = (x + 5 - y)(x + 5 + y)$

Therefore

$x^2 + 10x + 25 - y^2 = (x + 5 - y)(x + 5 + y)$

EXAMPLE 6 Factor $ax + ay + az + bx + by + bz$.

SOLUTION Note that the first three terms contain a as a common factor and the last three terms contain b as a common factor. So

$$ax + ay + az + bx + by + bz = a(x + y + z) + b(x + y + z) = (a + b)(x + y + z)$$

EXERCISES

Factor the following expressions:

1 $x^2 - xy - 6y^2$
2 $5a^2 + 25a + 30$
3 $6 - 11x + 4x^2$
4 $a^8 - b^8$
5 $x^7 + y^7$
6 $a^3b^3 - c^3$
7 $64 + c^3$
8 $a^{3n} - y^{3n}$
9 $21x^2 - 22xy - 24y^2$
10 $64a^2 - 36b^2$
11 $4x^6 - 9y^6$
12 $x^6 - 1$
13 $0.01a^2 - 0.01b^2$
14 $x^2 + 6x + 9 - y^2$
15 $3y^3 + 6y^2 + 3y$
16 $x^2 - 3x - xy + 3y$
17 $x^4 - 2x^3 + x^2 - 4$
18 $ax^3 - ay^3$
19 $x^2 - y^2 + 2y - 2x$
20 $(a + b)^2 - (c - d)^2$
21 $a^2bc + ac^2d - ab^2d - bcd^2$
22 $4a^5b^2 + 4a^2b^5$
23 $x^6 - 2x^3 + 1$
24 $(x^3 + y^3) - 2xy(x + y)$
25 $(a - b)^2 - 10(a - b) + 25$
26 $x^2 + 4xy + 4y^2 - z^2$
27 $(x - 3)^3 - 8y^3$
28 $40a^3b - 52a^2b^2 + 12ab^3$
29 $2x^4 + 6x^3 - 8x^2 - 24x$

In Exercises 30 to 35, write the given expression as the product of linear factors.

30 $x^3 - 5x^2 - x + 5$
31 $24x^3 - 96x^2 + 24x + 144$
32 $x^4 - 6x^2 + 8$
33 $x^3 - 3x^2 + 2$
34 $y^4 + y^3 - 39y^2 - 27y + 324$
35 $x^4 - x^3 - 3x^2 + 5x - 2$

In Exercises 36 to 39, simplify the given fraction.

36 $\dfrac{\dfrac{x^2 + y^2}{y} - x}{\dfrac{x^2}{y} + \dfrac{y^2}{x}}$

37 $\dfrac{1 - \dfrac{8}{x^3}}{\dfrac{2}{x} - 1}$

38 $\dfrac{\dfrac{x^3 - x^2 + 4x - 4}{2x^2 - x}}{\dfrac{x^2 + 4}{2x^2 + x - 1}}$

39 $\dfrac{a^4 - b^4}{\dfrac{3a^2}{b^4} + \dfrac{1}{b^2} - \dfrac{2}{a^2}}$

Express the following as a product of linear factors. Introduce complex numbers as necessary. If complex numbers are introduced, express them in the form $a + bi$.

40 $x^2 + 16$

41 $x^3 - 9$

42 $x^3 - 3x^2 + x + 5$

43 $x^4 + x^3 - 2x^2 - 6x - 4$

In Exercises 44 to 47, find the linear factors of the left member of the given equation, and state the roots of the equation. Introduce complex numbers as necessary. If complex numbers are introduced, express them in the form $a + bi$.

44 $x^3 - x^2 + x - 1 = 0$

45 $y^3 - 4y^2 + 7y - 6 = 0$

46 $y^4 - y^3 + 2y^2 - 4y - 8 = 0$

47 $x^5 + x^3 - 2x^2 - 12x - 8 = 0$

6.9
EQUATIONS INVOLVING RADICALS

RADICAL EQUATION

A *radical equation* is an equation in which the variable appears under a radical sign or with a fractional exponent. To illustrate,

$$\sqrt{9x^2 + 4} - 3x - 1 = 0 \qquad \text{and} \qquad x^{1/4} + 5 = 7$$

are radical equations.

Frequently a radical equation can be transformed to a linear or a quadratic equation. In this procedure the first step is to change the given equation to a form in which one side is a radical with a constant multiplier. The next step is to raise both sides to the power which is equal to the index of the radical.

When each side of an equation is raised to the same power, new roots may be introduced. Therefore, if this operation has been performed, the roots of the new equation, called the *derived equation,* must be tested by substitution in the original equation, and any value not satisfying the given equation should be rejected.

EXAMPLE 1 Solve the equation $\sqrt{y + 2} - 2y = 1$.

SOLUTION On adding $2y$ to both sides of the given equation, we obtain

$$\sqrt{y + 2} = 2y + 1 \tag{1}$$

After squaring both sides of equation (1), we have

$$y + 2 = 4y^2 + 4y + 1$$

or

$$4y^2 + 3y - 1 = 0 \tag{2}$$

Now recall that if two numbers are equal, their squares must be equal; however, the squares of two numbers may be equal without those numbers being equal. Therefore, a value of y may satisfy (2) and not satisfy (1). We find that equation (2) has the roots $\frac{1}{4}$ and -1. We test these values to determine if they are roots of equation (1).

TESTING $\frac{1}{4}$ Does $\sqrt{\frac{1}{4} + 2} - 2 \cdot \frac{1}{4} = 1$? Does $\frac{3}{2} - \frac{1}{2} = 1$? Yes. So $\frac{1}{4}$ is a root of the given equation.

TESTING -1 Does $\sqrt{-1 + 2} - 2(-1) = 1$? Does $1 + 2 = 1$? No. Consequently -1 is *not* a root of the original equation. Therefore, the given equation has one root, namely, $\frac{1}{4}$.

EXAMPLE 2 Solve $\sqrt{9x^2 + 4} - 3x - 1 = 0$.

SOLUTION Adding $3x + 1$ to both sides in order that the radical will stand alone on the left side, we get

$$\sqrt{9x^2 + 4} = 3x + 1$$

Raising both sides to the second power, we get

$$9x^2 + 4 = 9x^2 + 6x + 1 \qquad \text{or} \qquad 6x = 3 \qquad \text{or} \qquad x = \tfrac{1}{2}$$

TESTING $\frac{1}{2}$ Does $\sqrt{9(\frac{1}{2})^2 + 4} - 3(\frac{1}{2}) - 1 = 0$? Does $\sqrt{\frac{25}{4}} - \frac{3}{2} - 1 = 0$? Does $\frac{5}{2} - \frac{5}{2} = 0$? Yes. Hence $\frac{1}{2}$ is a root of the given equation.

EXAMPLE 3 Solve the equation $\sqrt{x + 1} - \sqrt{x} = -2$.

SOLUTION Adding $\sqrt{x}$ to both sides of the equation, we obtain

$$\sqrt{x + 1} = \sqrt{x} - 2$$

On squaring both sides, we get

$$x + 1 = x - 4\sqrt{x} + 4 \qquad \text{or} \qquad 4\sqrt{x} = 3$$

Squaring both sides of the latter equation, we have

$$16x = 9 \qquad \text{or} \qquad x = \tfrac{9}{16}$$

TESTING $\frac{9}{16}$ Does $\sqrt{\frac{9}{16} + 1} - \sqrt{\frac{9}{16}} = -2$? Does $\sqrt{\frac{25}{16}} - \sqrt{\frac{9}{16}} = -2$? Does $\frac{5}{4} - \frac{3}{4} = -2$? No. Hence $\frac{9}{16}$ does *not* satisfy the given equation, and this equation has no root.

EXAMPLE 4 Solve $\sqrt{3y - 5} - \sqrt{2y + 3} = -1$.

SOLUTION First we add $\sqrt{2y + 3}$ to both sides of the given equation so that $\sqrt{3y - 5}$

will stand alone on the left side. We then have

$$\sqrt{3y-5} = \sqrt{2y+3} - 1$$

On squaring both sides, we get

$$3y - 5 = 2y + 3 - 2\sqrt{2y+3} + 1$$

or

$$y - 9 = -2\sqrt{2y+3}$$

Squaring both sides, we get

$$y^2 - 18y + 81 = 4(2y+3) \qquad \text{or} \qquad y^2 - 18y + 81 = 8y + 12$$
$$y^2 - 26y + 69 = 0$$

The roots of the derived equation are 3 and 23.

TESTING 3 Does $\sqrt{9-5} - \sqrt{9} = -1$? Does $2 - 3 = -1$? Yes. Therefore 3 is a root of the given equation.

TESTING 23 Does $\sqrt{69-5} - \sqrt{46+3} = -1$? Does $\sqrt{64} - \sqrt{49} = -1$? Does $8 - 7 = -1$? No. Hence 23 is *not* a root of the original equation.

In summary, the given equation has one root, 3.

EXAMPLE 5 Solve $\sqrt[3]{x^2-1} = 2$.

SOLUTION Raising both sides to the third power, we get

$$x^2 - 1 = 8 \qquad \text{or} \qquad x^2 = 9 \qquad \text{or} \qquad x = \pm 3$$

TESTING 3 Does $\sqrt[3]{3^2-1} = 2$? Does $\sqrt[3]{8} = 2$? Yes.

TESTING -3 Does $\sqrt[3]{(-3)^2-1} = 2$? Does $\sqrt[3]{8} = 2$? Yes. Therefore 3 and -3 are roots of the given equation.

EXERCISES

Find the roots of the following equations. In each case be sure to test the roots of the derived equation in the given equation.

1. $\sqrt{x+10} - \sqrt{2x-5} = 0$
2. $\sqrt{x^2+27} + x = 9$
3. $\sqrt{y^2-5} + 1 = y$
4. $\sqrt{x-3} - \sqrt{x} = 3$
5. $\sqrt{10+x} - \sqrt{10-x} = 2$
6. $\sqrt[3]{y+2} = 3$
7. $\sqrt{20+x-x^2} + 5 = x$
8. $\sqrt{2x+5} + \sqrt{x-2} = 3$
9. $x^{3/2} = 8$
10. $\sqrt{2x+3} - \sqrt{x-3} = 3$
11. $(x-3)^{1/4} = 2$
12. $\sqrt[3]{x^2-1} + 1 = x$

13 A stone is dropped into a well and 6 seconds later is heard to strike the water surface. If sound travels 1,100 feet per second, how deep is

the well? *Hint:* Let x = depth of the well in feet. Then $\frac{\sqrt{x}}{4}$ = number of seconds required for the stone to reach the water surface (by using $d = 16t^2$). Further, $\frac{x}{1{,}100}$ = number of seconds required for the sound to reach the top of the well (by using $d = rt$). Then $\frac{\sqrt{x}}{4} + \frac{x}{1{,}100} = 6$.

14 A stone is dropped from a cliff and is heard to strike the ground below in $3\frac{1}{2}$ seconds. How high is the cliff?

7

Relations, Functions, and Graphs

7.1
RELATIONS AND FUNCTIONS

Recall from Sec. 4.9 that we have an *ordered pair* of numbers when we have two numbers, one of which is designated as the first and the other as the second. The two numbers are usually called, respectively, the *first entry* and the *second entry* of the ordered pair, and the symbol (a, b) is used to denote the ordered pair with a as the first entry and b as the second entry. A very common way in which sets of ordered pairs of numbers occur is in connection with open sentences with two variables.

FIRST ENTRY

SECOND ENTRY

An open sentence with two variables is defined by a generalization of the definition of an open sentence with one variable, given in Sec. 2.1. Let x and y be variables whose universe is U. A sentence, denoted by S_{xy}, with the variables x and y is called an **open sentence with two variables** if it has the property that whenever the variables are replaced by any members of U, we obtain a statement that is either true or false, but not both. We most often encounter open sentences with two variables in the form of equations or inequalities; for example, the equation $x - y = 2$ and the inequality $2x - 3y > 5$, where the universe of the variables may be taken as the set Re of real numbers.

AN OPEN SENTENCE WITH TWO VARIABLES

Consider the equation

$$x - y = 2 \tag{7.1}$$

in which the universe is Re. If x is replaced by 6 and y is replaced by 4, we have $6 - 4 = 2$, which is a *true* statement. We say that the ordered pair (6, 4) **satisfies** equation (7.1); on the other hand, the ordered pair (5, 9) does not satisfy the equation since $5 - 9 = 2$ is a *false* statement. The *solution* of equation (7.1) is the set of all the ordered pairs of real numbers that satisfy the equation. We denote the solution of (7.1) by

AN ORDERED PAIR SATISFIES A SENTENCE

$$\{(x, y) \mid x - y = 2\} \tag{7.2}$$

We read equation (7.2) as "the set of all ordered pairs (x, y) of real numbers that satisfy the equation $x - y = 2$ (or, for which $x - y = 2$ is true)," and we say that the equation *specifies* this set of ordered pairs.

Similarly,

$$\{(x, y) \mid 2x - 3y > 5\} \tag{7.3}$$

denotes the *solution* of the inequality

$$2x - 3y > 5 \tag{7.4}$$

that is, (7.3) is the set of all ordered pairs of real numbers that satisfy the inequality (7.4). The ordered pair (5, 1) is a member of the solution (7.3) since when x is replaced by 5 and y by 1, we obtain $2 \cdot 5 - 3 \cdot 1 > 5$, which is true; the ordered pair (4, 9) is not a member of the solution since $2 \cdot 4 - 3 \cdot 9 > 5$ is false. We say that the inequality (7.4) *specifies* the set of ordered pairs (7.3).

SOLUTION OF A SENTENCE S_{xy}

In general, the **solution of a sentence** S_{xy} with two variables is the set of all ordered pairs of elements in the universe U that satisfy the sentence. The solution is denoted by

$$\{(x, y) \mid S_{xy}\}$$

S_{xy} SPECIFIES A SET OF ORDERED PAIRS

and we say that the sentence S_{xy} **specifies** this set of ordered pairs.

Notice that we cannot tabulate the solution sets (7.2) and (7.3) since each consists of an unlimited number of members. As examples of sets of ordered pairs which can be tabulated, consider the following sets specified by sentences with two variables x and y, each of whose universe is $U = \{1, 2, 3\}$:

$$R_1 = \{(x, y) \mid y = x\} = \{(1, 1), (2, 2), (3, 3)\}$$
$$R_2 = \{(x, y) \mid y > x\} = \{(1, 2), (1, 3), (2, 3)\}$$
$$R_3 = \{(x, y) \mid x + y = 4\} = \{(1, 3), (2, 2), (3, 1)\}$$
$$R_4 = \{(x, y) \mid x^2 + y^2 = 5\} = \{(1, 2), (2, 1)\}$$
$$R_5 = \{(x, y) \mid y = 2\} = \{(1, 2), (2, 2), (3, 2)\}$$

Contrast the set R_5 with the set

$$B = \{y \mid y = 2\} = \{2\}$$

Both R_5 and B are specified by the same equation $y = 2$, but in R_5 we are considering $y = 2$ as a sentence with two variables and in B we have $y = 2$ as a sentence with one variable. This indicates one of the advantages of using the set notation to designate solutions of open sentences.

As a sentence with one variable specifies a *condition* on that variable, it is natural to think of a sentence with two variables as specifying a *relation* between the two variables. The concept of a relation between objects or between sets of objects is a very common and basic idea in human experience. We say that there is a relation among the members of a family, and there is a relation between the number of industries in a city and the amount of air pollution. In mathematics we refer to the "greater than" relation and to the relation between the numbers x and y expressed by the equation $x - y = 2$. An examination of the situations in which we use the word "relation" reveals that inherent in all these uses is the idea of a "pairing." There are several choices of just what we shall say a "relation" is. We could say that a relation is a rule which establishes a pairing (or a correspondence) between the members of two sets, or that it is the correspondence itself, or we could say that the relation is the set of ordered pairs which the rule or correspondence establishes. These different ways of thinking about a relation are of course equivalent.

Each of the ways of thinking about a relation (a rule, a pairing, a correspondence, a set of ordered pairs) has its particular advantages. Since one of our major interests is in the graph of a relation, we elect to consider a relation to be a set of ordered pairs, and we use the following definition:

A RELATION IN U

If a universe U is given, a **relation in** U is a set of ordered pairs whose entries are members of U.

So the sets R_1, R_2, R_3, R_4, and R_5 tabulated above are relations in the set $\{1, 2, 3\}$, and the sets symbolized, respectively, by (7.2) and (7.3) are relations in the set of real numbers.

DOMAIN

RANGE

The **domain** of a relation R in U is the set of those members of U which appear as first entries of the ordered pairs belonging to R. The **range** of a relation R in U is the set of those members of U which appear as second entries of the ordered pairs belonging to R. To illustrate, for the relations R_1 to R_5 given above for which it is stated that $U = \{1, 2, 3\}$,

domain of $R_1 = \{1, 2, 3\}$ range of $R_1 = \{1, 2, 3\}$
domain of $R_2 = \{1, 2\}$ range of $R_2 = \{2, 3\}$
domain of $R_3 = \{1, 2, 3\}$ range of $R_3 = \{1, 2, 3\}$
domain of $R_4 = \{1, 2\}$ range of $R_4 = \{1, 2\}$
domain of $R_5 = \{1, 2, 3\}$ range of $R_5 = \{2\}$

If the universe U is not specifically stated, we assume that it is the set Re of real numbers.

While a relation is frequently specified by a sentence S_{xy} (with two variables) in the form of an equation or an inequality, and symbolized by writing

$$R = \{(x, y) \mid S_{xy}\}$$

a relation may be specified by means of a table which gives the corresponding members of the domain and range. For example, the relation R_3 in the universe $U = \{1, 2, 3\}$, which we specified by $R_3 = \{(x, y) \mid x + y = 4\}$, could be specified by the following table:

WHEN $x =$	1	2	3
THEN $y =$	3	2	1

The relation R_6 specified by the table

WHEN $x =$	1	1	2	3	3	4
THEN $y =$	1	2	2	1	3	1

consists of the ordered pairs (1, 1), (1, 2), (2, 2), (3, 1), (3, 3), and (4, 1); its domain is $\{1, 2, 3, 4\}$, and its range is $\{1, 2, 3\}$.

Note that each member of the domain of a relation R is paired with *at least one* member of the range of R. In particular, it may happen that a member of the domain of a relation is paired with *two or more* members of the range. For example, in the relation

$$R_7 = \{(x, y) \mid y^2 = x\}$$

with the universe the set of real numbers, to the number 4 in the domain of R_7 there correspond *two* members of the range, namely, 2 and -2. That is, $(4, 2) \in R_7$ and $(4, -2) \in R_7$. Notice that the domain of R_7 is the set of nonnegative real numbers and the range of R_7 is the set of all real numbers.

A relation with the special property that to each member of its domain there is paired *only one* member of its range is called a

FUNCTION **function.** In other words, a function is a relation with the property that no two of its ordered pairs have the same first entry. To illustrate, of the relations R_1 to R_7 which we have discussed above, R_1, R_3, R_4, and R_5 are functions, while R_2, R_6, and R_7 are not functions.

Since a function is a particular kind of relation, we can think of a function as a rule, or a pairing, or a correspondence, or a set of ordered pairs. These ways of thinking about a function are equivalent. No matter in what form we state the definition of a function, it will be true that a function associates with each member of its domain a *unique* member of its range. In the next section we introduce a symbol for this unique member of the range that is associated with a given member of the domain.

EXERCISES

1 If $U = \{1, 2\}$, list all the ordered pairs whose entries are members of U.

2 If $U = \{1, 2\}$, tabulate each of the relations

$$R_1 = \{(x, y) \mid y = x\} \qquad R_2 = \{(x, y) \mid y > x\} \qquad R_3 = \{(x, y) \mid y < x\}$$

Give the domain and the range of each of these relations.

3 If $U = \{1, 2, 3, 4\}$, list all the ordered pairs whose entries are members of U.

4 If $U = \{1, 2, 3, 4\}$, tabulate each of the relations

$$R_1 = \{(x, y) \mid y = x\} \qquad R_2 = \{(x, y) \mid y > x\}$$

$$R_3 = \{(x, y) \mid y < x\} \qquad R_4 = \{(x, y) \mid x + y = 5\}$$

Give the domain and the range of each of these relations.

5 Find five ordered pairs which are members of the relation $R = \{(x, y) \mid y = 2x\}$.

6 Find five ordered pairs which are members of the relation $R = \{(x, y) \mid x^2 + y^2 = 25\}$, where the universe is the set of rational numbers.

7 Find five ordered pairs which are members of the relation $R_7 = \{(x, y) \mid y^2 = x\}$, where the domain is the set of positive integers.

8 Which of the relations in Exercise 2 above are functions? Why?

9 Which of the relations in Exercise 4 are functions? Why?

10 Is the relation in Exercise 5 a function? Why?

11 Is the relation in Exercise 6 a function? Why?

12 Is the relation in Exercise 7 a function? Why?

7.2 THE FUNCTIONAL NOTATION

If F is a function, the definition in Sec. 7.1 tells us that F associates with each member of its domain *exactly one* member of its range. If x is a member of the domain of a function F, we shall denote the member of the range which the function associates with x by the symbol

$$F(x)$$

$F(x)$ and we read this symbol as "F of x." In other words, $F(x)$ denotes the (unique) second entry of the ordered pair whose first entry is x. Thus, the open sentence $y = F(x)$ is a sentence that specifies the function F, that is,

$$F = \{(x, y) \mid y = F(x)\}$$

We have seen that a relation, and hence a function, can be specified by a sentence in two variables with the domain either explicitly designated or understood. Frequently a function will be specified by giving a formula for $F(x)$ with the domain either explicitly designated or understood. To illustrate,

$$F(x) = x^2 - 4x + 3 \qquad 0 < x < 5 \tag{7.5}$$

specifies the function whose domain is the set of real numbers greater than zero and less than 5 and which associates with any value x in this domain the member $x^2 - 4x + 3$ of the range. The same function could be specified of course by the sentence

$y = x^2 - 4x + 3$, with $0 < x < 5$. We note that, for the function specified by (7.5),

$$F(2) = 2^2 - 4(2) + 3 = -1$$
$$F(3) = 3^2 - 4(3) + 3 = 0$$
$$F(4) = 4^2 - 4(4) + 3 = 3$$
$$F(\tfrac{1}{2}) = (\tfrac{1}{2})^2 - 4(\tfrac{1}{2}) + 3 = \tfrac{5}{4}$$
$F(7)$ is not defined
$$F(a) = a^2 - 4a + 3 \qquad \text{if } a \text{ is a member of the domain}$$

We emphasize that if F is a function and if a is a member of the domain of F, then the symbol $F(a)$ means one and only one thing, the member of the range of F that is associated with the domain member a; in other words,

$$F(a) = b \quad \text{if and only if} \quad (a, b) \in F$$

IMAGE OF a

For this reason, $F(a)$ is sometimes called the **image of** a (under the function F).

In connection with a *function*

$$F = \{(x, y) \mid S_{xy}\}$$

it is customary to call the variable x, whose values are in the domain of the function, the **independent variable,** and the variable y, whose values are in the range of the function, the **dependent variable.**

INDEPENDENT VARIABLE

DEPENDENT VARIABLE

In addition to specifying a function by an equation in two variables, or by a formula for $F(x)$, a function can be specified by a table, as we saw in Sec. 7.1. A table for a function may be obtained from an equation $y = F(x)$, which specifies the function, by the following procedure:

1 Assign to the independent variable x some selected values.
2 Substitute each of these values in turn in the given equation which specifies the function.
3 Perform the operations called for by this equation, and thus calculate corresponding values of y.
4 Finally arrange the corresponding values of x and y in juxtaposition in a table.

The table obtained in this manner for a given function usually will not list all the ordered pairs which make up the function, because only selected values are assigned to x.

EXAMPLE 1 Construct a table for the function specified by the equation

$$y = 2x + 3 \qquad (1)$$

to give values of y corresponding to $x = -4, -3, -2, -1, 0, 1, 2, 3, 4$.

SOLUTION When $x = -4$, then $y = 2(-4) + 3 = -8 + 3 = -5$; when $x = -3$, then $y = 2(-3) + 3 = -6 + 3 = -3$. Notice that we do not assign a value to x and then assign a value to y. We *assign* a value to x and then *compute* the corresponding value of y by performing the arithmetic operations called for in the given equation (1). To find the value of y when $x = 0$, put 0 in place of x in (1) and perform the operations called for. When $x = 0$, we obtain $y = 2 \cdot 0 + 3 = 0 + 3 = 3$. In this manner we construct the following table:

WHEN $x =$	-4	-3	-2	-1	0	1	2	3	4
THEN $y =$	-5	-3	-1	1	3	5	7	9	11

Notice from this table that whenever x is increased by 1, y is increased by 2.

EXAMPLE 2 Construct a table for the function F specified by $F(x) = 2 - x^2$ to give values of $F(x)$ corresponding to $x = -3, -2, -1, 0, 1, 2, 3$.

SOLUTION $F(-3) = 2 - (-3)^2 = -7$ $\quad F(-2) = 2 - (-2)^2 = -2$
$F(-1) = 2 - (-1)^2 = 1$ $\quad F(0) = 2 - (0)^2 = 2$
$F(1) = 2 - (1)^2 = 1$ $\quad F(2) = 2 - (2)^2 = -2$
$F(3) = 2 - (3)^2 = -7$

We arrange these results in the following table:

WHEN $x =$	-3	-2	-1	0	1	2	3
THEN $F(x) =$	-7	-2	1	2	1	-2	-7

The functions appearing in Examples 1 and 2 are examples of a very important class of functions known as the *polynomial functions*. Polynomial functions are defined in this section, and graphs of some of these functions are discussed in Sec. 7.4.

The function F which is specified by the first-degree equation in x and y,

$$y = ax + b \qquad a \neq 0$$

GENERAL FIRST-DEGREE POLYNOMIAL FUNCTION

is called the **general first-degree polynomial function.** That is, the general first-degree polynomial function is the function F for which

$$F(x) = ax + b$$

where a and b are constants with $a \neq 0$.

To illustrate,

$$F = \{(x, y) \mid y = x - 2\}$$

is a first-degree polynomial function with

$$F(x) = x - 2$$

and consequently

$$F(-2) = -4 \qquad F(-1) = -3 \qquad F(0) = -2$$
$$F(1) = -1 \qquad F(2) = 0 \qquad F(k) = k - 2$$

The function F which is specified by the second-degree equation in x and y,

$$y = ax^2 + bx + c \qquad a \neq 0$$

GENERAL SECOND-DEGREE POLYNOMIAL FUNCTION

is called the **general second-degree polynomial function.** That is, the general second-degree polynomial function is the function F for which

$$F(x) = ax^2 + bx + c$$

where a, b, and c are constants with $a \neq 0$.

For example,

$$F = \{(x, y) \mid y = 2x^2 - x + 4\}$$

is a second-degree polynomial function with

$$F(x) = 2x^2 - x + 4$$

and

$$F(-2) = 14 \qquad F(-1) = 7 \qquad F(0) = 4$$
$$F(1) = 5 \qquad F(2) = 10 \qquad F(k) = 2k^2 - k + 4$$

GENERAL nth-DEGREE POLYNOMIAL FUNCTION

The **general nth-degree polynomial function** is the function F for which

$$F(x) = a_0x^n + a_1x^{n-1} + a_2x^{n-2} + \cdots + a_{n-1}x + a_n$$

where n is a positive integer and $a_0, a_1, a_2, \ldots, a_{n-1}$, and a_n are constants with $a_0 \neq 0$.

Thus,

$$F = \{(x, y) \mid y = 6x^4 - 2x^2 + 10x - 1\}$$

is a fourth-degree polynomial function with

$$F(x) = 6x^4 - 2x^2 + 10x - 1$$

Also

$$F = \{(x, y) \mid y = 3x^7 + x^6 - 15x^2 + 4\}$$

is a seventh-degree polynomial function with

$$F(x) = 3x^7 + x^6 - 15x^2 + 4$$

EXERCISES

1 If $F(x) = 6 - 3x$, find $F(-2)$, $F(0)$, $F(\frac{1}{3})$, $F(2)$.

2 If $F = \{(x, y) \mid y = 5x + 7\}$, what is $F(x)$? Find $F(0)$, $F(2)$, $F(\frac{5}{2})$.

3 If $F(x) = x^2 - 4x + 6$, find $F(-2)$, $F(-1)$, $F(0)$, $F(1)$, $F(2)$, $F(3)$, $F(5)$.

4 If $F(x) = \dfrac{x^2 + 3x + 2}{x + 5}$, find $F(1)$, $F(\frac{1}{2})$, $F(-1)$, $F(0)$. If you were asked to find $F(-5)$ for this function, what would be your reply? Why?

5 Construct a table for the function specified by the equation $y = 3x - 2$ to indicate the values of y corresponding to $x = -4, -3, -2, -1, 0, 1, 2, 3, 4$. From this table what can you say about the effect upon y when x is increased by 1?

6 Construct a table for the function specified by $y = -2x + 4$ to indicate the values of y corresponding to $x = -4, -3, -2, -1, 0, 1, 2, 3, 4$. From this table what can you say about the effect upon y when x is increased by 1?

7 In Exercise 6, replace $y = -2x + 4$ by $y = 2x$, and work as before.

8 In Exercise 6, replace $y = -2x + 4$ by $y = -3x$, and work as before.

9 Construct a table to show the values of y corresponding to $x = -4, -3, -2, -1, 0, 1, 2, 3, 4$ when $y = x^2 + 4$.

10 In Exercise 9, replace $y = x^2 + 4$ by $y = -x^2 + 4$, and work as before.

11 In Exercise 9, replace $y = x^2 + 4$ by $y = x^2 - 4$, and work as before.

12 In Exercise 9, replace $y = x^2 + 4$ by $y = -x^2 - 4$, and work as before.

7.3 COORDINATE SYSTEMS

In Chap. 3 we saw that there is a point on the number scale which corresponds to every rational number. Even though there is always another rational number between any two unequal rational numbers, the points which correspond to the rational numbers do not "fill up" a straight line. It was not until the rational number system was extended to the real number system (Sec. 5.2) that we could say that there is a one-to-one correspondence between the points of a line and a set of numbers. That is, given a line L, to every real number there corresponds a point on the line, and to every point on that line there corresponds a real number. When such a correspondence has been established, we call the line L a *number line* for the real numbers or a **one-dimensional coordinate system.** Such a coordinate system is illustrated in Fig. 7.1. In the coordinate system in Fig. 7.1, O is an arbitrarily chosen point corresponding to

ONE-DIMENSIONAL COORDINATE SYSTEM

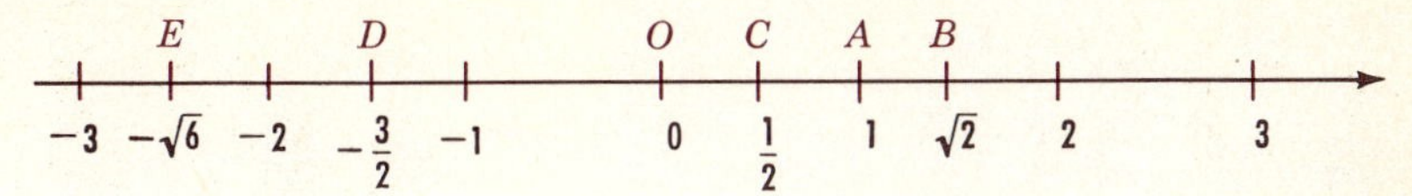

FIGURE 7.1

ORIGIN

the number 0, and it is called the **origin** of the coordinate system. A point A is chosen to represent the number 1, and the segment OA is the *unit of measure*. A may be any point distinct from O. For convenience we usually represent a one-dimensional coordinate system by a *horizontal line*. It could just as well be a vertical line or a line with some other orientation. Also, we customarily take the direction to the right on a horizontal line to be the positive direction (as indicated by the arrow in Fig. 7.1) and the direction to the left to be the negative direction. This is an arbitrary choice; the positive direction can be taken to the left if we wish.

In Fig. 7.1 the point B represents the number $\sqrt{2}$, the point C represents the number $\frac{1}{2}$, the point D represents the number $-\frac{3}{2}$, and the point E represents the number $-\sqrt{6}$. The real number a that corresponds to a point P in a one-dimensional coordinate system is called the **coordinate** of P, and P is called the **graph** of a. We write $P(a)$ to mean "the point P with coordinate a." Thus $\sqrt{2}$ is the coordinate of B, and we write $B(\sqrt{2})$; C is the graph of $\frac{1}{2}$, and we write $C(\frac{1}{2})$; and so on.

COORDINATE

GRAPH OF A NUMBER

Section 4.8 concerning solutions of inequalities of the first degree and the geometrical representation of these solutions should now be reread in the light of the above discussion, with the set Ra of rational numbers being replaced by the set Re of real numbers.

We now extend the ideas of the preceding paragraphs to set up a one-to-one correspondence between the set of points in a plane and the set of all ordered pairs of real numbers. Such a correspondence is made by means of a **two-dimensional coordinate system** (for real numbers), which we shall call simply a coordinate system. Figure 7.2 shows such a coordinate system. In this figure the lines OX and OY are one-dimensional coordinate systems placed so that they are perpendicular to each other and so that their origins coincide. OX is horizontal and is called the x axis; OY is vertical and is called the y axis; the point O of intersection of these axes is called the **origin.** The x axis and y axis together are spoken of as the **coordinate axes.** These coordinate axes divide the plane into four parts which are called *quadrants* and which are numbered I, II, III, and IV as shown in Fig. 7.2; quadrant I is the upper right-hand one, and the others are numbered in counterclockwise order.

TWO-DIMENSIONAL COORDINATE SYSTEM

x AXIS

y AXIS

ORIGIN

COORDINATE AXES

QUADRANTS

We agree that *horizontal distances will be considered positive*

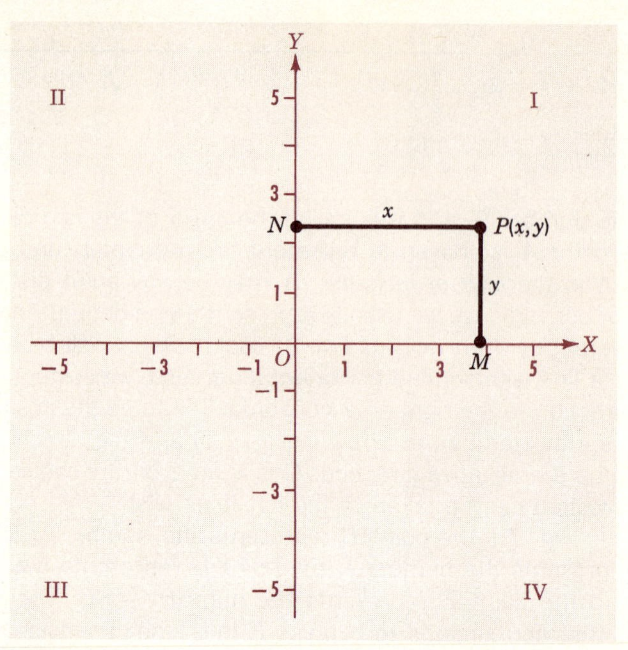

FIGURE 7.2

if measured from left to right and negative if measured from right to left. Vertical distances will be considered positive if measured upward and negative if measured downward.

Let P be any point in a plane on which coordinate axes have been drawn. A unique ordered pair of real numbers can be associated with the point P as follows: Construct two lines through P, one line perpendicular to the x axis and the other perpendicular to the y axis, as shown in Fig. 7.2. Let M be the point at which the line perpendicular to the x axis intersects that axis; let N be the point at which the line perpendicular to the y axis intersects that axis. The distance from N to P, measured by the scale on the x axis is the **abscissa,** or x coordinate of P, and the distance from M to P measured by the scale on the y axis is the **ordinate,** or y coordinate of P. The x coordinate is positive if P is to the right of the y axis and negative if P is to the left of the y axis. The y coordinate is positive if P is above the x axis and negative if P is below the x axis. The abscissa and ordinate together are called the rectangular coordinates, or simply **coordinates,** of the point P. They are written in parentheses as an ordered pair of numbers (x, y) with the abscissa written first, and this is the unique ordered pair of real numbers associated with the point P. To illustrate, in Fig. 7.3:

ABSCISSA

ORDINATE

COORDINATES OF A POINT

For P_1 the coordinates are $(4\frac{1}{2}, 2)$

For P_2 the coordinates are $(-3, 4)$

For P_3 the coordinates are $(-4\frac{1}{2}, -3\frac{1}{2})$

For P_4 the coordinates are $(2, -3)$

In this figure the same unit of measurement (or scale) is used on the x axis and the y axis. However, the same unit does not have to be used for the two axes, and in some problems it is more convenient to use different units on the two axes.

By reversing the process used to determine the ordered pair of real numbers that corresponds to a given point P, we can begin with a given ordered pair of real numbers and determine a unique point in the plane. To illustrate, to locate the point corresponding to the ordered pair (3, 2), we proceed as follows: Start from the origin, and measure 3 units to the right (since the abscissa is positive) along the x axis. From the point at which we arrive, measure up (since the ordinate is positive) a distance of 2 units perpendicular to the x axis (parallel to the y axis). This locates the desired point P shown in Fig. 7.4. Similarly, to determine the point corresponding to the ordered pair $(-4, -3)$, start at the origin, and measure 4 units to the left (since the x coordinate is negative) along the x axis, and then measure 3 units downward (since the y coordinate is negative) perpendicular to the x axis (parallel to the y axis) to arrive at the point with coordinates $(-4, -3)$ shown in Fig. 7.4.

We have established a two-dimensional coordinate system in such a way that to each point in the plane there corresponds one and

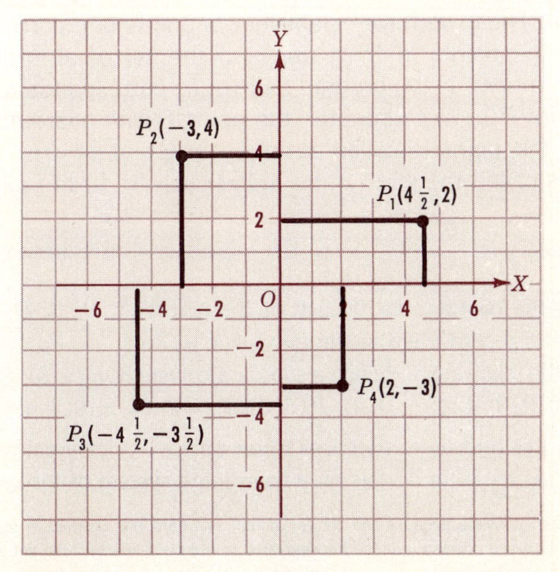

FIGURE 7.3

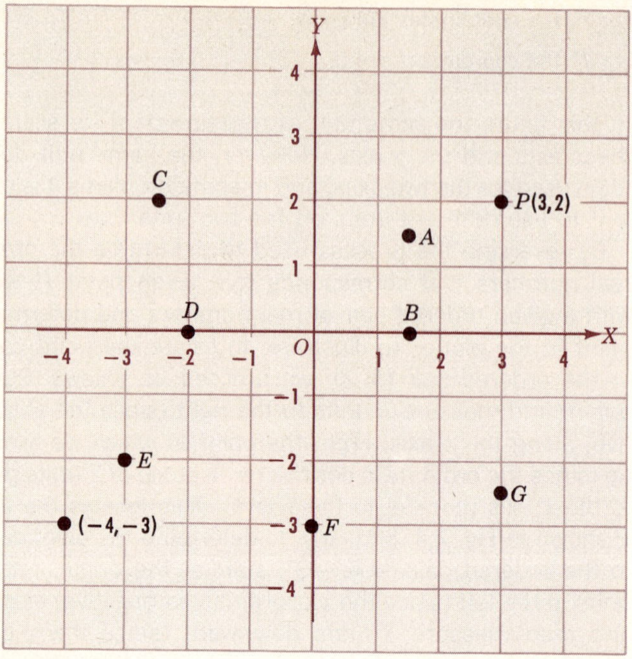

FIGURE 7.4

only one ordered pair of real numbers,* the first of these denoting the abscissa or x coordinate of the point, and the second denoting the ordinate or y coordinate of the point. Conversely, to each ordered pair of real numbers there corresponds one and only one point in the plane. Thus, we have established a one-to-one correspondence between the points in the plane and the set of all ordered pairs of real numbers. If the ordered pair (a, b) corresponds to the point P, we write $P(a, b)$ to denote "the point P with coordinates a and b," and the point is called simply "the point (a, b)." The point $P(a, b)$ is called the **graph of the ordered pair** (a, b) of real numbers.

GRAPH OF AN ORDERED PAIR

EXERCISES

1 Write the coordinates of each of the points A, B, C, D, E, F, G in Fig. 7.4, and record them in the form $A(x, y)$.

2 Plot the following points on a coordinate system: $P_1(5, 6)$; $P_2(-4, -5)$; $P_3(-4, 2)$; $P_4(3, -5)$; $P_5(0, 4)$; $P_6(3, 0)$; $P_7(-2, 0)$; $P_8(0, -4)$.

Note: In this and other exercises when we say, "Plot a point" or "Draw a graph," we mean to do this on a coordinate system of the type we have

*This has been possible because the set Re of real numbers was defined in such a way that there is a one-to-one correspondence between the points on a line and the set Re.

just discussed. In such a case you should carefully construct a coordinate system by drawing and labeling the x axis and y axis on "squared paper," or "graph paper." Also indicate the origin O, and label a convenient number of points of each axis.

3 In what quadrant is each of the following points located: $A(-5, 2)$; $B(3, 7)$; $C(-3, -4)$; $D(8, -3)$?

4 If a point is on the y axis, what is its x coordinate, or abscissa?

5 If a point is on the x axis, what is its y coordinate, or ordinate?

6 What are the coordinates of the origin?

7 What are the coordinates of a point on the x axis 5 units to the left of the origin?

8 What are the coordinates of the vertices of triangle ABC in Fig. 7.5?

9 On a coordinate system draw the rectangle whose vertices are $A(-5, 0)$, $B(-1, -6)$, $C(4, 6)$, $D(8, 0)$.

10 Find the length of AB in Fig. 7.6.

SOLUTION: Through A, draw a line parallel to the x axis, and through B, draw a line parallel to the y axis. These lines intersect at a point with the same abscissa as B and the same ordinate as A, that is, the point $C(-4, -3)$. Triangle ACB is a right triangle, in which

$CA = 6 \qquad CB = 7 \qquad AB = ?$

Applying the Pythagorean theorem (Sec. 5.8) to the right triangle ACB, we get

$\overline{AB}^2 = \overline{CA}^2 + \overline{CB}^2$ or $\overline{AB}^2 = 6^2 + 7^2$

So

$\overline{AB}^2 = 36 + 49 = 85$

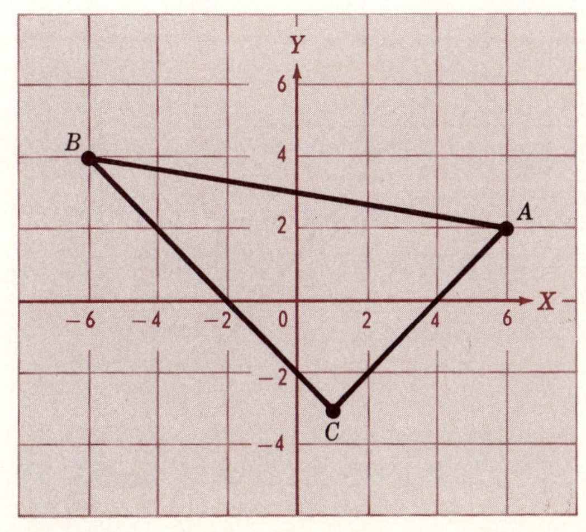

FIGURE 7.5

Then

$\overline{AB} = \sqrt{85}$

Expressed as a decimal correct to two decimal places, $AB = 9.22$.

11 Find the length of each side of the triangle in Exercise 8.

12 Find the length of each side of the rectangle in Exercise 9.

13 Let $P_1(x_1, y_1)$ and $P_2(x_2, y_2)$ be any two distinct points in the plane. Find the distance between these two points; that is, find the length of the line segment P_1P_2.

SOLUTION: Using Fig. 7.7 as a guide, we proceed as in Exercise 10. Through P_1, draw a line parallel to the x axis, and through P_2, draw a line parallel to the y axis. These two lines will intersect at a point C with the same y coordinate as P_1 and the same x coordinate as P_2, that is, the point $C(x_2, y_1)$. Triangle P_1CP_2 is a right triangle with legs P_1C and P_2C and hypotenuse P_1P_2. Applying the Pythagorean theorem to the triangle P_1CP_2, we have

$\overline{P_1P_2}^2 = \overline{P_1C}^2 + \overline{P_2C}^2$

The length $\overline{P_1C}$ of the segment P_1C is the absolute value of the difference of the x coordinates of the points P_1 and C; $\overline{P_1C} = |x_2 - x_1|$. The length

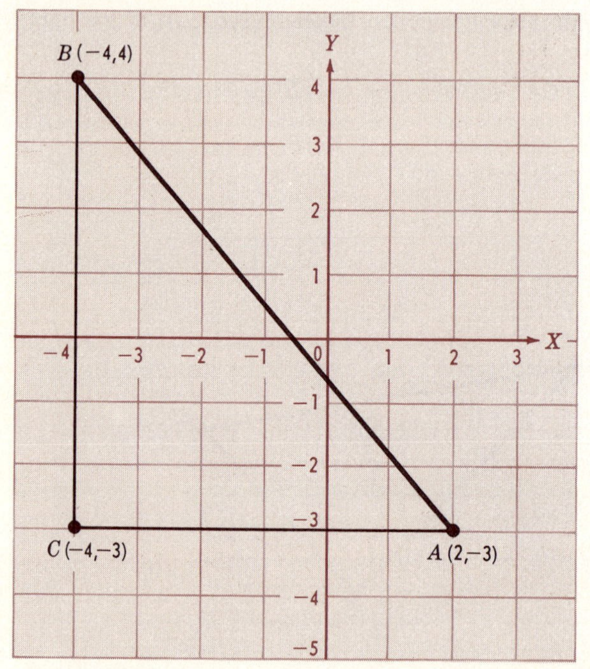

FIGURE 7.6

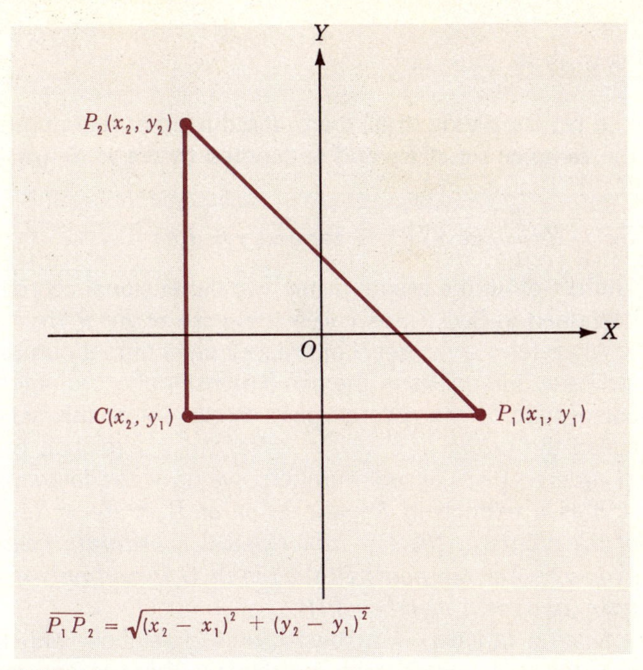

FIGURE 7.7

$\overline{P_2C}$ of segment P_2C is the absolute value of the difference of the y coordinates of P_2 and C; $\overline{P_2C} = |y_2 - y_1|$. Therefore,

$$\overline{P_1P_2}^2 = |x_2 - x_1|^2 + |y_2 - y_1|^2$$

and

DISTANCE FORMULA

$$\overline{P_1P_2} = \sqrt{(x_2 - x_1)^2 + (y_2 - y_1)^2}$$

This result, which gives the distance between two points in terms of the coordinates of the two points, is known as the **distance formula** (for two dimensions).

14 Use the distance formula given in Exercise 13 to find the distance between $P_1(6, 3)$ and $P_2(9, 7)$.

15 Use the distance formula to find the distance between $P_1(6, 3)$ and $P_2(-4, 8)$. Express the result in radical form.

16 Use the distance formula to find the distance between the points $(4, -7)$ and $(2, 3)$. Note that it makes no difference which point is designated P_1 and which is designated P_2, since $(x_2 - x_1)^2 = (x_1 - x_2)^2$ and $(y_2 - y_1)^2 = (y_1 - y_2)^2$.

17 Find the distance between $(2, -3)$ and $(-2, -4)$. Express the result correct to two decimal places.

7.4
GRAPHS IN $Re \times Re$

CARTESIAN SET OF Re

The set consisting of all the ordered pairs of real numbers is called the **cartesian set of** Re and is denoted by $Re \times Re$ (read "Re cross Re"). That is,

$$Re \times Re = \{(x, y) \mid x \in Re \text{ and } y \in Re\}$$

GRAPH OF $Re \times Re$

The set of all the points in the two-dimensional coordinate system described in Sec. 7.3 is called the **graph of** $Re \times Re$.

When for a given set S of ordered pairs of real numbers, we have indicated the points in the two-dimensional coordinate system that correspond to the ordered pairs in the set S, this set of points is called the *graph of S in $Re \times Re$*. Since a relation in Re is a set of ordered pairs of real numbers, we have the following definition:

GRAPH OF A RELATION

If R is a relation in Re, the **graph of** R in $Re \times Re$ *is the set G of all points in the two-dimensional coordinate system with the property that the point $P(a, b)$ is in G if and only if the ordered pair (a, b) is a member of R.*

Since a function is a special kind of relation, with the property that for each member of the domain there is one and only one member of the range, it follows that in the graph of a function there cannot be two points which have the same x coordinate. This means that the graph of a function F is intersected *at most once* by any line perpendicular to the x axis.

GRAPH OF AN EQUATION

GRAPH OF AN OPEN SENTENCE

Frequently we speak of the graph of the function $F = \{(x, y) \mid y = F(x)\}$ as the graph of the equation $y = F(x)$. More generally, the **graph of an equation** in two variables is the graph of the relation determined by that equation, and the **graph of an open sentence** with two variables is the graph of the relation determined by that sentence.

We shall be concerned primarily with the graphs of some simple functions. Consider a function

$$F = \{(x, y) \mid y = F(x)\}$$

The graph of an ordered pair (a, b) belonging to F is a point on the graph G of the function. The graph G is the set of all such points. For a given function specified by an equation $y = F(x)$, we can construct a (partial) table by assigning to x some selected values and computing the corresponding values of y (as explained in Sec. 7.2). The graphs of the ordered pairs in the table provide us with some points on the graph G of the function. For the simple functions which we shall consider, the graph G is obtained by joining the points so obtained by a smooth curve.

EXAMPLE 1 Construct the graph of the equation $y = 2x + 3$. In other words, construct the graph of the function $F = \{(x, y) \mid y = 2x + 3\}$.

SOLUTION From Example 1 in Sec. 7.2 we have the following table:

WHEN $x =$	-4	-3	-2	-1	0	1	2	3	4
THEN $y =$	-5	-3	-1	1	3	5	7	9	11

Each pair of corresponding values of x and y, such as $(-4, -5)$, are the coordinates of a point. If we plot these points and connect them by a smooth curve, we have the graph in Fig. 7.8.

Notice that for the equation in Example 1, when x is increased by 1, then y is increased by 2. Graphically this means that if we start at any point on the graph and go 1 unit to the *right*, it is necessary to go 2 units *up* to get back to the graph. Also note that the graph intersects the y axis at the point (0, 3), where the y coordinate is 3.

EXAMPLE 2 Construct the graph of the equation $y = -3x + 1$. In other words, construct the graph of $\{(x, y) \mid y = -3x + 1\}$.

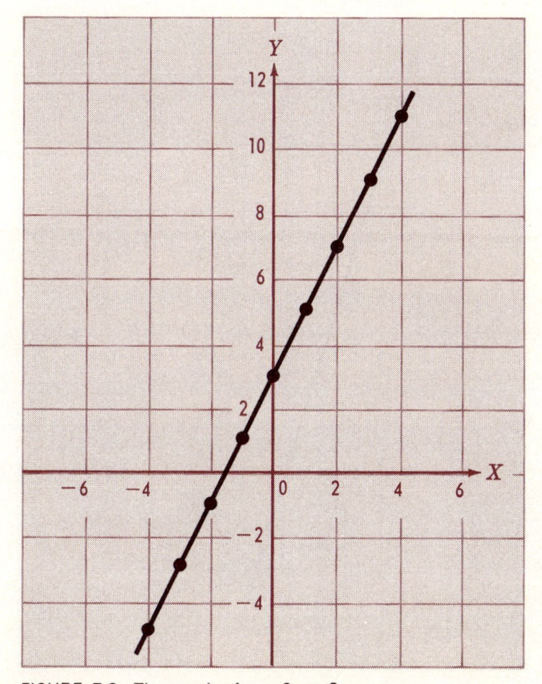

FIGURE 7.8 The graph of $y = 2x + 3$.

SOLUTION We can construct the following table for the equation $y = -3x + 1$:

WHEN $x =$	-3	-2	-1	0	1	2	3
THEN $y =$	10	7	4	1	-2	-5	-8

Plotting the points determined by the table and connecting them by a smooth curve will produce the graph in Fig. 7.9.

Notice that for the equation in Example 2, when x is increased by 1, then y is *decreased* by 3; if we start at any point on the graph in Fig. 7.9 and go 1 unit to the *right*, it is necessary to go 3 units *down* to get back to the graph. Also note that the graph intersects the y axis at (0, 1), whose y coordinate is 1.

EQUATION OF A LINE

In both of the above examples the graph of the given equation appears to be a *straight line*. This is indeed the case, for the graph of every equation of the form

$$y = ax + b \tag{7.6}$$

is a straight line. As illustrated by the examples, the constant a in equation (7.6) determines the "steepness" of the line. If a is *posi-*

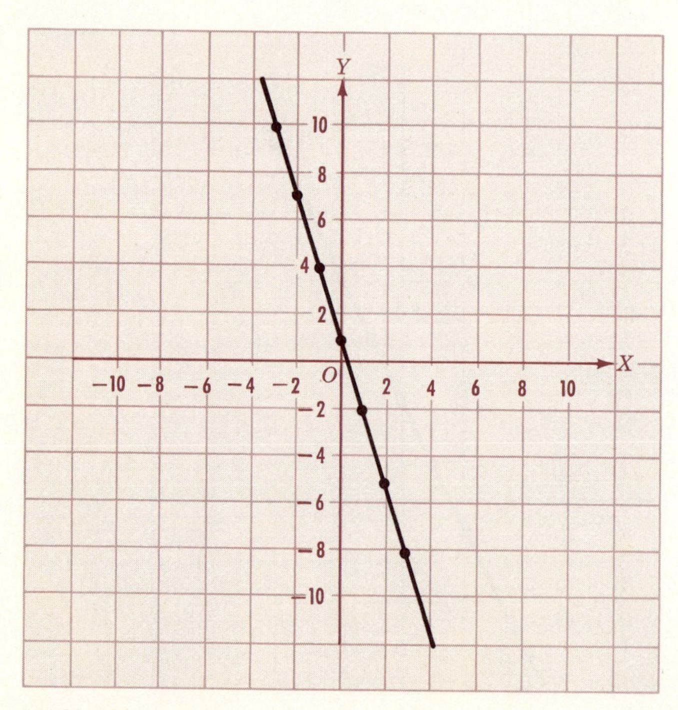

FIGURE 7.9 The graph of $y = -3x + 1$.

tive, increasing x by 1 will *increase* y by an amount determined by the absolute value of a, and the line will slope *up* to the right. If a is *negative,* increasing x by 1 will *decrease* y by an amount determined by the absolute value of a, and the line slopes *down* to the right. If a is *zero,* increasing x will produce no change in y, and the line is *horizontal.* For the line which is the graph of the equation (7.6), the constant a is called the **slope** of the line.

SLOPE OF A LINE

An examination of equation (7.6) shows that when $x = 0$, we have $y = b$. Thus the graph of equation (7.6) intersects the y axis at the point $(0, b)$, whose y coordinate is b. For the line which is the graph of equation (7.6), b is called the y **intercept** of the line.

y INTERCEPT

To illustrate, the graph of

$$y = 6x - 5$$

is a line with slope 6 and y intercept -5; the graph of

$$y = -4x - 2$$

is a line with slope -4 and y intercept -2.

The concepts of slope and intercept can be used to construct quickly the graph of an equation of the form $y = ax + b$.

EXAMPLE 3 Construct the graph of the equation $y = -4x + 2$.

SOLUTION The y intercept is 2; that is, the graph of the equation goes through the point (0, 2). We locate (0, 2) on the coordinate system and then note that the slope of the line is -4. This means that if we go 1 unit to the right from (0, 2), we must go *down* (since the slope is negative) 4 units to get back to the graph. Doing this, as indicated in Fig. 7.10, we obtain a second point of the desired graph and so can draw the line which is the graph of $y = -4x + 2$.

EXAMPLE 4 Find an equation of the straight line which passes through the points whose coordinates are given in the following table:

WHEN $x =$	0	2	4	6	8
THEN $y =$	1	7	13	19	25

SOLUTION Since x ranges from 0 to 8 in this table and y ranges from 1 to 25, it is inconvenient to use the same unit of measurement on the two coordinate axes. We elect to have the unit of measurement on the y axis two-fifths of the unit on the x axis (see Fig. 7.11).

Let us find an equation of the straight line through any two of the given points—say (0, 1) and (4, 13)—and then show that the coordinates of the

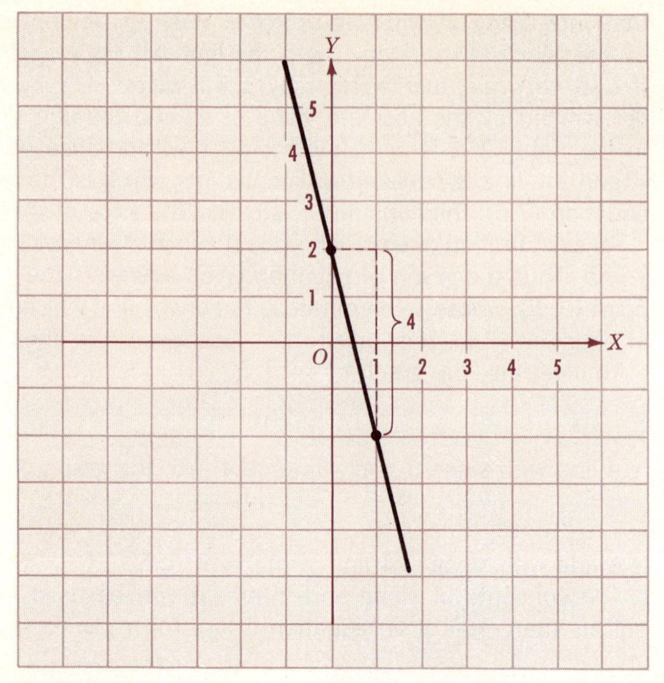

FIGURE 7.10 The graph of $y = -4x + 2$.

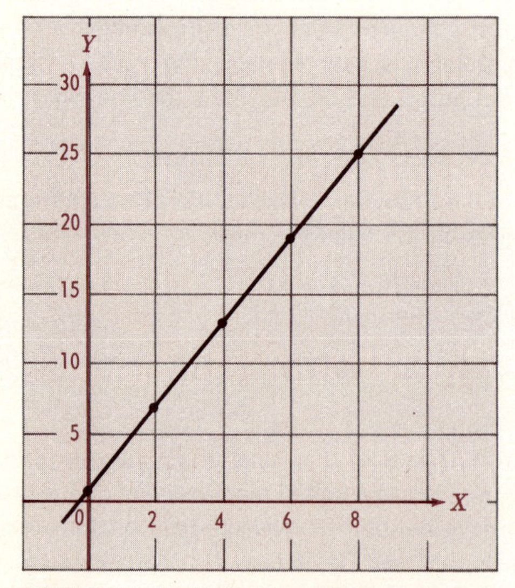

FIGURE 7.11 Graph of $y = 3x + 1$.

other given points also satisfy this equation. The general equation of a straight line is

$$y = ax + b \tag{1}$$

Since this equation is to be satisfied by the ordered pair (0, 1), we have

$$1 = a \cdot 0 + b$$

or

$$b = 1 \tag{2}$$

Similarly, since (1) is to be satisfied by the ordered pair (4, 13), we have

$$13 = a \cdot 4 + b$$

or

$$4a + b = 13 \tag{3}$$

From (2) we have $b = 1$; substituting this value in (3), we get

$$4a + 1 = 13 \qquad \text{or} \qquad 4a = 12 \qquad \text{or} \qquad a = 3$$

So $a = 3$ and $b = 1$ if the points (0, 1) and (4, 13) lie on the line whose equation is $y = ax + b$. That is,

$$y = 3x + 1 \tag{4}$$

is an equation of the straight-line graph through the points (0, 1) and (4, 13). This line has slope 3 and y intercept 1. That equation (4) is satisfied by the other pairs of values of x and y given in the table is readily verified. You should make those verifications.

EXAMPLE 5 Construct the graph of the equation $y = x^2 - 4x + 3$.

SOLUTION We first choose arbitrarily some values of x and compute from the equation the corresponding values of y. These values are recorded in the following table:

WHEN $x =$	-1	0	1	2	3	4	5
THEN $y =$	8	3	0	-1	0	3	8

The points corresponding to these pairs of values are plotted, and the points so obtained are connected by a smooth curve to give us the graph in Fig. 7.12.

EXAMPLE 6 Construct the graph of the equation $y = -2x^2 + 4x + 6$.

SOLUTION We choose for x the values indicated in the table on page 330, and by use of the given equation we compute the corresponding values of y that are shown in the table.

WHEN $x =$	-2	-1	0	$\frac{1}{2}$	1	$\frac{3}{2}$	2	3	4
THEN $y =$	-10	0	6	$\frac{15}{2}$	8	$\frac{15}{2}$	6	0	-10

Plotting the points corresponding to these pairs of values of x and y, and connecting them by a smooth curve, we obtain the curve shown in Fig. 7.13.

The equations in Examples 5 and 6 are examples of the general equation

$$y = ax^2 + bx + c \qquad a \neq 0$$

The graph of any equation of this type is a curve that is called a PARABOLA **parabola.** Notice that in the equation in Example 5, $a = 1$, which is positive, and there is a lowest point on the graph; this lowest point has coordinates $(2, -1)$. Also note that in the equation in Example 6, $a = -2$, which is negative, and there is a highest point on the graph; this highest point has coordinates $(1, 8)$.

If $a > 0$ in the equation $y = ax^2 + bx + c$, then it can be shown that the parabola which is the graph of this equation has a lowest point with coordinates $\left(-\frac{b}{2a}, c - \frac{b^2}{4a}\right)$. If $a < 0$ in the equation $y = ax^2 + bx + c$, then it can be shown that the parabola which is the graph of this equation has a highest point and that the coordinates of this highest point are $\left(-\frac{b}{2a}, c - \frac{b^2}{4a}\right)$. In each case the point

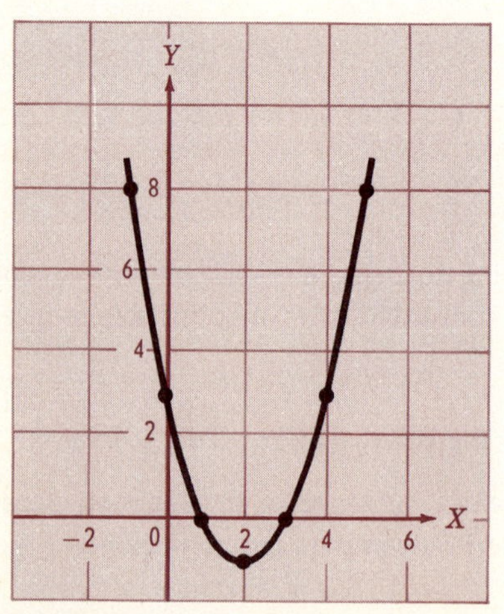

FIGURE 7.12 Graph of $y = x^2 - 4x + 3$.

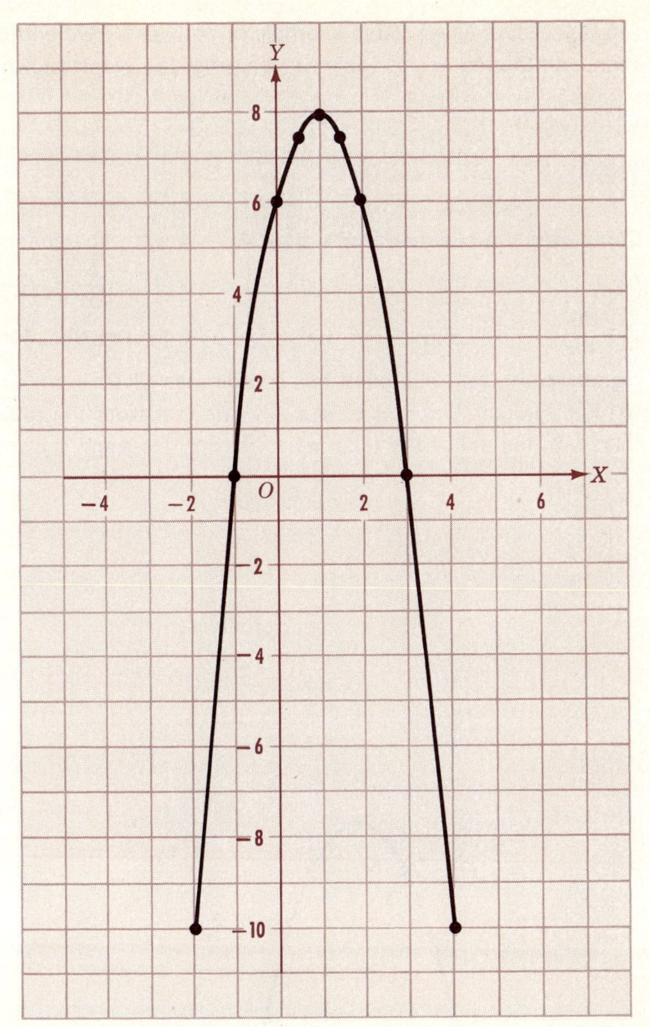

FIGURE 7.13 Graph of $y = -2x^2 + 4x + 6$.

VERTEX $V\left(-\frac{b}{2a}, c - \frac{b^2}{4a}\right)$ is called the **vertex** of the parabola, and the line

AXIS OF A PARABOLA through V parallel to the y axis is called the **axis** of the parabola.

EXAMPLE 7 For each of the following equations, construct the graph. State whether the graph has a highest point or a lowest point, and give the coordinates of that point.

(a) $y = -x^2 + 4x - 4$

(b) $y = x^2 + x + 1$

SOLUTION (a) Proceeding as we did in Example 6, we construct the table below, and then with the aid of this table we construct the graph shown in Fig. 7.14.

WHEN $x =$	-1	0	1	2	3	4	5
THEN $y =$	-9	-4	-1	0	-1	-4	-9

Comparing the equation $y = -x^2 + 4x - 4$ with the equation $y = ax^2 + bx + c$, we see that $a = -1$, $b = 4$, $c = -4$, $-\dfrac{b}{2a} = -\dfrac{4}{-2} = 2$, and $c - \dfrac{b^2}{4a} = -4 - \dfrac{16}{-4} = -4 + 4 = 0$. Since a is negative, the graph has a highest point, and this point has coordinates (2, 0).

(b) For the equation $y = x^2 + x + 1$, we construct the table below, and then with the aid of the table we construct the graph shown in Fig. 7.15.

WHEN $x =$	-3	-2	-1	$-\frac{1}{2}$	0	1	2
THEN $y =$	7	3	1	$\frac{3}{4}$	1	3	7

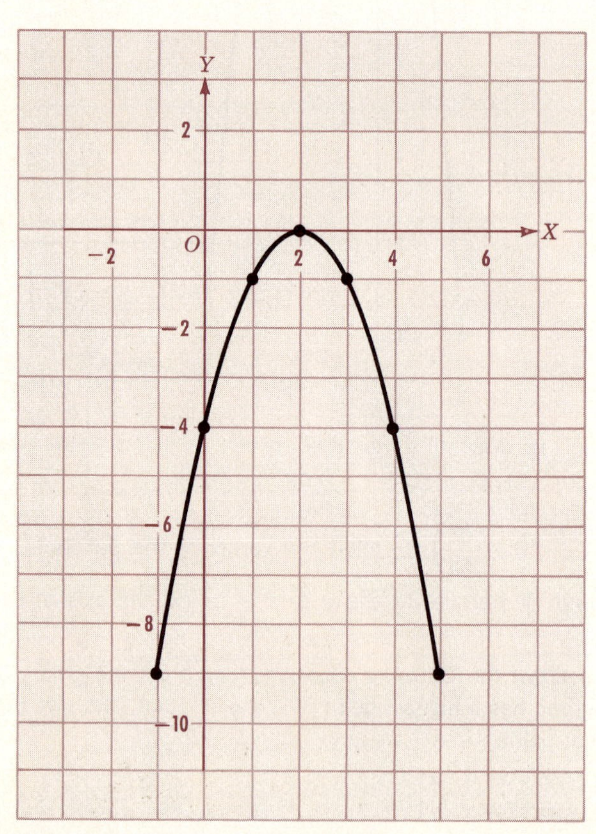

FIGURE 7.14 Graph of $y = -x^2 + 4x - 4$.

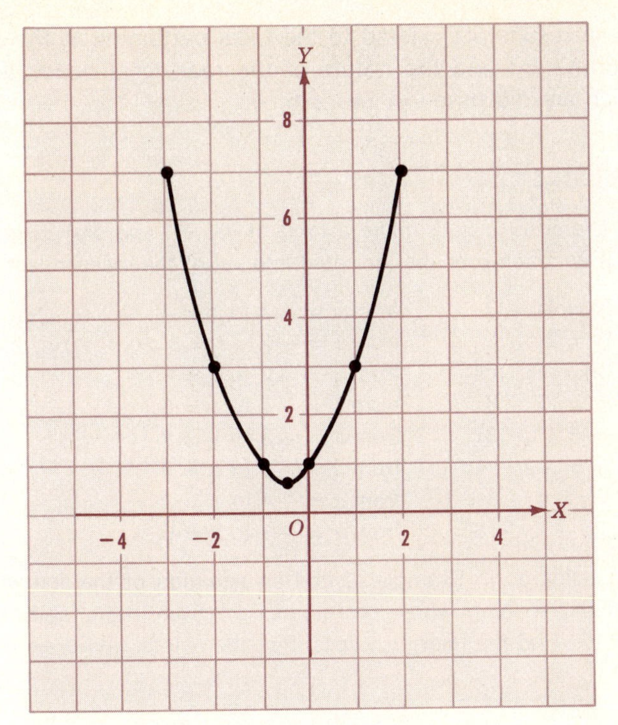

FIGURE 7.15 Graph of $y = x^2 + x + 1$.

For this equation $a = 1$, $b = 1$, and $c = 1$; so the graph has a lowest point with coordinates $(-\frac{1}{2}, \frac{3}{4})$.

Note that in each of Examples 5 and 6 the graph intersects the x axis at two distinct points. In both of these examples, the *discriminant* $b^2 - 4ac$ of the quadratic expression $ax^2 + bx + c$ has a positive value; in Example 5, $b^2 - 4ac = 16 - 12 = 4$, and in Example 6, $b^2 - 4ac = 16 + 48 = 64$. Also, in Example 7*a* note that the graph (Fig. 7.14) has only one point in common with the x axis, namely, (2, 0), and that the discriminant is zero: $b^2 - 4ac = 16 - 16 = 0$. Finally, in Example 7*b* note that the graph (Fig. 7.15) has no point in common with the x axis, and that the discriminant in this case is negative: $b^2 - 4ac = 1 - 4 = -3$. These examples illustrate the fact that the graph of

$$y = ax^2 + bx + c$$

intersects the x axis:

In two points if $b^2 - 4ac > 0$

In one point if $b^2 - 4ac = 0$

In no points if $b^2 - 4ac < 0$

These results are related to the facts pertaining to the sign of the discriminant and the nature of the roots of a quadratic equation which are discussed in Sec. 6.5.

EXERCISES

For Exercises 1 to 8, make a table of values, and draw a graph for the given equation over the indicated interval of the independent variable.

1 $y = 4x - 3$ from $x = -3$ to $x = 4$
2 $y = 2t + 1$ from $t = -4$ to $t = 3$
3 $2y = 3x + 4$ from $x = -2$ to $x = 6$
4 $y = 2x - 8$ from $x = -1$ to $x = 6$
5 $y = 10 - 3x$ from $x = -3$ to $x = 3$
6 $y = -2 - 4x$ from $x = -3$ to $x = 3$
7 $3x - y + 4 = 0$ from $x = -4$ to $x = 2$
8 $2x + 4y - 8 = 0$ from $x = -2$ to $x = 6$

Proceeding as in Example 4, find an equation of the line which passes through the points whose coordinates are given in the tables of Exercises 9 to 12. Find the missing values. Plot the points, and draw the line.

9

WHEN $x =$	1	2	4	5
THEN $y =$	4	6	10	12

10

WHEN $x =$	0	3	?	9
THEN $y =$	1	7	13	19

11

WHEN $r =$	0.5	1	2	?
THEN $s =$	0.2	0.5	1.1	−0.7

12

WHEN $x =$	−1	0	?	2	3
THEN $y =$	−5	−3	−1	1	?

For Exercises 13 to 16, make a table of values, and draw a graph for the given equation over an interval which you consider suitable.

13 $y = x^2 + 4x$
14 $y = x^2 - 4x$
15 $y = x^2 - 4x + 5$
16 $y = x^2 - 2x - 5$

In each of Exercises 17 to 20: (a) State whether the graph of the given equation has a lowest point or a highest point, and give the coordinates of this point. (b) How many points does the graph have in common with the x axis? What are the coordinates of these points? (c) Construct the graph of the equation, and verify your results in parts (a) and (b).

17 $y = x^2 + x - 6$

18 $y = 2x^2 - 6x + 3$

19 $y = -2x^2 - 2x + 6$

20 $y = -2x^2 - 2x - 2$

7.5
GRAPHICAL SOLUTION OF PAIRS OF EQUATIONS

In Sec. 4.9 there is a discussion of an algebraic method for finding the solution of a pair of equations of the first degree with two variables. Here we shall discuss a geometrical interpretation of the solution of a pair of equations.

Any equation of the first degree in x and y such as

$$Ax + By + C = 0 \qquad B \neq 0 \tag{7.7}$$

is an equation of a straight line, since it can be written in the form

$$y = ax + b$$

For this reason, any equation of the form (7.7), that is, any equation of the first degree in x and y, is called a **linear equation.**

LINEAR EQUATION

Consider the pair of equations,

$$\begin{aligned} 2x + 3y &= 7 \\ 3x - 2y &= 4 \end{aligned} \tag{7.8}$$

The solution of this pair of linear equations is the set of all ordered pairs of real numbers which satisfy both equations. Thus, the solution of the pair (7.8) can be written as

$$\{(x, y) \mid 2x + 3y = 7 \quad \text{and} \quad 3x - 2y = 4\}$$

which, as we have seen, is the same as

$$\{(x, y) \mid 2x + 3y = 7\} \cap \{(x, y) \mid 3x - 2y = 4\}$$

So the graph of the solution of the pair (7.8) is the *intersection* of the graph of $2x + 3y = 7$ and the graph of $3x - 2y = 4$. By constructing tables for each of these equations, we find that the points $(-1, 3)$ and $(5, -1)$ lie on the graph of $2x + 3y = 7$ and the points $(-2, -5)$ and $(4, 4)$ lie on the graph of $3x - 2y = 4$. The graph of $2x + 3y = 7$ is the line L_1 and the graph of $3x - 2y = 4$ is the line L_2 in Fig. 7.16. The graph of the solution of the given pair of equations (7.8) is the point P, the intersection of lines L_1 and L_2. The coordinates of P appear to be (2, 1). That (2, 1) does satisfy both equations in (7.8) can be verified by direct substitution in the

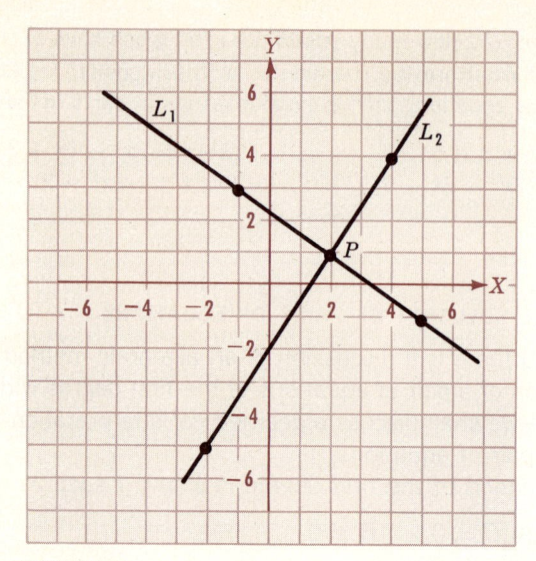

FIGURE 7.16

two equations, and since the two lines L_1 and L_2 have only one point of intersection, it follows that $\{(2, 1)\}$ is the solution of the pair (7.8).

EXAMPLE 1 Solve graphically the pair of linear equations

$$2x + y = 5 \qquad (1)$$

$$x + y = 2 \qquad (2)$$

SOLUTION For $2x + y = 5$, we construct the following table:

WHEN $x =$	-1	0	1	2
THEN $y =$	7	5	3	1

For $x + y = 2$, we construct the following table:

WHEN $x =$	-1	0	1	2
THEN $y =$	3	2	1	0

Since two points determine a straight line, we need to find only two points to determine the graph of each of equations (1) and (2). However, it is well to find one or two other points for checking purposes. With the tabular values as shown, we draw the graph for each equation (see Fig. 7.17), as we learned to do in the preceding section.

The two straight-line graphs intersect at a point whose coordinates appear from the graph to be $(3, -1)$. That $\{(3, -1)\}$ is the solution of the given pair of equations may be verified by direct substitution.

It should be apparent from the nature of graphs that the solution of a pair of linear equations obtained from a graph is usually an approximation. The accuracy of the approximation depends, in part, upon the care used in constructing the graph. Therefore, solutions obtained graphically may not check precisely.

It may happen that the graphs of two equations of the first degree are two parallel lines. In such a case the solution of the pair of linear equations is the empty set, and the equations are said to be *inconsistent.*

EXAMPLE 2 Find the solution of the equations

$$2x + 2y = 6 \qquad (1)$$

$$5x + 5y = 20 \qquad (2)$$

SOLUTION We note that in (1), when $x = 0$, then $y = 3$, and when $x = 3$, then $y = 0$, so that the graph of (1) is the line through the points (0, 3) and (3, 0).

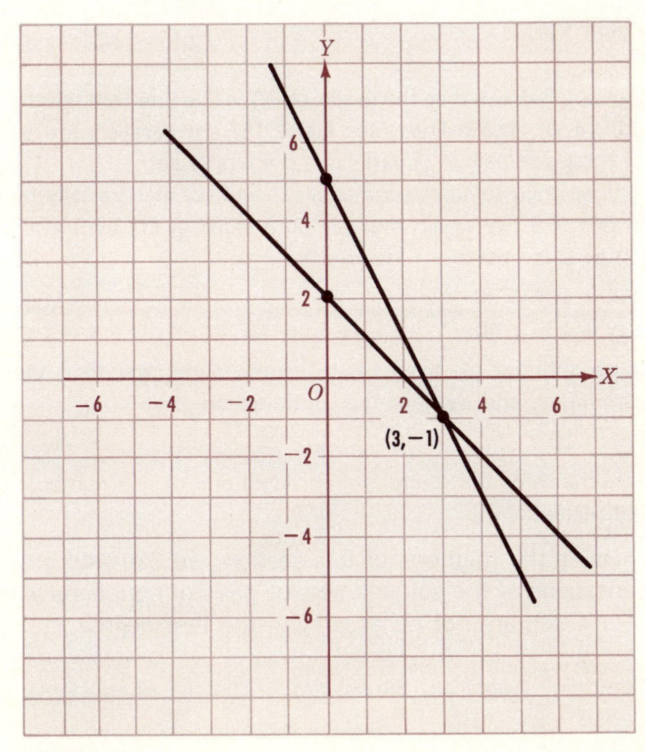

FIGURE 7.17

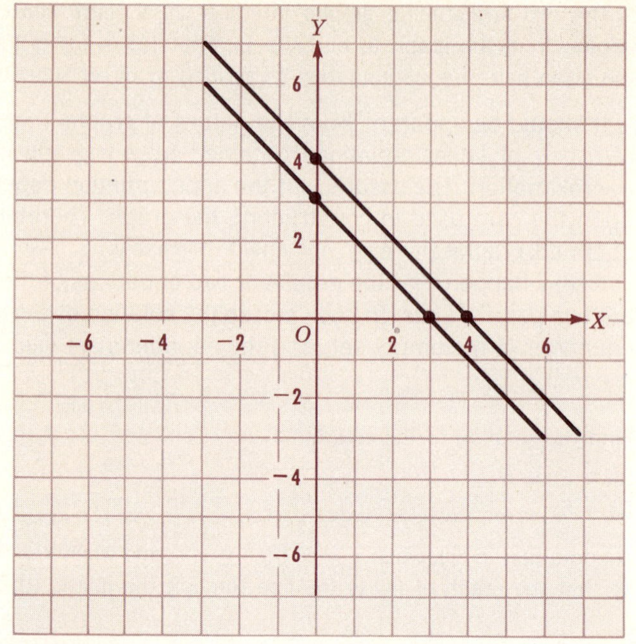

FIGURE 7.18

Similarly we see that the graph of (2) is the line through the points (0, 4) and (4, 0). These lines (see Fig. 7.18) are parallel, and so the solution of the given pair of equations is the empty set.

If we tried to solve equations (1) and (2) algebraically by the methods in Sec. 4.9, we would multiply both sides of (1) by 5 and both sides of (2) by 2 to obtain

$$10x + 10y = 30 \qquad (3)$$

$$10x + 10y = 40 \qquad (4)$$

If we subtract (4) from (3) to eliminate the x's, then the y's are also eliminated, and we have the contradiction

$$0 = -10$$

This indicates that there is no ordered pair of real numbers that satisfies both equations.

Using the methods of this section, we can find graphical representations of the solution sets of pairs of equations with two variables which are not necessarily of the first degree.

EXAMPLE 3 Determine graphically the solution of the pair of equations

$$\begin{aligned} y &= x^2 - 6x + 11 \\ y &= x + 1 \end{aligned} \qquad (1)$$

SOLUTION The solution of the pair of equations (1) is the set of all ordered pairs of real numbers that can be written

$$\{(x, y) \mid y = x^2 - 6x + 11 \quad \text{and} \quad y = x + 1\}$$

or

$$\{(x, y) \mid y = x^2 - 6x + 11\} \cap \{(x, y) \mid y = x + 1\}$$

So the graph of the solution of the pair (1) is the intersection of the graph of $y = x^2 - 6x + 11$ and the graph of $y = x + 1$.

Using the methods in Sec. 7.4, we find that the graph of $y = x^2 - 6x + 11$ is the parabola with vertex (3, 2) and opening upward as shown in Fig. 7.19. The graph of $y = x + 1$ is the line with slope 1 and y intercept 1 shown in Fig. 7.19. From the figure we see that the solution of the pair (1) consists of two ordered pairs of real numbers, the coordinates of points P_1 and P_2. The coordinates of P_1 appear to be (2, 3) and the coordinates of P_2 appear to be (5, 6). That $\{(2, 3), (5, 6)\}$ is the solution of the pair (1) can be verified by direct substitution in the two equations.

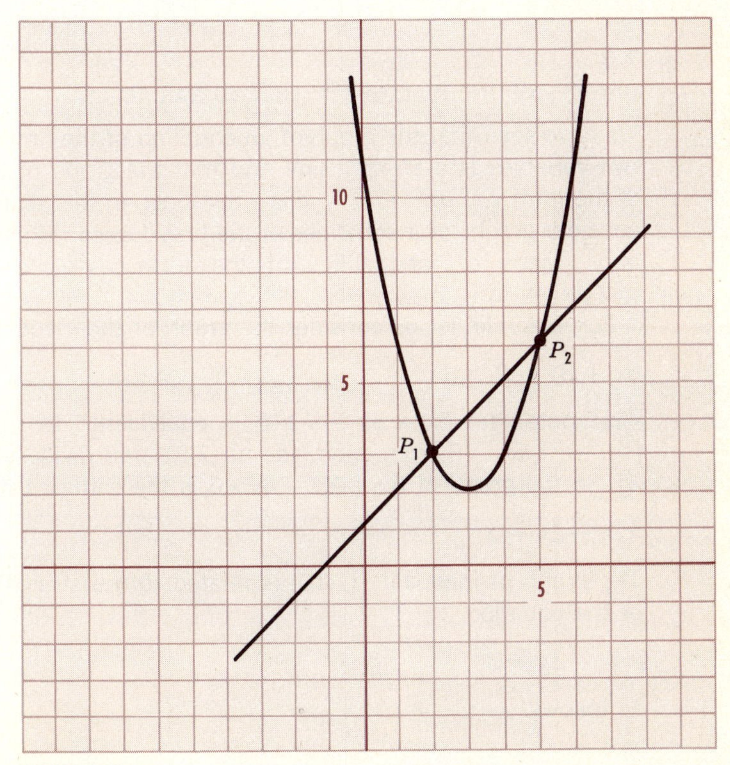

FIGURE 7.19

EXERCISES

Determine the solution of the following pairs of equations. In each exercise graph the equations, and on this graph indicate your solution. If the solution of some pairs of equations is the empty set, explain why.

1 $2x - 3y = 18$
$x + 4y = -13$

2 $x + y = 4$
$6x + 4y = -1$

3 $3x - y = -14$
$7x - 2y = -33$

4 $3x + 3y = -4$
$6x - 9y = 32$

5 $x + y = \sqrt{2}$
$2x - y = 2\sqrt{2}$

6 $x + y = 3\sqrt{2}$
$2x + 3y = 7\sqrt{2}$

7 $x + 2y = 9$
$3x + 4y = 13$

8 $2x + 3y = -1$
$2x - 6y = -7$

9 $x + y = \sqrt{2}$
$2\sqrt{2}x + 2\sqrt{2}y = 8$

10 $x + y = 3\sqrt{2}$
$\sqrt{2}x + \sqrt{2}y = 12$

11 $y = x^2 - 4x + 4$
$2x - y = 1$

12 $y = 10 + 2x - 3x^2$
$x + y = -8$

13 $4y = 2x^2 - 16$
$4y = 32 - x^2$

14 $3y = x^2 + 2x + 10$
$3y = 14 + 4x - x^2$

7.6 GRAPHS OF INEQUALITIES WITH TWO VARIABLES

We have seen that the graph of an equation of the first degree with two variables is a straight line and that the graph of an equation of the form $y = ax^2 + bx + c$ is a parabola. In this section we shall consider graphs of inequalities of the first degree with two variables and graphs of inequalities of the form $y < ax^2 + bx + c$ or $y > ax^2 + bx + c$.

For example, let us consider the graph of the inequality

$$6x + 3y - 9 > 0 \tag{7.9}$$

First note that $6x + 3y - 9 > 0$ is equivalent* to $y > -2x + 3$ [subtract $6x - 9$ from each side of (7.9), and divide each side by 3]. So the graph of $6x + 3y - 9 > 0$ is the same as the graph of

$$y > -2x + 3 \tag{7.10}$$

The graph of inequality (7.10) is related to the straight-line graph of the equation

$$y = -2x + 3 \tag{7.11}$$

as follows.

* The concept of equivalent inequalities is discussed in Sec. 4.8.

Suppose that the ordered pair (x_1, y_1) satisfies the *equation* (7.11) and that the ordered pair (x_1, y_2) satisfies the *inequality* (7.10). Then

$$y_1 = -2x_1 + 3 \qquad \text{and} \qquad y_2 > -2x_1 + 3$$

so that $y_2 > y_1$. Therefore, as shown in Fig. 7.20, the point corresponding to (x_1, y_2) lies *above* the point corresponding to (x_1, y_1). But the points corresponding to the ordered pairs that satisfy $y = -2x + 3$ lie on a straight line; so the points corresponding to the ordered pairs that satisfy $y > -2x + 3$ lie *above* the line determined by $y = -2x + 3$. The graph of $y > -2x + 3$, and hence the graph of $6x + 3y - 9 > 0$, is the region that lies *above* the graph of $y = -2x + 3$ and is indicated in Fig. 7.20 by the *shaded* region.

By the same type of argument we find that the graph of

$$y < -2x + 3 \tag{7.12}$$

is the region that lies *below* the graph of $y = -2x + 3$ and thus is the *unshaded* region in Fig. 7.20.

Thus, the line L which is the graph of $y = -2x + 3$ divides the plane into two parts—the graph of $y > -2x + 3$, which is the half plane lying above L, and the graph of $y < -2x + 3$, which is the half plane lying below L.

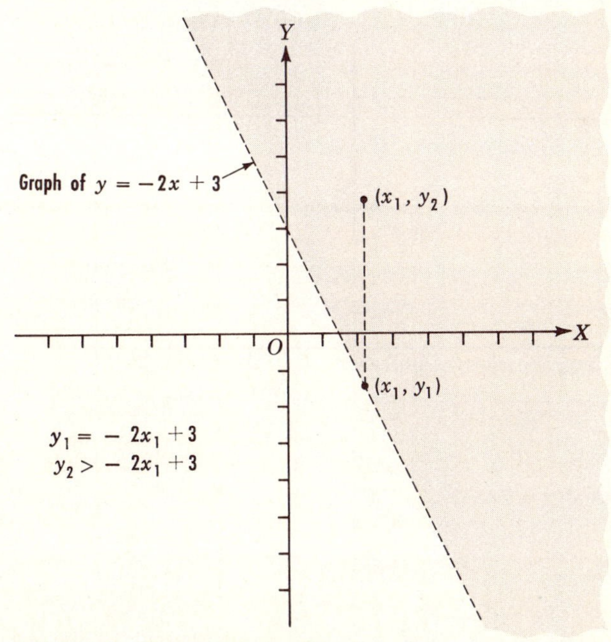

FIGURE 7.20 The graph of $y > -2x + 3$.

Using the methods illustrated in the preceding discussion, we can arrive at the following general statement:

The line L which is the graph of

$$y = ax + b$$

divides the plane into two parts—the graph of

$$y > ax + b$$

which is the half plane lying above the line L, and the graph of

$$y < ax + b$$

which is the half plane lying below the line L.

EXAMPLE 1 Determine the graph of

$$2x - y - 3 > 0 \tag{1}$$

SOLUTION We see that $2x - y - 3 > 0$ is equivalent to

$$y < 2x - 3 \tag{2}$$

[subtract $2x - 3$ from each side of (1), and divide each side by -1, reversing the sense of the inequality]. Since the graph of (2) is the region of the plane lying below the graph of $y = 2x - 3$, the graph of the given inequality (1) is the shaded region shown in Fig. 7.21.

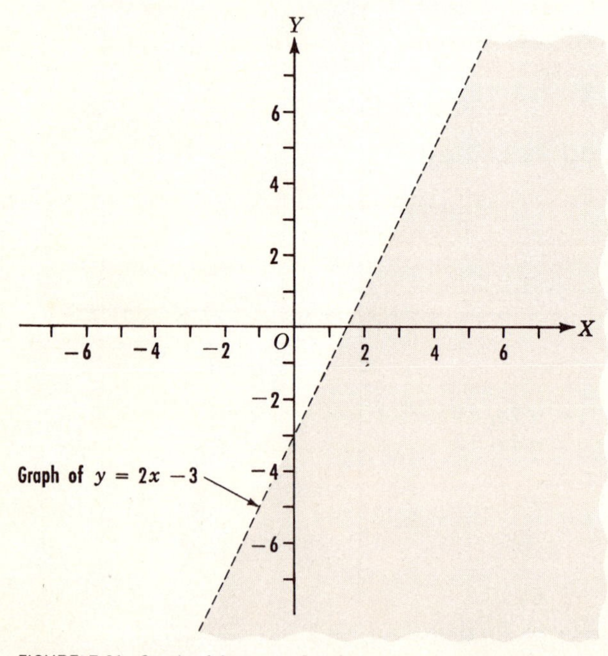

FIGURE 7.21 Graph of $2x - y - 3 > 0$.

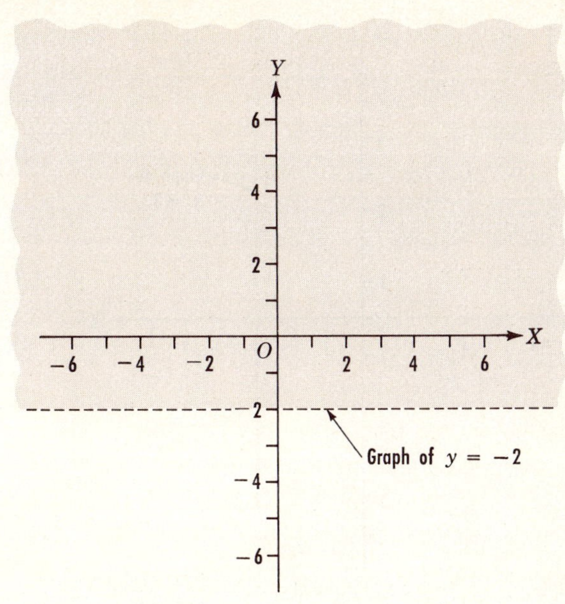

FIGURE 7.22 Graph of $\{(x, y) | y > -2\}$.

EXAMPLE 2 Determine the graph, *in the plane,* of $y > -2$. That is, determine the graph of the relation $\{(x, y) \mid y > -2\}$.

SOLUTION An ordered pair (x_1, y_1) will be a member of $\{(x, y) \mid y > -2\}$ if and only if the second entry of the pair is greater than -2. Therefore, the graph of $\{(x, y) \mid y > -2\}$ will consist of those points whose y coordinates are greater than -2. This region is shaded in Fig. 7.22 and is the region lying above the line which is the graph of $y = -2$.

EXAMPLE 3 Determine the graph, *in the plane,* of the inequality $x < 3$. That is, determine the graph of the relation $\{(x, y) \mid x < 3\}$.

SOLUTION An ordered pair (x_1, y_1) will satisfy the inequality $x < 3$ if and only if the first entry is less than 3. So the graph of $x < 3$ will consist of those points whose x coordinates are less than 3; that is, those points which lie on the left of the line which is the graph of $x = 3$. The graph is shown as the shaded region in Fig. 7.23.

EXAMPLE 4 Indicate the graph of $\{(x, y) \mid 2x + y > 5 \quad \text{and} \quad x - y > 3\}$.

SOLUTION Recall that

$$\{(x, y) \mid 2x + y > 5 \quad \text{and} \quad x - y > 3\} = \{(x, y) \mid 2x + y > 5\} \cap \{(x, y) \mid x - y > 3\}$$

So the graph of the given set is the intersection of the graphs of $2x + y > 5$ and $x - y > 3$. The inequality $x - y > 3$ is equivalent to $-y > -x + 3$

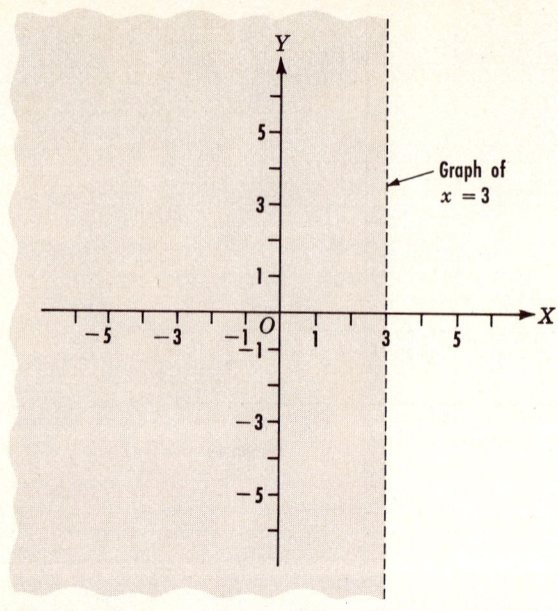

FIGURE 7.23 Graph of $\{(x, y) | x < 3\}$.

and hence to $y < x - 3$. The graph of $x - y > 3$ is indicated in Fig. 7.24. The graph of $2x + y > 5$ is the same as the graph of $y > -2x + 5$, and is indicated in Fig. 7.25. The intersection of the graph in Fig. 7.24 with that in Fig. 7.25 is indicated in Fig. 7.26.

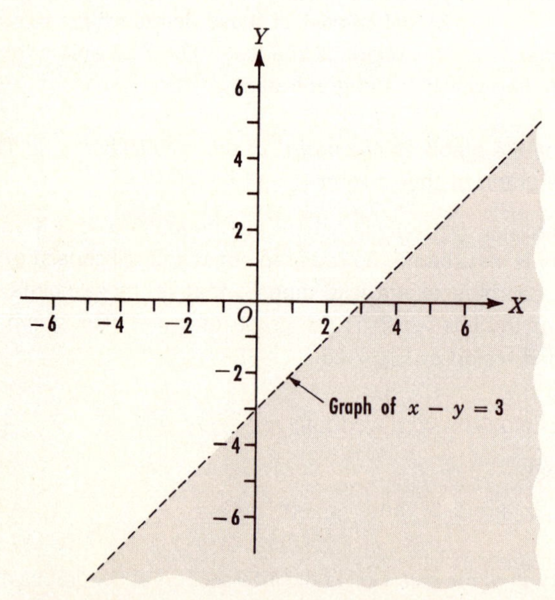

FIGURE 7.24 Graph of $x - y > 3$.

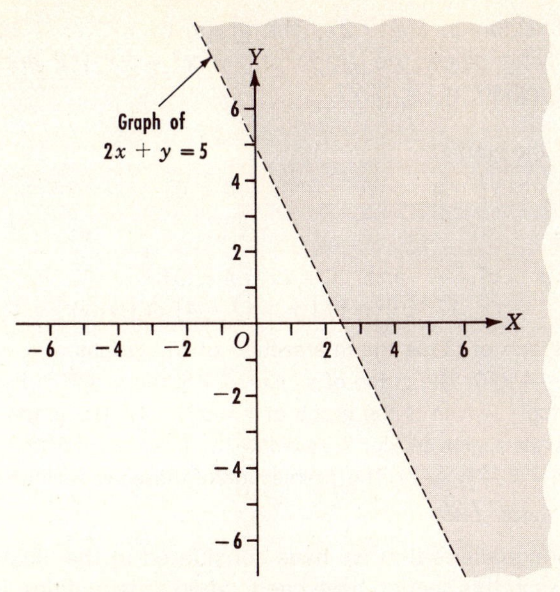

FIGURE 7.25 Graph of $2x + y > 5$.

Just as the graph of the inequality $y > ax + b$ is related to the graph of the equation $y = ax + b$, so is the graph of the inequality $y > ax^2 + bx + c$ related to the graph of the equation $y = ax^2 + bx + c$. For example, the graph of $y > x^2 - 2x + 2$ is the region which lies above the graph of $y = x^2 - 2x + 2$, as shown by the

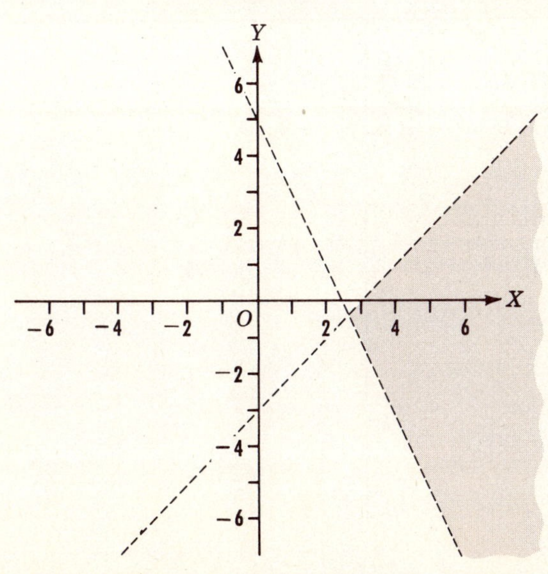

FIGURE 7.26 Graph of $\{(x, y) | 2x + y > 5 \text{ and } x - y > 3\}$.

shaded region in Fig. 7.27. The graph of $y < x^2 - 2x + 2$ is the region lying below the graph of $y = x^2 - 2x + 2$ and is the unshaded region in Fig. 7.27.

EXAMPLE 5 Indicate the graph of

$$\{(x, y) \mid y > x^2 - 4 \quad \text{and} \quad x + 2y - 4 < 0\} \tag{1}$$

SOLUTION We have

$$\{(x, y) \mid y > x^2 - 4 \quad \text{and} \quad x + 2y - 4 < 0\}$$
$$= \{(x, y) \mid y > x^2 - 4\} \cap \{(x, y) \mid x + 2y - 4 < 0\}$$

So the graph of (1) is the intersection of the graphs of $y > x^2 - 4$ and $x + 2y - 4 < 0$. The graph of $y > x^2 - 4$ is the region in Fig. 7.28 above the parabola (which is the graph of $y = x^2 - 4$). The graph of $x + 2y - 4 < 0$ is the region in Fig. 7.28 below the line (which is the graph of the equation $y = -\frac{1}{2}x + 2$). The intersection of these two regions is the shaded portion in Fig. 7.28.

The inequalities that we have considered in the illustrations and examples in this section have been "strict" inequalities. That is, they have contained only the symbols $<$ or $>$ and have not involved either $\leq$ or $\geq$, and the graphs have not included the line or curve bounding the region. If the symbol $<$ is replaced by $\leq$, or the

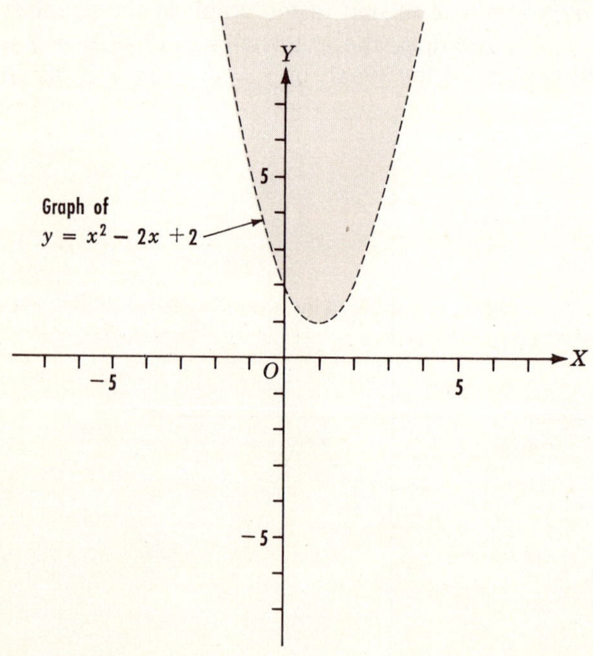

FIGURE 7.27

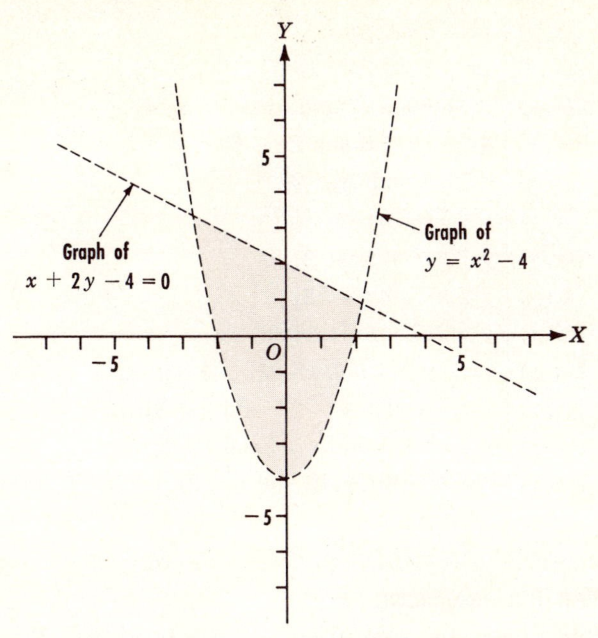

FIGURE 7.28 Graph of $\{(x, y) | y > x^2 - 4 \text{ and } x + 2y - 4 < 0\}$.

symbol $>$ is replaced by $\geq$, the corresponding graph of the inequality would include the boundary of the region. To illustrate, the inequality $2x + y \geq 5$ means "$2x + y > 5$ or $2x + y = 5$" and so

$$\{(x, y) \mid 2x + y \geq 5\} = \{(x, y) \mid 2x + y > 5 \quad \text{or} \quad 2x + y = 5\}$$

But this set can be written as

$$\{(x, y) \mid 2x + y > 5\} \cup \{(x, y) \mid 2x + y = 5\}$$

Thus the graph of $2x + y \geq 5$ is the union of the graphs of $2x + y > 5$ and $2x + y = 5$. The graph of $2x + y > 5$ is shown in Fig. 7.25, and to obtain the graph of $2x + y \geq 5$ we simply annex the line which is the boundary of the shaded region.

EXERCISES

Indicate the graphs of the following sets:

1 $\{(x, y) \mid 3x + 6y > 12\}$
2 $\{(x, y) \mid 6x - 3y > 18\}$
3 $\{(x, y) \mid 6x > 5\}$
4 $\{(x, y) \mid -2x > 6\}$
5 $\{(x, y) \mid 3y > 9\}$

6 $\{(x, y) \mid 4x + 2y < 8\}$
7 $\{(x, y) \mid 2x - 6y < 24\}$
8 $\{(x, y) \mid 3x + 6y > 12 \text{ and } 6x - 3y > 18\}$
9 $\{(x, y) \mid 4x + 2y < 8 \text{ and } 2x - 6y < 24\}$
10 $\{(x, y) \mid 6x > 5 \text{ and } 3y > 9\}$
11 $\{(x, y) \mid y < x^2 + 4\}$
12 $\{(x, y) \mid y < 2x^2 + 5x - 3\}$
13 $\{(x, y) \mid y > x^2 + 4x + 12\}$
14 $\{(x, y) \mid y > 3x^2 - 6x + 1\}$
15 $\{(x, y) \mid 2x - y > 4 \text{ and } x - 2y > 6 \text{ and } 2x - 3y > 12\}$
16 $\{(x, y) \mid x > 4 \text{ and } x + y < 5 \text{ and } y < 3\}$
17 $\{(x, y) \mid y < x^2 + 4 \text{ and } y > 2x + 1\}$
18 $\{(x, y) \mid y < x^2 - 6x + 10 \text{ and } x - 2y + 4 > 0\}$

7.7 LINEAR PROGRAMMING

In this section we give a very brief introduction to some of the simpler aspects of a relatively new and quite important branch of mathematics called *linear programming.*

CONVEX SET

Basic to the study of linear programming is the concept of convex sets. A set S of points in a plane is a **convex set** if whenever P_1 and P_2 are points in S, every point of the *line segment* P_1P_2 is in S. In other words, S is convex if and only if whenever $P_1 \in S$ and $P_2 \in S$, it is true that $P_1P_2 \subseteq S$. The concept of convex set is illustrated in Fig. 7.29.

We shall be concerned with the intersection of convex sets, and in this connection a fundamental question arises: Is the intersection

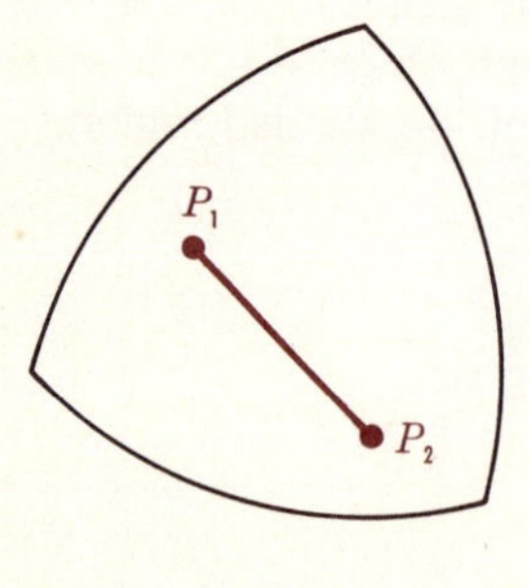

(a) A convex set

P_1
P_2

(b) Not a convex set

FIGURE 7.29

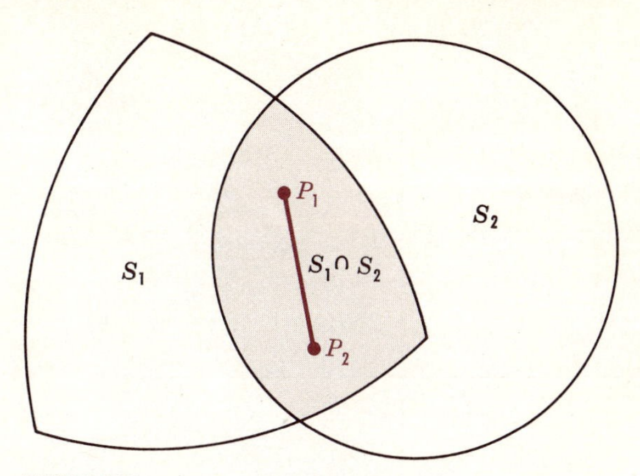

FIGURE 7.30

of convex sets also convex? To answer this question, let us suppose that S_1 and S_2 are convex sets, and let P_1 and P_2 be any two points in $S_1 \cap S_2$. Can we conclude that $P_1P_2 \subseteq S_1 \cap S_2$? The point P_1 is in both S_1 and S_2, and the point P_2 is in both S_1 and S_2 (see Fig. 7.30). Since P_1 and P_2 are both in the convex set S_1, we know that $P_1P_2 \subseteq S_1$; also, since P_1 and P_2 are both in the convex set S_2, we know that $P_1P_2 \subseteq S_2$. So the segment P_1P_2 lies in *both* S_1 and S_2, and therefore,

$$P_1P_2 \subseteq S_1 \cap S_2$$

and $S_1 \cap S_2$ is convex. We have proved the following proposition:

The intersection of two convex sets is a convex set. From this it follows:

The intersection of a finite number of convex sets is a convex set.

As we saw in the preceding section, a nonvertical line L which is the graph of

$$R_1 = \{(x, y) \mid y = ax + b\}$$

divides the coordinate plane into two parts which are the graphs of

$$R_2 = \{(x, y) \mid y > ax + b\}$$

and

$$R_3 = \{(x, y) \mid y < ax + b\}$$

as shown in Fig. 7.31. The graph of R_2 consists of points lying *above* L, and the graph of R_3 consists of points lying *below* L. We call the graph of each of R_2 and R_3 a **half plane.**

HALF PLANE

If L is a vertical line which is the graph of $\{(x, y) \mid x = k\}$, then

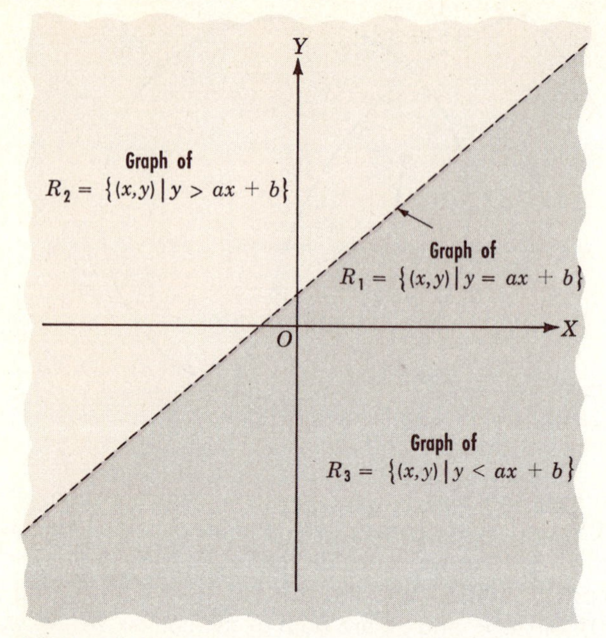

FIGURE 7.31 Half planes determined by a nonvertical line.

the half plane to the *right* of L is the graph of $\{(x, y) \mid x > k\}$, and the half plane to the *left* of L is the graph of $\{(x, y) \mid x < k\}$ (see Fig. 7.32). The graphs of the relations

$$R_1 \cup R_2 = \{(x, y) \mid y \geq ax + b\}$$

and

$$R_1 \cup R_3 = \{(x, y) \mid y \leq ax + b\}$$

CLOSED HALF PLANES

OPEN HALF PLANES

are called **closed half planes,** and therefore the graphs of R_2 and R_3 are sometimes called **open half planes.** Similarly, the graphs of $\{(x, y) \mid x \geq k\}$ and $\{(x, y) \mid x \leq k\}$ are closed half planes.

It can be shown that *any half plane (open or closed) is a convex set.* Then from the proposition proved at the beginning of this section it follows that:

The intersection of any finite number of half planes is a convex set.

POLYGONAL CONVEX SET

The intersection of a finite number of closed half planes is called a **polygonal convex set.**

EXAMPLE 1 Construct the graph of the set $S = \{(x, y) \mid x \geq 0 \text{ and } y \geq 0 \text{ and } x + y - 4 \geq 0 \text{ and } 5x + 9y - 45 \leq 0\}$.

SOLUTION We can write

$$S = S_1 \cap S_2 \cap S_3 \cap S_4$$

where

$$S_1 = \{(x, y) \mid x \geq 0\} \qquad S_2 = \{(x, y) \mid y \geq 0\}$$

$$S_3 = \{(x, y) \mid x + y - 4 \geq 0\} \qquad S_4 = \{(x, y) \mid 5x + 9y - 45 \leq 0\}$$

The graph of S is shown in Fig. 7.33, and is an illustration of a polygonal convex set.

BOUNDARY LINES Any polygonal convex set has as *boundary lines* a set of lines

$$L_1:\quad A_1x + B_1y + C_1 = 0$$
$$L_2:\quad A_2x + B_2y + C_2 = 0$$
$$\cdots\cdots\cdots\cdots\cdots\cdots\cdots\cdots\cdots$$
$$L_n:\quad A_nx + B_ny + C_n = 0$$

To illustrate, in Example 1 the boundary lines are

$$L_1:\quad x = 0$$
$$L_2:\quad y = 0$$
$$L_3:\quad x + y - 4 = 0$$
$$L_4:\quad 5x + 9y - 45 = 0$$

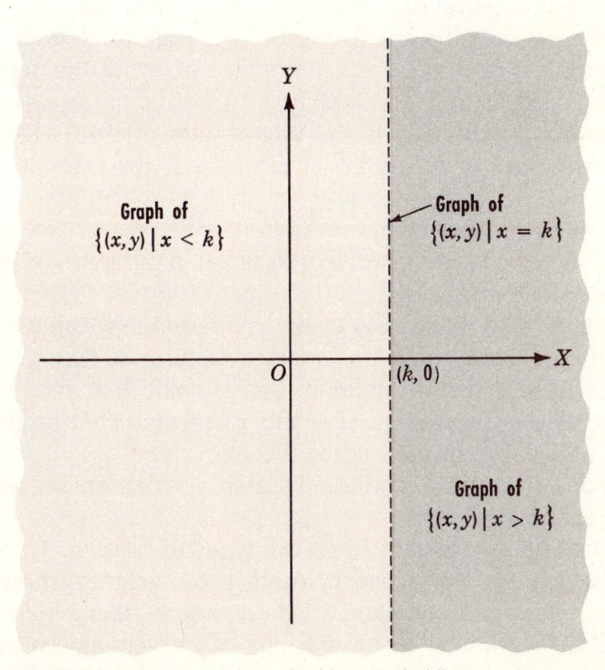

FIGURE 7.32 Half planes determined by a vertical line.

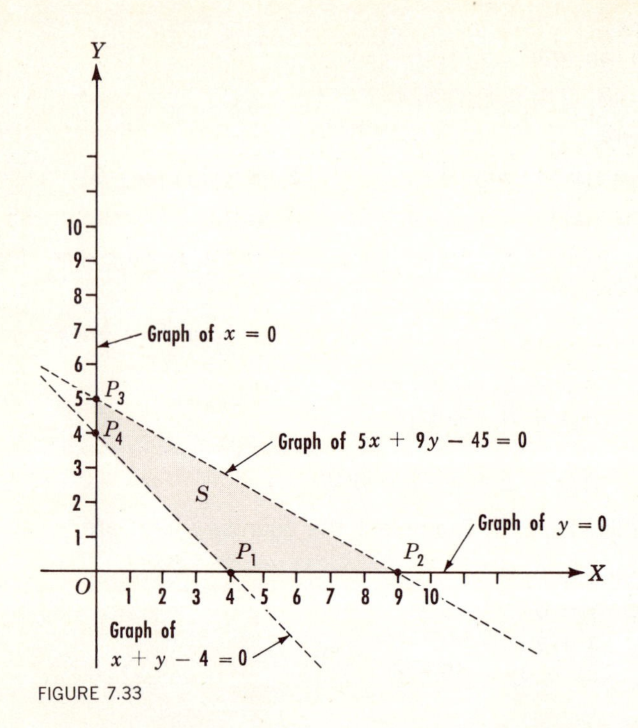

FIGURE 7.33

If S is a polygonal convex set, the intersection of any boundary line of S with the set S is called an **edge** of S. With reference to Example 1 and Fig. 7.33, the edges of S are the segments P_1P_2, P_3P_4, P_4P_1, and P_2P_3. A point in a polygonal convex set which is common to (that is, is the intersection of) two edges of S is an **extreme point.** In the set S of Example 1, the extreme points are

EDGE

EXTREME POINT

$P_1(4, 0) \qquad P_2(9, 0) \qquad P_3(0, 5) \qquad P_4(0, 4)$

The basic type of problem in linear programming (in the plane) is the following:

Let $E(x, y) = ax + by$ be an expression in which a and b are real numbers and x and y are variables. Determine the maximum (greatest) value and the minimum (least) value that $E(x, y) = ax + by$ can have if the values of x and y are restricted so that the point (x, y) lies in a given polygonal convex set S.

For a situation that will lead to this type of problem, let us consider an example.

A company manufactures two types of bicycles, type A and type B, which are assembled by using two machines, machine M_1 and machine M_2. Assembling a type A bicycle requires 2 hours on machine M_1 and 1 hour on machine M_2; assembling a type B bicycle requires 1 hour on machine M_1 and 3 hours on machine M_2. The profit made on the sale of a type A bicycle is \$12, and the profit on the sale of a type B bicycle is \$18. If each machine is to be operated no more than 40 hours per week, how many bicycles of each type should be made per week in order to produce the largest total profit?

Let x be the number of type A bicycles and y be the number of type B bicycles to be made in one week. The profit on the sale of x type A and y type B bicycles is

$$E(x, y) = 12x + 18y$$

and we wish to determine the values of x and y which will produce the maximum profit. Clearly x and y must be nonnegative. The number of hours that machine M_1 is operated to produce x type A and y type B bicycles is

$$2x + y$$

and the number of hours that machine M_2 is operated to produce x type A and y type B bicycles is

$$x + 3y$$

So, the values that x and y can take on must satisfy the four inequalities

$$\begin{array}{ll} x \geq 0 & y \geq 0 \\ 2x + y \leq 40 & x + 3y \leq 40 \end{array}$$

Thus the values of x and y are restricted so that the point (x, y) lies in the polygonal convex set S that is the graph of

$$\{(x, y) \mid x \geq 0 \text{ and } y \geq 0 \text{ and } 2x + y \leq 40 \text{ and } x + 3y \leq 40\}$$

This set S is shown in Fig. 7.34, and the linear programming problem is to determine the maximum value of $E(x, y) = 12x + 18y$ for (x, y) in the set S.

A famous theorem, known as the *Weierstrass theorem* on continuous functions, assures us that $E(x, y)$ will attain a maximum value and a minimum value at points in the polygonal convex set S, and a basic theorem in linear programming tells us that the maximum value and the minimum value will each be attained as some *extreme point* of S.

EXAMPLE 2 Find the number of type A bicycles and the number of type B bicycles that the company described above must produce in a week to produce the largest profit, and determine what this maximum profit is.

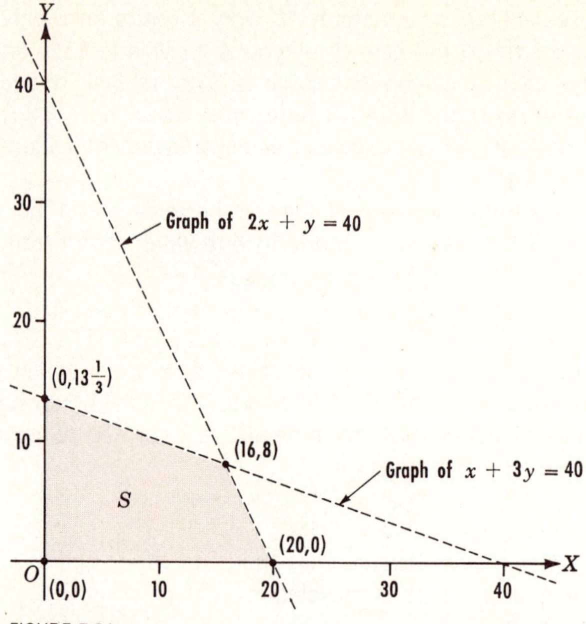

FIGURE 7.34

SOLUTION Referring to Fig. 7.34, we find that the extreme points of the polygonal convex set S are (0, 0), (20, 0), (16, 8), and $(0, 13\frac{1}{3})$. Evaluating the profit function $E(x, y) = 12x + 18y$ at each of these extreme points, we find

$$E(0, 0) = 12 \cdot 0 + 18 \cdot 0 = 0$$
$$E(20, 0) = 12 \cdot 20 + 18 \cdot 0 = 240$$
$$E(16, 8) = 12 \cdot 16 + 18 \cdot 8 = 336$$
$$E(0, 13\tfrac{1}{3}) = 12 \cdot 0 + 18 \cdot \tfrac{40}{3} = 240$$

So the maximum profit of $336 occurs when 16 type A and 8 type B bicycles are produced each week.

EXAMPLE 3 Find the maximum value and the minimum value of the expression $E(x, y) = 2x + 3y$ when the values of x and y are restricted so that the point (x, y) lies in the polygonal convex set S of Example 1.

SOLUTION We evaluate $E(x, y)$ at each of the extreme points of S, and find

$E(4, 0) = 8 \qquad E(9, 0) = 18$
$E(0, 5) = 15 \qquad E(0, 4) = 12$

We conclude that the maximum value of $E(x, y)$ on S is 18 and that the minimum value of $E(x, y)$ on S is 8.

EXERCISES

In each of Exercises 1 to 5, graph S, and find the extreme points of S. Then find the maximum and minimum values of $E(x, y)$ for $(x, y) \in S$.

1 $S = \{(x, y) \mid x \geq 0 \text{ and } y \geq 0 \text{ and } y \geq -\frac{4}{3}x + 4 \text{ and } y \leq -\frac{12}{5}x + 12\}$; $E(x, y) = x + y$

2 $S = \{(x, y) \mid y \geq 0 \text{ and } y \leq -\frac{1}{2}x + \frac{5}{2} \text{ and } y \geq -\frac{2}{3}x - 2\}$; $E(x, y) = 2x + 3y$

3 $S = \{(x, y) \mid y \geq 0 \text{ and } x + y \leq 7 \text{ and } -x + 5 \leq 10\}$; $E(x, y) = x - y$

4 $S = \{(x, y) \mid x \geq 0 \text{ and } y \geq 0 \text{ and } x + y - 5 \geq 0 \text{ and } x + 3y - 18 \leq 0\}$; $E(x, y) = x + 2y$

5 $S = \{(x, y) \mid x \geq 0 \text{ and } y \geq 4 \text{ and } x + y \geq 6 \text{ and } x \leq 5 \text{ and } y \leq 8\}$; $E(x, y) = x + 2y$

6 The Big Company makes two products, X_1 and X_2. These must be processed on machines m_1 and m_2, X_1 requiring 1 hour of time on m_1 and 3 hours of time on m_2, X_2 requiring 2 hours of time on m_1 and 1 hour on m_2. To describe the quantities of X_1 and X_2 which may be produced within a week, where x_1 is the quantity of X_1 and x_2 is the quantity of X_2, it is necessary and sufficient that

$$x_1 + 2x_2 \leq 100 \qquad 3x_1 + x_2 \leq 100 \qquad x_1 \geq 0 \qquad x_2 \geq 0$$

The company's earnings are \$4 per unit on X_1 and \$3 per unit on X_2. If $E(x_1, y_1)$ denotes the earnings, then $E(x_1, y_1) = 4x_1 + 3x_2$. Maximize E subject to the restraints listed above.

7 Customer K_1 wants to purchase from 400 to 1,200 units of a certain commodity at \$2 each, and customer K_2 wants to purchase from 600 to 1,000 units at \$2.20 each. Delivery charges are 8 cents per unit to K_1 and 16 cents per unit to K_2. If the seller has 2,000 units, find how many he should sell to each of K_1 and K_2 to obtain the maximum proceeds. What is the amount of the maximum proceeds?

8

Ratio, Proportion, and Variation

8.1
RATIO AND MEASUREMENT

RATIO By the **ratio** of the number a to the number b we mean the quotient $a \div b$. This ratio is written variously as

$$a \div b \quad \text{or} \quad \frac{a}{b} \quad \text{or} \quad a:b$$

To illustrate, the ratio of 3 to 4 may be written

$$3 \div 4 \quad \text{or} \quad \frac{3}{4} \quad \text{or} \quad 3:4$$

The use of ratio occurs most often in the comparison of two measurements. When we say that a man is 6 feet tall, we mean that the ratio of his height to the unit of measure, 1 foot, is 6:1; when we say that an object weighs 8 pounds, we mean that the ratio of its weight to the unit of weight, 1 pound, is 8:1. However, in such ratios it is not necessary to have one of the measurements the unit measurement. For example, if a man weighs 180 pounds and his son weighs 120 pounds, the ratio of the weight of the man to the weight of the son is 3:2, or $\frac{3}{2}$. In the examples which we have considered, the measurements compared were measurements of the same kinds of quantities, and these measurements were expressed in the same unit. *Whenever the measurements of the same kinds of quantities are compared, these measurements must be expressed in terms of a common unit.* Thus the ratio of 1 gallon to 3 quarts is not 1:3, but 4:3.

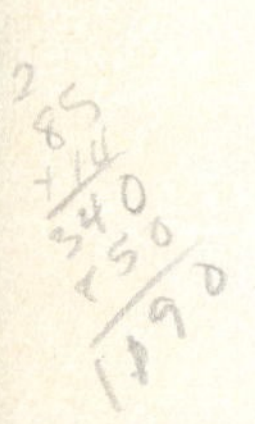

We can have ratios of measurements of different kinds of quantities. If a car travels 120 miles in 3 hours, the ratio 120:3 (or 40:1) is its average speed. This is written as 40 miles per hour and is often expressed more briefly as 40 mi/h. To say that an automobile averages 12 miles per gallon of gasoline means that the ratio of miles traveled to gallons of gasoline used is 12:1.

Some ratios are so important that they are given special names and have become a part of our everyday speech. We are all familiar with the constant π, which is the ratio of the circumference of a circle to its diameter. Engineers speak of a roadbed as having a 3 percent grade. By this they mean that a car going up the road will rise 3 feet vertically for each 100 feet of horizontal motion (Fig. 8.1). Hence, the ratio of rise to forward motion is $\frac{3}{100}$, or 3:100.

EXAMPLE 1 If a car can go 85 miles on 5 gallons of gasoline, how many miles can it go on 14 gallons of gasoline?

SOLUTION The car will travel $\frac{85}{5}$ miles per gallon; it will therefore travel $\frac{85}{5} \cdot 14 = 238$ miles on 14 gallons.

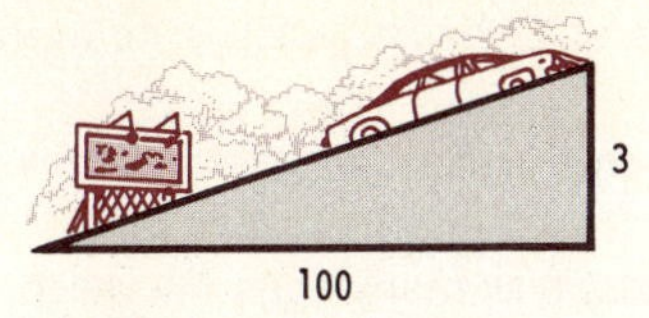

FIGURE 8.1

Sometimes we wish to express the speed of an automobile, or the rate of fuel consumption, or some other ratio in units which are different from the ones given. Consider the following examples.

EXAMPLE 2 If a car is traveling 50 miles per hour, what is its speed expressed as feet per second?

SOLUTION Since there are 5,280 feet in 1 mile and 3,600 seconds in 1 hour, the car is traveling $50 \cdot 5{,}280$ feet in 3,600 seconds, or $\frac{50 \cdot 5{,}280}{3{,}600}$ feet in 1 second. So its speed is $73\frac{1}{3}$ feet per second.

EXAMPLE 3 A road is being paved at the rate of 6.6 feet per minute. Express the rate in miles per 8 hours.

SOLUTION Since there are 60 minutes in 1 hour, there will be (6.6)(60) feet of road built in 1 hour or (6.6)(60)(8) feet of road built in 8 hours. Since 5,280 feet make a mile, there will be $\frac{(6.6)(60)(8)}{5{,}280} = 0.6$ mile of road built in 8 hours.

EXERCISES

1 A high school has 54 students, 24 of whom are girls. What is the ratio of the number of girl students to the number of boy students?

2 What is the ratio of the length of the circumference of a circle to the length of its radius?

3 Two boys divide 30 marbles in the ratio 5 : 1. How many marbles does each boy have?

4 A baseball team wins 3 games out of every 5 games it plays. What is the ratio of the number of the games it wins to the number of games it plays? What is the ratio of the number of the games it wins to the number of games it loses?

5 Two boys divide $4.20 in the ratio 5 : 2. How much does each boy receive?

6 A boy hits a target 3 out of every 10 shots. What is the ratio of his hits to his misses?

7 What is the ratio of 7 quarts to 2 gallons? Of 6 pounds to 48 ounces? Of 2 miles to 2,640 yards? Of y gallons to z pints?

8 If a car travels 110 miles in $2\frac{3}{4}$ hours, what is its average speed in miles per hour?

9 If a car can go 81 miles on 6 gallons of gasoline, how far can it go on 28 gallons of gasoline?

10 If 28 eggs weigh $2\frac{1}{3}$ pounds, how many eggs will weigh 5 pounds?

11 If a car travels 88 feet per second, what is its speed in miles per hour?

12 How many feet per second are equivalent to 30 miles per hour?

13 A road is being paved at the rate of 1 mile in 8 hours. What is the rate in feet per minute?

14 A conveyor belt carries away 18 cubic feet of coal per minute. How many cubic yards are carried per hour?

15 Convert 720 gallons per hour into quarts per second.

16 If 2 lemons give enough juice to fill $\frac{4}{5}$ of a glass, how many lemons are necessary to fill 2 glasses with juice?

17 An 87-pound bag of fertilizer contains 18 pounds of limestone. How many pounds of limestone are contained in 217.5 pounds of the same type of fertilizer?

18 Convert 50 pounds per second into tons per hour.

19 An automobile will go 44 miles on $2\frac{1}{2}$ gallons of gasoline. How far can this car travel on 7 gallons of gasoline? How many gallons of gasoline would be required for a 25-mile trip?

20 A grocery store advertises meat at 78 cents per pound. How many pounds of meat could be bought for \$1.95?

21 If 28 identical bolts weigh $2\frac{1}{3}$ pounds, how many pounds will 20 of these bolts weigh?

22 If 3 pounds of pecans are sold for \$1.10, how much will 5 pounds of pecans cost? How many pounds of pecans can be bought for \$2.50?

23 Dress material is priced at 3 yards for \$1.98. How much will $4\frac{1}{2}$ yards of this material cost?

24 If 2 oranges will give $\frac{3}{4}$ of a glass of juice, how many oranges will be needed to provide enough juice for 14 people if $1\frac{1}{2}$ glasses of juice are allowed for each person?

8.2 PROPORTION

PROPORTION A **proportion** is the statement that two ratios are equal. Thus $3:6 = 1:2$ or $\frac{3}{6} = \frac{1}{2}$ is a proportion. The proportion $\frac{a}{b} = \frac{c}{d}$ is read "a is to b as c is to d." Proportions enable us to solve quickly and easily many of the problems that arise in everyday living.

EXAMPLE 1 Solve the proportion $\frac{3x}{3x^2 - 2} = \frac{2}{2x + 1}$

SOLUTION The definition of equality of fractions states that

$$\frac{a}{b} = \frac{c}{d} \quad \text{if and only if} \quad ad = bc$$

Therefore, if neither $3x^2 - 2$ nor $2x + 1$ is zero, then the given proportion is equivalent to

$$3x(2x + 1) = 2(3x^2 - 2) \quad \text{or} \quad 6x^2 + 3x = 6x^2 - 4$$

which is equivalent to

$$3x = -4 \quad \text{or} \quad x = -\tfrac{4}{3}$$

Since neither $3x^2 - 2$ nor $2x + 1$ is zero when $x = -\frac{4}{3}$, then $\{-\frac{4}{3}\}$ is the solution of the given proportion.

EXAMPLE 2 A mixture of fertilizers contains 3 parts nitrogen, 2 parts potash, and 2 parts phosphate by weight. How many pounds of the mixture will contain 60 pounds of nitrogen?

SOLUTION Notice that the ratio of pounds of nitrogen to pounds of mixture is $\frac{3}{7}$. So, using x for the number of pounds of mixture, we have

$$\frac{3}{7} = \frac{60}{x} \quad \text{or} \quad 3x = 60 \cdot 7 \quad \text{or} \quad x = 140$$

Hence 140 pounds of the mixture will contain 60 pounds of nitrogen.

EXERCISES

Solve the following proportions. Recall from Sec. 4.7 the caution regarding the solving of a fractional equation.

1 $\frac{x}{4} = \frac{3}{6}$

2 $\frac{9}{2x} = \frac{12}{4}$

3 $\frac{27}{36} = \frac{x}{38}$

4 $\frac{63}{121} = \frac{9}{11x}$

5 $\frac{x - 1}{7} = \frac{16}{28}$

6 $\frac{x - 1}{x - 2} = \frac{6}{5}$

7 $\frac{2x}{x - 1} = \frac{4}{5}$

8 $\frac{7}{3x} = \frac{5}{x - 2}$

9 $\frac{4x - 1}{2} = \frac{3x + 7}{21}$

10 $\frac{2x - 3}{2} = \frac{x^2 - 7}{x}$

11 $\frac{x^2 - 5}{2x - 1} = \frac{x}{2}$

12 $\frac{x - 3}{x + 2} = \frac{x + 7}{x - 1}$

13 $\frac{5}{1 - x} = \frac{5x - 1}{1 - x^2}$

14 $\frac{2 - x}{x^2 - 1} = \frac{3}{4 - 3x}$

15 $\frac{3x}{x^2 - 1} = \frac{3}{x - 2}$

16 $\frac{1 - 2x}{3x - 4} = \frac{3 - 4x}{6x}$

17 A paving mixture contains 2 parts cement, 2 parts sand, and 3 parts gravel by weight. How many pounds of the mixture can be made with 60 pounds of cement?

18 A mixture consisting of 3 parts of alcohol to 5 parts of water by volume is used to protect an automobile radiator against freezing. How many quarts of alcohol are needed to produce 20 quarts of the mixture?

19 A sirup recipe calls for 3 pounds of sugar for every 5 pounds of berries. How many pounds of sugar are needed for 8 pounds of berries? How many pounds of sugar are needed for 4 pounds of berries?

8.3
DIRECT VARIATION

An alert person is constantly on the lookout for changes. Undoubtedly you would be concerned if there were an appreciable change within the next few hours in your body temperature or in the weather. Over a longer period you would be interested in noting any significant change in your weight, or a change in your standing in this class, or a change in your pay if you have a part-time job. Not only would you be interested in such changes, but you probably would ask about each one: What caused it?

It is by making observations, asking related questions, and testing tentative conclusions that our knowledge of quantitative relationships is developed. Many changes, or variations, fall into one of the several basic types which we now study.

Suppose that it costs 4 cents per mile to travel on a certain train. Then the cost of a trip by train will depend on the number of miles in the trip. More particularly, if

m = number of miles traveled
C = cost in cents to travel m miles by train

then

$$C = 4m$$

Here the ratio of C to m is always 4; that is,

$$\frac{C}{m} = 4$$

To describe this situation, we say that C varies directly with m.

VARIES DIRECTLY WITH

When two variables x and y are related so that the ratio $\frac{y}{x}$ is always the same (remains constant), we say that y **varies directly with** x. That is,

y varies directly with x means $\frac{y}{x} = k$

where k is some nonzero constant.

Since $\frac{y}{x} = k$ ($x \neq 0$) is equivalent to $y = kx$, either of the equations

$$\frac{y}{x} = k \tag{8.1}$$

or

$$y = kx \tag{8.2}$$

DIRECT VARIATION

CONSTANT OF VARIATION

represents **direct variation.** In these equations x is the independent variable, y is the dependent variable, and k is a constant. We call k the *constant of variation.*

For example, the circumference C of a circle varies directly with the radius r, for $C = 2\pi r$. Here 2π is the constant of variation.

If y varies directly with x, then x varies directly with y. To prove this, note that if

$$y = kx$$

then

$$x = \frac{1}{k}y$$

If k is a constant, then $\frac{1}{k}$ is likewise some constant, and this shows that x varies directly with y.

If y varies directly with x and if the constant of variation is positive, then y increases as x increases and y decreases as x decreases.

The following is an outline of a procedure which you will find useful for working many problems in direct variation and also problems in other types of variation to be considered in subsequent sections of this chapter:

1 Assign letters to the variables which appear in the problem (if no letters have already been assigned), and translate the statement of variation into an algebraic equation.
2 Use the given data to compute k.
3 Substitute the computed value of k in the equation connecting the variables.
4 Use the result obtained in the preceding step as a formula to solve the problem.

EXAMPLE 1 The distance that A travels at a uniform rate varies directly with the time. A distance of 60 miles is traveled in 2 hours. Find a formula for the distance in terms of time, and find the distance traveled in $3\frac{1}{2}$ hours.

SOLUTION Let $d =$ the distance and $t =$ the time. Then

$$d = kt \tag{1}$$

Substituting in (1) the values $d = 60$ and $t = 2$, we obtain

$60 = 2k$

So

$k = 30$

Replacing k by 30 in (1), we get

$$d = 30t \tag{2}$$

To answer the final question, we replace t by $3\frac{1}{2} = \frac{7}{2}$ in (2) and obtain

$d = 30 \cdot \frac{7}{2} = 15 \cdot 7 = 105$

So the distance traveled in $3\frac{1}{2}$ hours is 105 miles. Note that the constant of variation in this problem is the familiar (uniform) rate of change.

EXAMPLE 2 Suppose that y varies directly with x and $y = 35$ when $x = 14$. Find a formula for y in terms of x, and find y when $x = 12$.

SOLUTION Since the problem states that y varies directly with x, we write

$$y = kx \tag{1}$$

We are told that when $x = 14$, then $y = 35$. Placing these values in (1), we get

$35 = 14k$ or $5 = 2k$ or $k = \frac{5}{2}$

Hence the formula for y in terms of x is

$$y = \tfrac{5}{2}x \tag{2}$$

Finally, substituting 12 for x in (2), we obtain $y = \frac{5}{2} \cdot 12$, or $y = 30$.

Write the formula for direct variation in the form

$$\frac{y}{x} = k \tag{8.3}$$

Let x_1 and y_1 be corresponding values of x and y in equation (8.3). Then

$$\frac{y_1}{x_1} = k$$

Similarly, suppose that x_2 and y_2 are another pair of corresponding values of x and y in the same equation, so that

$$\frac{y_2}{x_2} = k$$

Then, since things equal to the same thing are equal to each other,

$$\frac{y_1}{x_1} = \frac{y_2}{x_2} \tag{8.4}$$

Note that (8.4) is equivalent to

$$\frac{y_1}{y_2} = \frac{x_1}{x_2} \tag{8.5}$$

This shows that if y varies directly with x, then the ratio of two values of y is equal to the ratio of the corresponding values of x. Since the ratio of y_1 to y_2 equals the ratio of x_1 to x_2, the relation (8.5) is a proportion. This accounts for our sometimes saying "*y is directly proportional to x*" for "y varies directly with x."

The observations we have just made show that a problem in direct variation, as Example 2 above, can be worked by using a proportion. As we have noted above, the equation

$$y = kx \tag{8.6}$$

implies, or is equivalent to, the proportion

$$\frac{y_1}{x_1} = \frac{y_2}{x_2} \tag{8.7}$$

The data given and the unknown to be found in Example 2 may be represented in the following table:

WHEN $x =$	14	12
THEN $y =$	35	y_2

Substituting the values from the table in (8.7), we get

$$\frac{35}{14} = \frac{y_2}{12} \qquad \text{or} \qquad y_2 = \frac{12 \cdot 35}{14} = 30$$

Observe that we do not have to find k when we use proportion to solve problems such as Example 2. Some prefer to use proportion in such problems. You should be able to work problems in direct variation either by working with the constant of variation k or by use of proportion. In more complex types of variation to be considered later, it will be necessary that we work with the constant of variation k.

EXERCISES

Work Exercises 1 to 4 by use of the constant of variation k. First write a formula for the dependent variable in terms of the independent variable and the unknown constant k. Then find k.

1 If y varies directly with x, and $y = 8$ when $x = 2$, find y when $x = 4$.

2 U varies directly with V, and $U = 34$ when $V = 2$. Find U when $V = 4$.

3 Y varies directly with X, and $Y = 20$ when $X = 5$. Find Y when $X = \frac{7}{2}$.
4 R varies directly with X, and $R = 15$ when $X = 3$. Find R when $X = 6$.

Using the letters indicated, write the statements in Exercises 5 to 8 as an equation, with k as the constant of variation.

5 The area A of a rectangle of constant width varies directly with its length l.
6 The cost C of a roast varies directly with its weight w.
7 The number N of gallons of gasoline used by a car varies directly with the distance d it travels.
8 The income I of a workman varies directly with the number d of days he works.

Work Exercises 9 to 14 by use of the constant of variation k. First write a formula for the dependent variable in terms of the independent variable and the unknown constant k. Then find k.

9 If y varies directly with x, and $y = 12$ when $x = 2$, find y when $x = 4$.
10 If r varies directly with s, and $r = 35$ when $s = 42$, find r when $s = 72$.
11 The wages W of a workman vary directly with the number H of hours he works. If he earns $21.75 when he works 15 hours, what will he earn when he works 30 hours?
12 The weight that can be lifted by an automobile jack varies directly with the force exerted downward on the handle of the jack. If a force of 15 pounds will lift 1,500 pounds, what force is required to lift 2,500 pounds?
13 The current I in an electric circuit varies directly with the applied voltage E. When $E = 220$ volts, then $I = 6$ amperes. Find I when $E = 110$ volts; when $E = 100$ volts. Find E when $I = 4$ amperes; when $I = 3.5$ amperes.
14 A submarine travels a distance of 42 miles in 2.5 hours. Assuming that it is traveling at a uniform rate, how far will it travel in $1\frac{1}{2}$ hours?

15 Work Exercise 9 by use of proportion.
16 Work Exercise 12 by use of proportion.
17 Work Exercise 13 by use of proportion.
18 The distance that B travels at a uniform rate varies directly with the time. A distance of 84 miles is traveled in 3 hours. Find a formula for distance d in terms of time t. With the aid of this formula complete the following table:

WHEN t, HOURS, =	$\frac{1}{2}$	1	2	$2\frac{1}{2}$	3	4	5	6
THEN d, MILES, =					84			

Notice that doubling the time doubles the distance. Show algebraically that when t is multiplied by any constant m, then d is multiplied by the same constant.

8.4 DIRECT VARIATION EXTENDED

When two variables x and y are so related that $\frac{y}{x^2}$ is constant, we say that y *varies directly with the square of* x. That is,

$$y \text{ varies directly with } x^2 \qquad \text{means} \qquad \frac{y}{x^2} = k$$

where k is some constant. Since $\frac{y}{x^2} = k$ $(x \neq 0)$ is equivalent to $y = kx^2$, either of the equations

$$\frac{y}{x^2} = k \tag{8.8}$$

or

$$y = kx^2 \tag{8.9}$$

indicates that y varies directly with the square of x.

If x_1 and y_1 are one pair of corresponding values of x and y in (8.8) and x_2 and y_2 are another pair of corresponding values of x and y, then

$$\frac{y_1}{x_1^2} = k \qquad \frac{y_2}{x_2^2} = k$$

So

$$\frac{y_1}{x_1^2} = \frac{y_2}{x_2^2} \tag{8.10}$$

or

$$\frac{y_1}{y_2} = \frac{x_1^2}{x_2^2} \tag{8.11}$$

Thus if y varies directly with the square of x, the ratio of y_1 to y_2 equals the ratio of x_1^2 to x_2^2. That is why we sometimes say "y is directly proportional to x^2" for "y varies directly with x^2."

EXAMPLE 1 For a body falling freely from rest under gravity, the number of feet d which the body falls varies directly with the square of the time t in seconds. If the body falls 144 feet in 3 seconds, find a formula for d in terms of t. How far will the body fall in 6 seconds? In 9 seconds?

SOLUTION Since d varies directly with the square of t, we have

$d = kt^2$

Setting $d = 144$ and $t = 3$, we get

$144 = k \cdot 3^2 \qquad \text{or} \qquad 9k = 144$

So $k = 16$; hence

$$d = 16t^2 \tag{1}$$

is the desired formula.

Substituting $t = 6$ in (1), we obtain $d = 16 \cdot 6^2 = 16 \cdot 36 = 576$ feet. Similarly, for $t = 9$ seconds, $d = 16 \cdot 9^2 = 16 \cdot 81 = 1{,}296$ feet.

You should now be able to extend the language of direct variation to direct variation with respect to the cube or a higher power. Thus

y varies directly with x^3 means $\dfrac{y}{x^3} = k$

EXAMPLE 2 If y varies directly with x^3, and $y = 4$ when $x = 2$, find y when $x = 40$.

SOLUTION Since y varies directly with x^3, we have

$y = kx^3$

We are given that when $x = 2$, then $y = 4$. Substituting these values in the above equation, we get

$4 = 8k$

and so

$k = \frac{1}{2}$

Hence

$y = \frac{1}{2}x^3$

For $x = 40$,

$y = \frac{1}{2} \cdot (40)^3 = \frac{1}{2} \cdot 40 \cdot (40)^2 = 20 \cdot 1{,}600 = 32{,}000$

EXERCISES

Work Exercises 1 to 4 by use of the constant of variation k. First write a formula for the dependent variable in terms of the independent variable and the unknown constant k. Then find k.

1 Given that y varies directly with the square of x, and $y = 6$ when $x = 6$, find y when $x = 12$.

2 The weight of a piece of wire of a given length varies directly with the square of its diameter. If a piece of wire 0.02 inch in diameter weighs 5 ounces, how much will another piece of wire of the same length made from the same material, but 0.05 inch in diameter, weigh?

3 The pay load P of a transport airplane varies directly with the cube of its length L. When $L = 40$ feet, $P = \$3{,}500$. Find P when $L = 50$ feet; when $L = 100$ feet.

4 The weight W of silt carried by a stream is proportional to the sixth power of its velocity V. When $V = 2$ feet per second, then $W = 10$ pounds per cubic yard. Find W when $V = 4$ feet per second.

5 The volume of a right-circular cylinder with a fixed height h varies directly with the square of its radius r. If V_1 and V_2 are the volumes of two such cylinders and r_1 and r_2 are the corresponding radii, what is the proportion that is the equivalent of the above statement? Using this proportion, show that if the radius of one cylinder is twice the radius of the other, then the volume of the first is four times the volume of the second.

6 The formula for the volume V of a sphere in terms of its radius r is

$$V = \tfrac{4}{3}\pi r^3$$

If V_1 and V_2 are the volumes of two spheres and r_1 and r_2 are their corresponding radii, use the above formula to obtain a proportion connecting V_1, V_2, r_1, and r_2. If two spheres are such that the radius of one is twice the radius of the other, find the ratio of the volume of the larger to the volume of the smaller.

7 The cost per hour for fuel required to run a given steamer varies directly with the cube of its speed. If the cost for fuel is \$40 per hour for a speed of 10 miles per hour, find the cost per hour for fuel if the steamer is run 15 miles per hour.

8 The volume V of water carried by a pipe which is running full of water varies directly with the square of the diameter d of the pipe. Express this statement as a proportion. If the diameter is doubled, how is the volume affected? If the diameter is trebled, how is the volume affected?

8.5 INVERSE VARIATION

Consider the formula for distance d traveled at the uniform rate r in time t,

$$d = rt$$

Suppose that $d = 120$ miles and that t is measured in hours. Then

$$rt = 120 \qquad \text{or} \qquad t = 120\,\frac{1}{r}$$

Note that the product of r and t is constant and t varies as the reciprocal of r.

VARIES INVERSELY WITH

When two variables x and y are related so that the product of x and y is constant, we say that y **varies inversely with** x. That is,

$$y \textit{ varies inversely with } x \qquad \text{means} \qquad xy = k$$

where k is some nonzero constant. Since $xy = k$ is equivalent to $y = \frac{k}{x}$ ($x \neq 0$), either of the equations

$$xy = k \tag{8.12}$$

or

$$y = \frac{k}{x} \tag{8.13}$$

represents *inverse variation.*

If y varies inversely with x and if the constant of variation k is positive, then y decreases as x increases, and y increases as x decreases.

If x_1 and y_1 are one pair of corresponding values of x and y in (8.13), then

$$y_1 = \frac{k}{x_1} \tag{8.14}$$

Similarly, if x_2 and y_2 are another pair of corresponding values of x and y in (8.13), we have

$$y_2 = \frac{k}{x_2} \tag{8.15}$$

Dividing equals by equals, we obtain

$$\frac{y_1}{y_2} = \frac{\frac{k}{x_1}}{\frac{k}{x_2}} = \frac{k}{x_1} \cdot \frac{x_2}{k} = \frac{x_2}{x_1}$$

or

$$\frac{y_1}{y_2} = \frac{x_2}{x_1} \tag{8.16}$$

Notice that the ratio of y_1 to y_2 is equal to the ratio of x_2 to x_1, which is the inverted ratio of x_1 to x_2. This is why we sometimes say "*y is inversely proportional to x*" for "y varies inversely with x."

You should be able to work problems on inverse variation either by working with the constant of variation k or by use of proportions like (8.16).

EXAMPLE 1 Boyle's law states that the volume V of a gas varies inversely with the pressure P when the temperature is kept constant. If the volume is 10 cubic inches when the pressure is 30 pounds, find the formula for V in terms of P. What is the volume when the pressure is 10 pounds? When the pressure is 15 pounds?

SOLUTION Since we are told that V varies inversely with P, we know that

$$VP = k$$

So

$$k = 30 \cdot 10 = 300$$

Hence

$$VP = 300 \qquad \text{or} \qquad V = \frac{300}{P}$$

Substituting 10 for P in the latter formula, we get $V = \frac{300}{10} = 30$ cubic inches. Similarly, when the pressure is 15 pounds, the volume is 20 cubic inches.

EXAMPLE 2 Suppose in Example 1 that V is 10 cubic inches when P is 30 pounds and we are asked only to find V when $P = 15$ pounds. Proportions may be used conveniently in this example.

SOLUTION Since V varies *inversely* with P, we know from (8.16) that the ratio of two values of V is equal to the *inverted ratio* of the corresponding values of P. That is,

$$\frac{V_1}{V_2} = \frac{P_2}{P_1}$$

Now $V_1 = ?$ when $P_1 = 15$. $V_2 = 10$ when $P_2 = 30$. Substituting these values in the above ratio, we get

$$\frac{V_1}{10} = \frac{30}{15} \qquad \text{or} \qquad V_1 = \frac{10 \cdot 30}{15} = 20 \text{ cubic inches}$$

EXERCISES

1 The resistance R of a wire of a given length varies inversely with the square of its diameter d. If a copper wire 10 feet in length and 0.001 of an inch in diameter has a resistance of 100 ohms, find the resistance of 10 feet of copper wire 0.002 of an inch in diameter.

2 The volume V of a gas varies inversely with the pressure P when the temperature is kept constant. If $V = 100$ when $P = 2$, find P when $V = 25$.

3 The illumination I received from a source of light varies inversely with the square of its distance d from the source. The illumination received from a given source at a distance of 10 feet from that source is in what ratio to the illumination from the same source at a distance of 20 feet?

4 The number of days required to finish a job varies inversely with the number of men employed. If 5 men can finish a job in 16 days, how many days will it take 8 men to do it?

5 The weight of an object above the surface of the earth varies inversely with the square of its distance from the center of the earth. If a man weighs 200 pounds at the earth's surface, how much would he weigh if he were 240 miles above the surface? In this problem take the radius of the earth to be 3,960 miles.
6 If y varies inversely with x, and if $y = 4$ when $x = 5$, what is the value of y when $x = 12$?
7 If y varies inversely with $2x + 4$, and $y = 32$ when $x = 2$, find y when $x = 3$.
8 Show that, if y varies inversely with x, then x varies inversely with y.

8.6
JOINT VARIATION

Consider the formula for simple interest

$$I = Prt$$

where I stands for the interest, P the principal, r the rate, and t the time. Let $P = 100$. Then

$$I = 100rt \tag{8.17}$$

In (8.17) I varies directly with the product of r and t. If r increases, then I increases; similarly, if t increases, then I increases. We say that I varies directly with r and t.

VARIES DIRECTLY WITH TWO OR MORE VARIABLES

One variable **varies directly with two or more other variables** when the first variable equals a constant times the product of the other variables:

z varies directly with x and y means $z = kxy$
u varies directly with x, y, and z means $u = kxyz$
V varies directly with r^2 and h means $V = kr^2h$

EXAMPLE If u varies directly with x, y, and z, and if $u = 6$ when $x = 3$, $y = 5$, and $z = 7$, find the value of u when $x = 1$, $y = 4$, and $z = 5$.

SOLUTION The algebraic equation which translates the statement of variation is

$$u = kxyz \tag{1}$$

Using the fact that $u = 6$ when $x = 3$, $y = 5$, and $z = 7$, we get

$$6 = k \cdot 3 \cdot 5 \cdot 7 \qquad \text{or} \qquad k = \tfrac{2}{35}$$

Substituting $\frac{2}{35}$ for k in (1), we obtain the formula

$$u = \tfrac{2}{35}\, xyz$$

If we substitute $x = 1$, $y = 4$, and $z = 5$ in this formula, we get

$$u = \tfrac{2}{35} \cdot 1 \cdot 4 \cdot 5 = \tfrac{8}{7}$$

8.7
COMBINED VARIATION

The various types of variation may be combined in the same problem. To illustrate, the statement

v varies directly with x and w and inversely with y

means

$$v = \frac{kxw}{y}$$

and

v varies directly with x and w and inversely with z^2

means

$$v = \frac{kxw}{z^2}$$

EXAMPLE The quantity v varies directly with x and w and inversely with y. When $x = 2$, $w = 6$, and $y = 4$, then v has the value 9. What value will v have when $x = 3$, $w = 1$, and $y = 5$?

SOLUTION From the statement of variation we have

$$v = \frac{kxw}{y} \tag{1}$$

To determine k, we note that $v = 9$ when $x = 2$, $w = 6$, and $y = 4$; so

$$9 = \frac{k \cdot 2 \cdot 6}{4} \qquad \text{or} \qquad k = 3$$

Substituting 3 for k in (1), we have the formula

$$v = \frac{3xw}{y}$$

When $x = 3$, $w = 1$, and $y = 5$, this formula gives

$$v = \frac{3 \cdot 3 \cdot 1}{5} = \frac{9}{5}$$

EXERCISES

Write each of the statements in Exercises 1 to 4 as an equation containing a constant of variation, k.

1 R varies directly with S and T.

2 The absolute temperature of a gas varies directly with the pressure and the volume.

3 H varies directly with the square of m and inversely with the cube of p.

4 The gravitational force F between any two bodies varies directly with their masses m_1 and m_2 and inversely with the square of the distance d between them.

For the equations in Exercises 5 to 8, give a verbal statement using the language of variation.

5 $S = kr^2h^2$

6 $V = \frac{4}{3}\pi r^3$

7 $I = \dfrac{kP}{d^2}$

8 $P = \dfrac{kT}{V}$

9 The load which a horizontal beam of a given length will safely carry varies directly with its breadth and the cube of its depth. If a timber 3 inches wide and 4 inches deep and of a given length will safely carry a load of 480 pounds, what is the safe load for a beam of the same length which is 3 inches wide and 12 inches deep?

10 The wage of a worker paid by the piece is directly proportional to his output. The constant of variation is called the *piece rate*. A certain worker gets $10 for a day in which he turns out 125 units. Let W be the wage, T the output, and r the piece rate. Find r. At the same rate, what would he earn for 148 units?

11 z varies directly with x and the square of y. Also $z = 6$ when $x = 9$ and $y = \frac{1}{2}$. Find the formula for z in terms of x and y. What is z when $x = 3$ and $y = 4$?

12 The distance required to stop a certain car varies directly with the square of its velocity. If the car can stop in 45 feet from a speed of 35 miles per hour, how many feet will be required to stop it from a speed of 60 miles per hour?

9

Basic Trigonometry

9.1
SIMILAR TRIANGLES AND INDIRECT MEASUREMENT

Two triangles are *similar* if the three angles of one triangle are equal, respectively, to the three angles of the other triangle. For example, in the triangles ABC and DEF in Fig. 9.1,

angle BAC = angle EDF
angle ACB = angle DFE
angle ABC = angle DEF

So the triangles are similar. In two similar triangles a side of one triangle is said to *correspond* to a side of the other triangle if these sides are opposite equal angles. To illustrate, in Fig. 9.1, side a corresponds to side d, since angle BAC = angle EDF.

We could describe the two triangles of Fig. 9.1 as having the same shape although they are not the same size.

One of the most useful properties of similar triangles is the fact that *if two triangles are similar, the ratios of the lengths of corresponding sides are equal.* In the triangles of Fig. 9.1, side a corresponds to side d, side b corresponds to side e, side c corresponds to side f, and we have

$$\frac{a}{d} = \frac{b}{e} = \frac{c}{f}$$

The following examples illustrate the use of similar figures in indirect measurement.

EXAMPLE 1 Find the height of a flagpole which casts a shadow 20 feet long at the same time that a 6-foot post casts a shadow $2\frac{1}{2}$ feet long.

SOLUTION Let x be the height in feet of the flagpole. Construct a figure showing the given relations.

We see that triangle ABC is similar to triangle DEF (Fig. 9.2) and hence that

$$\frac{x}{6} = \frac{20}{2.5} \qquad \text{or} \qquad x = 48$$

So the flagpole is 48 feet high.

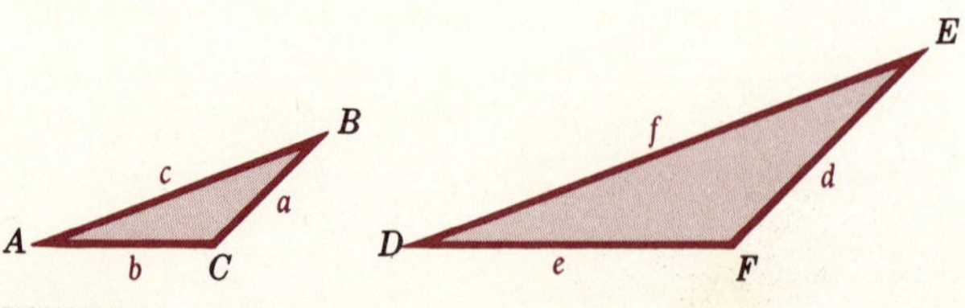

FIGURE 9.1

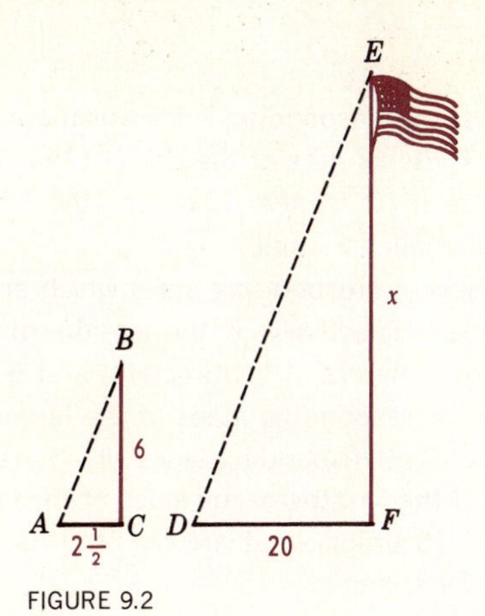

FIGURE 9.2

EXAMPLE 2 Find the distance AB across the pond shown in Fig. 9.3.

SOLUTION To determine this distance, we drive stakes at A, B, and C, where C is a point chosen so that either AC or BC skirts the pond. The points D and E are chosen so that the line DE is parallel to the line AB and so that it skirts the pond. Then triangle ABC is similar to triangle CDE (Fig. 9.3) so that

$$\frac{AB}{DE} = \frac{CB}{CE}$$

Hence if measurements show that $DE = \dot{6}0$ yards,* $BC = \dot{7}0$ yards, and $EC = 55$ yards, we know that

$$\frac{AB}{60} = \frac{70}{55} \quad \text{or} \quad \frac{AB}{60} = \frac{14}{11}$$

Therefore $AB = 76$ yards (approximately).

* The notation $\dot{6}0$ is explained in Appendix C.

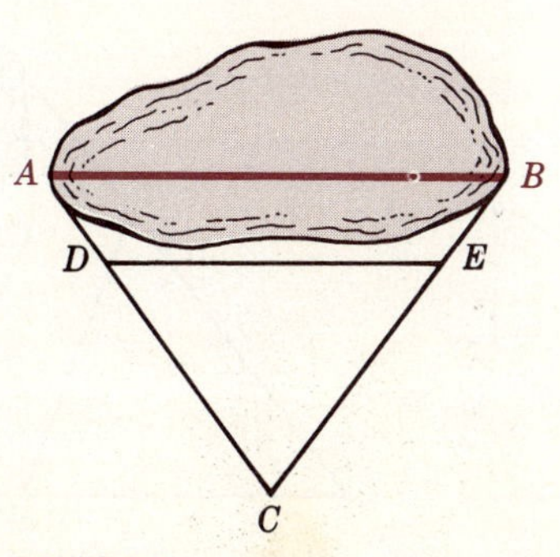

FIGURE 9.3

EXERCISES

1 Two similar triangles have corresponding sides which are 2 inches and 5 inches long, respectively. If the lengths of the sides of the larger triangle are 5 inches, 7.5 inches, and 10 inches, what are the lengths of the corresponding sides of the smaller triangle?

2 Two similar triangles have corresponding sides which are 3 centimeters and 7 centimeters long, respectively. If the lengths of the sides of the smaller triangle are 3 centimeters, 4 centimeters, and 6 centimeters, what are the lengths of the corresponding sides of the larger triangle?

3 Two similar triangles have corresponding sides which are 6 units and 11 units long, respectively. If the lengths of the sides of the larger triangle are 11 units, 13 units, and 15 units, what are the lengths of the corresponding sides of the smaller triangle?

4 A man casts a shadow 4.4 feet long at the same time his son casts a shadow 2.1 feet long. If the son is 3 feet tall, how tall is the man?

5 What is the height of a building which casts a 30-foot shadow at the same time a 9-foot post casts a 2-foot shadow?

6 A man 6 feet tall casts a shadow $2\frac{1}{2}$ feet long at the same time his wife casts a shadow $2\frac{1}{4}$ feet long. How tall is his wife?

7 How long will the shadow of a 5-foot post be when the shadow of a 47-foot steeple is 8 feet long?

8 A man 5 feet 9 inches tall casts a shadow 6 feet 3 inches long when he stands 20 feet from a street light. How high is the light?

9 To find the height of a tree, a boy drove a stake ED into the ground and then took the following measurements: $ED = 3.5$ feet, $AD = 8.2$ feet, and $AB = 31.8$ feet (see Fig. 9.4). How high was the tree?

10 To find the length of a pond, a surveyor drove stakes at A, B, and C. He then laid off DE parallel to AB and took the following measurements: $DE = 982$ feet, $CE = 875$ feet, and $EB = 425$ feet (see Fig. 9.5). What is the length of AB?

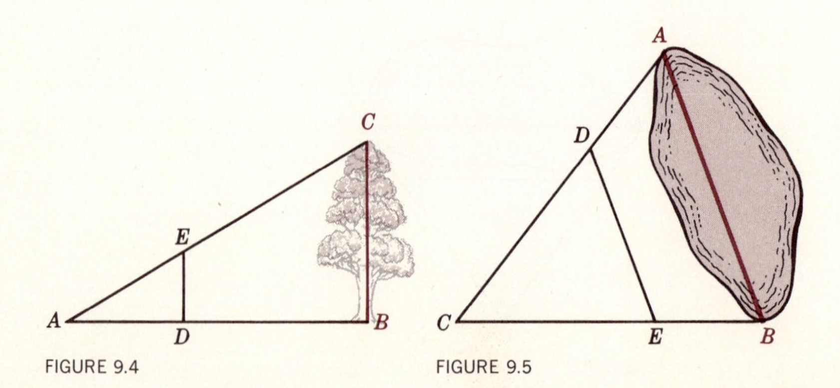

FIGURE 9.4

FIGURE 9.5

9.2
SCALE DRAWINGS

An important application of ratio and proportion is the construction and use of scale drawings. We do not need to see a full-scale drawing of a house or a boat in order to get a clear idea of what the object is. We can usually understand what we want to know about the object if a drawing shows the shape, even though the object is many times larger (or smaller) than the drawing. If in addition to showing the shape of an object, a drawing also indicates the size of the object, then the drawing becomes a scale drawing. Regardless of the way in which a scale may be shown on a scale drawing, its purpose is always to show how distances on the drawing and on the object are related. The scale shows how the length of a line on the drawing compares with the length of the corresponding line on the object.

Maps and charts are similar examples of scale drawings. Navigational maps and charts are drawn with extreme care, and the scale is carefully checked, for a navigator must be able to rely on the accuracy of his measurements. On the other hand, a road map does not require a high degree of accuracy and attention to minute details. Hence, measurements taken from most road maps usually give nothing more than fair approximations to actual distances. Suppose we want to know the distance between two cities which appear $2\frac{1}{8}$ inches apart on a road map which has a scale of $\frac{1}{4}$ inch = 50 miles. Let x be the distance in miles between the two cities. Then, approximately,

$$\frac{x}{2\frac{1}{8}} = \frac{50}{\frac{1}{4}} \qquad \text{or} \qquad x = 425$$

Therefore the distance between the two cities is approximately 425 miles.

Other examples of scale drawings are plans and blueprints which are used as guides in building and manufacturing. Magazines, books, and newspapers often contain the plans for homes which the publisher believes will interest his readers. Before a contractor can estimate the cost of erecting a building, he must have a carefully worked-out scale drawing, or blueprint, which gives all details of the construction. Many commercial positions require that employees be able to read and to understand blueprints.

If the scale on a blueprint is $\frac{1}{8}'' = 1'$ ($\frac{1}{8}$ inch = 1 foot), a line on the blueprint 2 inches long will represent a length of 16 feet. This can be obtained from the proportion

$$\frac{x}{2} = \frac{1}{\frac{1}{8}}$$

Likewise, an area on the blueprint which is $2\frac{1}{2}$ by $3\frac{1}{4}$ inches would represent an area of 520 square feet. This can be shown by finding the lengths in feet which correspond to $2\frac{1}{2}$ and $3\frac{1}{4}$ inches and multiplying them together, or it can be obtained by solving the proportion

$$\frac{y}{\frac{5}{2}\cdot\frac{13}{4}} = \frac{1}{(\frac{1}{8})^2}$$

You should notice that this proportion states that y square feet on the object is to $\frac{5}{2}\cdot\frac{13}{4}$ square inches on the blueprint as 1 square foot on the object is to $(\frac{1}{8})^2$ square inch on the blueprint, or that $\frac{1}{64}$ square inch on the blueprint corresponds to 1 square foot on the object.

EXERCISES

1 If the scale on a map is $\frac{1}{4}$ inch = 25 miles, how far apart are two cities which are $4\frac{3}{4}$ inches apart on the map?

2 Two cities are 120 kilometers apart. How far apart will they be on a map which has a scale of 1 centimeter = 50 kilometers?

3 The scale on the plans for a house is $\frac{1}{5}$ inch = 1 foot. What is the area of the floor of a room which measures $3\frac{1}{10}$ inches by $4\frac{3}{10}$ inches on the plans?

4 The house in Exercise 3 is 65 feet long and 22 feet wide. What are the corresponding lengths on the plans?

5 If the scale on a blueprint is $\frac{1}{16}$ inch = 1 foot, what is the area represented by a rectangle which is $\frac{3}{4}$ inch by $2\frac{1}{8}$ inches on the blueprint?

6 The scale on a map is marked as 1:10,000,000, which means that one unit on the map is equivalent to 10,000,000 of the same units on the earth. For example, two points which are 1 inch apart on the map would actually be 10,000,000 inches apart. How many feet apart (to the nearest foot) would these two points be? How many miles apart (to the nearest 10 miles) would they be?

7 If the scale on a map is marked as 1:500,000, one centimeter on the map is equivalent to how many kilometers?

8 If the scale on a map is marked as 1:100,000, one inch on the map is equivalent to how many miles?

9 The scale on a map is given as 1:500,000. If two cities are 26 miles apart, how many inches apart would they be on the map?

10 The scale on a map is 1:1,000,000. How many miles apart are two cities which are 4.6 inches apart on the map?

9.3
THE TANGENT OF AN ANGLE

In any triangle the sum of the three angles is 180°. It is customary to designate the angles of a triangle by capital letters. Let the angles

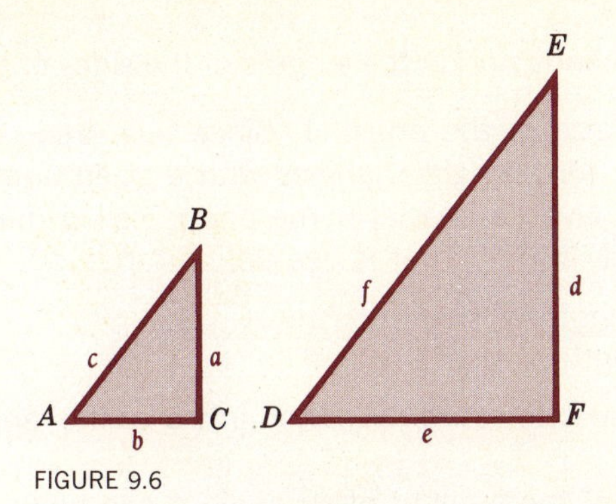

FIGURE 9.6

of a triangle be denoted by A, B, and C. Then

$$A + B + C = 180°$$

RIGHT ANGLE
RIGHT TRIANGLE
ACUTE ANGLE

An angle of 90° is called a *right angle,* and if one of the angles of a triangle is 90°, the triangle is called a *right triangle.* An angle less than 90° is called an *acute angle.* If one of the angles of a triangle is a right angle, the other two angles must be acute angles and the sum of these two acute angles is 90°. Consequently, when one of the acute angles in a right triangle is known, the other acute angle can be determined by subtracting the known angle from 90°. Let C denote the right angle in triangle ACB of Fig. 9.6. Then $A + B = 90°$, and $B = 90° - A$.

Suppose that an acute angle of one right triangle is equal to an acute angle of another right triangle. Let the two triangles be represented by Fig. 9.6, with $A = D$. Then

$$B = 90° - A \qquad E = 90° - D$$

So

$$B = E$$

since

$$A = D$$

Therefore, corresponding angles of the two triangles are equal, and the two triangles are similar. Since the two triangles are similar, the ratio of the lengths of any two sides of one triangle is equal to the ratio of the lengths of the corresponding sides of the other triangle. So

$$\frac{a}{b} = \frac{d}{e}$$

In right triangle ACB of Fig. 9.6, side a is called the **side opposite angle** A and side b is called the **side adjacent to angle** A. Side c, which is opposite the right angle C, is called the **hypotenuse.**

The ratio $\frac{a}{b}$ depends not on the lengths of the sides of the triangle but only on the size of the angle A. Since this ratio is constant, or fixed in value, for all right triangles with a given angle A as an acute angle, it is given a special name and is called the **tangent** of angle A, or simply *tan A*. That is, we define tan A by

TANGENT OF AN ANGLE

$$\tan A = \frac{a}{b} = \frac{\text{length of side opposite } A}{\text{length of side adjacent to } A}$$

Notice that we use a to stand for either side a or the length of that side.

This definition of the tangent of an angle A can be used to compute the values of the tangents of selected angles. Suppose that we want to determine tan 20° correct to two decimal places. We construct, as in Fig. 9.7, a right triangle ACB sufficiently large that

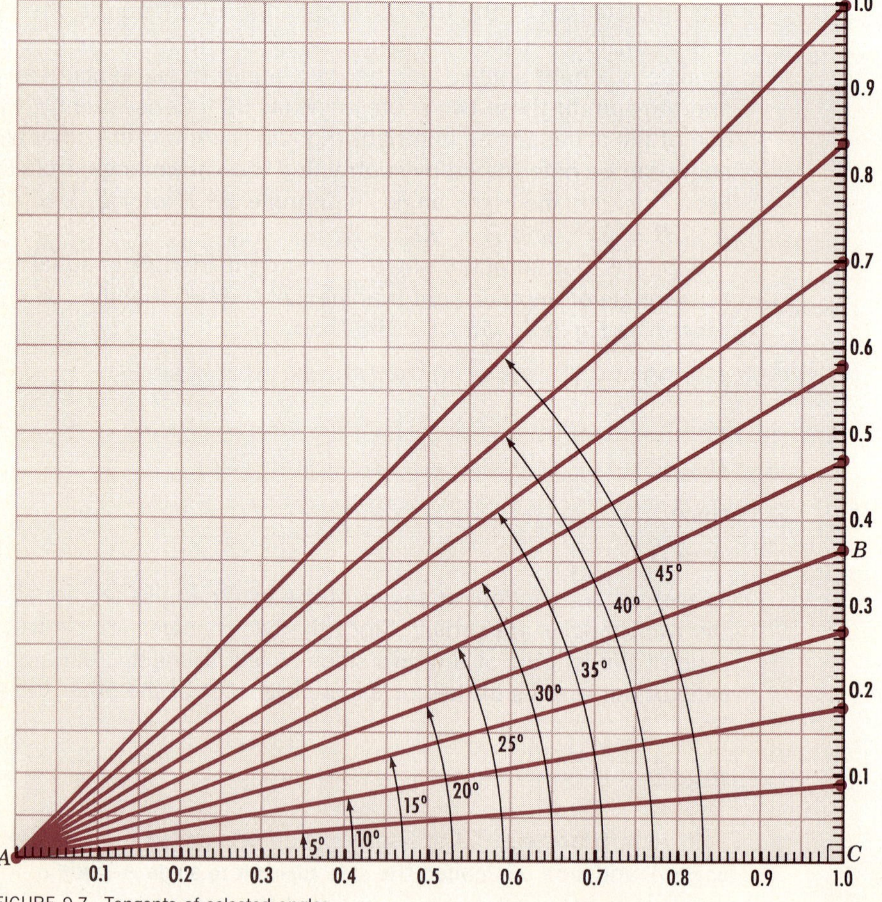

FIGURE 9.7 Tangents of selected angles.

if the length of the side AC adjacent to A is taken to be the unit length, then we can determine the length of the side BC opposite to A correct to two decimal places. With angle $A = 20°$, we find from Fig. 9.7 that

$$\tan 20° = \frac{0.36}{1.00} = 0.36$$

Proceeding in this manner, we can construct, with the aid of Fig. 9.7, the following table:

ANGLE A, DEGREES	SIDE BC OPPOSITE A	SIDE AC ADJACENT TO A	TAN $A = \frac{BC}{AC}$
5	0.09	1.00	0.09
10	0.18	1.00	0.18
15	0.27	1.00	0.27
20	0.36	1.00	0.36
25	0.47	1.00	0.47
30	0.58	1.00	0.58
35	0.70	1.00	0.70
40	0.84	1.00	0.84
45	1.00	1.00	1.00

With the use of more advanced mathematics (infinite series), mathematicians have prepared tables of tangents which are more accurate and more extensive than that given above. In the table on page 385 there is a column headed "Tangent." To find the tangent of a given angle, find the number of degrees of the angle in the angle column; then the number on the horizontal line to the right in the tangent column is the tangent of the angle. To illustrate,

$$\tan 29° = 0.554 \qquad \tan 82° = 7.12$$

In this table the values of $\tan A$ (and also those given for $\sin A$ and $\cos A$, which are discussed in the next section) are given correct to three significant figures.* In all the problems which are to be solved with the aid of this table, the numbers are given correct to three significant figures; so the answers should be rounded off to three significant figures.

ANGLE OF ELEVATION

If the point B is above the horizontal line AC, the **angle of elevation** of the point B as seen from the point A is the angle which the line of sight AB makes with the horizontal line AC (Fig. 9.8). If the point B is below the horizontal line AC, the **angle of depression** of the point B as seen from the point A is the angle which the line of sight AB makes with the horizontal line AC (Fig. 9.9).

ANGLE OF DEPRESSION

* See Appendix C.

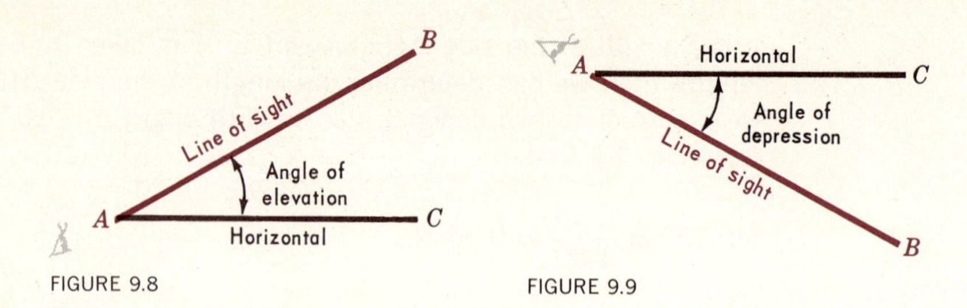

FIGURE 9.8

FIGURE 9.9

EXAMPLE Find the distance a of Fig. 9.10 if the angle of elevation of the point B as seen from A is 34° when b is 125 feet.

SOLUTION From the definition of the tangent of an angle we have

$$\tan 34^\circ = \frac{a}{125}$$

From the table on page 385 we have

$$\tan 34^\circ = 0.675$$

Therefore

$$\frac{a}{125} = 0.675 \qquad \text{or} \qquad a = 125(0.675)$$

or $a = 84.4$ feet to three significant figures.

EXERCISES

From the table on page 385, find the tangent of the following angles:

1 22° 2 37° 3 45° 4 79°

5 The angle of elevation of the top of the wall of a building as seen from a point 130 feet from the foot of the wall is 47°. How tall is the building?

6 A wire is stretched from the top of a vertical pole standing on level ground. The wire reaches a point 14.6 feet from the foot of the pole and makes an angle of 74° with the horizontal. Find the height of the pole.

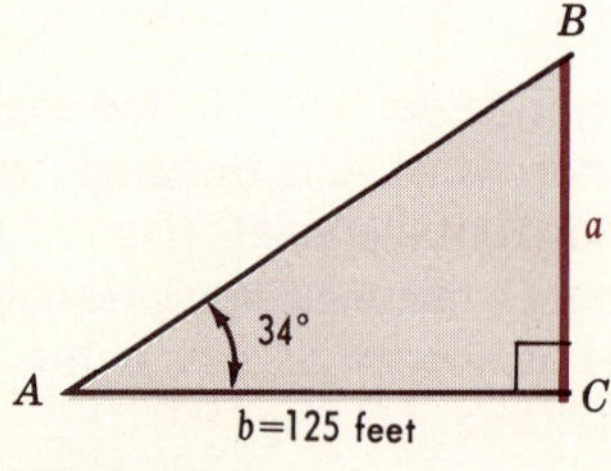

FIGURE 9.10

Table of values of sines, cosines, and tangents

ANGLE, DEGREES	SINE	COSINE	TANGENT	ANGLE, DEGREES	SINE	COSINE	TANGENT
1	0.0175	1.00	0.0175	45	0.707	0.707	1.00
2	0.0349	0.999	0.0349	46	0.719	0.695	1.04
3	0.0523	0.999	0.0524	47	0.731	0.682	1.07
4	0.0698	0.998	0.0699	48	0.743	0.669	1.11
5	0.0872	0.996	0.0875	49	0.755	0.656	1.15
6	0.105	0.995	0.105	50	0.766	0.643	1.19
7	0.122	0.993	0.123	51	0.777	0.629	1.23
8	0.139	0.990	0.141	52	0.788	0.616	1.28
9	0.156	0.988	0.158	53	0.799	0.602	1.33
10	0.174	0.985	0.176	54	0.809	0.588	1.38
11	0.191	0.982	0.194	55	0.819	0.574	1.43
12	0.208	0.978	0.213	56	0.829	0.559	1.48
13	0.225	0.974	0.231	57	0.839	0.545	1.54
14	0.242	0.970	0.249	58	0.848	0.530	1.60
15	0.259	0.966	0.268	59	0.857	0.515	1.66
16	0.276	0.961	0.287	60	0.866	0.500	1.73
17	0.292	0.956	0.306	61	0.875	0.485	1.80
18	0.309	0.951	0.325	62	0.883	0.469	1.88
19	0.326	0.946	0.344	63	0.891	0.454	1.96
20	0.342	0.940	0.364	64	0.899	0.438	2.05
21	0.358	0.934	0.384	65	0.906	0.423	2.14
22	0.375	0.927	0.404	66	0.914	0.407	2.25
23	0.391	0.921	0.424	67	0.921	0.391	2.36
24	0.407	0.914	0.445	68	0.927	0.375	2.48
25	0.423	0.906	0.466	69	0.934	0.358	2.61
26	0.438	0.899	0.488	70	0.940	0.342	2.75
27	0.454	0.891	0.510	71	0.946	0.326	2.90
28	0.469	0.883	0.532	72	0.951	0.309	3.08
29	0.485	0.875	0.554	73	0.956	0.292	3.27
30	0.500	0.866	0.577	74	0.961	0.276	3.49
31	0.515	0.857	0.601	75	0.966	0.259	3.73
32	0.530	0.848	0.625	76	0.970	0.242	4.01
33	0.545	0.839	0.649	77	0.974	0.225	4.33
34	0.559	0.829	0.675	78	0.978	0.208	4.70
35	0.574	0.819	0.700	79	0.982	0.191	5.14
36	0.588	0.809	0.727	80	0.985	0.174	5.67
37	0.602	0.799	0.754	81	0.988	0.156	6.31
38	0.616	0.788	0.781	82	0.990	0.139	7.12
39	0.629	0.777	0.810	83	0.993	0.122	8.14
40	0.643	0.766	0.839	84	0.995	0.105	9.51
41	0.656	0.755	0.869	85	0.996	0.0872	11.4
42	0.669	0.743	0.900	86	0.998	0.0698	14.3
43	0.682	0.731	0.933	87	0.999	0.0523	19.1
44	0.695	0.719	0.966	88	0.999	0.0349	28.6
45	0.707	0.707	1.00	89	1.00	0.0175	57.3

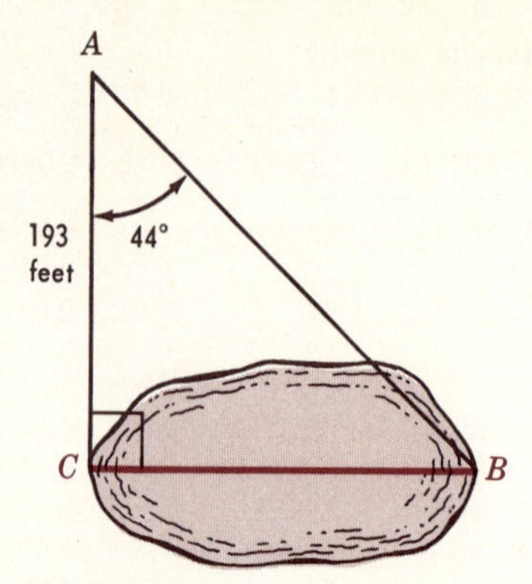

FIGURE 9.11

7 To find the distance CB across a pond, a surveyor measured the distance AC, with angle BCA equal to 90°, and found AC to be 193 feet (see Fig. 9.11). He then measured angle A and found it to be 44°. Find the width CB of the pond.

8 A pole is 67.2 feet high. Find the length of its shadow when the angle of elevation of the sun is 49°.

9 An observer in a helicopter which is directly above a point C and at an altitude of 6,320 feet notes that the angle of depression of an enemy fortification A is 12°. If the points A and C are on the same horizontal line, what is the distance AC?

10 From the top of a vertical cliff 212 feet high (see Fig. 9.12) the angle of depression of a boat at sea is 26°. How far is the boat from the foot of the cliff?

11 From the top of a tower 114 feet high the angle of depression of an object on a level with the base of the tower is 29°. What is the distance of the object from the base of the tower?

12 The angle of elevation of an airplane from a point A due west of it is 54°, and from a point B due east of it the angle of elevation is 49°. The distance between A and B is 1,430 feet. Find the height of the airplane.

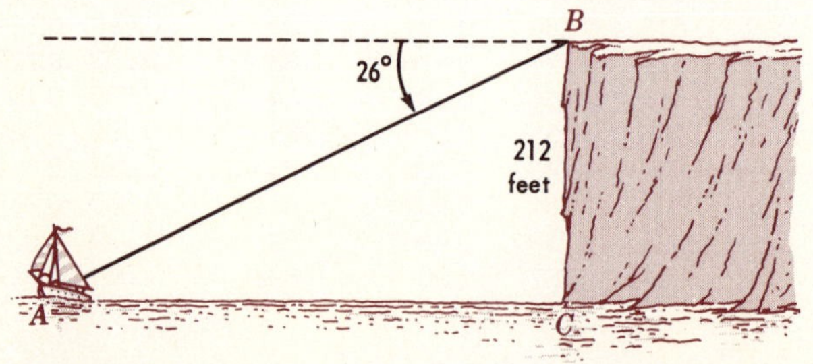

FIGURE 9.12

9.4
THE SINE AND COSINE OF AN ANGLE

In addition to the tangent there are other ratios connected with a right triangle that have constant values for any particular value of an acute angle of the triangle. For example, returning to the similar right triangles of Fig. 9.6, we see that

$$\frac{a}{c} = \frac{d}{f} \qquad \text{and} \qquad \frac{b}{c} = \frac{e}{f}$$

SINE
COSINE

These constant ratios have the special names **sine** of angle A, or simply $\sin A$, and **cosine** of angle A, or simply $\cos A$, respectively. That is, we define $\sin A$ and $\cos A$ by the following inequalities:

$$\sin A = \frac{a}{c} = \frac{\text{length of the side opposite } A}{\text{length of the hypotenuse}}$$

$$\cos A = \frac{b}{c} = \frac{\text{length of the side adjacent to } A}{\text{length of the hypotenuse}}$$

Just as we could use the definition of the tangent of an angle to compute values of the tangents of selected angles, so we can use these definitions of $\sin A$ and $\cos A$ to compute values of sines and cosines of selected angles. Suppose that we want to determine the values of $\sin 20°$ and $\cos 20°$ correct to two decimal places. We construct, as in Fig. 9.13, a right triangle ACB sufficiently large so that if the length of the hypotenuse is the unit length, then the lengths of the sides of the triangle can be determined correct to two decimal places. With angle $A = 20°$, we find from Fig. 9.13 that

$$\sin 20° = \frac{\text{length of the side opposite } A}{\text{length of the hypotenuse}} = \frac{0.34}{1.00} = 0.34$$

$$\cos 20° = \frac{\text{length of the side adjacent to } A}{\text{length of the hypotenuse}} = \frac{0.94}{1.00} = 0.94$$

In this manner, we can find the sine and the cosine of any given acute angle correct to two decimal places, and in Exercise 31 we ask the student to use this method to find $\sin 5°$, $\sin 10°$, and $\sin 15°$.

Tables have been constructed, by using methods from more advanced mathematics, which give numerical values of sines and cosines that are more accurate than can be computed by geometrical means from the definitions. In the table on page 385 values of $\sin A$ and $\cos A$ are given correct to three decimal places. To illustrate,

$$\sin 42° = 0.669 \qquad \cos 68° = 0.375$$

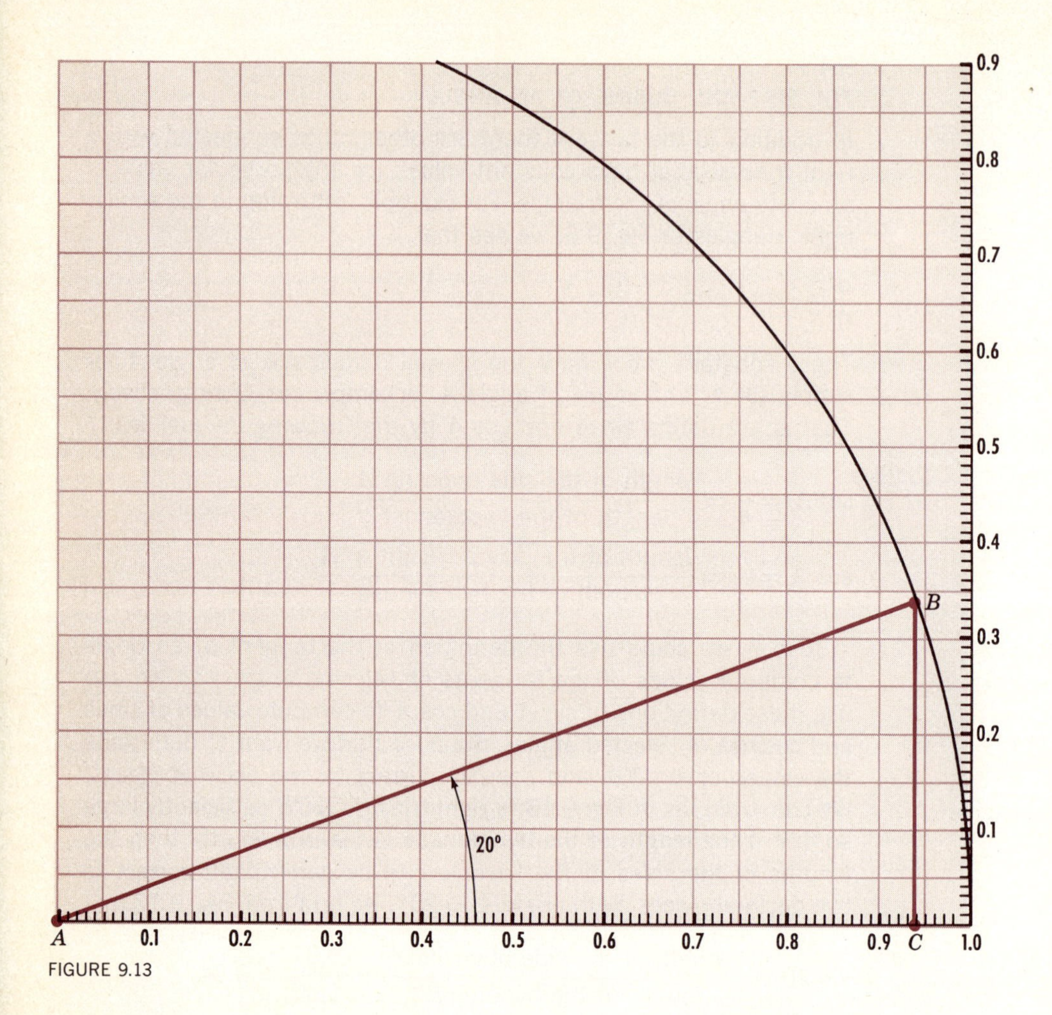

FIGURE 9.13

EXAMPLE The length of a kite string is 234 yards, and the angle of elevation of the kite is 52°. Find the height of the kite above the ground, assuming that the kite string is straight. Since a kite string is not actually straight but is concave upward, the true height of the kite will be different from the one we calculate. Is the calculated height less than, or greater than, the actual height?

SOLUTION Let a represent the height of the kite in yards. Assuming that the string is straight, the situation is as pictured in Fig. 9.14. Using this figure, we see that

$$\sin 52^\circ = \frac{a}{234}$$

From the table we read $\sin 52^\circ = 0.788$. Hence

$\frac{a}{234} = 0.788$ and $a = 184$ yards to three significant figures

Since the kite string would actually sag downward, the true height of the kite would be less than the height calculated above.

EXERCISES

From the table on page 385, find the sine of each of the following angles:

1 19° 2 28° 3 45° 4 90°

From the table on page 385, find the cosine of each of the following angles:

5 31° 6 49° 7 45° 8 81°

9 In a right triangle ACB, the hypotenuse AB is 10 units long and the side BC is 5 units long. What is the value of $\sin A$? Use the table on page 385 to find the number of degrees in angle A.

10 A wire is stretched from the top of a pole to a point on the ground 112 feet from the base of the pole. If the angle between the wire and the ground is 51°, what is the length of the wire? What is the height of the pole?

11 A steel cable is to run from a point B on top of a tall steel chimney to a point A on the ground (see Fig. 9.15). If $AC = 134$ feet and angle $CAB = 52°$, find the length of the cable.

12 A balloon is anchored to a point A with a cable 1,980 feet in length. If the angle of elevation of the balloon at A is 44°, find the height of the balloon.

13 A ladder 32.0 feet long leans against a building and makes an angle of 65° with the ground. What is the distance from the base of the building to the foot of the ladder?

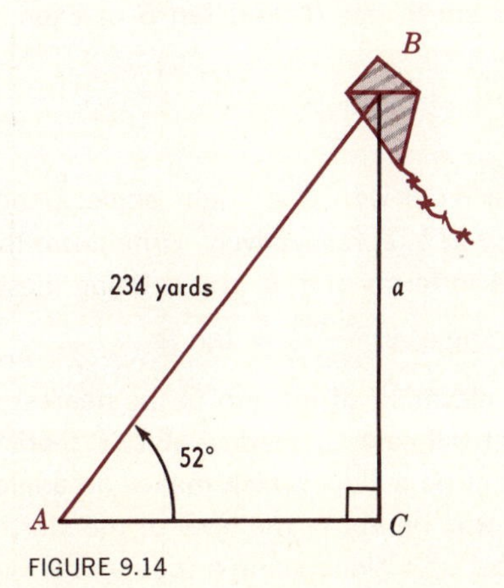

FIGURE 9.14

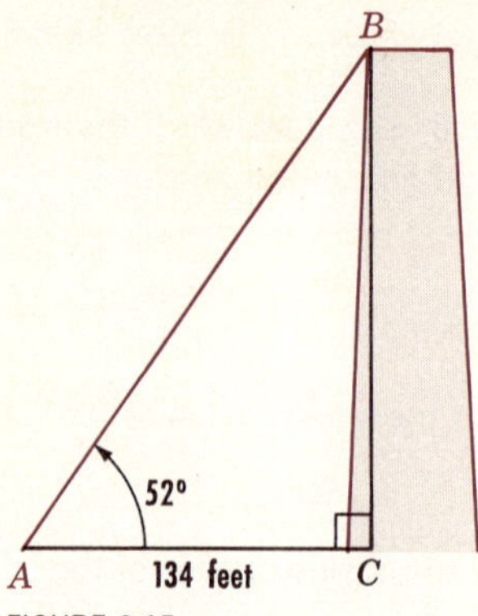

FIGURE 9.15

14 The distance from the earth to the underside of the clouds is called the *ceiling*. An observer at an airport sent up a vertical beam of light. At a point 765 feet from the searchlight, and at the same elevation as the searchlight, the observer found the angle of elevation of the spot of light on the cloud to be 68°. Find the ceiling.

15 In right triangle ACB with right angle at C, $a = 4$, $b = 3$, and $c = 5$. Find $\sin A$, $\cos A$, and $\tan A$. Find $\sin B$, $\cos B$, and $\tan B$.

16 Using the values for $\sin A$ and $\cos A$ in Exercise 15, verify that

$$(\sin A)^2 + (\cos A)^2 = 1$$

17 Using the values for $\sin B$ and $\cos B$ in Exercise 15, verify that

$$(\sin B)^2 + (\cos B)^2 = 1$$

18 Using the values for $\sin A$, $\cos A$, and $\tan A$ in Exercise 15, verify that

$$\frac{\sin A}{\cos A} = \tan A$$

19 Using the values for $\sin B$, $\cos B$, and $\tan B$ in Exercise 15, verify that

$$\frac{\sin B}{\cos B} = \tan B$$

20 Draw a right triangle ACB with C as right angle. Denote the sides opposite angles A, B, C by a, b, c, respectively. Write down the definitions of $\sin A$, $\cos A$, and $\tan A$ in terms of a, b, and c. Using these definitions, prove that for any acute angle A, $\dfrac{\sin A}{\cos A} = \tan A$.

21 What is the angle of elevation of the sun to the nearest degree when a telephone pole 30.0 feet tall casts a shadow of 12.0 feet?

22 A man walked 256 feet up a slope which makes an angle of 28° with the horizontal. How high was he above the level of the starting point?

23 A man walked 352 feet up a slope along a road and arrived at a point

47.0 feet above the level of his starting point. Find to the nearest degree the angle the road makes with the horizontal.

24 From one bank of a river the angle of elevation of the top of a cliff rising vertically from the opposite bank is 35°. The cliff is known to be 128 feet high. Find the width of the river.

25 When the angle of elevation of the sun is 52°, what is the length of the shadow cast by a flagpole 46.0 feet tall?

26 An observer in a helicopter at a height of 6,930 feet finds that the angle of depression of an enemy gun emplacement is 37°. What is the distance of the gun from a point directly beneath the observer? How far is the gun from the observer?

27 From a point 315 feet above sea level two boats are seen directly east of the observer. The angle of depression of one boat is 19°, and the angle of depression of the other boat is 24°. How far apart are the boats?

28 An airplane in leaving a runway flies along a straight line at an angle of 17° to the horizontal. What will be the altitude of the airplane after it has traveled 4,890 feet along this line of flight?

29 A flagpole DE is fastened on the edge of the top of a building. At a horizontal distance of 385 feet from a point at the foot of the building directly beneath the flagpole, the angle of elevation of the top of the pole is 57°. From the same point of observation the angle of elevation of the bottom of the pole is 53°. What is the length of the flagpole?

30 The angle of elevation of an airplane from a point A due west of it is 61°, and from a point B due east of it the angle of elevation is 54°. The distance between A and B is 1,390 feet. Find the height of the airplane.

31 With the aid of appropriate figures, find the sine and the cosine of each of 5°, 10°, and 15° correct to two decimal places.

9.5 THE TRIGONOMETRIC RATIOS—TRIGONOMETRIC IDENTITIES

In Secs. 9.3 and 9.4 we have defined and studied three ratios involving the sides of a right triangle that depend only on the size of one of the acute angles, A, namely (with reference to Fig. 9.16),

$$\sin A = \frac{a}{c} = \frac{\text{length of side opposite } A}{\text{length of hypotenuse}}$$

$$\cos A = \frac{b}{c} = \frac{\text{length of side adjacent to } A}{\text{length of hypotenuse}}$$

$$\tan A = \frac{a}{b} = \frac{\text{length of side opposite } A}{\text{length of side adjacent to } A}$$

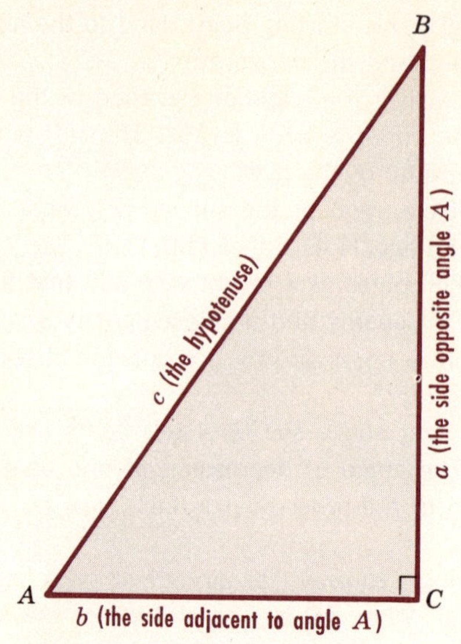

FIGURE 9.16 Right triangle ACB.

There are three other ratios involving sides of triangle ACB which depend only on the size of angle A. These are called

cotangent of A secant of A and cosecant of A

and have the respective abbreviations

$\cot A \qquad \sec A \qquad \csc A$

They are *defined* by the following equalities:

$$\cot A = \frac{b}{a} = \frac{\text{length of side adjacent to } A}{\text{length of side opposite } A}$$

$$\sec A = \frac{c}{b} = \frac{\text{length of hypotenuse}}{\text{length of side adjacent to } A}$$

$$\csc A = \frac{c}{a} = \frac{\text{length of hypotenuse}}{\text{length of side opposite } A}$$

THE TRIGONOMETRIC RATIOS

The six ratios $\sin A$, $\cos A$, $\tan A$, $\cot A$, $\sec A$, and $\csc A$, whose values depend only on the size of the acute angle A, are called the **trigonometric ratios** of A. These six ratios are connected by several equalities, one of them being

$$\sin A \csc A = 1 \qquad \text{or} \qquad \sin A = \frac{1}{\csc A} \qquad \text{or} \qquad \csc A = \frac{1}{\sin A}$$

To prove this equality, we use the above definitions of $\sin A$ and $\csc A$ and observe that

$$\sin A \csc A = \frac{a}{c} \cdot \frac{c}{a} = \frac{ac}{ca} = 1$$

Since $\sin A \csc A = 1$ holds for all values of A greater than 0° and less than 90°, we call the equality $\sin A \csc A = 1$ a *trigonometric identity.*

EXERCISES

In the following exercises, A denotes an acute angle in a right triangle (like that in Fig. 9.16).

1 Show that $\cos A \sec A = 1$.
2 Show that $\tan A \cot A = 1$.
3 Show that $\sin^2 A + \cos^2 A = 1$.
4 Show that $\tan^2 A + 1 = \sec^2 A$.
5 Show that $1 + \cot^2 A = \csc^2 A$.
6 Show that $\tan A = \dfrac{\sin A}{\cos A}$.
7 Show that $\cot A = \dfrac{\cos A}{\sin A}$.

FUNDAMENTAL TRIGONOMETRIC IDENTITIES

The equality $\sin A \csc A = 1$, which we established in this section, and the equalities which you established in Exercises 1 to 7 above are commonly referred to as the *eight fundamental trigonometric identities.* These identities can be used to verify other trigonometric identities, and they can be used to simplify a given expression that involves trigonometric ratios.

8 Show that $\tan A + \cot A = \csc A \sec A$. *Hint:* First write $\tan A + \cot A = \dfrac{\sin A}{\cos A} + \dfrac{\cos A}{\sin A}$, and then simplify using Exercise 3.

In Exercises 9 to 14, prove that the given equality is an identity for all values of A between 0 and 90°, by use of one or more of the eight fundamental trigonometric identities.

9 $\cos A \csc A = \cot A$
10 $\sin A \sec A = \tan A$
11 $\dfrac{\cos^2 A}{1 - \sin A} = 1 + \sin A$
12 $\dfrac{1 + \tan A}{1 + \cot A} = \dfrac{\sec A}{\csc A}$
13 $\dfrac{\tan^2 A + 1}{\cot^2 A + 1} = \tan^2 A$
14 $\dfrac{\sec A}{\cot A + \tan A} = \sin A$

In Exercises 15 to 18, use one or more of the eight fundamental trigonometric identities to show that the given expression has the value indicated for all values of A between 0 and 90°.

15 $(1 - \cos^2 A)\csc^2 A$; 1
16 $(\sin A + \cos A)^2 + (\sin A - \cos A)^2$; 2
17 $\sin A(\cot A + \csc A) - \cos A$; 1
18 $\dfrac{\tan^2 A}{\sec^2 A} + \dfrac{\cot^2 A}{\csc^2 A}$; 1

9.6
RADIAN MEASURE—TRIGONOMETRIC RATIOS OF AN ANGLE OF ANY SIZE

ONE DEGREE

We are accustomed to measuring angles in degrees, where an **angle of one degree** may be defined as follows. Place the vertex of an angle at the center of a circle. If the length of the arc of the circle cut off (or *subtended*) by the angle is $\frac{1}{360}$ of the circumference of the circle, then the angle is said to have measure of one degree, or to be an angle of one degree (see Fig. 9.17). The size of the circle has no effect on the measurement of the angle.

ONE RADIAN

Another common unit for measuring the size of an angle is the unit known as a *radian*. This unit of angle measure is used a great deal in mathematics. An angle has measure of **one radian,** or is an angle of one radian, if when its vertex is placed at the center of a circle, the length of the arc subtended by the angle is equal to the length of the radius of the circle (see Fig. 9.18). The size of the circle has no effect on the measurement of the angle.

Recall that the circumference of a circle is $2\pi r$, where r is the length of the radius of the circle. An angle of 90° placed with its vertex at the center of a circle subtends an arc whose length is one-fourth the circumference of the circle or $\frac{1}{4}(2\pi r) = \frac{\pi}{2}r$; an angle of 180° subtends an arc whose length is one-half the circumference of the circle, or $\frac{1}{2}(2\pi r) = \pi r$. So an angle of 90° is an angle of $\frac{\pi}{2}$ radians, and an angle of 180° is an angle of π radians. Therefore we write

$$\pi \text{ radians} = 180°$$

or

$$1 \text{ radian} = \left(\frac{180}{\pi}\right)°$$

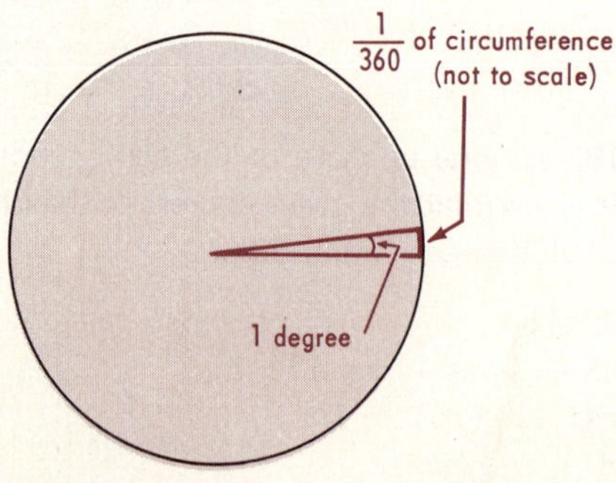

FIGURE 9.17

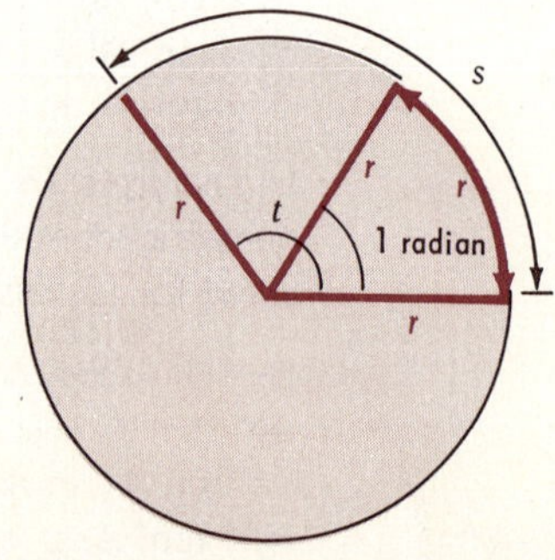

FIGURE 9.18

or

$$1^\circ = \frac{\pi}{180} \text{ radian}$$

If we use the approximation 3.1416 for π, we find that 1 radian is approximately 57.3°, and 1° is approximately 0.0175 radian.

The student is asked to verify that

$$30^\circ = \frac{\pi}{6} \text{ radian} \qquad 45^\circ = \frac{\pi}{4} \text{ radian} \qquad 60^\circ = \frac{\pi}{3} \text{ radians}$$

$$90^\circ = \frac{\pi}{2} \text{ radians} \qquad 120^\circ = \frac{2\pi}{3} \text{ radians} \qquad 150^\circ = \frac{5\pi}{6} \text{ radians}$$

$$210^\circ = \frac{7\pi}{6} \text{ radians} \qquad 270^\circ = \frac{3\pi}{2} \text{ radians} \qquad 360^\circ = 2\pi \text{ radians}$$

If an angle t is placed with vertex at the center of a circle, it follows from the definition of radian measure of an angle that

$$\text{radian measure of } t = \frac{s}{r} \tag{9.1}$$

where r is the radius of the circle and s is the length of the arc of the circle subtended by the angle t. (It is understood that s and r are measured in the same units of length.) Since we use the words "an angle of x radians" to mean "an angle whose measure is x radians," we write the above equality simply as

$$t = \frac{s}{r} \qquad \text{for } 0 \leq t \leq 2\pi \tag{9.2}$$

If in (9.2) we set $r = 1$, we get $t = s$; that is, if the radius of the circle of Fig. 9.18 is 1, then an angle of t radians subtends an arc whose length is t (linear) units.

Let us consider a circle whose radius is 1 unit in length and whose center is designated by O. As in Figs. 9.19 to 9.22, we place O at the origin of a rectangular coordinate system and designate by A the point at which the circle intersects the positive x axis. Let OA be rotated *counterclockwise* so that the point at A moves a distance s along the circumference of the circle and comes to rest at a point P on the circle C. We call this rotation a *positive* rotation, and we call the angle swept out by OA a *positive angle;* the radian measure of this angle is s. If OA is rotated *clockwise* so that the point at A moves a distance s along the circumference, the rotation is said to be *negative* and the angle swept out by OA is called a *negative angle;* the radian measure of this angle is $-s$. In this way we can define positive and negative angles whose radian measure can be any real number. Figures 9.19 to 9.22 illustrate four positive angles

POSITIVE ANGLE

NEGATIVE ANGLE

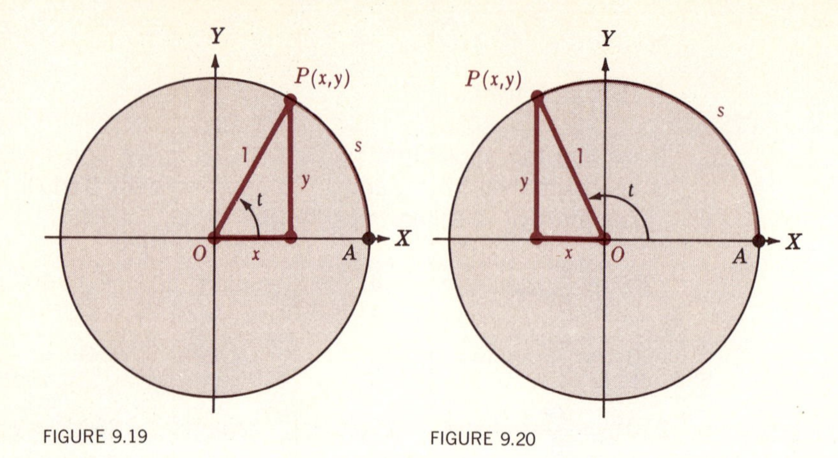

FIGURE 9.19

FIGURE 9.20

with measure less than 2π (or 360°). In this chapter we shall not encounter angles that are negative, and we shall seldom use an angle whose measure is greater than 2π radians (or 360°).

INITIAL SIDE

TERMINAL SIDE

STANDARD POSITION OF AN ANGLE

The side OA of the angle t is called the *initial side* of t, and the side OP is called the *terminal side* of t (Figs. 9.19 to 9.22). When the angle t is placed with its vertex at the origin and its initial side along the positive x axis, it is said to be in **standard position.**

The point P, which is reached by moving the point at A a distance s along the circle C (whose radius is 1), will have coordinates in the rectangular coordinate system; let these coordinates be (x, y). Thus, corresponding to any angle t in standard position (whose measure is s) there is a unique pair of coordinates (x, y). For any angle t we *define* the six trigonometric ratios of t as follows (remem-

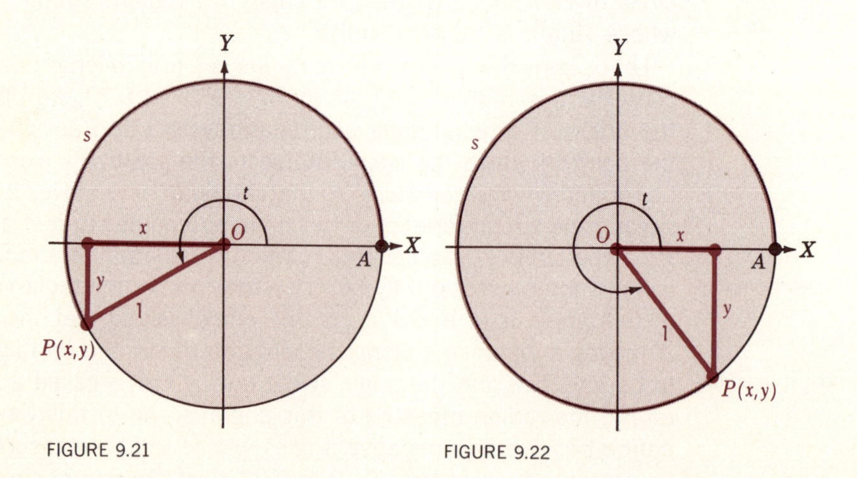

FIGURE 9.21

FIGURE 9.22

ber that the radius r of the circle C is 1):

SIX TRIGONOMETRIC RATIOS OF t

$$\sin t = y \qquad \cos t = x$$

$$\tan t = \frac{y}{x} \quad x \neq 0 \qquad \cot t = \frac{x}{y} \quad y \neq 0 \tag{9.3}$$

$$\csc t = \frac{1}{y} \quad y \neq 0 \qquad \sec t = \frac{1}{x} \quad x \neq 0$$

We shall direct our attention mainly to the ratios $\sin t$ and $\cos t$.

From the definitions

$$\sin t = y \tag{9.4}$$

and

$$\cos t = x \tag{9.5}$$

we can make numerous observations and deductions. From (9.3) and a consideration of appropriate figures in connection with the unit circle and the coordinate system, the trigonometric ratios of some angles can be determined. To illustrate, for the angle of 0 radians in standard position the coordinates of P are (1, 0); hence

$$\sin 0 = 0 \qquad \cos 0 = 1 \qquad \tan 0 = 0$$

In like manner the student is asked to verify that

$$\sin \frac{\pi}{2} = 1 \qquad \cos \frac{\pi}{2} = 0 \qquad \sin \pi = 0 \qquad \cos \pi = -1$$

$$\sin \frac{3\pi}{2} = -1 \qquad \cos 2\pi = 1$$

Let t be an angle in standard position with $\frac{\pi}{2} < t < \pi$, so that the terminal side of t falls in the second quadrant. Also, let

$$t_1 + t = \pi \qquad 0 < t_1 < \frac{\pi}{2}$$

so

$$t = \pi - t_1$$

We want to show that

$$\sin(\pi - t_1) = \sin t_1 \tag{9.6}$$

$$\cos(\pi - t_1) = -\cos t_1 \tag{9.7}$$

In Fig. 9.23 let OP_1 be the terminal side of t_1 and OP be the terminal side of t; also let (x_1, y_1) be the coordinates of P_1 and (x, y) be the coordinates of P. Since $t = \pi - t_1$, $\angle P_1OM_1 = \angle POM$, and right triangle OMP is congruent to right triangle OM_1P_1. Therefore,

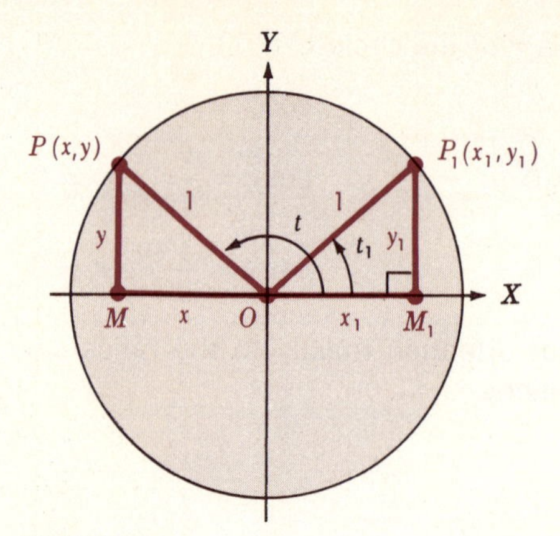

FIGURE 9.23

the length of MP is the same as the length of M_1P_1 and the length of OM is the same as the length of OM_1. But this requires that

$$y = y_1 \qquad \text{and} \qquad x = -x_1$$

From these equalities it follows that

$$\sin(\pi - t_1) = \sin t = y = y_1 = \sin t_1$$
$$\cos(\pi - t_1) = \cos t = x = -x_1 = -\cos t_1$$

and formulas (9.6) and (9.7) are proved.

If α is any acute angle measured in degrees, formulas (9.6) and (9.7) tell us that

$$\sin(180° - \alpha) = \sin \alpha \tag{9.8}$$

and

$$\cos(180° - \alpha) = -\cos \alpha \tag{9.9}$$

To illustrate,

$$\sin 145° = \sin(180° - 35°) = \sin 35°$$

and

$$\cos 145° = \cos(180° - 35°) = -\cos 35°$$

We shall need formulas (9.8) and (9.9) in the next section.

EXERCISES

Express each of the following in radians:

1 $10°$ 2 $-12°$ 3 $40°$ 4 $18°$

Express each of the following in terms of degrees:

5 $\frac{\pi}{8}$ 6 $\frac{3\pi}{5}$ 7 $-\frac{\pi}{2}$ 8 $\frac{11\pi}{2}$

Using formula (9.8) or (9.9) together with the table on page 385, find a decimal approximation for the following:

9 sin 100° 10 sin 164° 11 cos 119° 12 cos 173°

13 Using the definitions (9.3) of the trigonometric ratios of an angle of any size, verify that the eight fundamental trigonometric identities hold for all values of t for which the trigonometric ratios concerned are defined.

14 If $\pi < t < \frac{3\pi}{2}$, let $t - t_1 = \pi$ and $t = \pi + t_1$, where $0 < t_1 < \frac{\pi}{2}$. Prove that $\sin t = \sin(\pi + t_1) = -\sin t_1$ and $\cos t = \cos(\pi + t_1) = -\cos t_1$.

15 If $\frac{3\pi}{2} < t < 2\pi$, let $t_1 + t = 2\pi$, so $t = 2\pi - t_1$ with $0 < t_1 < \frac{\pi}{2}$. Show that $\sin t = \sin(2\pi - t_1) = -\sin t_1$ and $\cos t = \cos(2\pi - t_1) = \cos t_1$.

16 If $0 < t_1 < \frac{\pi}{2}$, show that $\tan(\pi - t_1) = -\tan t_1$, $\tan(\pi + t_1) = \tan t_1$ and $\tan(2\pi - t_1) = -\tan t_1$.

9.7 THE LAW OF COSINES AND THE LAW OF SINES

We now establish some formulas of trigonometry which are needed to solve certain types of problems not previously considered.

Let α, β, and γ denote the angles of triangle ABC, and let a, b, and c denote the sides opposite α, β, and γ, respectively (see Figs. 9.24 and 9.25). Let α' be the angle MAC of Fig. 9.24, so

$$\alpha + \alpha' = 180°$$

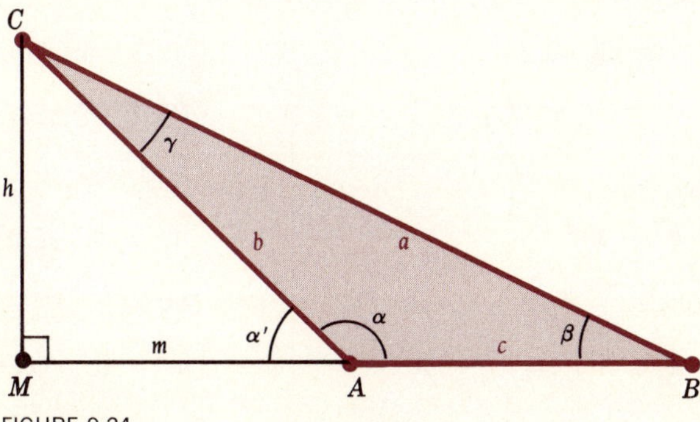

FIGURE 9.24

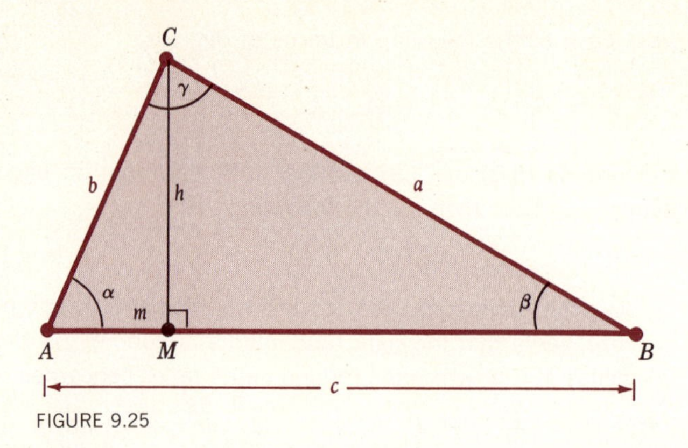

FIGURE 9.25

Here $\alpha > 90°$ and $\alpha' < 90°$. Using formulas (9.8) and (9.9), we get

$$\sin \alpha = \sin (180° - \alpha') = \sin \alpha' \tag{9.10}$$

and

$$\cos \alpha = \cos (180° - \alpha') = -\cos \alpha' \tag{9.11}$$

We now concern ourselves with the statement and proof of the law of cosines, which is a kind of generalization of the theorem of Pythagoras.

THE LAW OF COSINES Let α, β, and γ denote the angles of triangle ABC, and let a, b, and c be the sides opposite α, β, and γ, respectively. Then

LAW OF COSINES

$$a^2 = b^2 + c^2 - 2bc \cos \alpha \tag{9.12}$$

$$b^2 = a^2 + c^2 - 2ac \cos \beta \tag{9.13}$$

$$c^2 = a^2 + b^2 - 2ab \cos \gamma \tag{9.14}$$

PROOF Triangle ABC may be oblique* with α obtuse as in Fig. 9.24, or it may be as in Fig. 9.25 with α acute. In Figs. 9.24 and 9.25, construct a perpendicular from the vertex C to the side AB (extended if necessary), meeting AB at M.

In Fig. 9.24	In Fig. 9.25
$h^2 = a^2 - (c + m)^2$	$h^2 = a^2 - (c - m)^2$
and	and
$h^2 = b^2 - m^2$	$h^2 = b^2 - m^2$
So	So
$a^2 - (c + m)^2 = b^2 - m^2$	$a^2 - (c - m)^2 = b^2 - m^2$

* An oblique triangle is a triangle with one angle larger than 90°.

In Fig. 9.24	In Fig. 9.25
and	and
$a^2 = b^2 + c^2 + 2cm$	$a^2 = b^2 + c^2 - 2cm$
Now	Now
$m = b \cos (180° - \alpha) = -b \cos \alpha$	$m = b \cos \alpha$
Hence	Hence
$a^2 = b^2 + c^2 - 2bc \cos \alpha$	$a^2 = b^2 + c^2 - 2bc \cos \alpha$

We have proved (9.12) for the two possible cases (α acute or α obtuse). The proof of (9.13) and (9.14) is left to the student. This may be done either by drawing appropriate figures like Figs. 9.24 and 9.25, or by relettering Figs. 9.24 and 9.25 suitably.

Notice that the law of cosines may be stated as follows: In any triangle the square of the length of a side is equal to the sum of the squares of the lengths of the other two sides, minus twice the product of the lengths of these sides and the cosine of the angle between these sides.

In carrying out the numerical computations in solving a triangle, we assume here that all angle measurements are correct to the nearest degree, that all linear measurements are correct to three significant figures, and that all trigonometric ratios are given to three significant figures. To illustrate, if in a problem we write $a = 1.7$, it is understood that a is 1.7 to the nearest hundredth, that is, $a = 1.70$.

EXAMPLE 1 In triangle ABC, $\alpha = 60°$, $b = 15$, and $c = 7$. Find a.

SOLUTION By the law of cosines,

$$\begin{aligned} a^2 &= b^2 + c^2 - 2bc \cos \alpha \\ &= (15)^2 + 7^2 - 2(15)(7) \cos 60° \\ &= 225 + 49 - 210(\tfrac{1}{2}) = 274 - 105 = 169 \end{aligned}$$

Therefore $a = \sqrt{169}$, or $a = 13$.

EXAMPLE 2 If the sides of a triangle are $a = 7$, $b = 3$, and $c = 5$, find angle α.

SOLUTION Using (9.12), we get

$$\cos \alpha = \frac{b^2 + c^2 - a^2}{2bc} = \frac{3^2 + 5^2 - 7^2}{2 \cdot 3 \cdot 5} = \frac{9 + 25 - 49}{30} = -\frac{1}{2}$$

Since $\cos \alpha = -\frac{1}{2}$, and since $0 < \alpha < 180°$ (for it is an angle in a triangle), we conclude that $\alpha = 120°$.

THE LAW OF SINES Let α, β, and γ denote the angles of a triangle, and let a, b, c be the sides opposite α, β, γ, respectively. Then

LAW OF SINES
$$\frac{a}{\sin \alpha} = \frac{b}{\sin \beta} = \frac{c}{\sin \gamma} \tag{9.15}$$

PROOF The triangle ABC may be like that shown in Fig. 9.24 or like that shown in Fig. 9.25.

In the calculations below, we show that

$$\frac{a}{\sin \alpha} = \frac{b}{\sin \beta} \tag{9.16}$$

for the two types of triangles shown in Figs. 9.24 and 9.25.

In Fig. 9.24	In Fig. 9.25
$h = a \sin \beta$	$h = a \sin \beta$
and	and
$h = b \sin (180^\circ - \alpha) = b \sin \alpha$	$h = b \sin \alpha$
Hence	Hence
$a \sin \beta = b \sin \alpha$	$a \sin \beta = b \sin \alpha$
and	and
$\frac{a}{\sin \alpha} = \frac{b}{\sin \beta}$	$\frac{a}{\sin \alpha} = \frac{b}{\sin \beta}$

Similarly, by constructing a line segment from B perpendicular to the opposite side and proceeding as above, we can show that

$$\frac{a}{\sin \alpha} = \frac{c}{\sin \gamma} \tag{9.17}$$

Combining (9.16) and (9.17), we get (9.15), and the law of sines is proved.

If we are given two angles and a side of a triangle, then the remaining angle may be found by using the formula $\alpha + \beta + \gamma = 180^\circ$; after that the remaining sides may be found by using the law of sines (see Example 3).

If two sides and the angle opposite one of them is given, then there may be no triangle, one triangle, or two triangles. In this book we limit our consideration to situations of this type in which precisely one triangle is determined (see Example 4).

EXAMPLE 3 In triangle ABC, $\alpha = 66^\circ$, $\beta = 43^\circ$, and $c = 12$ feet. Find a and b.

SOLUTION From $\alpha + \beta + \gamma = 180^\circ$, we get

$$\gamma = 180° - (66° + 43°) = 180° - 109° = 71°$$

To find a, we use $\dfrac{a}{\sin \alpha} = \dfrac{c}{\sin \gamma}$, so

$$a = \frac{c \sin \alpha}{\sin \gamma} = \frac{12 \sin 66°}{\sin 71°}$$

Using the table on page 385, we get

$$a = \frac{12(0.914)}{0.946}$$

so $a = 11.6$ feet to three significant figures.

To find b, we use $\dfrac{b}{\sin \beta} = \dfrac{c}{\sin \gamma}$, so

$$b = \frac{c \sin \beta}{\sin \gamma} = \frac{12 \sin 43°}{\sin 71°}$$

Using the table on page 385, we get

$$b = \frac{12(0.682)}{0.946}$$

so $b = 8.65$ feet to three significant figures.

EXAMPLE 4 In triangle ABC, $a = 7$, $b = 8.6$, and $\beta = 28°$. Find α, γ, and c.

SOLUTION As shown in Fig. 9.26, lay off $a = BC = 7$, and with C as center, construct an arc with radius $b = CA = 8.6$. At B we construct angle $\beta = 28°$, and we see that only one triangle is possible.

Using the law of sines, we have

$$\frac{\sin \alpha}{a} = \frac{\sin \beta}{b}$$

or

$$\frac{\sin \alpha}{7} = \frac{\sin 28°}{8.6}$$

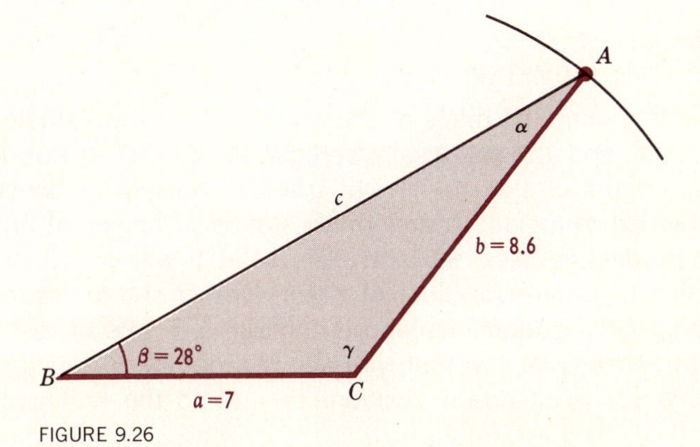

FIGURE 9.26

so

$$\sin \alpha = 7 \frac{\sin 28°}{8.6}$$

Hence

$$\sin \alpha = 7 \left(\frac{0.469}{8.6}\right) = 0.382$$

and

$$\alpha = 22°$$

Then $\gamma = 180° - (\alpha + \beta) = 180° - 50° = 130°$.

Again, using the law of sines, we have

$$\frac{c}{\sin \gamma} = \frac{b}{\sin \beta} \qquad \frac{c}{\sin 130°} = \frac{8.6}{\sin 28°}$$

$$c = 8.6 \frac{\sin 130°}{\sin 28°} = 8.6 \frac{\sin 50°}{\sin 28°} = \frac{8.6(0.766)}{0.469}$$

and $c = 14.0$ to three significant figures, which we write simply as 14.

EXERCISES

1 In triangle ABC, $a = 6$, $b = 10$, and $c = 14$. Find γ.

2 In triangle ABC, $a = 5$, $b = 3$, $c = 7$. Find α, β, and γ.

3 A triangular plot of ground has sides of length 430, 370, and 250 feet. Find the three angles of the triangle.

4 In triangle ABC, $a = 20$, $\beta = 30°$, $\gamma = 40°$. Find α, b, and c.

5 In triangle ABC, $b = 407$, $c = 568$, and $\gamma = 107°$. Find α, β, and a.

6 In triangle ABC, $\beta = 73°$, $\gamma = 65°$, $a = 11.1$ feet. Find α, b, and c.

7 Given triangle ABC with $\alpha = 40°$, $\beta = 80°$, and $a = 20$ feet; find γ, b, and c.

8 In triangle ABC, $\gamma = 66°$, $b = 200$ feet, and $a = 800$ feet. Find c, β, and α.

9.8
THE TRIGONOMETRIC FUNCTIONS

The trigonometric ratios of acute angles (which we studied in Secs. 9.3, 9.4, and 9.5) historically were developed first as aids in calculating lengths of sides of a right triangle. Next these concepts were extended to include trigonometric ratios of angles of any size (as we studied in Secs. 9.6 and 9.7). Later it was found desirable to define trigonometric ratios of a real number and to use these ratios for specifying functions whose domains are sets of real numbers.

In Sec. 9.6 we saw that with any real number t, we can associate the angle in standard position, relative to the unit circle, whose radian measure is t. Recall from Sec. 7.1 that a function is a set of ordered pairs, no two of which have the same first entry. Given any real number t, there is precisely *one* value of sin t and *one* value of cos t. Thus, if we associate with the real number x the unique angle in standard position whose radian measure is x, the sets of ordered pairs, $\{(x, y) \mid y = \sin x,\ x \in Re\}$ and $\{(x, y) \mid y = \cos x,\ x \in Re\}$, are functions. These functions are called, respectively, the *sine function* and the *cosine function* and are denoted by "Sin" and "Cos." That is, we define the functions Sin and Cos by

$$\text{Sin} = \{(x, y) \mid y = \sin x,\ x \in Re\}$$
$$\text{Cos} = \{(x, y) \mid y = \cos x,\ x \in Re\}$$

Since for any real number x, the terminal side of an angle of $x + 2\pi$ radians coincides with the terminal side of an angle of x radians, it follows from the definitions (9.3) of the trigonometric ratios that

$$\sin (x + 2\pi) = \sin x \qquad (9.18)$$

and

$$\cos (x + 2\pi) = \cos x \qquad (9.19)$$

These properties (9.18) and (9.19) tell us that each of $\sin x$ and $\cos x$ "repeats" itself when x is increased by the amount 2π. A function with this "repetitive" character is said to be "periodic"; that is, a function F is **periodic** if there is a number p such that $F(x + p) = F(x)$ for all x in the domain of the function. The smallest positive number p for which this is true is called the **period** of the function. Thus, the functions Sin and Cos are periodic, and from an examination of the definitions of the trigonometric ratios, we can see that the period of each of these functions is 2π.

PERIODIC FUNCTION

PERIOD OF A FUNCTION

As a consequence of the periodic character of Sin, as expressed in equality (9.18), the graph of the function Sin for any values of $x \in Re$ can be constructed provided the graph of Sin is known for $0 \leq x \leq 2\pi$. In Fig. 9.27 we give the graph of Sin for $0 \leq x \leq 2\pi$ as the *solid* curve. In this same figure the *dashed* curve is the graph of Sin for $2\pi \leq x \leq 3\pi$. Note that the dashed curve can be obtained simply by shifting to the right 2π units the portion of the solid curve for which $0 \leq x \leq \pi$.

In the construction of the graphs of the trigonometric functions it is convenient to have at hand the following table which gives two-decimal place approximations for some rational multiples of π (with $\pi = 3.14$).

0	$\frac{\pi}{6}$	$\frac{\pi}{4}$	$\frac{\pi}{3}$	$\frac{\pi}{2}$	$\frac{2\pi}{3}$	$\frac{3\pi}{4}$	$\frac{5\pi}{6}$	π
0	0.52	0.79	1.05	1.57	2.09	2.36	2.62	3.14

$\frac{7\pi}{6}$	$\frac{5\pi}{4}$	$\frac{4\pi}{3}$	$\frac{3\pi}{2}$	$\frac{5\pi}{3}$	$\frac{7\pi}{4}$	$\frac{11\pi}{6}$	2π
3.66	3.93	4.19	4.71	5.23	5.50	5.76	6.28

The graph of Sin as shown in Fig. 9.27 can be constructed with the use of the table on page 385, along with the conversion of radian measure into degree measure. Similarly, the graph of Cos for $0 \leq x \leq 2\pi$ can be constructed (and then extended to any values of $x \in Re$ by the use of the periodic property). The student is asked to make such a construction in Exercise 2.

An easy way of visualizing the behavior of the Sin and Cos functions is through the use of a circle of radius 1 with its center at the origin of a two-dimensional coordinate system, as shown in Fig. 9.28. When an angle of t radians is placed in standard position, the definitions (9.3) tell us that $\sin t$ is the y coordinate of point P and is thus represented by the length of the segment MP (this length is positive if P lies above the x axis, and is negative if P lies below the x axis). Similarly, $\cos t$ is the x coordinate of the point P and is represented by the length of the segment OM (this length is positive if M lies to the right of the y axis, and it is negative if M lies to the left of the y axis). As we visualize the angle with radian

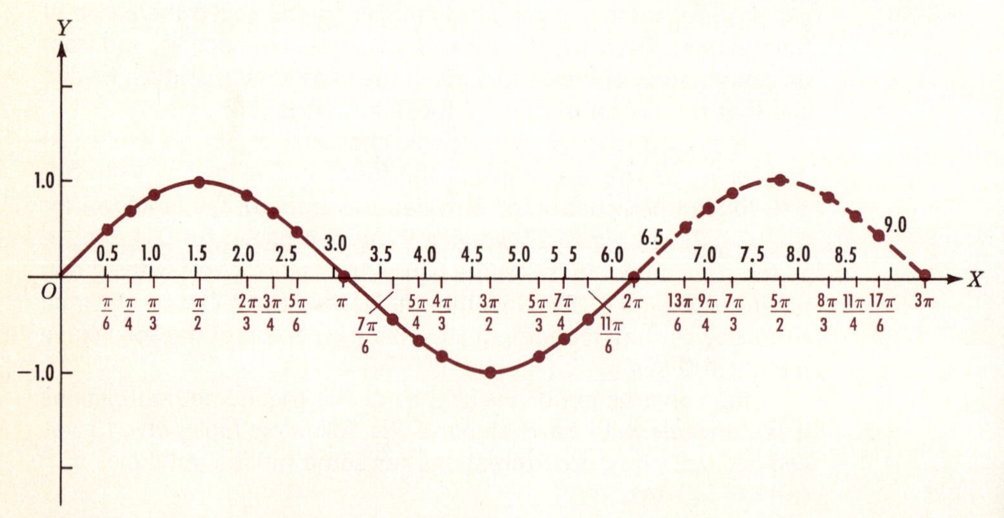

FIGURE 9.27 The graph of Sin for $0 \leq x \leq 3\pi$.

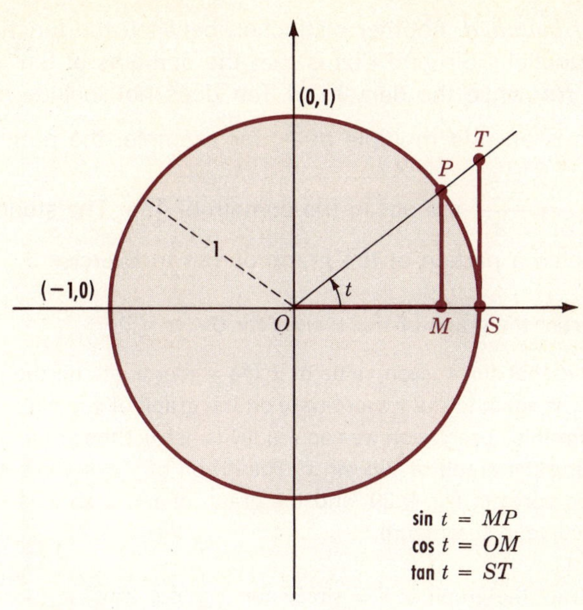

FIGURE 9.28

measure t increasing and the point P moving around the circle, the way in which sin t and cos t vary can be visualized. On this figure the behavior of tan t can also be observed. The definition of tan t given by (9.3) tells us that

$$\tan t = \frac{MP}{OM}$$

Since the triangles OMP and OST are similar, we have $\dfrac{MP}{OM} = \dfrac{ST}{OS}$; or since $OS = 1$, then $\dfrac{MP}{OM} = ST$ and tan t is represented by the length of the segment ST (this length is positive if T lies above the x axis, and negative if the point T lies below the x axis). Using this idea and Fig. 9.28, we see that tan 0 = 0 and that as t increases from 0 and approaches $\frac{\pi}{2}$, tan t increases and in fact can be made as large as we wish (the segment ST can be made any length) by choosing t sufficiently near $\frac{\pi}{2}$. This behavior of tan t is very different from the behavior of sin t and cos t. The values of sin t and cos t are restricted to lie between 1 and -1 (inclusive). We say that the functions Sin and Cos are *bounded,* while the function Tan, defined by

$$\text{Tan} = \{(x, y) \mid y = \tan x\}$$

is *not bounded*. Another distinction between the function Tan and the functions Sin and Cos is that the domains of Sin and Cos are both *Re*, while the domain of Tan does not include any number which is an *odd* multiple of $\frac{\pi}{2}$; for example, the numbers $\frac{\pi}{2}, \frac{3\pi}{2}, \frac{5\pi}{2}, \frac{-\pi}{2}, \frac{-3\pi}{2}$ are not in the domain of Tan. The student is asked to sketch a portion of the graph of Tan in Exercise 3.

EXAMPLE 1 Construct the graph of $y = 2 \sin x$ for $0 \leq x \leq 2\pi$.

SOLUTION Observe that for a given value of x the y coordinate on the graph of $y = 2 \sin x$ is equal to the y coordinate on the graph of $y = \sin x$ multiplied by 2. From this observation we can readily construct the graph of $y = 2 \sin x$ by using the graph of $y = \sin x$. The graph of $y = \sin x$ is shown as the dashed curve in Fig. 9.29, and the graph of $y = 2 \sin x$ is shown as the solid curve in that figure.

EXAMPLE 2 Construct the graph of $y = \sin 2x$ for $0 \leq x \leq 2\pi$.

SOLUTION Observe that when x is increased by π, then $2x$ is increased by 2π, and therefore the period of $\sin 2x$ is π rather than 2π. Also observe that when $x = \frac{\pi}{2}$, then $2x = \pi$; when $x = \pi$, then $2x = 2\pi$; and when $x = \frac{3\pi}{2}$, then

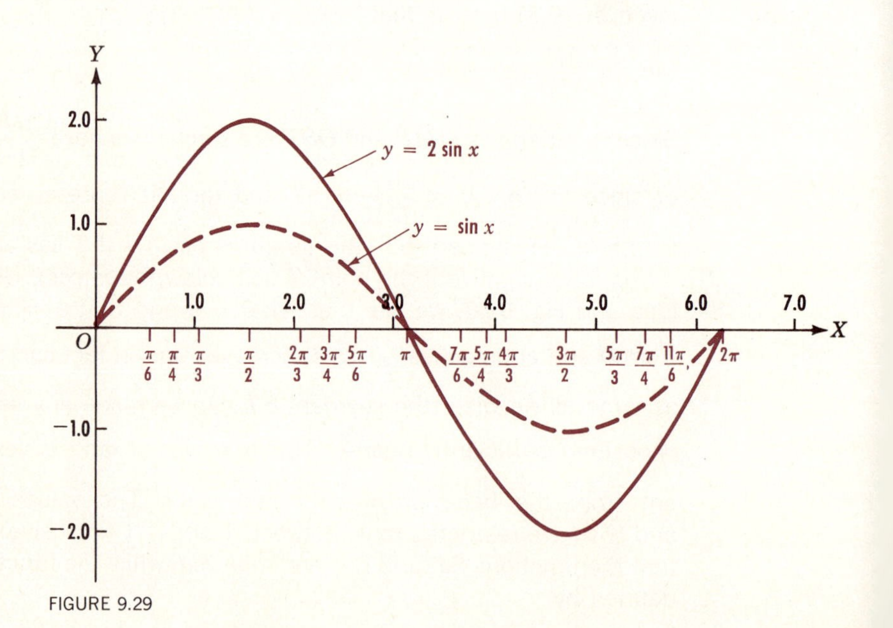

FIGURE 9.29

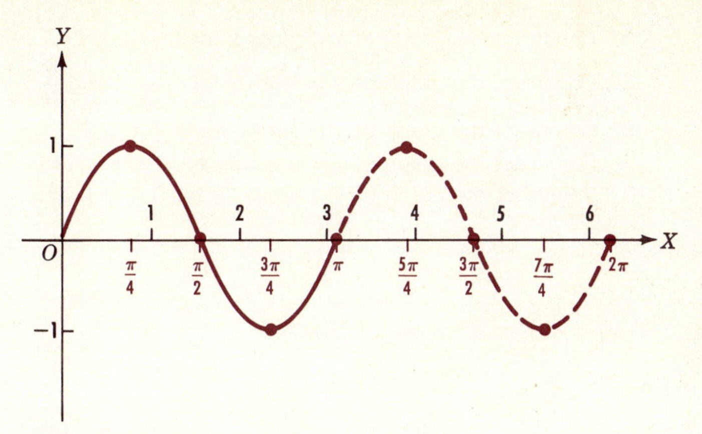

FIGURE 9.30 Graph of $y = \sin 2x$ for $0 \leq x \leq 2\pi$.

$2x = 3\pi$. Using these observations and the graph of $y = \sin x$ given in Fig. 9.27, we can obtain the graph of $y = \sin 2x$ shown in Fig. 9.30. Comparing Figs. 9.27 and 9.30, we see that instead of one "oscillation" over the interval $0 \leq x \leq 2\pi$ (as $y = \sin x$ has), the graph of $y = \sin 2x$ has two oscillations.

In addition to the functions Sin, Cos, and Tan, which we have defined, there are three other trigonometric functions which are obtained by the use of the trigonometric ratios cot t, sec t, and csc t defined in (9.3). These functions are denoted by Cot, Sec, and Csc respectively and are defined by

$$\text{Cot} = \{(x, y) \mid y = \cot x\}$$
$$\text{Sec} = \{(x, y) \mid y = \sec x\}$$
$$\text{Csc} = \{(x, y) \mid y = \csc x\}$$

All the trigonometric functions are periodic. Sin, Cos, Sec, and Csc all have period 2π, while Tan and Cot have period π. Because of their periodic properties, the trigonometric functions find considerable use in situations in which we are dealing with oscillatory or repetitive phenomena.

EXERCISES

1 Construct the graph of $y = \sin x$ for $-2\pi \leq x \leq 0$.
2 Construct the graph of $y = \cos x$ for $0 \leq x \leq 2\pi$.
3 Construct the graph of $y = \tan x$ for $-\dfrac{\pi}{3} \leq x \leq \dfrac{\pi}{3}$.
4 Construct the graph of $y = 3 \sin x$ for $0 \leq x \leq 4\pi$.
5 Construct the graph of $y = \frac{3}{2} \sin x$ for $0 \leq x \leq 2\pi$.

6 Construct the graph of $y = \frac{3}{4} \cos x$ for $0 \leq x \leq 2\pi$.

7 Construct the graph of $y = \frac{1}{2} \tan x$ for $-\frac{\pi}{3} \leq x \leq \frac{\pi}{3}$.

8 Construct the graph of $y = \sin 4x$ for $0 \leq x \leq 2\pi$.

9 Construct the graph of $y = 2 \sin 3x$ for $0 \leq x \leq 2\pi$.

10 Construct the graph of $y = 4 \cos 2x$ for $0 \leq x \leq 2\pi$.

10

Logarithms

10.1
INTRODUCTION

Many of the problems which arise in the application of mathematics to science and engineering require arithmetic calculations that are long and tedious by use of the four fundamental operations of arithmetic. In this chapter we shall see how such computations can be done more conveniently with the use of logarithms.

To introduce the concept of logarithms and to indicate how logarithms may be used as an aid in computation, consider the following table, which gives some powers of 2:

WHEN $x =$	0	1	2	3	4	5	6	7	8	9	10
THEN $2^x =$	1	2	4	8	16	32	64	128	256	512	1,024

WHEN $x =$	11	12	13	14	15	16
THEN $2^x =$	2,048	4,096	8,192	16,384	32,768	65,536

Suppose that we have this table available and want to calculate $32 \cdot 512$. We note that $32 = 2^5$ and $512 = 2^9$. Therefore,

$$32 \cdot 512 = 2^5 \cdot 2^9 = 2^{5+9} = 2^{14} = 16{,}384$$

Thus with the use of the above table, the multiplication of the numbers 32 and 512 is reduced to the addition of the numbers 5 and 9.

Similarly, let us consider the determination of the quotient $\dfrac{65{,}536}{512}$.

We have, with the use of the table,

$$\frac{65{,}536}{512} = \frac{2^{16}}{2^9} = 2^{16-9} = 2^7 = 128$$

With the use of this table the division of 65,536 by 512 is effected by subtracting 9 from 16.

For another illustration let us calculate $(32)^3$ with the aid of the table. Now

$$(32)^3 = (2^5)^3 = 2^{5 \cdot 3} = 2^{15} = 32{,}768$$

The determination of $32 \cdot 32 \cdot 32$ is reduced, through use of the table of powers of 2, to finding the product of 5 and 3.

For $64 = 2^6$ we write $\log_2 64 = 6$. The latter expression is read "the logarithm of 64 to the base 2 is 6."

In general, if

$$N = 2^x$$

we write

$$\log_2 N = x$$

and read this as "the logarithm of N to the base 2 is x."

If $N = b^x$ (b positive and different from 1), then the exponent x is called the **logarithm** of N to the base b and we write $x = \log_b N$. The equations

LOGARITHM

$$N = b^x \tag{10.1}$$

and

$$x = \log_b N \tag{10.2}$$

are two different ways of expressing the same statement: (10.1) is called the *exponential form* of this statement, and (10.2) is called the *logarithmic form*.

EXPONENTIAL FORM

LOGARITHMIC FORM

To illustrate

$\log_6 36 = 2$, since $6^2 = 36$
$\log_5 125 = 3$, since $5^3 = 125$
$\log_{10} 100 = 2$, since $10^2 = 100$
$\log_{10} 1{,}000 = 3$, since $10^3 = 1{,}000$

$$\log_{10} 0.01 = -2, \text{ since } 10^{-2} = \frac{1}{10^2} = 0.01$$

$$\log_{10} 0.001 = -3, \text{ since } 10^{-3} = \frac{1}{10^3} = 0.001$$

EXERCISES

Express the following statements in logarithmic form:

1 $5^2 = 25$
2 $10^4 = 10{,}000$
3 $7^{-2} = \frac{1}{49}$
4 $8^{2/3} = 4$

Express the following statements in exponential form:

5 $\log_2 128 = 7$
6 $\log_{10} 0.0001 = -4$
7 $\log_{10} 1{,}000 = 3$
8 $\log_6 216 = 3$

9 What is the logarithm to the base 10 of each of the following numbers: 1; 10; 100; 1,000; 10,000; 100,000; 1,000,000? Of each of the numbers 0.1; 0.01; 0.001; 0.00001; 0.000001?

10 When the base is 10, what is the number whose logarithm is 7? 8? -4? -10?

10.2
COMMON LOGARITHMS

COMMON LOGARITHMS

Any positive number b except 1 may be used as the base of a system of logarithms. However, the system of logarithms most widely used for computation has the base 10 and is called the system of **common logarithms.** In the remainder of this book the only logarithms with which we shall be concerned will be common logarithms. For this reason we shall understand that the base is 10, and write

$\log N$ to mean $\log_{10} N$

The *common logarithm* of any number N is the exponent x of the power to which 10 must be raised to equal N. That is, if

$N = 10^x$

then

$\log N = x$

Study the accompanying table, which gives some powers of 10 and the corresponding logarithmic forms.

EXPONENTIAL FORM	LOGARITHMIC FORM
$10^6 = 1{,}000{,}000$	$\log 1{,}000{,}000 = 6$
$10^5 = 100{,}000$	$\log 100{,}000 = 5$
$10^4 = 10{,}000$	$\log 10{,}000 = 4$
$10^3 = 1{,}000$	$\log 1{,}000 = 3$
$10^2 = 100$	$\log 100 = 2$
$10^1 = 10$	$\log 10 = 1$
$10^0 = 1$	$\log 1 = 0$
$10^{-1} = 0.1$	$\log 0.1 = -1$
$10^{-2} = 0.01$	$\log 0.01 = -2$
$10^{-3} = 0.001$	$\log 0.001 = -3$
$10^{-4} = 0.0001$	$\log 0.0001 = -4$
$10^{-5} = 0.00001$	$\log 0.00001 = -5$
$10^{-6} = 0.000001$	$\log 0.000001 = -6$

From this table note that

$\log 1 = 0$ and $\log 10 = 1$

An important property of $\log N$ is that, as N increases, $\log N$ increases. From these facts it follows that the logarithm of any number between 1 and 10 is between 0 and 1, that is, the logarithm of any number between 1 and 10 can be expressed as a decimal.

If the result of a computation is to contain four significant* figures, each number used in that computation must have at least four

*See Appendix C.

significant figures. There is a method (using infinite series) developed in more advanced mathematics which enables one to calculate the logarithm of a number N when N is given. In the table on pages 416 and 417 the logarithms of some numbers with four significant figures are given to four significant figures. In using this table to find $\log N$, where N is between 1.000 and 9.990 and has four significant figures the last of which is zero, we locate the first two digits of N in the column which is headed by N. Log N is in the row which is determined by the first two digits of N and in the column headed by the last two digits of N.

To illustrate, from the table on pages 416 and 417 we find the following:

$$\log 1.830 = 0.2625 \tag{10.3}$$

which means that

$$1.830 = 10^{0.2625} \tag{10.4}$$

Also

$$\log 9.370 = 0.9717 \tag{10.5}$$

which means that

$$9.370 = 10^{0.9717} \tag{10.6}$$

SCIENTIFIC NOTATION

When writing very large or very small numbers, we frequently desire to express them in *scientific notation*. To express a number N in scientific notation, we write it in the form $A \cdot 10^k$, where $1 \leq A < 10$ and k is an integer, and A has the same number of significant figures as N. For example, light has a velocity of 186,000 miles per second, correct to three significant figures. Written in scientific notation, this is $1.86 \cdot 10^5$ miles per second. Had this figure been correct to four significant figures, we would have written $1.860 \cdot 10^5$. The number 0.00000018 written in scientific notation is $1.8 \cdot 10^{-7}$.

EXAMPLE 1 Write 273,000,000 in scientific notation.

SOLUTION We write 2.73 and observe that it is necessary to multiply 2.73 by $100{,}000{,}000 = 10^8$ to obtain 273,000,000. Hence 273,000,000 written in scientific notation is $2.73 \cdot 10^8$.

EXAMPLE 2 Write 0.0000000018 in scientific notation.

SOLUTION We write 1.8 and observe that we must multiply 1.8 by $0.000000001 = 10^{-9}$ to obtain 0.0000000018. Hence 0.0000000018 written in scientific notation is $1.8 \cdot 10^{-9}$.

Logarithms of numbers from 1.000 to 9.990

N	00	10	20	30	40	50	60	70	80	90
1.0	0.0000	0.004321	0.008600	0.01284	0.01703	0.02119	0.02531	0.02938	0.03342	0.03743
1.1	0.04139	0.04532	0.04922	0.05308	0.05690	0.06070	0.06446	0.06819	0.07188	0.07555
1.2	0.07918	0.08279	0.08636	0.08991	0.09342	0.09691	0.1004	0.1038	0.1072	0.1106
1.3	0.1139	0.1173	0.1206	0.1239	0.1271	0.1303	0.1335	0.1367	0.1399	0.1430
1.4	0.1461	0.1492	0.1523	0.1553	0.1584	0.1614	0.1644	0.1673	0.1703	0.1732
1.5	0.1761	0.1790	0.1818	0.1847	0.1875	0.1903	0.1931	0.1959	0.1987	0.2014
1.6	0.2041	0.2068	0.2095	0.2122	0.2148	0.2175	0.2201	0.2227	0.2253	0.2279
1.7	0.2304	0.2330	0.2355	0.2380	0.2405	0.2430	0.2455	0.2480	0.2504	0.2529
1.8	0.2553	0.2577	0.2601	0.2625	0.2648	0.2673	0.2695	0.2718	0.2742	0.2765
1.9	0.2788	0.2810	0.2833	0.2856	0.2878	0.2900	0.2923	0.2945	0.2967	0.2989
2.0	0.3010	0.3032	0.3054	0.3075	0.3096	0.3118	0.3139	0.3160	0.3181	0.3201
2.1	0.3222	0.3243	0.3263	0.3284	0.3304	0.3324	0.3345	0.3365	0.3385	0.3404
2.2	0.3424	0.3444	0.3464	0.3483	0.3502	0.3522	0.3541	0.3560	0.3579	0.3598
2.3	0.3617	0.3636	0.3655	0.3674	0.3692	0.3711	0.3729	0.3747	0.3766	0.3784
2.4	0.3802	0.3820	0.3838	0.3856	0.3874	0.3892	0.3909	0.3927	0.3945	0.3962
2.5	0.3979	0.3997	0.4014	0.4031	0.4048	0.4065	0.4082	0.4099	0.4116	0.4133
2.6	0.4150	0.4166	0.4183	0.4200	0.4216	0.4232	0.4249	0.4265	0.4281	0.4298
2.7	0.4314	0.4330	0.4346	0.4362	0.4378	0.4393	0.4409	0.4425	0.4440	0.4456
2.8	0.4472	0.4487	0.4502	0.4518	0.4533	0.4548	0.4564	0.4579	0.4594	0.4609
2.9	0.4624	0.4639	0.4654	0.4669	0.4683	0.4698	0.4713	0.4728	0.4742	0.4757
3.0	0.4771	0.4786	0.4800	0.4814	0.4829	0.4843	0.4857	0.4871	0.4886	0.4900
3.1	0.4914	0.4928	0.4942	0.4955	0.4969	0.4983	0.4997	0.5011	0.5024	0.5038
3.2	0.5051	0.5065	0.5079	0.5092	0.5105	0.5119	0.5132	0.5145	0.5159	0.5172
3.3	0.5185	0.5198	0.5211	0.5224	0.5237	0.5250	0.5263	0.5276	0.5289	0.5302
3.4	0.5315	0.5328	0.5340	0.5353	0.5366	0.5378	0.5391	0.5403	0.5416	0.5428
3.5	0.5441	0.5453	0.5465	0.5478	0.5490	0.5502	0.5514	0.5527	0.5539	0.5551
3.6	0.5563	0.5575	0.5587	0.5599	0.5611	0.5623	0.5635	0.5647	0.5658	0.5670
3.7	0.5682	0.5694	0.5705	0.5717	0.5729	0.5740	0.5752	0.5763	0.5775	0.5786
3.8	0.5798	0.5809	0.5821	0.5832	0.5843	0.5855	0.5866	0.5877	0.5888	0.5899
3.9	0.5911	0.5922	0.5933	0.5944	0.5955	0.5966	0.5977	0.5988	0.5999	0.6010
4.0	0.6021	0.6031	0.6042	0.6053	0.6064	0.6075	0.6085	0.6096	0.6107	0.6117
4.1	0.6128	0.6138	0.6149	0.6160	0.6170	0.6180	0.6191	0.6201	0.6212	0.6222
4.2	0.6232	0.6243	0.6253	0.6263	0.6274	0.6284	0.6294	0.6304	0.6314	0.6325
4.3	0.6335	0.6345	0.6355	0.6365	0.6375	0.6385	0.6395	0.6405	0.6415	0.6425
4.4	0.6435	0.6444	0.6454	0.6464	0.6474	0.6484	0.6493	0.6503	0.6513	0.6522
4.5	0.6532	0.6542	0.6551	0.6561	0.6571	0.6580	0.6590	0.6599	0.6609	0.6618
4.6	0.6628	0.6637	0.6646	0.6656	0.6665	0.6675	0.6684	0.6693	0.6702	0.6712
4.7	0.6721	0.6730	0.6739	0.6749	0.6758	0.6767	0.6776	0.6785	0.6794	0.6803
4.8	0.6812	0.6821	0.6830	0.6839	0.6848	0.6857	0.6866	0.6875	0.6884	0.6893
4.9	0.6902	0.6911	0.6920	0.6928	0.6937	0.6946	0.6955	0.6964	0.6972	0.6981
5.0	0.6990	0.6998	0.7007	0.7016	0.7024	0.7033	0.7042	0.7050	0.7059	0.7067
5.1	0.7076	0.7084	0.7093	0.7101	0.7110	0.7118	0.7126	0.7135	0.7143	0.7152
5.2	0.7160	0.7168	0.7177	0.7185	0.7193	0.7202	0.7210	0.7218	0.7226	0.7235
5.3	0.7243	0.7251	0.7259	0.7267	0.7275	0.7284	0.7292	0.7300	0.7308	0.7316
5.4	0.7324	0.7332	0.7340	0.7348	0.7356	0.7364	0.7372	0.7380	0.7388	0.7396

Logarithms of numbers from 1.000 to 9.990 (*Continued.*)

N	00	10	20	30	40	50	60	70	80	90
5.5	0.7404	0.7412	0.7419	0.7427	0.7435	0.7443	0.7451	0.7459	0.7466	0.7474
5.6	0.7482	0.7490	0.7497	0.7505	0.7513	0.7520	0.7528	0.7536	0.7543	0.7551
5.7	0.7559	0.7566	0.7574	0.7582	0.7589	0.7597	0.7604	0.7612	0.7619	0.7627
5.8	0.7634	0.7642	0.7649	0.7657	0.7664	0.7672	0.7679	0.7686	0.7694	0.7701
5.9	0.7709	0.7716	0.7723	0.7731	0.7738	0.7745	0.7752	0.7760	0.7767	0.7774
6.0	0.7782	0.7789	0.7796	0.7803	0.7810	0.7818	0.7825	0.7832	0.7839	0.7846
6.1	0.7853	0.7860	0.7868	0.7875	0.7882	0.7889	0.7896	0.7903	0.7910	0.7917
6.2	0.7924	0.7931	0.7938	0.7945	0.7952	0.7959	0.7966	0.7973	0.7980	0.7987
6.3	0.7993	0.8000	0.8007	0.8014	0.8021	0.8028	0.8035	0.8041	0.8048	0.8055
6.4	0.8062	0.8069	0.8075	0.8082	0.8089	0.8096	0.8102	0.8109	0.8116	0.8122
6.5	0.8129	0.8136	0.8142	0.8149	0.8156	0.8162	0.8169	0.8176	0.8182	0.8189
6.6	0.8195	0.8202	0.8209	0.8215	0.8222	0.8228	0.8235	0.8241	0.8248	0.8254
6.7	0.8261	0.8267	0.8274	0.8280	0.8287	0.8293	0.8299	0.8306	0.8312	0.8319
6.8	0.8325	0.8331	0.8338	0.8344	0.8351	0.8357	0.8363	0.8370	0.8376	0.8382
6.9	0.8388	0.8395	0.8401	0.8407	0.8414	0.8420	0.8426	0.8432	0.8439	0.8445
7.0	0.8451	0.8457	0.8463	0.8470	0.8476	0.8482	0.8488	0.8494	0.8500	0.8506
7.1	0.8513	0.8519	0.8525	0.8531	0.8537	0.8543	0.8549	0.8555	0.8561	0.8567
7.2	0.8573	0.8579	0.8585	0.8591	0.8597	0.8603	0.8609	0.8615	0.8621	0.8627
7.3	0.8633	0.8639	0.8645	0.8651	0.8657	0.8663	0.8669	0.8675	0.8681	0.8686
7.4	0.8692	0.8698	0.8704	0.8710	0.8716	0.8722	0.8727	0.8733	0.8739	0.8745
7.5	0.8751	0.8756	0.8762	0.8768	0.8774	0.8779	0.8785	0.8791	0.8797	0.8802
7.6	0.8808	0.8814	0.8820	0.8825	0.8831	0.8837	0.8842	0.8848	0.8854	0.8859
7.7	0.8865	0.8871	0.8876	0.8882	0.8887	0.8893	0.8899	0.8904	0.8910	0.8915
7.8	0.8921	0.8927	0.8932	0.8938	0.8943	0.8949	0.8954	0.8960	0.8965	0.8971
7.9	0.8976	0.8982	0.8987	0.8993	0.8998	0.9004	0.9009	0.9015	0.9020	0.9025
8.0	0.9031	0.9036	0.9042	0.9047	0.9053	0.9058	0.9063	0.9069	0.9074	0.9079
8.1	0.9085	0.9090	0.9096	0.9101	0.9106	0.9112	0.9117	0.9122	0.9128	0.9133
8.2	0.9138	0.9143	0.9149	0.9154	0.9159	0.9165	0.9170	0.9175	0.9180	0.9186
8.3	0.9191	0.9196	0.9201	0.9206	0.9212	0.9217	0.9222	0.9227	0.9232	0.9238
8.4	0.9243	0.9248	0.9253	0.9258	0.9263	0.9269	0.9274	0.9279	0.9284	0.9289
8.5	0.9294	0.9299	0.9304	0.9309	0.9315	0.9320	0.9325	0.9330	0.9335	0.9340
8.6	0.9345	0.9350	0.9355	0.9360	0.9365	0.9370	0.9375	0.9380	0.9385	0.9390
8.7	0.9395	0.9400	9.9405	0.9410	0.9415	0.9420	0.9425	0.9430	0.9435	0.9440
8.8	0.9445	0.9450	0.9455	0.9460	0.9465	0.9469	0.9474	0.9479	0.9484	0.9489
8.9	0.9494	0.9499	0.9504	0.9509	0.9513	0.9518	0.9523	0.9528	0.9533	0.9538
9.0	0.9542	0.9547	0.9552	0.9557	0.9562	0.9566	0.9571	0.9576	0.9581	0.9586
9.1	0.9590	0.9595	0.9600	0.9605	0.9609	0.9614	0.9619	0.9624	0.9628	0.9633
9.2	0.9638	0.9643	0.9647	0.9652	0.9657	0.9661	0.9666	0.9671	0.9675	0.9680
9.3	0.9685	0.9689	0.9694	0.9699	0.9703	0.9708	0.9713	0.9717	0.9722	0.9727
9.4	0.9731	0.9736	0.9741	0.9745	0.9750	0.9754	0.9759	0.9763	0.9768	0.9773
9.5	0.9777	0.9782	0.9786	0.9791	0.9795	0.9800	0.9805	0.9809	0.9814	0.9818
9.6	0.9823	0.9827	0.9832	0.9836	0.9841	0.9845	0.9850	0.9854	0.9859	0.9863
9.7	0.9868	0.9872	0.9877	0.9881	0.9886	0.9890	0.9894	0.9899	0.9903	0.9908
9.8	0.9912	0.9917	0.9921	0.9926	0.9930	0.9934	0.9939	0.9943	0.9948	0.9952
9.9	0.9956	0.9961	0.9965	0.9969	0.9974	0.9978	0.9983	0.9987	0.9991	0.9996

EXAMPLE 3 Write $5.8 \cdot 10^6$ in ordinary notation.

SOLUTION We observe that $10^6 = 1{,}000{,}000$. Therefore,

$$5.8 \cdot 10^6 = (5.8)(1{,}000{,}000) = 5{,}800{,}000$$

EXAMPLE 4 Write $4.32 \cdot 10^{-5}$ in ordinary notation.

SOLUTION We observe that $10^{-5} = 0.00001$. Hence

$$4.32 \cdot 10^{-5} = (4.32)(0.00001) = 0.0000432$$

Further examples are given in the table below.

NUMBER	NUMBER OF SIGNIFICANT FIGURES	NUMBER IN SCIENTIFIC NOTATION
175,000,000	3	$1.75 \cdot 10^8$
110,000	3	$1.10 \cdot 10^5$
26,000,000,000	2	$2.6 \cdot 10^{10}$
1,000,000	4	$1.000 \cdot 10^6$
1,000,000	1	10^6
0.000123	3	$1.23 \cdot 10^{-4}$
0.0000010	2	$1.0 \cdot 10^{-6}$
0.000001	1	10^{-6}
0.0000107	3	$1.07 \cdot 10^{-5}$
0.00009000	4	$9.000 \cdot 10^{-5}$

Any number N can be expressed in the scientific notation, that is, as the product of a number between 1 and 10 and an integral power of 10,

$$N = A \cdot 10^k \qquad 1 \leq A < 10 \tag{10.7}$$

In (10.7), N and A have the same number of significant figures, and k is a positive integer or a negative integer or zero. By use of (10.4), we have

$$1{,}830 = 1.830 \cdot 10^3 = 10^{0.2625} \cdot 10^3 = 10^{0.2625+3}$$

and

$$0.001830 = 1.830 \cdot 10^{-3} = 10^{0.2625} \cdot 10^{-3} = 10^{0.2625-3}$$

Consequently,

$$\log 1{,}830 = 0.2625 + 3 \qquad \text{and} \qquad \log 0.001830 = 0.2625 - 3$$

From these illustrations it should be clear that the (common) logarithm of a number may be written as the sum of a nonnegative decimal and a positive or negative integer. The nonnegative decimal part of a logarithm is called the **mantissa** of the logarithm, and the

MANTISSA

CHARACTERISTIC

integral part is called the **characteristic** of the logarithm. To illustrate, in $\log 1{,}830 = 0.2625 + 3$, we call 0.2625 the mantissa, and 3 the characteristic; in $\log 0.001830 = 0.2625 - 3$, the mantissa is 0.2625, and the characteristic is -3.

While the mantissa of the logarithm of a number is usually an approximation, the characteristic of the logarithm is always exact.

From (10.7) we see that any number with four significant figures may be written as the product of a number between 1 and 10 with four significant figures and an integral power of 10. From this fact it follows that the table of logarithms gives directly the mantissa of the logarithm of any number N, where N has four significant figures the last of which is zero.

Note that if k, l, and m are any digits, and if

$$10^x = k.lm0 \qquad \textit{to four significant figures}$$

then

$$10^x = k.lm \qquad \textit{to three significant figures}$$

Also, if k and l are any digits, and if

$$10^x = k.l00 \qquad \textit{to four significant figures}$$

then

$$10^x = k.l \qquad \textit{to two significant figures}$$

Further, if k is any digit, and if

$$10^x = k.000 \qquad \textit{to four significant figures}$$

then

$$10^x = k \qquad \textit{to one significant figure}$$

From the facts just stated it follows that

$$\log k.lm = \log k.lm0 \qquad \log k.l = \log k.l00 \qquad \log k = \log k.000$$

Therefore, we can find directly from the table on pages 416 and 417 the mantissa of the logarithm of any number N, where N has one, two, or three significant figures. To illustrate,

$634 = 6.34 \cdot 10^2$; so $\log 634 = 0.8021 + 2$, since
$\log 6.34 = \log 6.340 = 0.8021$

$63 = 6.3 \cdot 10^1$; so $\log 63 = 0.7993 + 1$, since
$\log 6.3 = \log 6.300 = 0.7993$

$6 = 6 \cdot 10^0$; so $\log 6 = 0.7782$, since
$\log 6 = \log 6.000 = 0.7782$

To determine the characteristic of a number N, think of the number as written in the scientific notation

$$N = A \cdot 10^k \qquad 1 \leq A < 10$$

where k is a positive or negative integer, or zero. *Then k is the characteristic of N.*

Study the illustrations in the following table, where the characteristics are found as indicated in the preceding paragraph. The mantissa is found from the table of logarithms.

GIVEN NUMBER N	N IN SCIENTIFIC NOTATION	CHARACTERISTIC OF LOG N	MANTISSA OF LOG N	LOG N
427	$4.27 \cdot 10^2$	2	0.6304	$0.6304 + 2$
42.7	$4.27 \cdot 10^1$	1	0.6304	$0.6304 + 1$
4.27	$4.27 \cdot 10^0$	0	0.6304	0.6304
0.427	$4.27 \cdot 10^{-1}$	-1	0.6304	$0.6304 - 1$
0.0427	$4.27 \cdot 10^{-2}$	-2	0.6304	$0.6304 - 2$
0.00427	$4.27 \cdot 10^{-3}$	-3	0.6304	$0.6304 - 3$

In practice it is customary to make use of the following rules for finding the characteristic (the reader should satisfy himself that these rules are consistent with the statement just made for finding the characteristic):

1 If a number is greater than 1, the characteristic of its logarithm is positive or zero and is 1 less than the number of digits to the left of the decimal point in the given number.
2 If a number is less than 1, the characteristic of its logarithm is negative and is numerically 1 more than the number of zeros between the decimal point and the first nonzero digit in the given number.
3 The characteristic of log 1 is zero, since log 1 = 0.

EXERCISES

Express the following numbers in scientific notation:

1 173,000,000
2 1,698,000
3 1,550,000
4 689,000
5 69,643,217
6 4,238,000,000
7 0.000000017432
8 0.0000192
9 0.0002950
10 0.000000003000
11 0.001543798
12 0.000000595

Write the following numbers in ordinary notation:

13 $7.364 \cdot 10^5$
14 $1.2 \cdot 10^{-4}$
15 $3.47 \cdot 10^{10}$
16 $5.678 \cdot 10^{-12}$
17 $4.6 \cdot 10^0$
18 $6.66 \cdot 10^{-1}$

Perform the indicated operations without using logarithms, and report your answers in scientific notation.

19 $(6.3 \cdot 10^{12})(1.9 \cdot 10^{4})$

20 $(9.81 \cdot 10^{3})(8.972 \cdot 10^{5})$

21 $(9.6 \cdot 10^{15})(3.2 \cdot 10^{-8})$

22 $(4.23 \cdot 10^{-10})(5.1 \cdot 10^{15})$

23 $(3.2 \cdot 10^{14}) \div (1.6 \cdot 10^{8})$

24 $(4.91 \cdot 10^{8}) \div (8.8 \cdot 10^{-7})$

25 $\dfrac{(9.7 \cdot 10^{12})(12.8 \cdot 10^{7})}{6.4 \cdot 10^{11}}$

26 $\dfrac{3.4 \cdot 10^{-8}}{(1.7 \cdot 10^{9})(4.0 \cdot 10^{4})}$

27 $\dfrac{9.9 \cdot 10^{-16}}{(-3.3 \cdot 10^{-11})(1.8 \cdot 10^{-2})}$

28 $\dfrac{2.5 \cdot 10^{4}}{(6.1 \cdot 10^{-3})(5.00 \cdot 10^{-7})}$

29 Express 48,000,000 in scientific notation.

30 A light-year is the distance that light travels in 1 year. Using the fact that light travels $1.86 \cdot 10^{5}$ miles per second, express in scientific notation the number of miles in a light-year.

31 A certain star is $1.18 \cdot 10^{15}$ miles from the earth. Using the number of miles in a light-year which you obtained in Exercise 30, express the distance from this star to the earth in light-years.

32 The star Sirius is approximately $5.1 \cdot 10^{13}$ miles from the earth. How many years are required for light to travel from Sirius to the earth?

33 It has been estimated that the mass of the earth is approximately $6.0 \cdot 10^{21}$ tons and that the mass of the sun is approximately $3.3 \cdot 10^{5}$ times the mass of the earth. Using these figures, find the mass of the sun.

34 The distances of the sun and moon from the earth are $9.3 \cdot 10^{7}$ and $2.4 \cdot 10^{5}$ miles, respectively. Find the number n, in scientific notation, such that the first of these distances is n times the second of these distances.

Find the logarithm of the following numbers:

35 384
36 86
37 4,070
38 0.00342
39 0.0792
40 230,000
41 52,900
42 42.20
43 $\frac{3}{4}$
44 1.92
45 0.000973
46 2,030

If the number N has the significant figures 567, give N as a decimal when the characteristic of its logarithm is:

47 4
48 −5
49 0
50 −3

10.3 INTERPOLATION

The table of logarithms may be used to determine the logarithm of any number N with four significant figures. If the last significant figure is not zero, a process called *interpolation* must be used. This process is illustrated in the following examples.

EXAMPLE 1 Find log 547.3.

SOLUTION Since 547.3 is three-tenths of the way between 547.0 and 548.0, its logarithm will be approximately three-tenths of the way between the logarithm of 547.0 and 548.0. We arrange our work as follows:

$$\left.\begin{array}{l}\log 547.0 = 0.7380 + 2 \\ \log 547.3 = ? \\ \log 548.0 = 0.7388 + 2\end{array}\right\} \text{difference} = 0.0008$$

Therefore,

$$\begin{aligned}\log 547.3 &= 0.7380 + 2 + \tfrac{3}{10}(0.0008) \\ &= 0.7380 + 2 + 0.0002 \\ \log 547.3 &= 0.7382 + 2\end{aligned}$$

In finding the logarithm of a number N, where N has four significant figures, we round off the mantissa of $\log N$ to four significant figures.

EXAMPLE 2 Find log 0.002424.

SOLUTION We note that 0.002424 is four-tenths of the way between 0.002420 and 0.002430, so its logarithm is approximately four-tenths of the way between log 0.002420 and log 0.002430. The work may be arranged as follows:

$$\left.\begin{array}{l}\log 0.002420 = 0.3838 - 3 \\ \log 0.002424 = ? \\ \log 0.002430 = 0.3856 - 3\end{array}\right\} \text{difference} = 0.0018$$

Therefore,

$$\begin{aligned}\log 0.002424 &= 0.3838 - 3 + \tfrac{4}{10}(0.0018) \\ &= 0.3838 - 3 + 0.0007 \\ \log 0.002424 &= 0.3845 - 3\end{aligned}$$

If a number has more than four significant figures we first round it off to four significant figures before we find its logarithm by use of the table of logarithms on pages 416 to 417. To illustrate, suppose we are given the number 439.682. We first round this number off to four significant figures, getting 439.7; then we find log 439.7 by interpolation from the table of logarithms.

EXERCISES

With the aid of the table of logarithms, find the logarithm of each of the following numbers. Give the mantissa to four significant figures.

1 789.6	2 1,608	3 37.21
4 0.06427	5 781,400	6 30,080

7 62.534 8 0.32735 9 0.004295
10 10,001,000 11 8,973,000 12 42.780

10.4 ANTILOGARITHMS

To use logarithms as an aid in a computational problem, we must be able to find the number N when we are given $\log N$. If $\log N = x$, then N is called the **antilogarithm** of x. The procedure for finding an antilogarithm is the reverse of that for finding a logarithm.

ANTILOGARITHM

EXAMPLE 1 If $\log N = 0.8432 + 3$, find N.

SOLUTION Since $\log N = 0.8432 + 3$, we know that $N = A \cdot 10^3$, where $\log A = 0.8432$.

We look in the table of logarithms for the mantissa 0.8432. It is found in the row with 6.9 on the left and in the column with 70 at the top. Therefore,

$$\log 6.970 = 0.8432$$

Since the characteristic is 3, we have

$$N = 6.970 \cdot 10^3 = 6{,}970$$

EXAMPLE 2 If $\log N = 0.7585 + 3$, find N.

SOLUTION In this case the mantissa is not listed in the table of logarithms, and interpolation is necessary. We search for two consecutive mantissas such that one is smaller and the other is larger than 0.7585. We find them to be

0.7582 and 0.7589

We note that

$$\left.\begin{array}{r} \left.\begin{array}{r}\log 5{,}730 = 0.7582 + 3 \\ \log N = 0.7585 + 3\end{array}\right\} 0.0003 \\ \log 5{,}740 = 0.7589 + 3 \end{array}\right\} 0.0007$$

with the differences in the logarithms as indicated. Since $\log N$ is three-sevenths of the way between log 5,730 and log 5,740, N will be approximately three-sevenths of the way between 5,730 and 5,740. Using this approximation, we get

$$\begin{aligned} N &= 5{,}730 + \tfrac{3}{7}(10) \\ &= 5{,}730 + 4 \\ N &= 5{,}734 \qquad \textit{to four significant figures} \end{aligned}$$

EXAMPLE 3 If $\log N = 0.5968 - 3$, find N.

SOLUTION Again we have a mantissa which is not listed in the table of logarithms. We look for two consecutive mantissas such that 0.5968 lies between them.

We find these mantissas to be

0.5966 and 0.5977

and arrange our work as follows:

$$\left.\begin{array}{r} \left.\begin{array}{r} \log 0.003950 = 0.5966 - 3 \\ \log N = 0.5968 - 3 \end{array}\right\} 0.0002 \\ \log 0.003960 = 0.5977 - 3 \qquad\qquad \end{array}\right\} 0.0011$$

Therefore,

$$N = 0.003950 + \tfrac{2}{11}(0.00001)$$
$$= 0.00395 + 0.000002$$
$$N = 0.003952 \qquad \textit{to four significant figures}$$

EXERCISES

In the following problems, find the number N to four significant figures:

1 $\log N = 0.8414 + 3$
2 $\log N = 0.5955 - 4$
3 $\log N = 0.4286 + 2$
4 $\log N = 0.9821 + 5$
5 $\log N = 0.4293 + 1$
6 $\log N = 0.5263 - 1$
7 $\log N = 0.3927 + 4$
8 $\log N = 0.8716 - 5$
9 $\log N = 0.5634$
10 $\log N = 0.9471 + 6$

10.5 LAWS OF LOGARITHMS

We now prove certain rules which are called the *laws of logarithms*. These laws are essentially the same as the laws of exponents. These laws of logarithms will be stated and proved for logarithms to the base 10; however, the laws are true for logarithms to any base.

Let

$$x = \log M \qquad \text{and} \qquad y = \log N$$

Then

$$M = 10^x \tag{10.8}$$

and

$$N = 10^y \tag{10.9}$$

Further,

$$M \cdot N = 10^x \cdot 10^y = 10^{x+y}$$

So $x + y = \log (M \cdot N)$. We have proved:

The logarithm of the product of two numbers is equal to the sum of the logarithms of the numbers,

$$\log M \cdot N = \log M + \log N$$

From (10.8) and (10.9) we have

$$\frac{M}{N} = \frac{10^x}{10^y} = 10^{x-y}$$

which means that $x - y = \log \frac{M}{N}$, and we have proved:

II *The logarithm of the quotient of two numbers is equal to the logarithm of the dividend minus the logarithm of the divisor,*

$$\log \frac{M}{N} = \log M - \log N$$

If we take the pth power of both sides of (10.9), we get

$$N^p = (10^y)^p = 10^{py}$$

So py is the logarithm of N^p, and we have proved:

III *The logarithm of the pth power of N is equal to p times the logarithm of N,*

$$\log N^p = p \log N$$

If we take the rth root of both sides of (10.9), we obtain

$$\sqrt[r]{N} = N^{1/r} = (10^y)^{1/r} = 10^{y/r}$$

This means that

$$\log \sqrt[r]{N} = \frac{y}{r} = \frac{1}{r} \log N$$

We have proved:

IV *The logarithm of the rth root of N is equal to the logarithm of N divided by r,*

$$\log \sqrt[r]{N} = \frac{1}{r} \log N$$

The following are illustrations of the use of laws I, II, III, and IV, respectively:

1 $\log (456.3)(6.824) = \log 456.3 + \log 6.824$

2 $\log \frac{456.3}{6.824} = \log 456.3 - \log 6.824$

3 $\log (963)^{14} = 14 \log 963$

4 $\log \sqrt[5]{8{,}764} = \frac{1}{5} \log 8{,}764$

EXERCISES

1 Using law I, write each of the following as the sum of two logarithms:

$\log (26)(35)$ $\quad \log (\frac{2}{3})(\frac{5}{7})$ $\quad \log (217)(314)$

2 Using law II, write each of the following as the difference of two logarithms:

$\log \frac{7}{9}$ $\log \frac{203}{8}$ $\log \frac{\frac{2}{3}}{109}$

3 Using law III, write each of the following as an integer times a logarithm:

$\log (17)^3$ $\log (219)^{-3}$ $\log (47)^8$

4 Using law IV, write each of the following as a fraction times a logarithm:

$\log (123)^{1/2}$ $\log (267)^{2/3}$ $\log (1{,}111)^{4/5}$

If $a = \log x$ and $b = \log y$, express each of the following in terms of a and b:

5 $\log xy$

6 $\log x^2y$

7 $\log xy^3$

8 $\log \frac{x}{y}$

9 $\log \frac{x^2}{y^3}$

10 $\log \frac{x^3}{y}$

10.6 COMPUTATION BY USE OF LOGARITHMS

If a problem in computation involves multiplication, division, finding a power, or finding a root, logarithms may be used to advantage. For by use of logarithms:

1 The solution of a problem in multiplication is based on the solution of a problem in addition.
2 The solution of a division problem is based on the solution of a problem in subtraction.
3 The solution of a problem of finding a power is based on the solution of a problem in multiplication.
4 The solution of a problem of finding a root is based on the solution of a division problem.

EXAMPLE 1 Find the value of $N = (437.0)(8.920)(67.80)$.

SOLUTION Applying law I, we get

$$\log N = \log 437.0 + \log 8.920 + \log 67.80$$

It is convenient to arrange the work as shown below:

$$\begin{aligned} \log 437.0 &= 0.6405 + 2 \\ \log 8.920 &= 0.9504 \\ \log 67.80 &= \underline{0.8312 + 1} \qquad \textit{add} \\ \log N &= 2.4221 + 3 \\ \log N &= 0.4221 + 5 \end{aligned}$$

We have found that $\log N = 0.4221 + 5$. Since the mantissa 0.4221 is not found in the table of logarithms, we interpolate as follows:

$$\log 264{,}000 = 0.4216 + 5$$
$$\log N = 0.4221 + 5$$
$$\log 265{,}000 = 0.4232 + 5$$

$$N = 264{,}000 + \tfrac{5}{16}(1{,}000)$$
$$= 264{,}000 + 300$$
$$N = 264{,}300 \qquad \textit{to four significant figures}$$

EXAMPLE 2 Calculate $N = \dfrac{(56.80)(892.0)}{72.60}$.

SOLUTION Application of laws I and II gives

$$\log N = \log 56.80 + \log 892.0 - \log 72.60$$

The work is shown below:

$$\begin{aligned} \log 56.80 &= 0.7543 + 1 \\ \log 892.0 &= \underline{0.9504 + 2} \qquad \textit{add} \\ & 1.7047 + 3 \\ \log 72.60 &= \underline{0.8609 + 1} \qquad \textit{subtract} \\ \log N &= 0.8438 + 2 \end{aligned}$$

To find N, we interpolate as follows:

$$\log 697 = 0.8432 + 2$$
$$\log N = 0.8438 + 2$$
$$\log 698 = 0.8439 + 2$$

So

$$N = 697 + \tfrac{6}{7}(1)$$
$$= 697 + 0.9$$
$$N = 697.9 \qquad \textit{to four significant figures}$$

EXAMPLE 3 Find the value of $N = \dfrac{45.60}{78.90}$.

SOLUTION Applying law II, we get

$$\log N = \log 45.60 - \log 78.90$$
$$= (0.6590 + 1) - (0.8971 + 1)$$
$$\log N = -0.2381$$

This procedure gives a negative decimal for $\log N$, whereas we have agreed that the logarithm of a number is to be written in such a form that its decimal part is nonnegative. The difficulty can be avoided by writing $\log 45.60$ as 1.6590 instead of $0.6590 + 1$ and proceeding with the calculation as shown below:

$$\begin{aligned} \log 45.60 &= 1.6590 \\ \log 78.90 &= \underline{0.8971 + 1} \qquad \textit{subtract} \\ \log N &= 0.7619 - 1 \end{aligned}$$

Therefore

$N = 0.5780$

EXAMPLE 4 Find the value of $N = (9{,}675)^4$.

SOLUTION By use of law III, we get

$$\begin{aligned}\log N &= 4 \log 9{,}675\\ &= 4(0.9856 + 3)\\ &= 3.9424 + 12\\ \log N &= 0.9424 + 15\end{aligned}$$

Since the mantissa 0.9424 does not appear in the table of logarithms, we interpolate and arrange our work as follows:

$$\begin{aligned}\log 8{,}750{,}000{,}000{,}000{,}000 &= 0.9420 + 15\\ \log N &= 0.9424 + 15\\ \log 8{,}760{,}000{,}000{,}000{,}000 &= 0.9425 + 15\end{aligned}$$

$$\begin{aligned}N &= 8{,}750{,}000{,}000{,}000{,}000 + \tfrac{4}{5}(10{,}000{,}000{,}000{,}000)\\ &= 8{,}750{,}000{,}000{,}000{,}000 + 8{,}000{,}000{,}000{,}000\\ N &= 8{,}758{,}000{,}000{,}000{,}000 \qquad \textit{to four significant figures}\end{aligned}$$

EXAMPLE 5 Evaluate $N = \sqrt[3]{85.30}$.

SOLUTION Applying law IV, we get

$$\begin{aligned}\log N &= \tfrac{1}{3} \log 85.30\\ &= \tfrac{1}{3}(0.9309 + 1)\\ &= \tfrac{1}{3}(1.9309)\\ \log N &= 0.6436\end{aligned}$$

The mantissa 0.6436 does not appear in the table of logarithms, but we find by interpolation

$$\begin{aligned}N &= 4.40 + \tfrac{1}{9}(0.01)\\ &= 4.40 + 0.001\\ N &= 4.401 \qquad \textit{to four significant figures}\end{aligned}$$

EXAMPLE 6 Evaluate $N = \dfrac{(32.10)(4.580)^5}{4{,}780\sqrt[3]{962.0}}$.

SOLUTION Application of law II gives

$$\log N = \log [32.10(4.580)^5] - \log (4{,}780\sqrt[3]{962.0})$$

Application of law I to this gives

$$\log N = \log 32.10 + \log (4.580)^5 - (\log 4{,}780 + \log \sqrt[3]{962.0})$$

Application of laws III and IV to the second and third terms, respectively, gives

$$\log N = \log 32.10 + 5 \log 4.580 - (\log 4{,}780 + \tfrac{1}{3} \log 962.0)$$

We arrange our work as follows:

$$\begin{aligned} \log 32.10 &= 0.5065 + 1 \\ 5 \log 4.580 &= \underline{3.3045} \qquad \textit{add} \\ & 3.8110 + 1 \end{aligned}$$

$$\text{or} \qquad 1.8110 + 3 \quad (a)$$

$$\begin{aligned} \log 4{,}780 &= 0.6794 + 3 \\ \tfrac{1}{3} \log 962.0 &= \underline{0.9944} \qquad \textit{add} \\ & 1.6738 + 3 \quad (b) \end{aligned}$$

Subtracting (b) from (a), we have

$\log N = 0.1372$

We interpolate as follows:

$$\begin{aligned} \log 1.37 &= 0.1367 \\ \log N &= 0.1372 \\ \log 1.38 &= 0.1399 \end{aligned}$$

$$\begin{aligned} N &= 1.37 + \tfrac{5}{32}(0.01) \\ &= 1.37 + 0.002 \\ N &= 1.372 \qquad \textit{to four significant figures} \end{aligned}$$

EXAMPLE 7 Evaluate $N = \sqrt[2]{0.463}$.

SOLUTION Applying law IV, we get

$$\begin{aligned} \log N &= \tfrac{1}{2} \log 0.463 \\ &= \tfrac{1}{2} \log (4.63 \cdot 10^{-1}) \\ &= \tfrac{1}{2}[0.6656 + (-1)] \\ \log N &= 0.3328 - 0.5 \end{aligned}$$

Having $\log N$ expressed in this form does not permit us to find N by use of the table of logarithms. In order to use this table to find N, $\log N$ needs to be expressed as a nonnegative decimal fraction plus a positive or negative integer. So we return to

$\log N = \tfrac{1}{2}(0.6656 - 1)$

and write this as

$\log N = \tfrac{1}{2}(1.6656 - 2)$

in order to get

$\log N = 0.8328 - 1$

With the aid of the table of logarithms, we find that $N = 6.805 \cdot 10^{-1} = 0.6805$.

EXERCISES

Use logarithms to calculate the following to four significant figures:

1 $(436.2)(67.85)$

2 $\dfrac{856.4}{48.32}$

3 $(37.82)^3$

4 $\dfrac{(42.63)(6.135)}{(25.60)(91.82)}$

5 $\sqrt[5]{42{,}700}$

6 $\dfrac{(367.2)(1.760)^5}{9{,}750}$

7 $(99.90)(0.1070)(611.7)$

8 $\sqrt[4]{9{,}283}$

9 $\dfrac{(96.73)(8.420)^3}{\sqrt[5]{6{,}872}}$

10 Find the number of cubic feet in the volume of a cube if the edge is 41.35 feet long.

11 The formula for the volume V of a sphere may be written as $V = 4.189R^3$, where R is the radius. If $R = 20.44$ inches, find V.

12 The time t for a complete swing of a pendulum of length l in feet is given by the formula $t = 2\pi\sqrt{\dfrac{l}{g}}$. Find t when $l = 4.682$ feet. Use $\pi = 3.142$ and $g = 32.16$.

13 Use logarithms to find N, given that $N = \sqrt[3]{0.02314}$.

14 Use logarithms to find N, given that $N = \sqrt[2]{\dfrac{2{,}463}{5{,}892}}$.

15 Recall from Exercise 11 of Sec. 5.8 that the area A of an equilateral triangle of side s is given by $A = \frac{1}{4}s^2\sqrt{3}$. With the aid of logarithms, find the area of the equilateral triangle for which the length of the side is 378.0 inches.

16 The weight P, in tons, which will crush a solid cast-iron cylindrical column is given by the formula $P = \dfrac{50d^{3.6}}{L^{1.7}}$, when d is the diameter in inches and L is the length in feet. Find the weight which will crush such a column for which $d = 4.780$ inches and $L = 14.60$ feet.

10.7
THE EXPONENTIAL AND LOGARITHMIC FUNCTIONS

In Sec. 10.1 we constructed a table giving values of 2^x for positive integer values of x from 1 to 16 inclusive. Using the definition of zero and negative integer exponents given in Sec. 3.9, we can compute values of 2^x for $x = 0$ and for negative integer values of x. Setting $y = 2^x$, let us construct such a table for $x = -4, -3, -2, -1, 0, 1, 2, 3, 4$.

x	-4	-3	-2	-1	0	1	2	3	4
$y = 2^x$	$\frac{1}{16}$	$\frac{1}{8}$	$\frac{1}{4}$	$\frac{1}{2}$	1	2	4	8	16

Clearly the table could be extended to give values of 2^x for any *integral* values of x. From Sec. 3.10 we know the meaning of 2^x for x, any *rational number;* for example, if $x = \frac{3}{2}$, then $2^x = 2^{3/2} = (\sqrt{2})^3$. By the use of logarithms we can find the (approximate) value of 2^x for x *any rational number;* for example, if $x = \frac{3}{2}$, we have

$$\log 2^{3/2} = \tfrac{3}{2} \log 2 = \tfrac{3}{2}(0.3010) = 0.4515$$

so

$$2^{3/2} = 2.828$$

Let us consider what the graph of the equation

$$y = 2^x$$

might look like. We can begin by plotting the points whose coordinates are given by the corresponding values of x and y in the table above. These points are shown as the unlettered points in Fig. 10.1.

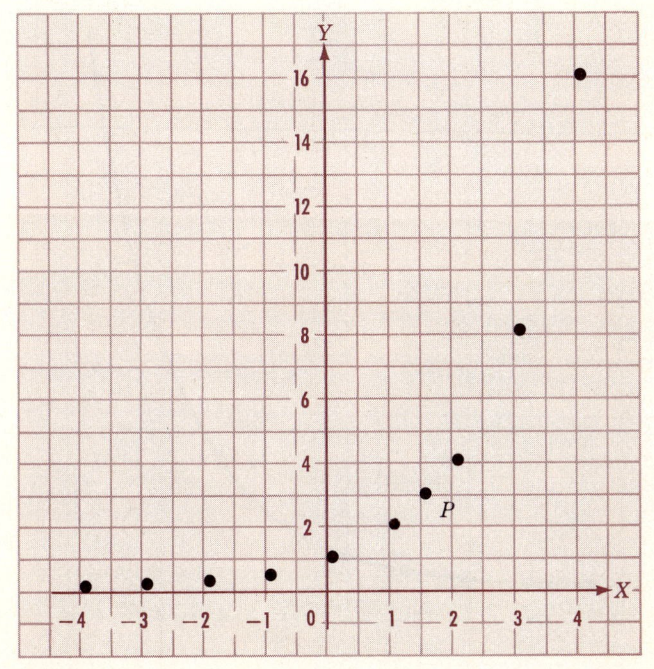

FIGURE 10.1

Next we could plot points corresponding to *rational* values of x; for example, the point $(\frac{3}{2}, 2.828)$ is shown as point P in Fig. 10.1. In this way we could obtain points that are arbitrarily close together (since the points on the x axis that correspond to rational numbers are arbitrarily close together). Of course we could not produce in this way any points on the graph of $y = 2^x$, whose x coordinates were irrational. However, we shall *assume* that such points with irrational x coordinates will lie on the smooth curve drawn through the points we can obtain by using rational values for x. With this assumption, the graph of $y = 2^x$ is defined to be the graph we obtain by joining the points with rational x coordinates with a smooth curve. The result is shown in Fig. 10.2. The graph of $y = 2^x$ shown in Fig. 10.2 is called an *exponential curve* and the function $F = \{(x, y) \mid y = 2^x\}$ is called the *exponential function with base 2*. Proceeding in a similar way, we can construct the graph of any equation of the form

EXPONENTIAL FUNCTION WITH BASE a

$$y = a^x \qquad \text{where} \qquad a > 0$$

EXPONENTIAL CURVE

and hence the graph of any function F for which $F(x) = a^x$, with $a > 0$. Such a function is an **exponential function with base a**, and its graph is an **exponential curve.** If $a > 1$, the graph of $y = a^x$ is

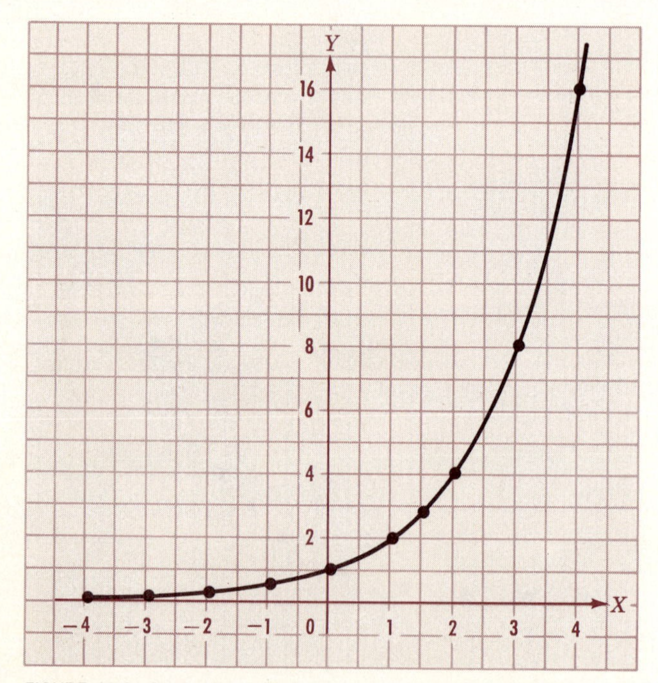

FIGURE 10.2 Graph of $y = 2^x$.

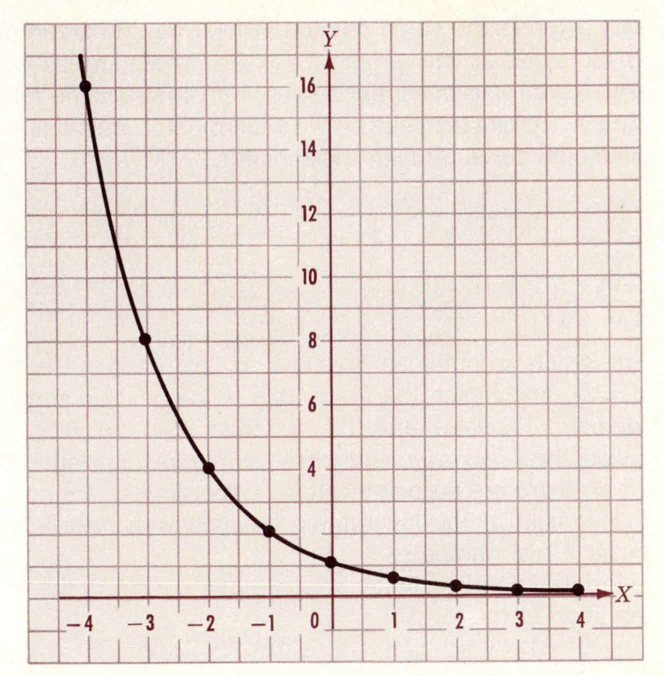

FIGURE 10.3 Graph of $y = (\frac{1}{2})^x$.

similar to the graph of $y = 2^x$ in Fig. 10.2. If $0 < a < 1$, the graph of $y = a^x$ has the appearance of the graph in Fig. 10.3. If $a = 1$, the graph of $y = a^x$ is the horizontal line passing through the point (0, 1). Note that the graph of $y = a^x$ passes through the point (0, 1) regardless of the value of a, since $a^0 = 1$ for all $a > 0$.

Exponential functions and exponential curves have many applications connected with natural phenomena. For example, under certain conditions the growth of a population is described by an exponential curve with base greater than 1; for instance, a portion of the curve in Fig. 10.2 lying to the right of $x = 0$ could show the behavior of a population when the population doubles for each unit increase in x. Also, the decay of radioactive material is described by an exponential curve with base less than 1.

Closely associated with exponential functions and exponential curves are logarithmic functions and logarithmic curves. Consider the graph of the equation

$$y = \log_2 x$$

From the definition in Sec. 10.1 we know that the equations

$$y = \log_2 x \qquad \text{and} \qquad x = 2^y$$

both express the same relation. Therefore, the graph of $y = \log_2 x$ is the same as the graph of $x = 2^y$. Proceeding as we did in the discussion of exponential curves, we construct the following table for $x = 2^y$, plot the points whose coordinates are obtained, and draw a smooth curve through these points.

$x = 2^y$	$\frac{1}{16}$	$\frac{1}{8}$	$\frac{1}{4}$	$\frac{1}{2}$	1	2	4	8	16
$y =$	-4	-3	-2	-1	0	1	2	3	4

The graph so obtained is shown in Fig. 10.4 and is the graph of $x = 2^y$, and hence the graph of $y = \log_2 x$. This curve is called a *logarithmic curve* and the function $G = \{(x, y) \mid y = \log_2 x\}$ is called the *logarithmic function with base 2*. Notice that on the graph there are no points whose x coordinates are not positive; so the domain of the logarithmic function with base 2 is the set of positive real numbers.

Proceeding in a similar way, we can construct the graph of any equation of the form

$$y = \log_a x \qquad \text{where} \qquad a > 0$$

If $a \neq 1$, the graph of this equation will be the graph of a function.

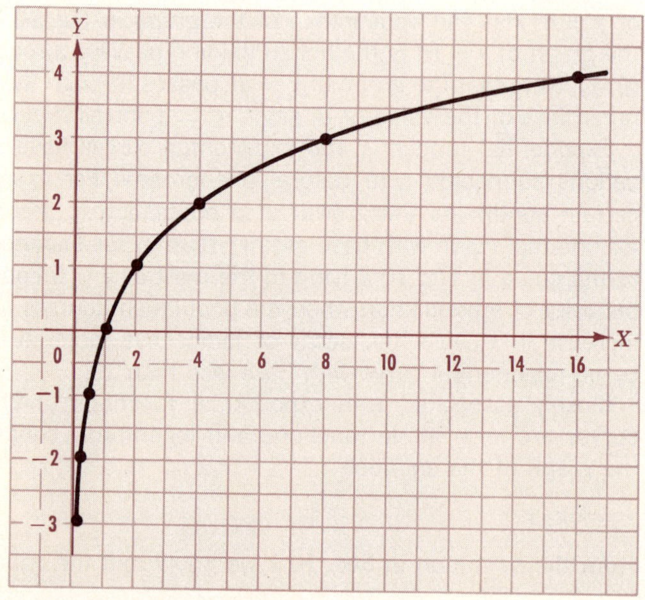

FIGURE 10.4 Graph of $y = \log_2 x$.

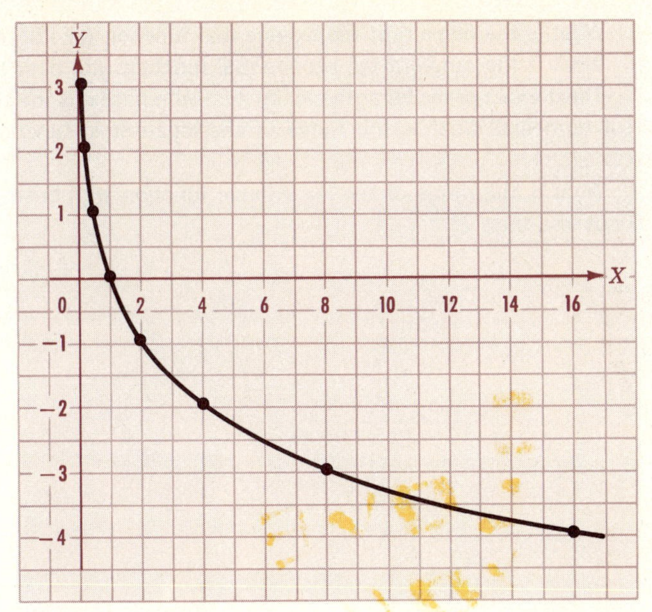

FIGURE 10.5 Graph of $y = \log_{1/2} x$.

The function

$$G = \{(x, y) \mid y = \log_a x,\ a > 0,\ a \neq 1\}$$

for which

$$G(x) = \log_a x \qquad a > 0 \quad a \neq 1$$

LOGARITHMIC FUNCTION WITH BASE a

LOGARITHMIC CURVE

is called the **logarithmic function with base** a, and its graph is called a **logarithmic curve.**

If $a = 1$, the graph of $y = \log_a x$ is a vertical line and therefore is not the graph of a function. Hence, there is no logarithmic function with base 1.

If $a > 1$, the graph of $y = \log_a x$ has the appearance of the graph in Fig. 10.4. If $0 < a < 1$, the graph of $y = \log_a x$ has the appearance of the graph in Fig. 10.5.

EXERCISES

1 Draw the graph of the equation $y = 3^x$ for $-3 \leq x \leq 4$.

2 Draw the graph of the equation $y = (\frac{1}{4})^x$ for $-4 \leq x \leq 3$.

3 Draw the graph of the function

$$F = \{(x, y) \mid y = \log_{10} x,\ 0 < x \leq 1{,}000\}$$

4 Draw the graph of the function
$H = \{(x, y) \mid y = \log_3 x,\ 0 < x \leq 27\}$

5 What is the domain of the exponential function with base a?

6 What is the range of the exponential function with base a?

7 The domain of the logarithmic function with base a is the set of positive real numbers. What is the range of the logarithmic function with base $a > 1$?

8 What is the range of the logarithmic function with base greater than 0 but less than 1?

11

Compound Interest and Annuities

11.1
COMPOUND INTEREST

In some business transactions when the interest becomes due at the end of a given period of time, it is added to the original principal. This second principal earns interest during the second period, and this interest is added at the end of the second period to the second principal to form a third principal. If this process is repeated for a given number of periods, the total amount accumulated at the end of the stated time is called the **compound amount** for the given period of time and the given rate of interest. The difference between the compound amount and the original principal is called the **compound interest.**

COMPOUND AMOUNT

COMPOUND INTEREST

When interest is added to the principal at the end of each period, we say that the interest is *compounded,* or *converted,* into principal. The time between two successive conversions of interest into principal is called the **conversion period.** Interest is usually quoted as so much per year, even though the conversion period is other than a year. In such a case the rate per conversion period is found by dividing the given annual rate by the number of conversion periods per year. For example, if the stated rate is 6% compounded semiannually for 4 years, there are 2 conversion periods per year, or a total of 8 conversion periods, and the interest rate per conversion period is 3%.

CONVERSION PERIOD

To show how compound interest is computed, and to contrast compound interest and simple interest, we give the following illustrations:

EXAMPLE 1 For a principal of $100 and an interest rate of 8%, prepare a table showing the principal at the beginning of each year and the interest at the end of each year, for 4 years, when (a) the interest is simple, (b) the interest is compounded annually.

SOLUTION (a) From our previous study of simple interest the student should verify the entries which are given in the table on page 439 for this case. (b) The computation for the compound amount and compound interest may be arranged as follows:

Original principal	$100.00
Add 8% interest	8.00
Principal at beginning of 2d year	$108.00
Add 8% interest	8.64
Principal at beginning of 3d year	$116.64
Add 8% interest	9.33
Principal at beginning of 4th year	$125.97
Add 8% interest	10.08
Compound amount at end of 4th year	$136.05

The compound interest is obtained by subtracting the original principal from the compound amount. So the compound interest = \$136.05 − \$100.00 = \$36.05. The compound interest exceeds the simple interest by \$4.05. For a given rate the compound interest always exceeds the simple interest if the number of conversion periods is greater than 1.

	(a) SIMPLE INTEREST		(b) COMPOUND INTEREST	
YEAR	PRINCIPAL AT BEGINNING OF YEAR	INTEREST AT END OF YEAR	PRINCIPAL AT BEGINNING OF YEAR	INTEREST AT END OF YEAR
1	\$100	\$8	\$100.00	\$ 8.00
2	100	8	108.00	8.64
3	100	8	116.64	9.33
4	100	8	125.97	10.08
PRINCIPAL PLUS INTEREST	\$132		\$136.05	
INTEREST	\$32		\$36.05	

EXAMPLE 2 For a principal of \$100 and an interest rate of 8% compounded semiannually, compute the compound amount at the end of 2 years.

SOLUTION In this case the interest is converted into principal at the end of each 6-month period, the rate per conversion period being 4%. The computations are as follows:

Original principal	\$100.00
Add 4% interest	4.00
Principal at beginning of 2d six months	\$104.00
Add 4% interest	4.16
Principal at beginning of 3d six months	\$108.16
Add 4% interest	4.33
Principal at beginning of 4th six months	\$112.49
Add 4% interest	4.50
Compound amount at end of 2d year	\$116.99

In this case the compound interest for the 2-year period is \$16.99. In contrast, we see from the results tabulated for the preceding example that \$100 compounded annually at 8% would yield interest of \$16.64 in 2 years.

NOMINAL ANNUAL RATE

EFFECTIVE ANNUAL RATE

As pointed out earlier, it is customary to state interest rates on a yearly basis even though the period of conversion is other than a year. This stated yearly rate is called the **nominal annual rate.** In Example 2 above the nominal annual rate is 8%. The **effective annual rate** i corresponding to a given nominal annual rate j is the rate i such that \$1 invested at the rate i for 1 year would amount to the same as $1 invested for 1 year at the nominal annual rate j. In Example 2 the effective annual rate is 8.16%, because $100 invested at the nominal rate of 8% compounded semiannually amounts to $108.16 in 1 year. When interest is converted annually, the effective annual rate is the same as the nominal annual rate, but when the interest is converted more than once a year the effective annual rate is greater than the nominal annual rate.

EXERCISES

Using the method of Example 1 above, compare the simple and compound interest for the situations in Exercises 1 to 4.

1 $100 at 3% for 5 years.
2 $250 at 8% for 4 years.
3 $3,800 at 5% for 3 years.
4 $7,500 at 6% for 4 years.

In Exercises 5 to 8, state the conversion period, the total number of conversion periods, and the interest rate per conversion period.

5 7% compounded semiannually for 2 years.
6 8% compounded quarterly for 3 years.
7 6% compounded monthly for 1 year.
8 $6\frac{1}{2}$% compounded semiannually for 4 years.

9 Find the amount of $1,000 invested for 2 years at 6% simple interest; at 6% interest compounded annually; at 6% interest compounded semiannually.
10 What is the effective annual rate in the last part of Exercise 9?
11 Find the amount of $1,234 invested for 2 years at 5% simple interest; at 5% interest compounded annually; at 5% interest compounded semiannually.

In Exercises 12 and 13, give the answer to two decimal places.

12 Find the effective annual rate in the last part of Exercise 11.
13 Find the effective annual rate when the nominal annual rate is 4% compounded semiannually; the nominal annual rate is 6% compounded quarterly.

11.2
FORMULA FOR COMPOUND AMOUNT

Compound interest may always be computed as indicated in the preceding section, but the arithmetic work becomes increasingly tedious as the number of conversion periods becomes larger. In actual practice it is customary to use a formula for the compound amount A at the end of t conversion periods in connection with a table to be explained later.

Let P be the original principal, r the interest rate per conversion period, and t the number of conversion periods.

The interest for the first period will be Pr, and the amount at the end of the *first* period (and also the principal at the beginning of the second period) will be

$$P + Pr = P(1 + r)$$

The interest for the second period will be $P(1 + r)r$, and the amount at the end of the *second* period will be

$$P(1 + r) + P(1 + r)r = P(1 + r)(1 + r) = P(1 + r)^2$$

The interest for the third period will be $P(1 + r)^2r$, and the amount at the end of the *third* period will be

$$P(1 + r)^2 + P(1 + r)^2r = P(1 + r)^2(1 + r) = P(1 + r)^3$$

If we continue this process for t periods, we see that the compound amount at the end of t periods is $P(1 + r)^t$. Therefore,

The compound amount A of a principal P at the end of t conversion periods at an interest rate r per period is given by

$$A = P(1 + r)^t \tag{11.1}$$

It is customary to denote by s the compound amount of a principal of 1 for t periods at the rate r per period. Since $1(1 + r)^t = (1 + r)^t$, we have

$$s = (1 + r)^t \tag{11.2}$$

as a special case of the formula (11.1). Since the compound amount of P is the product of P and $(1 + r)^t$, we call $(1 + r)^t$ in formula (11.1) the **accumulation factor.**

ACCUMULATION FACTOR

Finding the compound amount of 1 is time-consuming, and it is customary to use specially prepared tables for common rates of interest. Such a table appears on page 442. The entries in it are given to six decimal places. Logarithms to the appropriate number of significant figures were used in constructing this table. Logarithms can be used to compute the compound amount of 1 when available tables do not contain entries that apply to the specific problem at hand.

The compound amount A of any principal P is found by multiplying the principal by the compound amount of 1 for the given rate r and time t,

$$A = P(1 + r)^t \tag{11.3}$$

The compound interest I may be found by subtracting the principal from the amount,

$$I = A - P \tag{11.4}$$

For given values of r, t, and A, the corresponding principal, or *present value,* P may be found by using the formula

$$P = \frac{A}{(1 + r)^t} \tag{11.5}$$

which is obtained by solving (11.3) for P.

EXAMPLE 1 With the use of the table on this page, find the compound amount of $1,000 at 4% compounded semiannually for 10 years. What is the compound interest?

SOLUTION Here $P = \$1{,}000$, $r = 0.02$, and $t = 20$. Substituting these values in formula (11.3), we get

$$A = 1{,}000(1 + 0.02)^{20} = 1{,}000(1.02)^{20}$$

From the table we find that

$$(1.02)^{20} = 1.485947$$

Hence we have for the compound amount

$$A = 1{,}000(1.485947) = \$1{,}485.95$$

Amount of 1 at compound interest, $s = (1 + r)^t$

NO OF PERIODS, t	RATE OF INTEREST PER PERIOD, r					
	0.005 (½%)	0.015 (1½%)	0.02 (2%)	0.03 (3%)	0.04 (4%)	0.06 (6%)
1	1.005000	1.015000	1.020000	1.030000	1.040000	1.060000
2	1.010025	1.030225	1.040400	1.060900	1.081600	1.123600
3	1.015075	1.045678	1.061208	1.092727	1.124864	1.191016
4	1.020150	1.061364	1.082432	1.125509	1.169859	1.262477
5	1.025251	1.077284	1.104081	1.159274	1.216653	1.338226
6	1.030378	1.093443	1.126162	1.194052	1.265319	1.418519
7	1.035529	1.109845	1.148686	1.229874	1.315932	1.503630
8	1.040707	1.126493	1.171659	1.266770	1.368569	1.593848
9	1.045911	1.143390	1.195093	1.304773	1.423312	1.689479
10	1.051140	1.160541	1.218994	1.343916	1.480244	1.790848
11	1.056396	1.177949	1.243374	1.384234	1.539454	1.898299
12	1.061678	1.195618	1.268242	1.425761	1.601032	2.012196
15	1.077683	1.250232	1.345868	1.557967	1.800944	2.396558
20	1.104896	1.346855	1.485947	1.806111	2.191123	3.207135
25	1.132796	1.450945	1.640606	2.093778	2.665836	4.291871
50	1.283226	2.105242	2.691588	4.383906	7.106683	18.420154
100	1.646669	4.432046	7.244646	19.218632	50.504948	339.302084

The compound interest I is given by

$$I = A - P = \$1{,}485.95 - \$1{,}000 = \$485.95$$

During recent years there has developed considerable competition among banks and savings and loan institutions for funds from depositors. As a result of this competition some savings institutions now compound interest monthly, and some even compound it daily. Examples 2 and 3 are intended to help the student gain insight into the comparative effects of several methods of compounding interest.

EXAMPLE 2 For a principal of \$10,000 and an interest rate of 6%, find the amount A at the end of 1 year:

(*a*) When the interest is compounded annually.

(*b*) When the interest is compounded semiannually.

(*c*) When the interest is compounded quarterly.

(*d*) When the interest is compounded monthly.

SOLUTION In all these situations $P = \$10{,}000$.

(*a*) Here $t = 1$ and $r = 0.06$, so

$$A = \$10{,}000(1.06) = \$10{,}600.00$$

(*b*) In this case $t = 2$ and $r = 0.03$, so we have

$$A = \$10{,}000(1.03)^2 = \$10{,}000(1.0609) = \$10{,}609.00$$

(*c*) For interest compounded quarterly, $t = 4$ and $r = 0.015$, so

$$A = \$10{,}000(1.015)^4 = \$10{,}000(1.061364) = \$10{,}613.64$$

(*d*) For interest compounded monthly, $t = 12$ and $r = 0.005$, so

$$A = \$10{,}000(1.005)^{12} = \$10{,}000(1.061678) = \$10{,}616.78$$

From the results of Example 2, we see that, for the situations described in the example, the effective annual rates of interest, correct to four decimal places, are as follows:

6% when the interest is compounded annually

6.0900% when the interest is compounded semiannually

6.1364% when the interest is compounded quarterly

6.1678% when the interest is compounded monthly

EXAMPLE 3 If a principal of \$10,000 is invested at an interest rate of 6% compounded daily, find the amount A at the end of 1 year.

SOLUTION Here $t = 365$ (for a non-leap year), $r = \frac{0.06}{365}$, and the compound amount A is given by

$$A = \$10{,}000\left(1 + \frac{0.06}{365}\right)^{365}$$

In this case the brief table on page 442 is of no help to us. However, we may determine A by the use of logarithms. The result, which we exhibit without verification, is that

$(1 + \frac{0.06}{365})^{365} = 1.061831$ correct to six decimal places

Consequently,

$A = \$10{,}000(1.061831) = \$10{,}618.31$

For this situation the effective annual rate of interest is 6.1831%.

From an examination of Examples 2 and 3, we note that the effective annual rate of interest i increases as the number of conversion periods t increases. However, as t continues to increase, the increase in i becomes less rapid. Indeed it can be shown that if we pass from daily conversion ($t = 365$) to hourly conversion ($t = 24 \cdot 365$) for a nominal annual rate of 6%, the effective annual rate i remains 6.183%, correct to three decimal places. By letting the number of conversion periods t become larger and larger, we could conceive of what is referred to as *continuous conversion* of interest. By advanced mathematics it can be shown that if the nominal annual rate of interest of 6% is converted continuously, then the effective annual rate of interest is 6.1837%, correct to four decimal places.

EXAMPLE 4 Find the principal which will amount to \$3,581.70 in 10 years at 6% compounded annually. What is the compound interest?

SOLUTION In this case $A = \$3{,}581.70$, $r = 0.06$, and $t = 10$. On substituting these values in formula (11.5), we have

$$P = \frac{3{,}581.70}{(1.06)^{10}}$$

From the table on page 442, we find that $(1.06)^{10} = 1.790848$. So the principal is

$$P = \frac{3{,}581.70}{1.790848} = \$2{,}000$$

For the compound interest, we have

$I = A - P = \$3{,}581.70 - \$2{,}000 = \$1{,}581.70$

EXERCISES

In Exercises 1 to 12, find the compound amount and the compound interest.

	PRINCIPAL	INTEREST RATE	NO. OF YEARS AT INTEREST
1	\$1,000	6% compounded annually	5
2	1,000	6% compounded annually	10

	PRINCIPAL	INTEREST RATE	NO. OF YEARS AT INTEREST
3	1,000	6% compounded annually	20
4	1,000	6% compounded annually	50
5	1,000	6% compounded every 4 months	5
6	1,000	8% compounded semiannually	5
7	1,000	8% compounded quarterly	5
8	1,875	4% compounded annually	5
9	1,875	4% compounded semiannually	5
10	963	8% compounded semiannually	25
11	963	8% compounded quarterly	25
12	267	16% compounded quarterly	5

13 A building and loan association pays interest on deposits at the rate of 4% compounded annually. Find the amount of a deposit of $5,000 at the end of 15 years.

14 At the birth of a son the father invests $2,000 at 4% compounded semiannually, to accumulate until the boy's twenty-fifth birthday. What will be the amount at that time?

15 A invested $1,500 at 4% simple interest for 25 years. B invested $1,500 at 4% compounded semiannually for 25 years. At the end of this period, which has the greater amount, and how much greater is it?

16 A invested $1,000 at 5% compounded annually for 15 years. At what rate would B have to invest $1,000 at simple interest in order to have the same amount as A at the end of the 15 years?

17 A dividend of $56.78 is left with the Good Life Insurance Company to accumulate at 4% compounded semiannually. How much will it amount to in 10 years?

18 The First Savings Bank pays interest on savings accounts at 4% compounded annually, the Second Savings Bank pays 4% compounded semiannually. What will be the difference in the amounts in accounts at these banks at the end of 10 years after $1,234.56 is deposited in each bank?

19 An acreage of timber has been found to increase in value each year at the rate of 5% of the value in the preceding year. If its value at the beginning of the period is $50,000, what will be its value at the end of 5 years? 10 years? 15 years? 50 years?

20 What sum invested now at 4% compounded semiannually will amount to $5,000 ten years from now?

21 How much will $2,000 amount to in 12 years if it is compounded semiannually at 5% per year?

22 How much will $1,870 amount to in 18 years if it is compounded annually at the rate of 4% per year?

23 How long (to the nearest quarterly period) will it take for $2,400 to amount to $4,000 if interest is 4% compounded quarterly?

24 What amount should be deposited now at 3% compounded annually to amount to \$1,000 in 14 years?

25 For a principal of \$10,000 and a nominal annual rate of 8%, find the compound amount at the end of 1 year (a) when the interest is compounded annually; (b) when the interest is compounded semiannually; (c) when the interest is compounded quarterly.

26 Determine the effective annual rate of interest, correct to four decimal places, for a nominal annual interest rate of 8% when (a) the interest is compounded annually; (b) the interest is compounded semiannually; (c) the interest is compounded quarterly.

11.3 ARITHMETIC PROGRESSIONS

ARITHMETIC PROGRESSION

An *arithmetic progression* is an ordered set of numbers each of which, after the first, is obtained by adding a fixed number to the preceding number of the set. The numbers which are members of the set are called *terms* of the progression, and the fixed number which is added to each term to obtain the following term is called the *common difference*.

TERMS

COMMON DIFFERENCE

To illustrate,

$$5, 8, 11, 14, 17, 20, 23$$

are the terms of an arithmetic progression for which the first term is 5, the common difference is 3, the number of terms is 7, and the last (or seventh) term is 23;

$$10, 8, 6, 4, 2, 0, -2, -4, -6$$

are the terms of an arithmetic progression for which the first term is 10, the common difference is -2, the number of terms is 9, and the last (or ninth) term is -6.

Note that the common difference d for an arithmetic progression may be found by subtracting any term of the progression from the term which follows it. In the first illustration above, $d = 8 - 5 = 3$.

Consider the arithmetic progression for which a denotes the first term and d the common difference. We may then write the first n terms of this progression in the form

$$a, a + d, a + 2d, a + 3d, \dots, a + (n - 1)d$$

Observe that for each term the number of d's added to a is 1 less than the number of the term. If l stands for the nth term, then we have

$$l = a + (n - 1)d \tag{11.6}$$

Associated with an arithmetic progression of n terms are five quantities of special importance. They are

a, the first term

d, the common difference

n, the number of terms

l, the nth term

S, the sum of the first n terms

We shall see that these five quantities are connected by three fundamental equalities (any one of which may be derived from the other two). One of these is the equality (11.6) above, which is a formula for l in terms of a, n, and d.

To obtain a formula for S in terms of a, n, and l, we first write the sum of the terms in the regular order. We have

$$S = a + (a + d) + (a + 2d) + \cdots + (l - 2d) + (l - d) + l \tag{11.7}$$

Secondly, writing the sum of the terms of the progression in reverse order, we get

$$S = l + (l - d) + (l - 2d) + \cdots + (a + 2d) + (a + d) + a \tag{11.8}$$

Adding (11.7) and (11.8) and combining corresponding terms so that the d's will be eliminated, we obtain

$$2S = (a + l) + (a + l) + (a + l) + \cdots + (a + l) + (a + l) + (a + l)$$

Note that there are n terms of the form $a + l$ in the latter sum. Therefore, we can write $2S = n(a + l)$, or

$$S = \frac{n}{2}(a + l) \tag{11.9}$$

If we substitute in (11.9) the expression for l given by (11.6), we get another formula for S,

$$S = \frac{n}{2}\{a + [a + (n - 1)d]\}$$

or

$$S = \frac{n}{2}[2a + (n - 1)d] \tag{11.10}$$

When any three of the quantities a, d, n, l, and S are given, the other two may be found by use of some two of the equalities (11.6), (11.9), and (11.10).

EXAMPLE 1 Find the tenth term and the sum of the first 10 terms of the arithmetic progression whose first 4 terms are 25, 21, 17, 13.

SOLUTION For the given arithmetic progression, $a = 25$ and $d = 21 - 25 = -4$. From the formula

$$l = a + (n - 1)d$$

we get for the tenth term

$$l = 25 + (10 - 1)(-4) = -11$$

From the formula

$$S = \frac{n}{2}(a + l)$$

we get for the sum of the first 10 terms

$$S = \frac{10}{2}(25 - 11) = 70$$

EXAMPLE 2 Find the common difference d and the number of terms n in an arithmetic progression whose first term is -11, whose last term is 103, and whose sum is 1,794.

SOLUTION We are given that $a = -11$, $l = 103$, and $S = 1{,}794$. Substituting these values in the formula

$$S = \frac{n}{2}(a + l)$$

we get

$$1{,}794 = \frac{n}{2}(-11 + 103) \qquad \text{or} \qquad 1{,}794 = \frac{n}{2} \cdot 92 \qquad \text{or} \qquad 1{,}794 = 46n$$

Therefore, $n = \frac{1{,}794}{46}$; so $n = 39$.

On substituting the values $a = -11$, $n = 39$, and $l = 103$ in the formula

$$l = a + (n - 1)d$$

we obtain $103 = -11 + 38d$ or $38d = 114$; so $d = 3$.

EXAMPLE 3 A man borrowed $360 from a loan company with the agreement that he would repay the loan (principal and interest) in 12 monthly payments of $40.80, $39.90, $39.00, $38.10, and so forth.

(a) Find the total amount he has to pay the loan company.

(b) He could have borrowed the $360 from another source and repaid the entire amount 1 year later with 6% simple interest. What is the total amount he would have had to pay the lender at the end of a year?

SOLUTION (a) The monthly payments form an arithmetic progression for which $a = \$40.80$, $d = -\$0.90$, and $n = 12$. By using the formula,

$$S = \frac{n}{2}[2a + (n - 1)d]$$

$$S = \frac{12}{2}[81.60 + 11(-0.90)] = 6(81.60 - 9.90) = 6(71.70) = \$430.20$$

(b) Recall from Sec. 4.5 the formula for amount A at simple interest,

$$A = P(1 + rt)$$

Here $P = \$360$, $r = 0.06$, and $t = 1$; therefore

$$A = 360(1 + 0.06) = 360(1.06) = \$381.60$$

So $381.60 in principal and interest must be paid the lender at the end of the year.

EXERCISES

In Exercises 1 to 6, determine whether or not the given set of numbers constitute the first three terms of an arithmetic progression. If the given set is the first three terms in an arithmetic progression, give the common difference d, and find the next two terms.

1 10, 16, 22
2 7, 10, 14
3 2, 8, 32
4 $-3, -6, -9$
5 $a - d, a, a + d$
6 $4k, k, -3k$

7 Find the twelfth term and the sum of the first 12 terms of the arithmetic progression for which the first four terms are 2, 5, 8, 11.

8 Find the twenty-sixth term and the sum of the first 26 terms of the arithmetic progression for which the first four terms are 14, $12\frac{1}{2}$, 11, $9\frac{1}{2}$.

9 The number -73 is what term in the arithmetic progression whose first four terms are 8, 5, 2, -1?

10 If in an arithmetic progression $a = 7$, $d = \frac{7}{2}$, $l = 42$, find n and S.

11 Find the sum of the integers from 1 to 100 inclusive.

12 Find the sum of the integers from 1 to 1,000 inclusive.

13 Find the sum of the even integers from 2 to 100 inclusive.

14 Find the sum of the odd integers from 1 to 99 inclusive.

15 Show that the sum of the first n positive odd integers is n^2.

16 Show that the sum of the first n positive even integers is $n(n + 1)$.

17 Show that the sum of the first n positive integers is $\frac{n}{2}(n + 1)$.

18 A child saves 5 cents the first week, 10 cents the second week, 15 cents the third week, and so forth. At this rate in how many weeks will he save $66.30.

19 A man bought a house priced at $10,000. He paid $2,000 cash and agreed to pay $500 at the end of each year to reduce the principal, plus 6% interest on the unpaid balance at the beginning of that year. How long will it take him to pay for the house? What is the total amount of interest paid?

11.4 GEOMETRIC PROGRESSIONS

GEOMETRIC PROGRESSION

A *geometric progression* is an ordered set of numbers each of which, after the first, is obtained by multiplying the preceding number by a fixed number. The numbers which are members of the set are called *terms* of the geometric progression, and the fixed number by which each term is multiplied to obtain the following term is called the *common ratio*.

TERMS

COMMON RATIO

To illustrate,

3, 6, 12, 24, 48

are the terms of a geometric progression for which the first term is 3, the common ratio is 2, the number of terms is 5, and the last (or fifth) term is 48; as another example,

$36, -18, 9, -\frac{9}{2}, \frac{9}{4}, -\frac{9}{8}$

are the terms of a geometric progression for which the first term is 36, the common ratio is $-\frac{1}{2}$, the number of terms is 6, and the last (or sixth) term is $-\frac{9}{8}$.

Observe that the common ratio r of a geometric progression may be found by dividing any term of the progression by the term which precedes it. In the second illustration above, $r = -18 \div 36 = -\frac{1}{2}$.

Consider the geometric progression for which a signifies the first term and r the common ratio. The first n terms of this progression may be written in the form

$$a, ar, ar^2, ar^3, \ldots, ar^{n-1}$$

Note that for each term, the number of times r occurs as a factor is 1 less than the number of the term. If we let l represent the nth term, we have

$$l = ar^{n-1} \tag{11.11}$$

Associated with a geometric progression of n terms are five basic quantities. They are

a, the first term

r, the common ratio

n, the number of terms

l, the nth term

S, the sum of the first n terms

We shall see that these five quantities are connected by three fundamental equalities (any one of which may be derived from the other two). One of these is the equality (11.11) above, which is a formula for l in terms of a, r, and n.

To obtain a formula for S in terms of a, n, and r, we write

$$S = a + ar + ar^2 + \cdots + ar^{n-2} + ar^{n-1} \tag{11.12}$$

Multiplying both sides of equation (11.12) by r, we obtain

$$rS = ar + ar^2 + ar^3 + \cdots + ar^{n-1} + ar^n \tag{11.13}$$

Subtracting (11.13) from (11.12), we get

$$S - rS = a - ar^n \qquad \text{or} \qquad S(1 - r) = a(1 - r^n)$$

If we assume that $r \neq 1$, we can write

$$S = a \cdot \frac{1 - r^n}{1 - r} \qquad \text{or} \qquad S = a \cdot \frac{r^n - 1}{r - 1} \tag{11.14}$$

For the special case in which $r = 1$, we obtain $S = na$ directly from (11.12).

We may write (11.14) in the form

$$S = \frac{a - ar^n}{1 - r}$$

Now, from (11.11), $ar^n = rl$. Therefore we may write

$$S = \frac{a - rl}{1 - r} \tag{11.15}$$

which is a formula for S in terms of a, r, and l.

If the values of some three of the quantities a, r, n, l, and S for a geometric progression are given, the remaining two of these quantities may be found by use of some two of the three equalities (11.11), (11.14), and (11.15).

EXAMPLE 1 Find the twelfth term and the sum of the first 12 terms of the geometric progression for which $\frac{5}{2}$, 5, 10, 20 are the first four terms.

SOLUTION For the given progression $a = \frac{5}{2}$ and $r = 5 \div \frac{5}{2} = 2$. Substituting these values and the value $n = 12$ in

$$l = ar^{n-1}$$

we get

$$l = \tfrac{5}{2} \cdot 2^{12-1} = \tfrac{5}{2} \cdot 2^{11} = 5 \cdot 2^{10} = 5 \cdot 1{,}024 = 5{,}120$$

Substituting the appropriate values in

$$S = a \cdot \frac{1 - r^n}{1 - r}$$

we get

$$S = \frac{5}{2} \cdot \frac{1 - 2^{12}}{1 - 2} = \frac{5}{2} \cdot \frac{1 - 4{,}096}{-1} = \frac{5}{2} \cdot 4{,}095 = 10{,}237\tfrac{1}{2}$$

EXAMPLE 2 For a geometric progression $a = 162$, $l = \frac{2}{9}$, and $S = 242\frac{8}{9}$. Find r and n.

SOLUTION Substituting the given values of a, l, and S in $S = \dfrac{a - rl}{1 - r}$, we get

$$242\tfrac{8}{9} = \frac{162 - r \cdot \frac{2}{9}}{1 - r} \quad \text{or} \quad 242\tfrac{8}{9}(1 - r) = 162 - \frac{2}{9}r$$

$$\frac{2{,}186}{9}(1 - r) = 162 - \frac{2}{9}r \quad \text{or} \quad 2{,}186(1 - r) = 1{,}458 - 2r$$

$$2{,}186 - 2{,}186r = 1{,}458 - 2r \quad \text{or} \quad 728 = 2{,}184r$$

Therefore, $r = \dfrac{728}{2{,}184}$, or $r = \dfrac{1}{3}$.

To find n, we put the values $a = 162$, $r = \frac{1}{3}$, and $l = \frac{2}{9}$ in $l = ar^{n-1}$ and get

$$\frac{2}{9} = 162\left(\frac{1}{3}\right)^{n-1} \quad \text{or} \quad \left(\frac{1}{3}\right)^{n-1} = \frac{1}{729}$$

Recall that $3^6 = 729$. Consequently,

$$\left(\frac{1}{3}\right)^6 = \frac{1}{729}$$

So

$$n - 1 = 6 \quad \text{and} \quad n = 7$$

In summary, for a geometric progression in which $a = 162$, $l = \frac{2}{9}$, and $S = 242\frac{8}{9}$, the common ratio r has the value $\frac{1}{3}$, and there are seven terms.

EXAMPLE 3 A man deposits \$100 in a savings bank at the end of each year for 5 years, and these deposits draw interest at 4% compounded annually. How much is in his account immediately after he has made the deposit of \$100 at the end of the fifth year?

SOLUTION We may represent the deposits on the following line diagram:

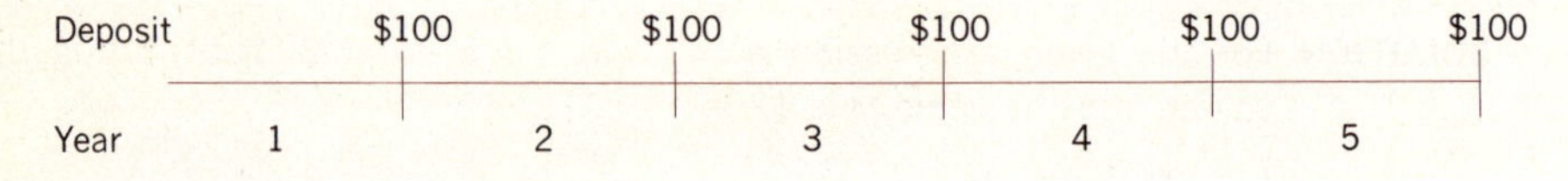

Using the formula for compound amount $A = P(1 + r)^t$, we find that at the end of the fifth year:

The first deposit of \$100 will amount to $\$100(1.04)^4$.

The second deposit of \$100 will amount to $\$100(1.04)^3$.

The third deposit of \$100 will amount to $\$100(1.04)^2$.

The fourth deposit of \$100 will amount to \$100(1.04).

The fifth deposit of \$100 will amount to \$100.

We could find the value of each of these amounts by use of the table on page 442 and add them to get the desired result. However, note that these amounts are the terms in a geometric progression for which

$a = 100 \qquad r = 1.04 \qquad n = 5$

Let S be the sum of the above amounts, which is the same as the sum of the geometric progression having the stated values of a, r, and n. Using the formula

$$S = a \cdot \frac{r^n - 1}{r - 1}$$

we get

$$S = 100 \cdot \frac{(1.04)^5 - 1}{1.04 - 1} = 100 \cdot \frac{(1.04)^5 - 1}{0.04}$$

$$= 100 \cdot \frac{1.216653 - 1}{0.04} = \$541.63$$

In the above calculation we use the above-mentioned table to get the value of $(1.04)^5$.

EXERCISES

In Exercises 1 to 6, determine whether or not the given set of numbers constitutes the first three terms in a geometric progression. If the given set is the first three terms in a geometric progression, give the common ratio r, and find the next two terms.

1 6, 12, 24
2 20, 15, $\frac{45}{4}$
3 8, 16, 24
4 10, 5, 1
5 -1, 15, -30
6 100, 10, 1

7 Find the eleventh term and the sum of the first 11 terms of the geometric progression for which the first four terms are 2, 6, 18, 54.

8 Find the tenth term and the sum of the first 10 terms of the geometric progression for which the first four terms are 3, -2, $\frac{4}{3}$, $-\frac{8}{9}$.

9 The number $\frac{243}{16}$ is what term in the geometric progression whose first five terms are $\frac{4}{3}$, 2, 3, $\frac{9}{2}$, $\frac{27}{4}$?

10 If in a geometric progression, $a = 405$, $r = \frac{2}{3}$, $l = 80$, find n and S.

11 If in a geometric progression, $a = 1$, $r = \frac{1}{2}$, $l = \frac{1}{128}$, find n and S.

In Exercises 12 to 14, find an expression for the sum of the n terms.

12 $1.03, (1.03)^2, (1.03)^3, \ldots, (1.03)^n$

13 $1.04, (1.04)^2, (1.04)^3, \ldots, (1.04)^n$

14 $1 + r, (1 + r)^2, (1 + r)^3, \ldots, (1 + r)^n$

15 A man deposits \$250 in a savings bank at the end of each year for 10 years, and these deposits draw interest at 4% compounded annually. How much is in his account immediately after he has made the deposit of \$250 at the end of the tenth year?

16 If for a geometric progression, $S = 171$, $r = -2$, and $l = 256$, find a and n.

17 If for a geometric progression, $a = 5$, $r = 2$, and $S = 315$, find n and l.

11.5 ANNUITIES

ANNUITY

An *annuity* is a set of equal payments of P dollars made so that there are equal intervals of time between the payments. A set of equal periodic payments on a debt and the periodic investment of equal amounts at a stated rate of interest are illustrations of annuities.

PAYMENT PERIOD OF AN ANNUITY

The **payment period** *of an annuity* is an interval of time equal to the time between two consecutive payments. We shall limit our consideration here to annuities in which the payment period is the same as the conversion period (see Sec. 11.1) of the interest rate. The number of payment periods of an annuity will be symbolized by t.

Also, we shall restrict our attention to annuities in which the periodic payments are made at the end of equal time intervals (these are called *ordinary* annuities). Note that for such an annuity the first payment is made at the end of the first payment period.

The term of an annuity is the time from the beginning of the first payment period to the end of the last payment period. Observe that for ordinary annuities the term begins one payment period before the first payment and ends with the last payment.

AMOUNT OF AN ANNUITY

The **amount A_t of an annuity** is the sum to which all the payments accumulate at the end of the term of the annuity, that is, at the end of t payment periods.

PRESENT VALUE OF AN ANNUITY

The **present value P_t of an annuity** is the sum of the present values (taken at the beginning of the term) of all the payments of the annuity, which is the same as the present value of the amount A_t of the annuity.

EXAMPLE 1 Find the amount and the present value of an annuity in which $400 is invested annually for 6 years at 4% interest compounded annually.

SOLUTION Here the term is 6 years. We represent the annuity on the following line diagram:

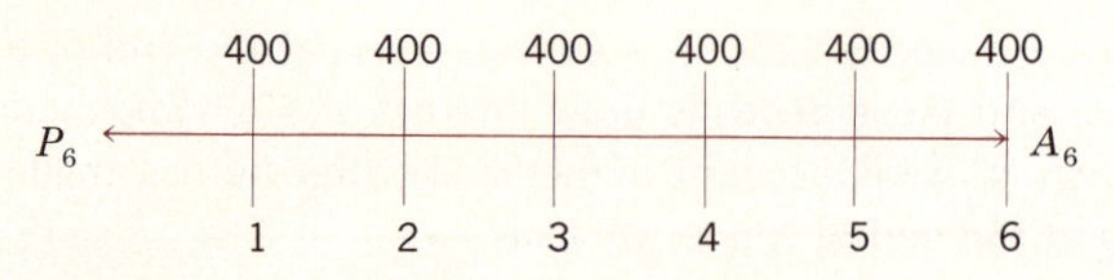

PAYMENT	AMOUNT OF PAYMENT AT END OF TERM	PRESENT VALUE OF PAYMENT AT BEGINNING OF TERM*
1	$400(1.04)^5$	$400(1.04)^{-1}$
2	$400(1.04)^4$	$400(1.04)^{-2}$
3	$400(1.04)^3$	$400(1.04)^{-3}$
4	$400(1.04)^2$	$400(1.04)^{-4}$
5	$400(1.04)$	$400(1.04)^{-5}$
6	400	$400(1.04)^{-6}$

*The values in this column are obtained by use of (11.5).

The amount A_6 of the annuity is given by

$$A_6 = 400 + 400(1.04) + 400(1.04)^2 + 400(1.04)^3 + 400(1.04)^4 + 400(1.04)^5$$

Note that this is the sum of a geometric progression for which $a = 400$, $r = 1.04$, and $n = 6$. Using the formula

$$S = a \cdot \frac{r^n - 1}{r - 1}$$

we get

$$A_6 = 400 \cdot \frac{(1.04)^6 - 1}{1.04 - 1} = 400 \cdot \frac{1.265319 - 1}{0.04} \quad \text{by use of the table on page 442}$$

$$= 400 \cdot \frac{0.265319}{0.04} = 10{,}000(0.265319) = \$2{,}653.19$$

or

$$A_6 = \$2{,}653.19 \tag{1}$$

The present value P_6 of the annuity is given by

$$P_6 = 400(1.04)^{-1} + 400(1.04)^{-2} + 400(1.04)^{-3} + 400(1.04)^{-4} + 400(1.04)^{-5} + 400(1.04)^{-6}$$

Note that this is the sum of a geometric progression for which $a = 400(1.04)^{-1}$, $r = (1.04)^{-1}$, and $n = 6$. Substituting these values in the formula

$$S = a \cdot \frac{1 - r^n}{1 - r}$$

we get

$$P_6 = 400(1.04)^{-1}\left[\frac{1 - (1.04)^{-6}}{1 - (1.04)^{-1}}\right] = \frac{400}{1.04}\left[\frac{1 - \frac{1}{(1.04)^6}}{1 - \frac{1}{1.04}}\right]$$

or

$$P_6 = \left[\frac{400}{1.04} \cdot \frac{(1.04)^6 - 1}{1.04 - 1}\right]\frac{1}{(1.04)^5} = \left[400 \cdot \frac{(1.04)^6 - 1}{1.04 - 1}\right]\frac{1}{(1.04)^6}$$

or

$$P_6 = A_6 \cdot \frac{1}{(1.04)^6} = A_6(1.04)^{-6} \tag{2}$$

Using the value of A_6 given by (1) and the value of $(1.04)^6$ given by the table on page 442, we get

$$P_6 = \frac{2{,}653.19}{1.265319} = \$2{,}096.85$$

If $2,096.85 were invested at 4% interest compounded annually, it would yield $2,653.19 at the end of a period of 6 years.

It is also true that if $2,096.85 were put in a savings account with interest at 4% compounded annually, one could withdraw from that account $400 at the end of each year for the next 6 years, as shown in the following table:

PERIOD	BALANCE AT BEGINNING OF PERIOD	INTEREST EARNED DURING PERIOD	BALANCE AT END OF PERIOD BEFORE WITHDRAWAL	WITHDRAWAL AT END OF PERIOD
1	$2,096.85	$83.87	$2,180.72	$400
2	1,780.72	71.23	1,851.95	400
3	1,451.95	58.08	1,510.03	400
4	1,110.03	44.40	1,154.43	400
5	754.43	30.18	784.61	400
6	384.61	15.39*	400.00	400

*In constructing such a table a small discrepancy may occur in the final figure, but this should never amount to very much. This discrepancy may arise because the individual amounts of interest, as well as the initial balance, are rounded off to the nearest cent. It is usual to correct this discrepancy in the last interest computation. For example, in the illustration above the interest during the last period is

$(0.04)(384.61) = 15.3844$

and we arbitrarily make this $15.39 since we see that this is the amount needed.

Now consider the general case of an annuity in which the periodic payment of D dollars is made at the end of each period for t periods with the compound interest rate r per period. We represent the annuity on the following line diagram:

$$P_t \longleftarrow \overset{D}{\underset{1}{+}} \text{---} \overset{D}{\underset{2}{+}} \text{---} \overset{D}{\underset{3}{+}} \cdots \overset{D}{\underset{t-1}{+}} \longrightarrow \overset{D}{\underset{t}{|}} \quad A_t$$

The first payment of D dollars will earn interest for $t - 1$ periods and will amount to $D(1 + r)^{t-1}$ at the end of the term. The second payment will earn interest for $t - 2$ periods, and the third for $t - 3$ periods. The next to the last payment will earn interest for 1 period, and the last payment will earn no interest. The amount A_t of the annuity is the sum of the accumulated amounts of all the payments to the end of the term. Therefore,

$$A_t = D + D(1 + r) + D(1 + r)^2 + \cdots + D(1 + r)^{t-2} + D(1 + r)^{t-1}$$

This is the sum of a geometric progression for which D is the first term, $1 + r$ is the common ratio, and there are t terms.

Substituting these values in the familiar formula for the sum of a geometric progression, we get

$$A_t = D \cdot \frac{(1 + r)^t - 1}{1 + r - 1} \quad \text{or} \quad A_t = D \cdot \frac{(1 + r)^t - 1}{r}$$

It should be clear that, when the periodic payment is 1, the amount of the annuity is

$$\frac{(1 + r)^t - 1}{r}$$

AMOUNT OF AN ANNUITY OF 1 PER PERIOD

This quantity we call the **amount of an annuity of 1 per period,** and we denote it by $a_{\overline{t}|r}$, that is,

$$a_{\overline{t}|r} = \frac{(1 + r)^t - 1}{r} \tag{11.16}$$

Hence we have

$$A_t = D \cdot a_{\overline{t}|r} \tag{11.17}$$

The following table gives the values of the amount of an annuity of 1 per period for selected values of r and t:

Amount of an annuity of 1 per period: $a_{\overline{t}|r} = \frac{(1 + r)^t - 1}{r}$

NO. OF PERIODS, t	RATE OF INTEREST PER PERIOD, r			
	0.02 (2%)	0.04 (4%)	0.05 (5%)	0.06 (6%)
1	1.000000	1.000000	1.000000	1.000000
2	2.020000	2.040000	2.050000	2.060000
3	3.060400	3.121600	3.152500	3.183600
4	4.121608	4.246464	4.310125	4.374616
5	5.204040	5.416323	5.525631	5.637093
6	6.308121	6.632975	6.801913	6.975319
7	7.434283	7.898294	8.142008	8.393838
8	8.582969	9.214226	9.549109	9.897468
9	9.754628	10.582795	11.026564	11.491316
10	10.949721	12.006107	12.577893	13.180795
15	17.293417	20.023588	21.578564	23.275970
20	24.297370	29.778079	33.065954	36.785591
25	32.030300	41.645908	47.727099	54.864512
50	84.579401	152.667084	209.347996	290.335904
100	312.232306	1,237.623705	2,610.025157	5,638.368059

For example, from this table we find that

$$a_{\overline{15}|0.05} = 21.578564 \qquad \text{or} \qquad a_{\overline{15}|0.05} = \$21.58$$

to the nearest cent which means that $1 invested at the end of each year for 15 years with interest at 5% compounded annually would accumulate at the end of the term of 15 years to $21.58 (to the nearest cent).

The entries in the preceding table are given to six decimal places. Logarithms to the appropriate number of significant figures were used in constructing the table.

EXAMPLE 2 With the use of the preceding table, find the amount of an annuity of $100 a year for 20 years if the interest rate is 4% compounded annually.

SOLUTION Here $D = \$100$, $r = 0.04$, and $t = 20$. Substituting these values in formula (11.17), we get

$$A_{20} = 100a_{\overline{20}|0.04}$$

From the table on page 457, we find that

$$a_{\overline{20}|0.04} = 29.778079$$

Therefore $A_{20} = 100(29.778079) = \$2{,}977.81$.

This result means that $100 invested at the end of each year for 20 years would amount to $2,977.81 at the end of the 20-year period if the interest rate were 4% compounded annually. The 20 payments total $20 \cdot \$100 = \$2{,}000$. The difference $\$2{,}977.81 - \$2{,}000.00 = \$977.81$ is the total compound interest on the 20 annual payments to the end of the term.

EXAMPLE 3 What equal payment should be made at the end of each year for 10 years to accumulate to $4,000 at the end of 10 years at the interest rate of 4% compounded annually?

SOLUTION In this case $t = 10$, $r = 0.04$, and $A_{10} = \$4{,}000$. Using formula (11.17), we have

$$4{,}000 = D \cdot a_{\overline{10}|0.04}$$

So

$$D = \frac{4{,}000}{a_{\overline{10}|0.04}}$$

From the table on page 457, we find that $a_{\overline{10}|0.04} = 12.006107$. Hence

$$D = \frac{4{,}000}{12.006107} \qquad \text{or} \qquad D = \$333.16$$

If the amount A_t of an annuity is known or can be found, then the present value P_t of the annuity can be calculated by use of the equality

$$P_t = A_t(1 + r)^{-t} \qquad \text{or} \qquad P_t = \frac{A_t}{(1 + r)^t} \tag{11.18}$$

where the value of $(1 + r)^t$ is found from the table on page 442.

EXAMPLE 4 Find the amount and the present value of an annuity of $1,000 a year for 6 years if the interest rate is 4% compounded annually.

SOLUTION Here $t = 6$, $r = 0.04$, and $D = \$1{,}000$. Substituting these values in formula (11.17), we have

$$A_6 = 1{,}000 \cdot a_{\overline{6}|0.04} = 1{,}000(6.632975) = \$6{,}632.98$$

Then by use of formula (11.18) we get

$$P_6 = \frac{A_6}{(1.04)^6} = \frac{6{,}632.98}{1.265319} \qquad \text{(using the table on page 442)}$$

So $P_6 = \$5{,}242.14$.

Note: As the student works a problem on annuities he should carefully interpret each result and then relate these results to the given data (and to each other, if possible). To illustrate, from the results of Example 4 we conclude that if money is worth 4% compounded annually, then $5,242.14 is equivalent to $6,632.98 six years hence; also $5,242.14 is equivalent to $1,000 at the end of each year for the next 6 years.

EXERCISES

In Exercises 1 to 8, compute the amount and present value of the annuity.

1 $500 payable annually for 20 years, interest at 2% compounded annually.

2 $500 payable semiannually for 25 years, interest at 4% compounded semiannually.

3 $500 payable annually for 20 years, interest at 4% compounded annually.

4 $250 payable every 4 months for 5 years, interest at 6% compounded every 4 months.

5 $875 payable quarterly for 25 years, interest at 8% compounded quarterly.

6 $10 payable annually for 50 years, interest at 5% compounded annually.

7 $100 payable semiannually for 25 years, interest at 4% compounded semiannually.

8 $845 payable annually for 8 years, interest at 5% compounded annually.

9 Verify the correctness of your answer in Exercise 8 for the present value of the annuity, by constructing a table like that on page 455.
10 What is the value of the equal annual payment P which \$1,200 invested now at 4% compounded annually will yield at the end of each year for the next 5 years?
11 Verify the correctness of your answer in Exercise 10 for the payment P, by constructing a table like that on page 456.
12 A man purchases a house by making a cash payment of \$2,500 and agreeing to pay \$500 at the end of each year for the next 20 years. If money is worth 6% compounded annually, what is the equivalent cash price of the house?

In Exercises 13 to 16, find the periodic payment of the annuity.

13 $A_t = \$5{,}000$, 8 annual payments at the end of each year, interest at 4% compounded annually.
14 $A_t = \$5{,}000$, 20 semiannual payments at the end of each 6 months, interest at 4% compounded semiannually.
15 $A_t = \$5{,}000$, 8 annual payments at the end of each year, interest at 6% compounded annually.
16 $P_t = \$4{,}000$, 20 annual payments at the end of each year, interest at 4% compounded annually.

17 The parents desire to provide an educational fund of \$6,000 for their son at a date 15 years from now. How much money should be invested at the end of each year at 4% compounded annually to provide for the fund?

12
Probability

12.1 INTRODUCTION

Aside from the concept of number, the mathematical concept that occurs most often in human activity is "probably" the concept of *probability*. We truly live surrounded by probabilities rather than certainties.

In predictions about an outcome that involves the action and reaction of human personalities, we speak in terms of "most likely" or "most probable" and *not* in terms of 100% certainty. A store owner must decide which of the possible items that he could stock are the most likely to be attractive to his customers. A person in deciding whether to announce as a candidate for public office (and prospective contributors to his campaign) will want to estimate the probability that he will win. When Mickey Mantle came to bat in a baseball game, the probability was better than 30% that he would get a hit.

We also encounter probability in situations that cannot be affected by human actions. There is a strong probability that the sun will rise tomorrow. We consider a coin to be "fair" if, when it is tossed, the probability of its coming up heads is one-half. The daily weather forecast lists the probability of rain or snow.

Historically, the study of probability grew out of questions about games of chance and gambling, and in the minds of the general public it is still closely associated with these activities. Witness such questions as: What is the probability of throwing a seven in a game of dice? What are the odds on the Super Bowl Game? What is the probability of being dealt a bridge hand with 13 spades?

In general, the use of the concept of probability in situations such as those in the preceding paragraphs relates to the degree of confidence one has that a particular event will occur (what is the probability that X will be elected?) or to what may be expected to happen in repeated trials (if Mickey Mantle repeatedly comes to bat he will get a hit about 310 or so times out of 1,000). It is the purpose of the theory of probability to give precise meaning to the concept of probability in terms of numbers and to provide systematic ways of determining these numbers. This enables us to deal more effectively with the uncertainties involved in many decisions which we face. It is the purpose of this chapter to consider some of the basic ideas and procedures in the theory and application of probability.

12.2 INTUITIVE CONCEPTS OF PROBABILITY

We want the definitions which we make in order to work systematically with probability to fit in as much as possible with our intuitive ideas. Let us examine some simple instances in which we have an intuitive feeling about what we want probability to mean.

When we toss a coin once, we assume that it must fall with either heads or tails upward. If it is an ordinary coin, we expect that it is just as likely that heads will turn up as it is that tails will turn up. If we toss a coin once, what is the probability that heads will turn up? Since there are *two* ways the coin can fall but only *one* way heads can occur, we say that there is one chance out of two that we will observe heads, and the probability that heads will come up is $\frac{1}{2}$.

Suppose we toss an ordinary cubical die (one member of a pair of dice) with faces numbered 1, 2, 3, 4, 5, 6, and observe the number that appears on the upper face. There are six possible outcomes, and if the die is not "loaded," we expect that any one of the numbers is just as likely to turn up as any other of the six numbers. If we toss a single die once, what is the probability that the number 3 will appear on the top face? Since there are six possible ways the die can fall and only one way in which a 3 can appear on the top face, we say that there is one chance out of six that we will observe a 3 on the top face, and the probability of throwing a 3 is $\frac{1}{6}$.

If we call tossing the single coin an *experiment* and the turning-up of heads an *event,* we have an experiment with two possible outcomes and an event that can occur in only one way. Assuming that each of the outcomes of the experiment is equally likely, we said that the probability of the event is $\frac{1}{2}$ (the number of ways the event can occur divided by the number of possible outcomes of the experiment).

If we call the tossing of the single die an *experiment* and the appearance of a 3 on the top face an *event,* we have an experiment with six possible outcomes and an event that can occur in only one way. Again assuming that each of the outcomes of the experiment is equally likely, we said that the probability of the event is $\frac{1}{6}$ (the number of ways the event can occur divided by the number of possible outcomes of the experiment).

For the experiment of tossing a die, let us consider another event, say the event that the number on the top face of the die is an even number. What shall we say is the probability that an even number will appear on the top face? Here the experiment still has six possible outcomes, each assumed to be equally likely, but the event can now occur in three ways: a 2 or a 4 or a 6 on the top face. We shall say that the probability of an even number appearing is $\frac{3}{6} = \frac{1}{2}$ (the number of ways the event can occur divided by the number of possible outcomes of the experiment).

We summarize the preceding examples by saying that if we perform an experiment each of whose outcomes is equally likely, then the probability that a given event will occur on a single trial of the experiment is given by

$$\text{probability of an event} = \frac{\text{number of ways the event can occur}}{\text{number of possible outcomes of the experiment}} \tag{12.1}$$

In the following sections we shall examine the concepts of experiment, trial, event, occurrence of an event, and probability of an event with more precision and in more detail.

EXAMPLE Suppose that three coins are tossed once and we record for each coin whether it is heads or tails. What is the probability of each of the following?

(a) There are exactly two tails showing.

(b) There is at least one tail showing.

SOLUTION First we determine the possible outcomes of the experiment of tossing three coins and recording for each coin whether it is heads or tails. One possibility is that the first coin is heads, the second coin is heads, and the third coin is heads; this outcome is recorded as HHH. Another possibility is that the first coin is heads, the second heads, and the third tails; HHT. By continuing this type of analysis, we can find that the possible outcomes are HHH, HHT, HTH, HTT, THH, THT, TTH, TTT. One way to obtain a listing of possible outcomes that will make certain that we have listed all of the possible outcomes is the use of a scheme called a *tree diagram* shown in Fig. 12.1. We have found that there are eight possible outcomes of the experiment.

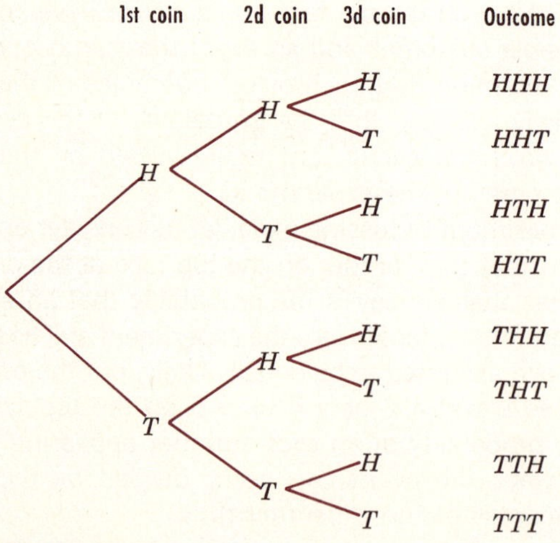

FIGURE 12.1

(a) An examination of the list of possible outcomes shows that there are three ways that exactly two tails can occur, HTT, THT, TTH. Therefore, the probability that exactly two tails will show is $\frac{3}{8}$.

(b) An examination of the list of possible outcomes shows that there are seven ways in which at least one tail can occur. (The only outcome that does not show at least one tail is HHH). So, the probability that at least one tail will be showing is $\frac{7}{8}$.

It is worthwhile for future reference to point out that the probability that three heads are showing (HHH) is the same as the probability that no tails are showing, and that this probability is $\frac{1}{8}$, so that

$$\text{probability that at least one tail shows} + \text{probability that no tail shows} = 1$$

EXERCISES

1 Two coins are tossed once. What is the probability (a) that at least one of the coins shows heads? (b) That exactly two of the coins show heads? (c) That none of the coins shows heads? *Hint:* The possible outcomes of the experiment are HH, HT, TH, TT.

2 An ordinary cubical die with faces numbered 1, 2, 3, 4, 5, 6 is tossed once, and the number on the top face is observed. What is the probability that the number on the top face is (a) either a 5 or a 6? (b) An even number less than 5? (c) Not a 2? *Hint:* In (a) the event can occur in two ways; in (b) the event can occur in two ways.

3 A cubical die with faces marked 1, 1, 2, 3, 4, 5 is tossed, and the number on the top face is observed. What is the probability that the number on the top face is (a) a 1? (b) A 3? (c) Not a 1. *Hint:* There are still six possible outcomes of the experiment since there are six faces, any one of which can be the top face.

4 Suppose that three pieces of paper, with the number 1 written on one piece, the number 2 written on another piece, and the number 3 written on the third piece, are put in a box. If two pieces of paper are drawn at random from the box, one after another without replacement, list the possible outcomes of the experiment. Each outcome will be an ordered pair (a, b) of numbers, where a is the number on the first piece of paper drawn and b is the number on the second piece of paper. Find the probability that the sum of the two numbers on the two pieces of paper is 5. Find the probability that the numbers are both less than 3.

12.3 SAMPLE SPACES, EVENTS, AND PROBABILITY OF AN EVENT

EXPERIMENT

In each of the illustrations, examples, and exercises of the preceding section, an experiment was described. By **experiment** we mean a process by which an observation or measurement is made. To determine an experiment fully, we must describe an action to be carried out, and tell what is to be observed or measured. To illustrate, suppose that we have a cubical die with faces numbered 1, 2, 3, 4, 5, 6 and with two faces colored red, two faces colored green, and two faces colored blue. To say simply, "Toss the die once," will not determine an experiment because we are not told what to observe or record. One possible experiment in connection with tossing such a die once is determined by the sentence, "Toss the die once, and observe the color of the top face." Another experiment would be "Toss the die once, and observe the number that appears on the top face."

TRIAL

Each performance of an experiment is called a **trial** of that experiment. Any given experiment has certain possible outcomes, and we have seen that the set of these possible outcomes plays a key role in the computation of probabilities. A set S of all possible outcomes of an experiment is called the **sample space** for the experiment provided it is *finite* and:

SAMPLE SPACE

1 Each member of S corresponds to an outcome of the experiment.
2 Any trial of the experiment produces an outcome that corresponds to one and only one member of S.

To illustrate, for the experiment of tossing a coin once and observing the face that is upward, the sample space is

$$S_1 = \{H, T\}$$

For the experiment of tossing one die and observing the number on the top face, the sample space is

$$S_2 = \{1, 2, 3, 4, 5, 6\}$$

For the experiment of tossing three coins once and observing for each coin whether it is heads or tails, the sample space is

$$S_3 = \{HHH, HHT, HTH, HTT, THH, THT, TTH, TTT\}$$

In the preceding section we called the occurrence of heads when a coin is tossed an event; the occurrence of an even number on the top face when a die is tossed was called an event; the occurrence of exactly two tails showing when three coins are tossed was called an event. In each of these cases the event can be considered a subset of the sample space:

For the occurrence of heads, $\{H\} \subseteq S_1$

For the occurrence of an even number, $\{2, 4, 6\} \subseteq S_2$

For the occurrence of exactly two tails, $\{HTT, THT, TTH\} \subseteq S_3$

Thus we are led to the following definition:

EVENT For a given experiment with sample space S, an **event** is a subset E of the sample space S, that is, $E \subseteq S$.

If we are looking for the event of the appearance of heads, $E = \{H\}$, we say that the event occurs if the coin comes up heads. If we are looking for the event of the appearance of an even number on the top face of a die $E = \{2, 4, 6\}$, we say that the event occurs if the die shows a 2 or a 4 or a 6. In either of these examples the event E occurs in a trial of the experiment if the outcome of the trial is an element of the event E. In general, if an event E is a subset of a sample space S, we say that the event E **occurs** in a trial of the experiment if the outcome of the trial is an element of E.

AN EVENT OCCURS

Each member of a sample space S for an experiment is called a **sample point.** For the sample space $S_1 = \{H, T\}$, H and T are sample points. Sometimes we denote the sample points of a sample space S by $e_1, e_2, e_3, \ldots, e_n$ and write

SAMPLE POINT

$$S = \{e_1, e_2, e_3, \ldots, e_n\}$$

SIMPLE EVENT A **simple event** is a subset of the sample space which has only one member; that is, a simple event is a set whose only member is a sample point. If $S = \{e_1, e_2, e_3, \ldots, e_n\}$, then each of the events $\{e_1\}, \{e_2\}, \{e_3\}, \ldots, \{e_n\}$ is a simple event. The event $\{H\}$ for the experiment of tossing a coin is a *simple event;* the event $\{2, 4, 6\}$ for the experiment of tossing a die is *not* a simple event.

EXAMPLE 1 Give the sample space, the sample points, and the simple events for the experiment of tossing two coins once and observing for each coin whether it is heads or tails. Indicate (in set notation) each of the events which are described in the following sentences:

i At least one head appears.

ii Two tails appear.

iii The number of heads equals the number of tails.

iv Three heads appear.

SOLUTION To determine the sample space, that is, to determine all the possible outcomes of the experiment, we may* use a tree diagram as shown in Fig. 12.2. So the sample space is $S = \{HH, HT, TH, TT\}$; the sample points are HH, HT, TH, TT; and the simple events are $\{HH\}$, $\{HT\}$, $\{TH\}$, $\{TT\}$.

i The event described by "at least one head appears" is $\{HH, HT, TH\}$.

ii The event described by "two tails appear" is $\{TT\}$.

* There are of course several other ways to determine the possible outcomes of this experiment, any one of which is acceptable.

1st coin	2d coin	Outcome
H	H	HH
	T	HT
T	H	TH
	T	TT

FIGURE 12.2

iii The event described by "the number of heads equals the number of tails" is $\{HT, TH\}$.

iv The event described by "three heads appear" cannot occur and is denoted by $\varnothing$ (the empty set). Recall from Sec. 1.3 that the empty set $\varnothing$ is a subset of every set, so $\varnothing \subseteq S$, and $\varnothing$ is an event.

As we pointed out in the above example, for *any* sample space S, the empty set $\varnothing$ is an event since $\varnothing \subseteq S$. We should also note that the sample space S itself is an event because $S \subseteq S$.

EXAMPLE 2 For the experiment in Example 1, assume that each of the outcomes is "equally likely", that is, assume that each coin is as likely* to turn up heads as it is to turn up tails. Determine the probability that each of the simple events in this experiment will occur on a single trial of the experiment, and determine the probability that each of the events described by sentences *i*, *ii*, *iii*, and *iv* in Example 1 will occur. Also, determine the probability that the sample space S, considered as an event, will occur on a single trial of the experiment.

SOLUTION Here $S = \{HH, HT, TH, TT\}$, and there are four possible outcomes of the experiment. Each of the simple events $\{HH\}$, $\{HT\}$, $\{TH\}$, and $\{TT\}$ can occur in only one way, so the probability that each will occur is $\frac{1}{4}$.

For the event $\{HH, HT, TH\}$, described by the sentence *i*, "At least one head appears," there are three ways for the event to occur (there are three members of the set), and the probability of this event is $\frac{3}{4}$.

The event $\{TT\}$, described by sentence *ii*, "Two tails appear," is one of the simple events whose probability we have seen to be $\frac{1}{4}$.

The event $\{HT, TH\}$, described by sentence *iii*, "The number of heads is equal to the number of tails," can occur in two ways (there are two members of the set), so the probability of this event is $\frac{2}{4} = \frac{1}{2}$.

The event $\varnothing$, described by sentence *iv*, "Three heads appear," can occur in *zero* ways (there are no members in the set), so the probability of this event is $\frac{0}{4} = 0$.

*Actually we are using "equally likely" as an undefined term. That is, we are taking the idea that "one happening is just as likely as another happening" as a primitive concept which we use without definition.

Considering the sample space S as an event, we see that there are four ways for the event to occur (the set has four members), so the probability of S is $\frac{4}{4} = 1$.

Using the examples and illustrations of this and the preceding sections as guides, we now give precise and formal definitions of some of the basic concepts used in probability and its applications.

In Example 2 each of the *simple* events $\{HH\}$, $\{HT\}$, $\{TH\}$, $\{TT\}$ associated with the sample space had a probability which was a nonnegative number, and the sum of the probabilities of the simple events was 1. This result is consistent with the following definition of the probability of a simple event.

Let S be a sample space for an experiment with sample points $e_1, e_2, e_3, \ldots, e_n$,

$$S = \{e_1, e_2, e_3, \ldots, e_n\}$$

PROBABILITY OF A SIMPLE EVENT

Suppose that to each simple event $\{e_i\}$, where $i = 1, 2, 3, \ldots, n$, there is assigned a number, denoted by $\Pr(e_i)$. The number $\Pr(e_i)$ is called the **probability of the simple event** $\{e_i\}$ *if* the following conditions are satisfied:

$$\begin{aligned} &(a)\ \Pr(e_i) \geq 0 \qquad \text{for } i = 1, 2, 3, \ldots, n \\ &(b)\ \Pr(e_1) + \Pr(e_2) + \Pr(e_3) + \cdots + \Pr(e_n) = 1 \end{aligned} \tag{12.2}$$

ACCEPTABLE ASSIGNMENT OF PROBABILITIES

An assignment of probabilities to the simple events of a given sample space S is **acceptable** if this assignment satisfies conditions (a) and (b) in (12.2), that is, if *the probability of each simple event is a nonnegative number and the sum of the probabilities of the simple events of S is 1.*

In Example 2 the probabilities of the *simple* events were all the same, namely, $\frac{1}{4}$. This came about because of the assumption that the coins were "fair" so that the outcomes of the experiment were equally likely, and it seemed reasonable to make the assignment

$$\Pr(HH) = \frac{1}{4} \qquad \Pr(HT) = \frac{1}{4}$$

$$\Pr(TH) = \frac{1}{4} \qquad \Pr(TT) = \frac{1}{4}$$

If the coins had been bent in some way so that, for example, the first coin was more likely to fall heads than tails, the probabilities of the simple events *might* have been assigned as

$$\Pr(HH) = \frac{3}{8} \qquad \Pr(HT) = \frac{3}{8}$$

$$\Pr(TH) = \frac{1}{8} \qquad \Pr(TT) = \frac{1}{8}$$

The assignment of acceptable probabilities is done either as a result of *experimentation* or by *hypothesis*. The assignments we have been making in our intuitive examples have been made by *hypothesis*. We have used the hypothesis of equally likely outcomes of the experiment and the following definition.

PROBABILITY OF EQUALLY LIKELY SIMPLE EVENTS

If an experiment has n *equally likely* outcomes, then we define the **probability of each of the n equally likely simple events** to be $\frac{1}{n}$.

Now consider an event that is *not* a simple event, for instance, in Example 2 the event $\{HH, HT, TH\}$ that at least one head appears. How shall we define the probability of such an event? Notice that the event, "At least one head appears," is the *union* of three simple events,

$$\{HH, HT, TH\} = \{HH\} \cup \{HT\} \cup \{TH\}$$

and notice that the probability of this event is $\frac{3}{4}$, which is the *sum* of the probabilities of the three simple events. This is consistent with the following definition.

PROBABILITY OF AN EVENT E

The **probability of an event E**, denoted by $\Pr(E)$, is the sum of the probabilities of the *simple* events of which E is the union; that is,

$$\text{if } E = \{e_1\} \cup \{e_2\} \cup \{e_3\} \cup \cdots \cup \{e_n\}, \text{ then} \\ \Pr(E) = \Pr(e_1) + \Pr(e_2) + \Pr(e_3) + \cdots + \Pr(e_n) \tag{12.3}$$

From this definition it follows that if we consider a sample space S as an event, then

$$\Pr(S) = 1$$

PROBABILITY OF $\varnothing$

We define the **probability of the empty set $\varnothing$** to be zero,

$$\Pr(\varnothing) = 0 \tag{12.4}$$

If each of the n outcomes of an experiment is *equally likely,* then an event E which is the union of k of the simple events will have its probability given by

$$\Pr(E) = \underbrace{\frac{1}{n} + \frac{1}{n} + \frac{1}{n} + \cdots + \frac{1}{n}}_{k \text{ terms in this sum}} = \frac{k}{n} \tag{12.5}$$

Each of the simple events, of which E is the union, is a way in which E can occur; so the *number* of simple events, of which E is the union, is the number of ways E can occur. Therefore, *if the outcomes of an experiment are equally likely,* then the definition (12.5) can be written as

$$\Pr(E) = \frac{\text{number of ways } E \text{ can occur}}{\text{number of possible outcomes of the experiment}} \tag{12.6}$$

This definition agrees with our intuitive concepts summarized in equality (12.1).

From this point on, unless specifically stated to the contrary, we shall assume that the outcomes of an experiment are *equally likely,* and therefore definition (12.6) can be used to determine the probability of an event E.

EXAMPLE 3 If three coins are tossed once and we observe for each coin whether it is heads or tails, we have seen that the sample space is $S = \{HHH, HHT, HTH, HTT, THH, THT, TTH, TTT\}$. Indicate in set notation each of the events described below, and determine the probability of each event:

E_1: There are exactly two heads.

E_2: The number of heads exceeds the number of tails.

E_3: The number of heads is less than three.

E_4: There are three heads.

E_5: There is at least one tail.

E_6: There are five heads.

SOLUTION There are eight sample points in the sample space, so there are eight possible outcomes of the experiment. The individual events can be tabulated by observing which of the sample points satisfy the description of that event. The number of elements so tabulated for a given event is the number of ways the event can occur. Then equality (12.6) can be used to compute the probability of the event. Proceeding in this fashion we find

$$
\begin{aligned}
&E_1 = \{HHT, HTH, THH\} && \Pr(E_1) = \tfrac{3}{8} \\
&E_2 = \{HHH, HHT, HTH, THH\} && \Pr(E_2) = \tfrac{4}{8} = \tfrac{1}{2} \\
&E_3 = \{HHT, HTH, HTT, THH, THT, TTH, TTT\} && \Pr(E_3) = \tfrac{7}{8} \\
&E_4 = \{HHH\} && \Pr(E_4) = \tfrac{1}{8} \\
&E_5 = \{HHT, HTH, HTT, THH, THT, TTH, TTT\} && \Pr(E_5) = \tfrac{7}{8} \\
&E_6 = \varnothing && \Pr(E_6) = 0
\end{aligned}
$$

Notice that $E_3 = E_5$, so an event may have more than one description. Also notice that $E_4 \cup E_5 = S$ and $E_4 \cap E_5 = \varnothing$, so that E_4 is the *complement* of E_5 *relative* to S. Further, $\Pr(E_4) + \Pr(E_5) = \tfrac{1}{8} + \tfrac{7}{8} = 1$. It is always true (as we shall see in Sec. 12.8) that if the union of two events is the sample space S and if their intersection is the empty set $\varnothing$, then the sum of the probabilities of the two events is 1. This fact is frequently useful in the computation of probabilities.

EXAMPLE 4 From a bag containing 10 green marbles and 7 red marbles, 1 marble is drawn at random. What is the probability that the marble drawn is red?

SOLUTION Here the experiment is drawing one marble from a bag and observing its color. The event is the occurrence of a red marble. There are 17 marbles in the bag, so there are 17 possible outcomes of the experiment. The words "at random" indicate that each of the 17 outcomes is equally likely. Since there are 7 red marbles in the bag, there are 7 ways the event can occur. So we use (12.6) to compute the probability of drawing a red marble, and we find that

$$\text{Pr(the marble is red)} = \tfrac{7}{17}$$

EXAMPLE 5 A set of numbered metal tags is made with no two tags having the same number. The number on each tag is a two-digit numeral with the tens digit different from zero. If the set contains the largest possible number of tags, what is the probability that a tag selected at random will be numbered with a number divisible by 5?

SOLUTION The experiment in this example is selecting a tag at random and observing the number on the tag. The event E is the occurrence of a number divisible by 5. Since the selection is made at random, the outcomes of the experiment are equally likely, and we can use (12.6) to determine the probability of the event. Therefore, we wish to determine (or *count*) the number of possible outcomes of the experiment and the number of ways the event can occur.

The number of outcomes of the experiment will be the number of two-digit numerals that can be formed with the tens digit different from zero. These numerals are

10, 11, 12, . . . , 19, 20, 21, 22, . . . , 29,
30, 31, 32, . . . , 39, 40, 41, 42, . . . , 49,
50, 51, . . . , 60, 61, . . . , 70, 71, . . . , 80,
81, 82, . . . , 90, 91, . . . , 99

and we see that there are 90 of them (9 groups of 10 numerals each) and thus there are 90 possible outcomes of the experiment.

The event E will occur if the tag selected has a number that is divisible by 5. An examination of the possible outcomes will reveal that 18 of the tags are so numbered (there are two numbers divisible by 5 in each of the 9 groups). So there are 18 ways the event E can occur and

$$\text{Pr}(E) = \frac{18}{90} = \frac{1}{5}$$

For the examples we have considered, it has been relatively easy to count the number of ways the events can occur and the number of possible outcomes of the experiments. However, we need to develop methods of counting that will enable us to handle more complicated situations. For example, consider the following problem.

If a set of four marbles is chosen at random from a sack containing five red and seven green marbles, what is the probability that two are red and two are green? To compute this probability, we need to know the number of ways that four marbles can be selected from a set of twelve marbles and how many of these ways will result in two red and two green marbles. In the next four sections we shall examine some methods of counting various types of outcomes and occurrences, among which will be methods for solving the marble problem.

EXERCISES

1 An ordinary cubical die with faces numbered 1, 2, 3, 4, 5, 6 and a coin are tossed once, and we observe the number on the top face of the die and whether the coin shows heads or tails. Tabulate the sample space for this experiment.

2 For the experiment in Exercise 1, indicate, in set notation, four different events which are unions of three simple events; that is, four events, each of which contains three sample points of the sample space. (As we shall see in Sec. 12.7, there are 220 such events.)

3 Let $S = \{e_1, e_2, e_3, e_4\}$ be the sample space for an experiment. Is the assignment $\Pr(e_1) = 0.3$, $\Pr(e_2) = 0.3$, $\Pr(e_3) = 0.3$, $\Pr(e_4) = 0.1$ an acceptable assignment of probabilities? Explain.

4 For the sample space of Exercise 3, is the assignment $\Pr(e_1) = \frac{2}{5}$, $\Pr(e_2) = \frac{1}{3}$, $\Pr(e_3) = \frac{2}{3}$, $\Pr(e_4) = \frac{1}{5}$ an acceptable assignment of probabilities? Explain.

5 For the sample space of Exercise 3, if $\Pr(e_1) = \frac{1}{3}$, $\Pr(e_2) = \frac{1}{3}$, and $\Pr(e_3) = \frac{1}{6}$, what value must be assigned to $\Pr(e_4)$ in order to produce an acceptable probability assignment?

6 For the sample space in Exercise 3, if $\Pr(e_1) = \frac{1}{3}$, $\Pr(e_2) = \frac{1}{6}$, and $\Pr(e_3) = \frac{1}{12}$, what value must be assigned to $\Pr(e_4)$ in order to produce an acceptable probability assignment?

7 If an ordinary cubical die, as in Exercise 1, is tossed once, what is the probability of throwing a number greater than 3? Greater than 4? Greater than 6?

8 From a bag containing 12 red and 18 green marbles, 1 marble is drawn at random. What is the probability that the marble drawn is red? Green? Either red or green? Purple?

9 What is the probability of drawing an ace when a single card is drawn at random from a pack of 52 bridge cards? (A pack of bridge cards contains 2, 3, 4, 5, 6, 7, 8, 9, 10, jack, queen, king, ace of each of four suits, spades, hearts, diamonds, clubs. The spades and clubs are black, the hearts and diamonds are red.)

10 What is the probability of drawing a red king when a single card is drawn at random from a pack of 52 bridge cards? What is the probability of drawing a jack? What is the probability of drawing the ace of hearts?

11 A coin is weighted so that tails is three times as likely to appear as heads. Determine an acceptable probability assignment for $\Pr(H)$ and $\Pr(T)$. *Hint:* Let $\Pr(H) = c$, then $\Pr(T) = 3c$, and we know that $\Pr(H) + \Pr(T) = 1$.

12 For three horses A, B, C in a race, A is twice as likely to win as B, and C is three times as likely to win as B. Determine an acceptable probability assignment for $\Pr(A \text{ wins})$, $\Pr(B \text{ wins})$, $\Pr(C \text{ wins})$.

12.4
THE FUNDAMENTAL COUNTING PRINCIPLE

As we pointed out in the preceding section, a basic problem in the computation of probabilities is the problem of counting the number of ways an event can occur and the number of possible outcomes of an experiment. In this and succeeding sections we shall examine some important methods of counting occurrences and outcomes.

A very common type of procedure encountered in probability problems is a procedure that consists of several acts to be performed one after the other. To illustrate, choosing a couple (one girl and one boy) from a group consisting of two girls, Susan and Mary, and three boys, Bob, Ralph, and Ray; or choosing a shirt, a tie, and a pair of trousers from a wardrobe containing 10 ties, 12 shirts, and 6 pairs of trousers, are such procedures. If we know, or can determine, the number of ways the individual acts can be performed in succession, then we can determine the total number of ways the entire procedure can be performed.

Consider the procedure of choosing a couple (one girl and one boy) from the group of two girls and three boys named in the preceding paragraph. This procedure can be considered to consist of two acts performed successively: (1) choose a girl, and then (2) choose a boy. The first act can be performed in two ways (either Susan or Mary can be chosen), and after the first act has been performed, the second act can be performed in three ways (Bob or Ralph or Ray can be chosen). The possible ways of choosing a couple by this procedure can be pictured by the *tree diagram* in Fig. 12.3. In the first column of the diagram we list the ways in which the first act can be performed, and then for *each* of these ways we list in the second column the ways in which the second act can be performed. The number of "branch endings" in the final column is the number of ways of carrying out the entire procedure.

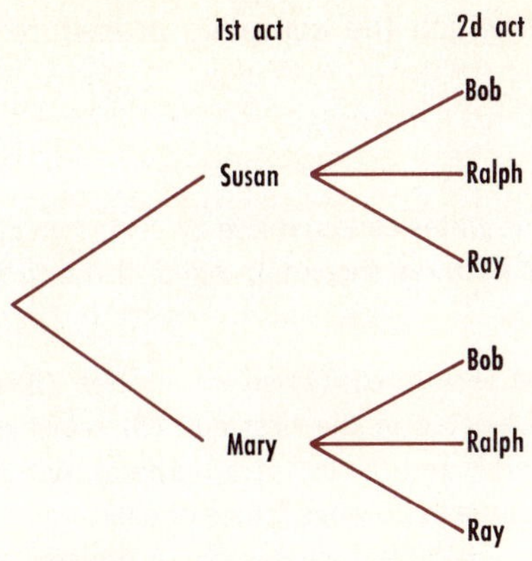

FIGURE 12.3

Here there are $2 \cdot 3 = 6$ ways the procedure of choosing a couple can be performed; the six possible couples can be read off of the tree diagram to be (Susan, Bob), (Susan, Ralph), (Susan, Ray), (Mary, Bob), (Mary, Ralph), (Mary, Ray).

Next, consider the procedure of choosing a shirt, a tie, and a pair of trousers from a wardrobe containing 12 shirts, 10 ties, and 6 pairs of trousers. This procedure can be considered to consist of three acts performed successively: (1) choose a shirt, then (2) choose a tie, and then (3) choose a pair of trousers. The first act can be performed in 12 ways; after the first act has been performed, the second act can be performed in 10 ways; after the second act has been performed, the third act can be performed in 6 ways. If we make a tree diagram for this procedure, the first column will have 12 entries; the second column will have 10 entries for *each* of the 12 entries in the first column, which makes a total of 120 entries in the second column; the third column will have 6 entries for *each* of the 120 entries in the second column, which makes a total of 720 entries in the third column. So, there are $12 \cdot 10 \cdot 6 = 720$ ways of performing the procedure of selecting a shirt, a tie, and a pair of trousers.

FUNDAMENTAL COUNTING PRINCIPLE

The two examples above illustrate the **fundamental counting principle** which can be stated as follows:

Suppose a procedure consists of k acts to be performed successively. If the first act can be performed in n_1 ways, and after the first act has been performed, the second act can be performed in n_2 ways, and after the second act has been performed, the third act can be performed in n_3 ways, and so on, for k acts, where n_k is the number of ways the kth act can be performed, then the number n of ways in which the complete procedure can be performed is given by

$$n = n_1 \cdot n_2 \cdot n_3 \cdots n_k$$

EXAMPLE 1 How many distinct license plates can be made by using two different letters of the alphabet followed by three (decimal) digits, if the first digit cannot be 0?

SOLUTION There are five acts to be performed: (1) select a letter, (2) select a letter different from the one selected in the first act, (3) select a nonzero numeral, (4) select a numeral, and (5) select a numeral. Act 1 can be done in 26 ways; act 2 can be done in 25 ways (since one letter has been chosen, there are only 25 letters available for use); there are ten digits, 0, 1, 2, 3, 4, 5, 6, 7, 8, 9, so act 3 can be done in 9 ways (0 cannot be used); act 4 can be done in 10 ways; act 5 can be done in 10 ways. The fundamental counting principle tells us that we can produce

$$26 \cdot 25 \cdot 9 \cdot 10 \cdot 10 = 585{,}000$$

distinct license plates in the manner described.

EXAMPLE 2 There are six girls and five boys in a club. Three members are selected to be president, vice president, and secretary, respectively. No person may hold more than one office. (a) How many different slates of officers are possible if no restrictions are placed on who may occupy the offices? (b) How many different slates consisting of all boys are possible? (c) If the selection is made at random, what is the probability that all three officers will be boys?

SOLUTION (a) There are three positions to be filled (three acts to perform). One position can be filled in any one of 11 ways. There are then 10 members left, so a second position can be filled in any one of 10 ways. There are then 9 members left, so the third position can be filled in 9 ways. Therefore the slate of officers can be selected in

$$11 \cdot 10 \cdot 9 = 990$$

ways.

(b) The three positions are all to be filled by boys. There are five boys who can be chosen for the first position, so one position can be filled in five ways. Then a second position can be filled in four ways, and the third position can be filled in three ways. Thus, there are

$$5 \cdot 4 \cdot 3 = 60$$

possible slates consisting of all boys.

(c) The experiment (select a president, a vice president, and a secretary) has 990 possible outcomes, as we saw in part (a). The event E: all the officers are boys, can occur in 60 ways, as we saw in part (*b*). Since the selection is made at random, all the outcomes are equally likely, and from (12.6) we have

$\Pr(E) = \frac{60}{990} = \frac{2}{33}$

As another example of the use of a tree diagram, consider the following example.

EXAMPLE 3 Ray and Leon play a tennis match. The first person to win either two games in a row or a total of three games wins the match. We shall assume that Ray and Leon are evenly matched. (*a*) In how many different ways can the match be played in order to have a winner? (*b*) What is the probability that the match will require *exactly* four games?

SOLUTION In Fig. 12.4, R represents a win by Ray and L represents a win by Leon. The first column represents what could happen in the first game, the

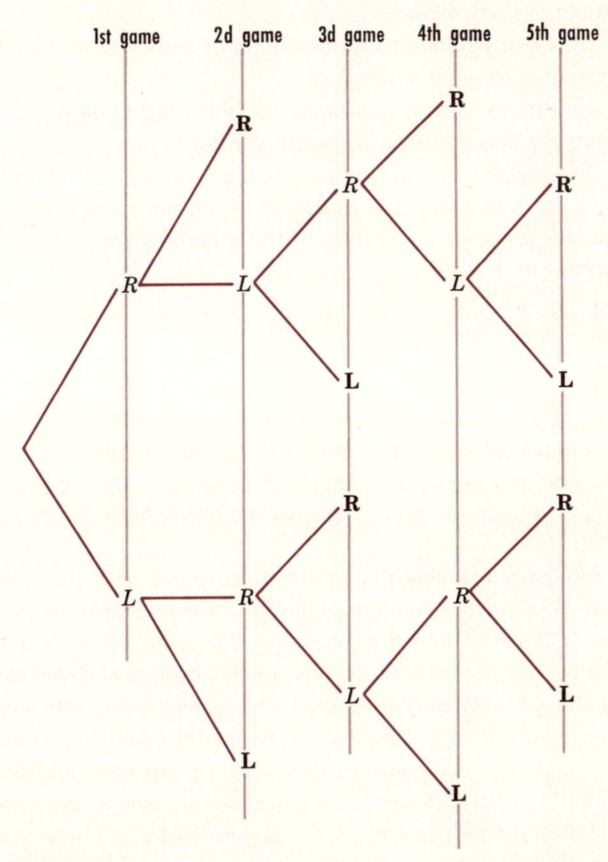

FIGURE 12.4

second column represents what could happen after each possible outcome of the first game, and so on. When a player has either won two games in a row or won a total of three games, the "branch" ends, which signifies that the match is over, and a boldface letter is placed at that point.

(*a*) In the figure there are 10 branch ends, so there are 10 different ways in which the match can be played.

(*b*) Of the 10 ways that the match can be played, 2 of them require exactly four games (there are two branch ends in the column headed "4th game"). Using (12.6), we find that the probability that exactly four games will be required is $\frac{2}{10} = \frac{1}{5}$.

EXERCISES

1 How many different three-digit numerals can be formed with the digits 1, 2, 3, 4, 5, 6 if no digit can be used twice? How many can be formed if repetitions are allowed?

2 How many different combinations of shirt and tie can a man select from four shirts and seven ties?

3 How many signals can be made by raising three flags on a flagpole if there are seven different flags to choose from?

4 Two cubical dice are thrown. If the faces of each die are numbered 1, 2, 3, 4, 5, 6, and if the numbers on the top faces of the two dice are observed, how many outcomes of the experiment are there? Tabulate the outcomes in the form

$$\begin{array}{llll} (1, 1) & (1, 2) & \cdots & (1, 6) \\ (2, 1) & (2, 2) & \cdots & (2, 6) \\ \cdots & \cdots & \cdots & \cdots \\ (6, 1) & (6, 2) & \cdots & (6, 6) \end{array}$$

5 If a person can travel from Tallahassee to Atlanta by either bus or plane and can go from Atlanta to Washington, D.C., by bus or train or plane, how many different ways can he travel from Tallahassee to Washington, D.C.?

6 Motorcycle license plates are to be made with three letters of the alphabet followed by two (decimal) digits. The middle letter must be a vowel (A, E, I, O, or U) and the other two letters cannot be vowels. The first of the two digits must not be zero. (*a*) How many such license plates can be made if no letter and no digit can appear more than once? (*b*) How many can be made if repetitions of both letters and numbers are allowed?

7 A club has 20 members of whom 12 are boys and 8 are girls. A president, a vice president, and a secretary are to be selected. If the president must be a girl and the vice president a boy, how many different slates of officers are possible?

8 For the club in Exercise 7, if the president can be either a girl or a boy, but the vice president must be of the opposite sex to the president and the secretary of the same sex as the president, how many different slates of officers are possible? *Hint:* Compute separately the number of slates with a girl for president and the number of slates with a boy for president.

9 Repeat Exercise 8 with the club enlarged to admit four more girls and three more boys.

10 How many three-digit numerals can be formed from the digits 0, 1, 2, 3, 4, 5 if the first digit cannot be zero and the numeral must represent an odd number? How many three-digit numerals can be formed if no digit can be used more than once? *Hint:* Select the *odd* units digit first, then the tens digit, then the *nonzero* hundreds digit; compute separately the number of numerals with a digit other than 0 as the tens digit.

11 Baseball teams A and B compete in the World Series. The first team to win a total of four games wins the series. If team A wins the first game, use a tree diagram to determine the number of different ways the series can be played in order to have a winner. If the teams are evenly matched and if team A wins the first game, what is the probability that A will win the series?

12 In Exercise 11, what is the probability that the series will end with the fifth game?

13 If two cubical dice are thrown (see Exercise 4), what is the probability that the *sum* of the numbers shown on the top faces of the dice is 5? *Hint:* In the tabulation of the outcomes of the experiment of tossing two dice, count the number of outcomes in which the sum of the numbers is 5.

14 If two cubical dice are thrown, what is the probability that the *sum* of the numbers shown on the top faces is greater than 8? (See Exercises 4 and 13.) What is the probability that the *sum* of the two numbers on the top faces is greater than 8 and the smaller of the two numbers is less than 5?

15 If two dice are thrown, what is the probability of obtaining (a) a sum of 11? (b) A sum of 7? (c) What is the "most probable" sum in a throw of two dice?

16 If two dice are thrown, what is the probability that (a) only one shows a 5? (b) At least one shows a 5?

12.5 PERMUTATIONS I

Many of the procedures or experiments we have discussed have as their outcomes arrangements (of objects) called permutations. For instance, forming a three-digit numeral by using three of the digits

1, 2, 3, 4, 5, 6 with no digit used twice (Exercise 1 of the preceding section) results in an arrangement of digits known as a *permutation*.

LINEAR PERMUTATION

A **linear permutation** of n *distinct* objects is an arrangement of these n objects in a row *in a given order*. Another way of saying the same thing is to say that a linear permutation of n distinct objects is an *ordered* set of these n distinct objects. *Ordering is essential in a permutation.* To illustrate, there are six permutations of the three letters a, b, c, namely,

$$abc \quad acb \quad bac \quad bca \quad cab \quad cba \tag{12.7}$$

Note that although each of these permutations contains the same letters, they form six different permutations because the *order* of the letters is not the same in any two of the arrangements.

Often, as in the illustration in the first paragraph of this section, we are interested in an arrangement of a certain number of objects from a larger set of distinct objects. When, for example, we form a three-digit numeral by using three of the digits 1, 2, 3, 4, 5, 6, with no digit used more than once, we have formed a permutation from six objects taken three at a time, or as we shall call it, *a 3-permutation of six objects.*

If we select r objects from a set of n *distinct* objects and arrange these r objects in a row *in a given order,* we call this arrangement a **linear r-permutation of n objects.***

LINEAR r-PERMUTATION OF n OBJECTS

To illustrate, the 2-permutations of the three letters a, b, c are

$$ab \quad ba \quad ac \quad ca \quad bc \quad cb \tag{12.8}$$

the 2-permutations of the four letters a, b, c, d are

$$\begin{matrix} ab & ba & ac & ca & ad & da \\ bc & cb & bd & db & cd & dc \end{matrix} \tag{12.9}$$

and the 3-permutations of the four letters a, b, c, d are

$$\begin{matrix} abc & acb & bac & bca & cab & cba \\ acd & adc & cad & cda & dac & dca \\ bcd & bdc & cbd & cdb & dbc & dcb \\ abd & adb & bad & bda & dab & dba \end{matrix} \tag{12.10}$$

Unless the numbers n and r are small, listing all the r-permutations of n objects can be a formidable task. However, in many instances we are primarily interested in the *number* of such permutations, and it is a relatively simple matter to determine the number of r-permutations of n objects. If n and r are nonnegative integers,

*Such an ordered arrangement is also called a *linear permutation of n objects taken r at a time;* however, this terminology fails to reflect the fact that the permutation is an arrangement of r objects rather than n objects.

chosen so that $0 \leq r \leq n$, we shall use the symbol $P(n, r)$ to denote the *number* of r-permutations of n objects.* For given values of n and r, let us consider the number of ways an r-permutation of n objects can be formed. The formation of an r-permutation of n objects can be considered to be a procedure that consists of r acts, namely, the filling of r spaces, in succession, with members of the set of n distinct objects. The first act (filling the 1st space) can be

n	$n-1$	$n-2$	...	$n-6$	...	$n-(r-1)$
1st space	2d space	3d space		7th space		rth space

performed in n ways. There are now $n - 1$ objects remaining to be used, so the second act (filling the 2d space) can be performed in $n - 1$ ways. Continuing in the same way, we see that the 3d space can be filled in $n - 2$ ways, the 4th space can be filled in $n - 3$ ways, and, finally, the rth space can be filled in

$$n - (r - 1) = n - r + 1$$

ways. Now we apply the fundamental counting principle and find that the procedure of forming an r-permutation of n objects can be performed in

$$n(n-1)(n-2)(n-3) \cdots (n-r+1)$$

NUMBER OF r-PERMUTATIONS OF n OBJECTS

ways. This is the *number* of r-permutations of n objects, denoted by $P(n, r)$, and we have derived the formula

$$P(n, r) = n(n-1)(n-2)(n-3) \cdots (n-r+1) \tag{12.11}$$

If $n = 3$ and $r = 2$, then $n - r + 1 = 2$, and we have

$$P(3, 2) = 3 \cdot 2 = 6$$

which agrees with the number of 2-permutations of the three letters a, b, c listed in (12.8). If $n = 4$ and $r = 2$, then $n - r + 1 = 3$, and we have

$$P(4, 2) = 4 \cdot 3 = 12$$

which agrees with the number of 2-permutations of the four letters a, b, c, d listed in (12.9). If $n = 4$ and $r = 3$, then $n - r + 1 = 2$, and we have

$$P(4, 3) = 4 \cdot 3 \cdot 2 = 24$$

which agrees with the tabulation (12.10).

* The symbol ${}_nP_r$ is also used by many writers to denote the *number* of r-permutations of n objects.

These illustrations and an examination of formula (12.11) show that $P(n, r)$ is expressed as the product of r factors; the largest factor is n, and the successive factors decrease by 1 until the factor $n - r + 1$ is reached. This observation enables us to use formula (12.11) very easily. To illustrate,

$$P(8,3) = \underbrace{8 \cdot 7 \cdot 6}_{3 \text{ factors}} \qquad P(9,1) = \underbrace{9}_{\substack{1 \\ \text{factor}}}$$

$$P(11,5) = \underbrace{11 \cdot 10 \cdot 9 \cdot 8 \cdot 7}_{5 \text{ factors}} \qquad P(20,4) = \underbrace{20 \cdot 19 \cdot 18 \cdot 17}_{4 \text{ factors}}$$

$$P(5,5) = \underbrace{5 \cdot 4 \cdot 3 \cdot 2 \cdot 1}_{5 \text{ factors}} \qquad P(3,3) = \underbrace{3 \cdot 2 \cdot 1}_{3 \text{ factors}}$$

The last two illustrations above show the use of formula (12.11) when $r = n$. In this case $P(n, n)$ is the number of n-permutations of n objects, or simply the *number of permutations of n objects,* and we have

NUMBER OF PERMUTATIONS OF n OBJECTS

$$P(n, n) = n(n-1)(n-2)(n-3) \cdots (3)(2)(1) \tag{12.12}$$

That is, the number of permutations of n objects is the product of the positive integers 1 through n.

The product on the right side of formula (12.12) is denoted by the symbol $n!$ and is called n factorial. That is, we define n **factorial** to be the product of the positive integers from 1 to n, inclusive, and denote it by $n!$. Thus

n FACTORIAL

$n!$

$$n! = n(n-1)(n-2)(n-3) \cdots (3)(2)(1) \tag{12.13}$$

Notice that (12.13) defines $n!$ only when n is a positive integer ($n \in N$). It is useful to have a meaning for the symbol 0!; let us consider how we might define 0! From equality (12.13) it follows that

$$n! = n(n-1)! \qquad \textit{if } (n-1) \in N \tag{12.14}$$

Thus

$$8! = 8(7!) \qquad 43! = 43(42!)$$

Suppose we want (12.14) to hold for $n = 1$. Then $n - 1 = 0$ and we should have

$$1! = 1(0!)$$

and for this to be true, we must have $0! = 1$. This leads us to make the *definition*

0! $0! = 1$ (12.15)

By using the factorial notation, formula (12.13) for $P(n, n)$ may be written

$$P(n, n) = n! \qquad (12.16)$$

Formula (12.11) also has a factorial form. We multiply the right side of (12.11) by

$$1 = \frac{(n-r)(n-r-1)(n-r-2)\cdots(3)(2)(1)}{(n-r)(n-r-1)(n-r-2)\cdots(3)(2)(1)}$$

to obtain

$$P(n, r) = \frac{n(n-1)(n-2)\cdots(n-r+1)(n-r)(n-r-1)(n-r-2)\cdots(3)(2)(1)}{(n-r)(n-r-1)(n-r-2)\cdots(3)(2)(1)}$$

We observe that the numerator of the fraction on the right side is $n!$ and that the denominator is $(n - r)!$ So we can write

$$P(n, r) = \frac{n!}{(n-r)!} \qquad (12.17)$$

and note that this formula holds when n and r are any nonnegative integers with $0 \leq r \leq n$. In particular, if $r = n$, formula (12.17) becomes $P(n, n) = \frac{n!}{(n-n)!} = \frac{n!}{0!}$, or by the use of definition (12.15), $P(n, n) = n!$, which is formula (12.16).

EXAMPLE 1 (*a*) Find the number of permutations of the letters a, b, c, d, e.
(*b*) Find the number of 3-permutations of the five letters a, b, c, d, e.

SOLUTION (*a*) Here $n = 5$, and by formula (12.16) we have

$$P(5, 5) = 5! = 5 \cdot 4 \cdot 3 \cdot 2 \cdot 1 = 120$$

(*b*) Here $n = 5$ and $r = 3$. Formula (12.11) gives

$$P(5, 3) = 5 \cdot 4 \cdot 3 = 60$$

If we elect to use formula (12.17), we find

$$P(5, 3) = \frac{5!}{(5-3)!} = \frac{5 \cdot 4 \cdot 3 \cdot 2 \cdot 1}{2 \cdot 1} = 5 \cdot 4 \cdot 3 = 60$$

So, there are 60 3-permutations of the five letters a, b, c, d, e.

EXAMPLE 2 How many three-digit numerals can be formed by using the digits 1 through 9 with *no repetitions permitted?*

SOLUTION Since no repetition of digits is permitted, each three-digit numeral to be formed is a 3-permutation of nine objects. Therefore, we wish to determine

$P(9, 3)$, the number of 3-permutations of nine objects. By use of formula (12.11) we find

$$P(9, 3) = 9 \cdot 8 \cdot 7 = 504$$

We emphasize that the permutation formula is applicable in this example *only* because no repetition of digits is allowed and therefore each three-digit numeral is a permutation (since it is an *ordered* arrangement of *distinct* objects in a row).

EXAMPLE 3 John, Ted, Pete, and Sam are to be seated in a row containing four seats. (*a*) How many seating arrangements are possible? (*b*) If John is always to occupy the seat on the left end of the row, how many seating arrangements are possible? (*c*) If the seating is made at random, what is the probability that Sam will occupy an end seat?

SOLUTION (*a*) Each seating arrangement is a linear permutation of four objects. The number of such permutations is

$$P(4, 4) = 4! = 4 \cdot 3 \cdot 2 \cdot 1 = 24$$

(*b*) If John is to be seated on the left end of the row, any change in the seating arrangement will involve only Ted, Pete, and Sam. Each seating arrangement of these three people is a linear permutation of three objects.

John			

So, the number of seating arrangements with John at the left end is

$$P(3, 3) = 3! = 3 \cdot 2 \cdot 1 = 6$$

(*c*) The experiment of seating four men in a row has 24 outcomes, as we saw in part (*a*). Since the seating is done at random, these outcomes are equally likely. To compute the probability that Sam occupies an end seat, we need to determine the number of ways this event can occur. Sam can sit in the left end seat or in the right end seat. If he sits on the left end, there are six possible seating arrangements, as we saw in part (*b*). Similarly, if he sits on the right end, there are six possible seating arrangements. Therefore, there are 12 seating arrangements that will have Sam in an end seat, and the probability of the event E: Sam occupies an end seat, is given by

$$\Pr(E) = \tfrac{12}{24} = \tfrac{1}{2}$$

EXERCISES

1 Evaluate each of the following: $7!$; $\dfrac{8!}{7!}$; $\dfrac{10!}{8!}$; $\dfrac{7!}{3!4!}$.

2 Evaluate each of the following: $\frac{12!}{8!}$; $\frac{7!}{4!}$; $\frac{8!}{2!2!4!}$; $\frac{16!}{(8!)^2}$.

3 Simplify each of the following: $\frac{n!}{(n-1)!}$; $\frac{(n+1)!}{(n-1)!}$; $\frac{(2n)!}{(n!)^2}$.

In Exercises 4 and 5, use formula (12.17) to make the evaluations.

4 Evaluate each of the following: $P(7, 3)$; $P(12, 7)$; $P(15, 4)$; $P(10, 4)$.

5 Evaluate each of the following: $P(9, 4)$; $P(15, 3)$; $P(40, 2)$; $\frac{P(8, 8)}{P(6, 6)}$.

6 How many three-letter designations of fraternities can be made by using the 24 letters of the Greek alphabet if no repetition of letters is allowed in a single designation?

7 How many three-letter designations of fraternities can be made by using the 24 letters of the Greek alphabet if repetition of letters is allowed in a single designation? *Hint:* Use the fundamental counting principle.

8 How many five-digit numerals can be made with the digits 1, 2, 3, 4, 5, 6, 7 if no repetitions are allowed in a numeral?

9 How many three-digit numerals can be made with the digits 1, 2, 3, 4, 5, 6 if no repetitions are allowed in a numeral?

10 How many arrangements of four letters can be made by using the letters of the word *factor* (*a*) if there are no repetitions in an arrangement? (*b*) If repetitions are permitted?

11 In how many ways can 10 distinct objects be arranged in a row (*a*) when a given object is at a specified end? (*b*) When a given object is at an end? (*c*) When a designated pair of the objects are side by side? (*d*) When a designated pair of the objects are never side by side?

12 In how many ways can three different geometry books and four different algebra books be arranged on a shelf so that the geometry books are always together?

13 If $P(n, 3) = 6P(n, 2)$ and $n > 1$, find n.

14 If $P(n, 5) = 72P(n, 3)$ and $n > 1$, find n.

12.6 PERMUTATIONS II

So far we have considered only permutations of objects in a row (linear permutations) when repetitions of the objects was not allowed. In this section we consider briefly *circular* permutations (in contrast to linear permutations) and permutations of objects which are *not all distinct* (in contrast to permutations of distinct objects).

Consider the problem: In how many different ways can four persons, denoted by a, b, c, and d, be arranged at a circular table, when any arrangement obtained from a given arrangement by a

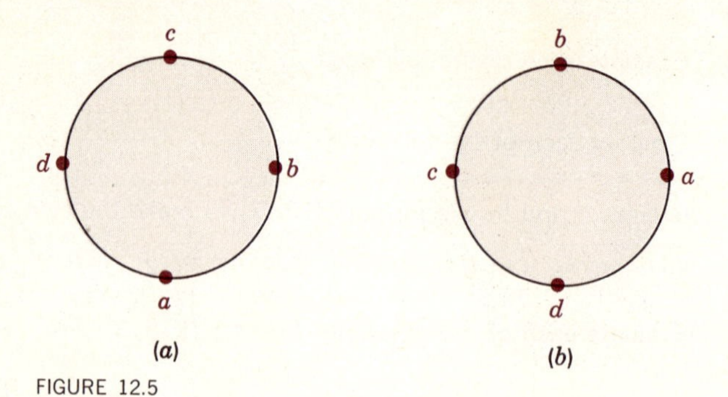

FIGURE 12.5

rotation of the table is considered the same as the given arrangement? To illustrate, the two arrangements shown in Fig. 12.5 are considered to be the same. To determine the number of possible arrangements under our agreement, we can fix the position of one person at the table. Then the chair at his right can be filled in three ways; the next chair to his right can be filled in two ways; the final chair can be filled in one way. So, the fundamental counting principle shows that after the first person is seated, the remaining three chairs can be filled in $3 \cdot 2 \cdot 1 = 3! = 6$ ways; that is, $3! = 6$ is the number of ways in which four people can be arranged at a circular table.

CIRCULAR PERMUTATION

An arrangement of objects about a circle (or any other simple closed curve) is called a **circular permutation.** Reasoning as in the preceding paragraph, we conclude that the *number* of circular permutations of n objects is $(n - 1)!$

We should note that the two circular permutations shown in Fig. 12.6 are different. In the next paragraph we discuss a situation in which these two arrangements are considered to be the same.

We now consider the problem: In how many ways can four distinct beads, denoted by a, b, c, and d, be arranged on a ring? Suppose

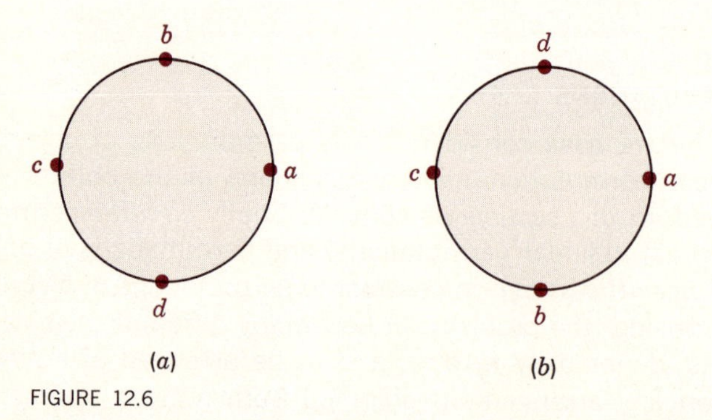

FIGURE 12.6

that the beads are arranged as shown in Fig. 12.6*a*. If the ring is picked up and turned over, the arrangement will appear as in Fig. 12.6*b*; therefore, *for beads on a ring,* the two arrangements shown in Fig. 12.6 are in effect the same. Thus, for beads on a ring that can be picked up and turned over, we conclude that there are half as many possible arrangements as there are circular permutations. Therefore, for four beads on a ring, there are $\frac{1}{2}[(4-1)!] = \frac{1}{2}(3!) = 3$ different arrangements. Similarly, the number of distinct arrangements of n distinct objects on a ring (that is free to be picked up and turned over) is $\frac{1}{2}[(n-1)!]$.

Suppose that we have n objects which are not all distinct. For example, we might have 10 marbles as follows: 4 white marbles which are indistinguishable from each other, 5 red marbles which are indistinguishable from each other, and 1 green marble. We shall describe this situation by saying that we have 10 objects of 3 different kinds; 4 of the first kind (white), 5 of the second kind (red), and 1 of the third kind (green). In general, we consider a situation in which we have n objects of k different kinds, n_1 of the first kind, n_2 of the second kind, and so on, with n_k of the last, or kth, kind. We consider all the objects of any one kind as indistinguishable from each other. For such a situation as this we wish to determine the number of (distinguishable) linear permutations of the n objects.

Let us look at the example of the 10 marbles (4 white, 5 red, 1 green). Two of the possible permutations of these 10 marbles are shown in Fig. 12.7. In Fig. 12.7 if we interchange the position of any of the marbles of the same color, we shall not produce permutations that are distinguishable from the ones in the figure. So, the number of distinguishable linear permutations that can be formed from these 10 marbles is *less* than the number of permutations of 10 distinct (distinguishable) objects. Let D represent the *number* of distinguishable linear permutations that can be formed from our 10 marbles, and suppose that these permutations are arrayed as in Fig. 12.7. For *each* of these D permutations there will be $P(4, 4)$

W R R G W W R W R R

(*a*)

W W R W W R G R R R

(*b*)

FIGURE 12.7

arrangements of the four white marbles that will not change the permutation, there will be $P(5, 5)$ arrangements of the five red marbles that will not change the permutation, and $P(1, 1)$ arrangements of the one green marble that will not change the permutation. Therefore, if the marbles *were* all distinguishable from each other, there would be (by the fundamental counting principle)

$$D \cdot P(4, 4) \cdot P(5, 5) \cdot P(1, 1)$$

different arrangements of the 10 marbles in a row. We know, however, that there are $P(10, 10)$ possible arrangements (permutations) of 10 distinct objects in a row, and so we have

$$D \cdot P(4, 4) \cdot P(5, 5) \cdot P(1, 1) = P(10, 10)$$

Solving this equation for D, we find

$$D = \frac{P(10, 10)}{P(4, 4) \cdot P(5, 5) \cdot P(1, 1)} = \frac{10!}{4!5!1!} \tag{12.18}$$

or

$$D = 1{,}260$$

For the general situation with n objects of k different kinds, n_1 of the first kind, n_2 of the second kind, . . . , n_k of the kth kind, if we let D denote the *number* of distinct linear permutations of the n objects, formula (12.18) generalizes to

$$D = \frac{n!}{n_1!n_2! \cdots n_k!} \tag{12.19}$$

EXAMPLE How many permutations can be made from the letters of the word *M i s s i s s i p p i ?*

SOLUTION There are 11 letters in the word *Mississippi*, 4 i's, 4 s's, 2 p's, and 1 m. So, here $n = 11$, $n_1 = 4$, $n_2 = 4$, $n_3 = 2$, $n_4 = 1$, and an application of (12.19) gives for the desired number D

$$D = \frac{11!}{4!4!2!1!} = 34{,}650$$

EXERCISES

1 In how many ways can seven persons be seated at a round table if only their relative positions are of interest?

2 In how many relative positions can seven keys be arranged on a ring?

3 In how many ways can six persons be seated at a round table if only their relative positions are considered but so that two particular persons always are seated side by side?

4 In how many ways can nine keys be arranged on a key ring?

5 In how many ways can 10 persons be seated at a round table if only relative positions are considered?

6 In how many ways can 10 persons be seated at a round table if only relative positions are considered but so that two particular persons are seated side by side?

7 In how many ways can 10 persons be seated at a round table if only relative positions are considered but so that two particular persons will never be seated side by side?

In Exercises 8 to 11, find the number of permutations that can be made using the letters of the given word.

8 Tell 9 Mathematics 10 College 11 Illinois

12 In how many distinguishable ways can 10 balls be arranged in a line if 3 are white (and alike), 5 are red, and 2 are blue?

13 How many distinct signals can be formed by hoisting 10 flags if 4 are red, 3 yellow, 2 green, and 1 blue?

12.7 COMBINATIONS

In Sec. 12.5 we observed that there are six 2-permutations of the three letters a, b, c, namely,

$$ab \quad ba \quad ac \quad ca \quad bc \quad cb \tag{12.20}$$

Here ab and ba are different permutations, although they contain the same letters, because *order is important in a permutation.* Let us consider a situation in which order is not important. Suppose that we wish to form a committee of two persons to be selected from a set A of three people. How many such committees can be formed? Let $A = \{a, b, c\}$. There are six 2-permutations of the three members a, b, c, and we might think that there are six committees of size two. However, we observe that the committee denoted by $\{a, b\}$ is indistinguishable from the committee denoted by $\{b, a\}$; the order in which committee members are listed has no influence on the membership of the committee. We conclude that there are three committees of two persons each that can be formed from a set of three people, namely,

$$\{a, b\} \quad \{a, c\} \quad \{b, c\} \tag{12.21}$$

Each of the sets in (12.21) is called a *combination,* or more specifically,* a 2-*combination of the three letters* a, b, c. The order in

* The terminology "a combination of the letters a, b, c taken two at a time" is also used.

which the members of each set are listed is not significant, and the members of each set are distinct.

COMBINATION

In general, a **combination** is an *unordered* set of *distinct* objects. We emphasize that *in a combination order is of no concern.* This characteristic of a combination contrasts with the requirement that in a permutation order *is* of primary concern.

If a combination consists of r elements chosen from a set with n members, it is called* an **r-combination of n objects.**

AN r-COMBINATION OF n OBJECTS

To illustrate, a poker hand consisting of 5 cards dealt from an ordinary pack of 52 cards is a 5-combination of 52 objects; if we select 3 marbles from a bag containing 12 marbles, we have produced a 3-combination of 12 objects.

The *number* of r-combinations of n objects that can be formed is denoted by†

$$C(n, r) \qquad \text{where} \qquad 0 \leq r \leq n$$

and this number is of considerable importance in the application of probability and in other areas of mathematics. Fortunately, for given values of the nonnegative integers n and r, with $r \leq n$, it is relatively simple to determine the value of this number $C(n, r)$.

First, let us consider the 3-combinations that can be formed from the members of the set $\{a, b, c, d\}$. We observe that there are four subsets of $\{a, b, c, d\}$ consisting of three members each, namely,

$$\{a, b, c\} \qquad \{a, b, d\} \qquad \{a, c, d\} \qquad \{b, c, d\} \tag{12.22}$$

Hence $C(4, 3) = 4$. For *each* of the four 3-combinations in (12.22) we can construct $P(3, 3) = 3! = 6$ permutations; so, the *total* number of 3-*permutations* that can be formed from the members of $\{a, b, c, d\}$ is $4 \cdot 6$ or

$$C(4, 3) \cdot 3!$$

But we know that the number of 3-permutations of four objects is given by $P(4, 3)$, and so we have the equalities

$$C(4, 3) \cdot 3! = P(4, 3)$$

and

$$C(4, 3) = \frac{P(4, 3)}{3!}$$

Now, for given values of n and r with $0 \leq r \leq n$, consider the general problem of determining the number $C(n, r)$ of r-combinations

* The terminology "a combination of n objects taken r at a time" is also used for this combination. However, this terminology fails to reflect the fact that the combination being described is a collection of r objects rather than n objects.

† The symbols ${}_nC_r$ and $\binom{n}{r}$ are also used to denote the *number* of r-combinations of n objects.

of n objects. For *each* of the $C(n, r)$ r-combinations we can construct $P(r, r) = r!$ *permutations.* So, the *total* number of *r-permutations* that can be formed from the set with n members is $C(n, r) \cdot r!$ But this number of r-permutations of n objects is given by $P(n, r)$, and so

$$C(n, r) \cdot r! = P(n, r)$$

or

NUMBER OF r-COMBINATIONS OF n OBJECTS

$$C(n, r) = \frac{P(n, r)}{r!} \tag{12.23}$$

Since we know, from formula (12.11) or (12.17), how to compute $P(n, r)$ for given values of n and r, formula (12.23) provides a way to compute $C(n, r)$ for given values of n and r, with $0 \leq r \leq n$.

To illustrate,

$$C(6, 3) = \frac{P(6, 3)}{3!} = \frac{6 \cdot 5 \cdot 4}{3 \cdot 2 \cdot 1} = 20$$

$$C(8, 5) = \frac{P(8, 5)}{5!} = \frac{8 \cdot 7 \cdot 6 \cdot 5 \cdot 4}{5 \cdot 4 \cdot 3 \cdot 2 \cdot 1} = 56$$

Formula (12.23) for $C(n, r)$ can be put in another form by the use of formula (12.17), $P(n, r) = \frac{n!}{(n-r)!}$. Substituting this expression for $P(n, r)$ in formula (12.23) gives

$$C(n, r) = \frac{n!}{r!(n-r)!} \tag{12.24}$$

If in formula (12.24) we replace r by $n - r$, we have

$$C(n, n-r) = \frac{n!}{(n-r)![n-(n-r)]!} = \frac{n!}{(n-r)!r!}$$

which is the same as $C(n, r)$. Thus, *the number of r-combinations of n objects is the same as the number of $(n - r)$-combinations of n objects;*

$$C(n, r) = C(n, n-r) \tag{12.25}$$

This result is useful when r is greater than $\frac{n}{2}$. For example, to compute $C(10, 8)$ by use of formula (12.23), we would write

$$C(10, 8) = \frac{10 \cdot 9 \cdot 8 \cdot 7 \cdot 6 \cdot 5 \cdot 4 \cdot 3}{8 \cdot 7 \cdot 6 \cdot 5 \cdot 4 \cdot 3 \cdot 2 \cdot 1}$$

but using (12.25) we can write

$$C(10, 8) = C(10, 2) = \frac{10 \cdot 9}{2 \cdot 1} = 45$$

EXAMPLE 1 How many different committees of three persons each can be selected from a set of nine persons?

SOLUTION Since the order in which the three members of a committee are selected is of no concern, each committee is a *combination* of three persons, chosen from a set of nine persons. Therefore, we are looking for the number of 3-combinations of nine objects. We find, by use of (12.23), that

$$C(9, 3) = \frac{9 \cdot 8 \cdot 7}{3 \cdot 2 \cdot 1} = 84$$

So, there are 84 three-member committees that can be selected from a group of nine persons.

EXAMPLE 2 In how many ways can a committee of three persons, consisting of a chairman and two other members, be selected from a set of nine persons?

SOLUTION We first select the chairman. This may be done in nine ways. The order in which the other two members may be chosen from the remaining *eight* persons is of no concern, and this choice can be made in $C(8, 2) = \frac{8 \cdot 7}{2 \cdot 1} = 28$ ways. Therefore, the desired number is $9 \cdot 28 = 252$.

EXAMPLE 3 In how many ways can a committee of 9 members be selected from a set of 12 persons?

SOLUTION The number of subsets with 9 members each that can be formed from a set of 12 people is $C(12, 9)$. Using formulas (12.25) and (12.23), we have

$$C(12, 9) = C(12, 3) = \frac{12 \cdot 11 \cdot 10}{3 \cdot 2 \cdot 1} = 220$$

EXAMPLE 4 A box contains four red, six green, and two white marbles. If three marbles are selected at random from the box, what is the probability that (*a*) the three marbles are red? (*b*) The three marbles are green? (*c*) The three marbles are white?

SOLUTION The experiment in this problem is selecting three marbles at random from a set of twelve marbles and observing the colors of the marbles. There are three events whose probabilities we wish to compute,

E_1: Three red marbles are selected.

E_2: Three green marbles are selected.

E_3: Three white marbles are selected.

Since the selection is made "at random," the outcomes are equally likely, and so we can use the result (12.6) to compute the probabilities of the events.

The number of ways in which three marbles can be selected from a set of twelve marbles is

$$C(12, 3) = \frac{12 \cdot 11 \cdot 10}{3 \cdot 2 \cdot 1} = 220$$

(a) For event E_1 to occur, three red marbles must be selected. These three red marbles must come from the four that are in the box. So the number of ways that E_1 can occur is the number of subsets with three members each that can be formed from a set of four objects; thus E_1 can occur in $C(4, 3) = C(4, 1) = 4$ ways, and

$$\Pr(E_1) = \tfrac{4}{220} = \tfrac{1}{55}$$

(b) For event E_2 to occur, the three marbles selected must all be green. The three green marbles must come from the six green marbles in the box, so the number of ways that E_2 can occur is $C(6, 3) = \dfrac{6 \cdot 5 \cdot 4}{3 \cdot 2 \cdot 1} = 20$. Therefore,

$$\Pr(E_2) = \tfrac{20}{220} = \tfrac{1}{11}$$

(c) Event E_3 cannot occur, since there are only two white marbles in the box. That is, $E_3 = \varnothing$ and

$$\Pr(E_3) = 0$$

EXAMPLE 5 If four marbles are selected from a box containing seven green and nine blue marbles, what is the probability that two of the marbles are green and the other two blue?

SOLUTION The experiment in this example is selecting 4 marbles from a set of 16 marbles and observing the colors; that is, selecting a subset having 4 members from a set with 16 members. The number of outcomes of this experiment is given by

$$C(16, 4) = \frac{16 \cdot 15 \cdot 14 \cdot 13}{4 \cdot 3 \cdot 2 \cdot 1} = 1{,}820$$

The event E that two green and two blue marbles are selected can be considered as the successive performance of two acts: (1) select two green marbles and (2) select two blue marbles. The two green marbles must come from the seven green marbles in the box and therefore two green marbles can be selected in $C(7, 2)$ ways. The two blue marbles must be selected from the nine blue marbles in the box and therefore two blue marbles can be selected in $C(9, 2)$ ways. Then, by the fundamental counting principle, the number of ways that the event E can occur is the product of the number of ways the first act can be performed and the number of ways the second act can be performed. Hence

$$\text{number of ways } E \text{ can occur} = C(7, 2) \cdot C(9, 2) = \frac{7 \cdot 6}{2 \cdot 1} \cdot \frac{9 \cdot 8}{2 \cdot 1} = 756$$

Then, by (12.6)

$\Pr(E) = \frac{756}{1{,}820} = \frac{189}{455}$

If in formula (12.23) we set $r = n$, we have

$$C(n, n) = \frac{P(n, n)}{n!} = \frac{n!}{n!} = 1$$

which symbolizes the fact that there is only one subset with n members that can be selected from a set A with n members, namely, the set A itself.

If, in formula (12.24) we set $r = 0$, we have (recalling that $0! = 1$)

$$C(n, 0) = \frac{n!}{0!n!} = 1$$

which symbolizes the fact that there is only one subset with no members that can be selected from a given set A of n objects, namely, the empty set $\varnothing$.

It is worthwhile to note that the number of subsets consisting of one member each that can be selected from a given set A with n members is n; that is, $C(n, 1) = n$, or since $C(n, n - 1) = C(n, 1)$

$$C(n, n - 1) = C(n, 1) = n$$

Note that $C(n, 1)$ is simply the number of ways that a single object can be selected from a given set with n members.

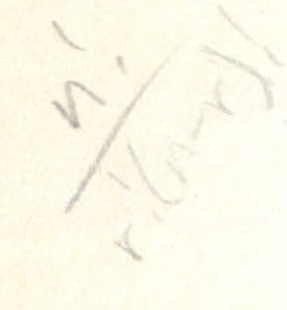

EXERCISES

1 Evaluate each of the following: $C(8, 3)$; $C(18, 16)$; $C(10, 5)$; $C(11, 2)$.

2 Evaluate each of the following:

$C(75, 73)$; $C(17, 14)$; $\dfrac{C(5, 3)}{C(6, 4)}$; $\dfrac{C(10, 5)}{C(5, 2)}$.

3 In how many ways can a set of 50 cards be chosen from a set of 52 playing cards?

4 How many straight lines can be drawn through six points, no three of which are on the same straight line?

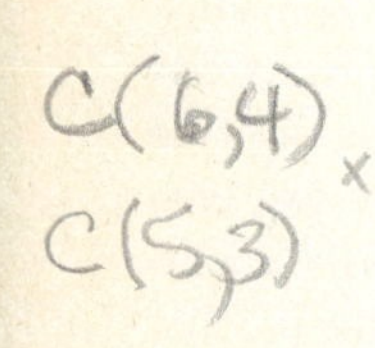

5 From a set of six men and a set of five women, how many ways can we choose a committee of four men and three women?

6 In how many ways can a committee consisting of two teachers and three students be selected from a set of eight teachers and a set of 15 students?

7 How many different sums can be made by using any combination of one or more of a penny, a nickel, a dime, a quarter, a half-dollar, and a dollar? *Hint:* Find how many sums can be made by using one coin, then how many can be made by using two coins, then how many with three, and so on.

8 From seven Russians and four Americans a committee of six is to be selected. In how many ways can this committee be chosen if it is to contain exactly two Americans?

9 From seven Russians and four Americans a committee of six is to be formed. In how many ways can this committee be chosen if it is to contain at least two Americans?

10 Four marbles are drawn at random from a box containing five red marbles and three green marbles. What is the probability that (a) all four are red? (b) All four are green? (c) Two are red and two are green? (d) One is red and three are green?

11 For the experiment in Exercise 10, what is the probability that (a) three marbles are red and one is green? (b) More than two are red? (c) More than two are green? (d) Fewer than three are green?

12 Six beads are selected at random from a box containing seven red, four green, three blue, and six yellow beads. What is the probability that (a) two are red, two are green, and two are blue? (b) One is red, two are green, two are blue, and one is yellow? (c) All are yellow? (d) At least one is not yellow? *Hint:* $C(20, 6) = 41{,}160$. The union of the events (c) and (d) is the entire sample space.

The remaining Exercises involve an ordinary pack of 52 cards containing four suits, spades, hearts, diamonds, and clubs. There are 13 different "kinds" in each suit; these kinds are labeled 2, 3, 4, 5, 6, 7, 8, 9, 10, jack, queen, king, ace. A bridge hand consists of 13 cards selected at random. There are $C(52, 13) = 635{,}013{,}559{,}600$ different 13-card hands that can be formed. A poker hand consists of 5 cards selected at random. There are $C(52, 5) = 2{,}598{,}960$ different poker hands that can be formed. Suppose we wish to determine the probability of being dealt a poker hand with all five of the cards in the same suit; such a hand is called a *flush*. We can consider the event of five cards in the same suit as a succession of two tasks: (1) select a suit and (2) select five cards from that suit. The fundamental counting principle tells us that the number of ways the event can occur is the product of the number of ways the first task can be performed and the number of ways the second task can be performed. There are four ways in which a suit can be selected. Then from the 13 cards of that suit we select 5 cards; this can be done in $C(13, 5)$ ways (since the order of the cards is not significant, we want the number of *combinations*). Thus the event of a flush can occur in

$$4 \cdot C(13, 5) = 4 \cdot \frac{13 \cdot 12 \cdot 11 \cdot 10 \cdot 9}{5 \cdot 4 \cdot 3 \cdot 2 \cdot 1} = 5{,}148$$

ways, and the probability of drawing a flush is

$$\frac{5{,}148}{2{,}598{,}960} = 0.0019$$

that is, 19 chances out of 10,000.

13 What is the probability of being dealt a bridge hand with all of the cards being spades? With all the cards in the same suit?

14 What is the probability of being dealt a bridge hand with no aces? With no face cards? (The face cards are J, Q, K, A.) Record each answer in the form $\frac{C(n, r)}{C(m, r)}$.

15 If a poker hand of 5 cards is dealt from an ordinary pack of 52 cards, determine the number of ways each of the following hands can occur: (a) A, K, Q, J, 10 all in the same suit (a *royal straight flush*). (b) Five cards all in the same suit arranged in numerical order, counting a jack as 11, a queen as 12, a king as 13, and an ace as *either* 1 or 14. (Examples: A, 2, 3, 4, 5; 6, 7, 8, 9, 10; 8, 9, 10, J, Q; 10, J, Q, K, A.) Such a hand is called a *straight flush*. *Hint:* How many suits are there? In each suit, how many sequences of five cards in numerical order are there? (Remember that the ace is either 1 or 14.) (c) Five cards arranged in numerical order but *not* all in the same suit. (d) Four of a kind (four 2s, four 3s, four 4s, and so on).

16 In how many ways can each of the following poker hands be dealt? (a) Three cards of one kind and two cards of another kind. Examples: three aces and two 5s; three 2s and two 10s. Such a hand is called a *full house*. *Hint:* The event of a full house can be considered to be a succession of tasks: (1) Select one kind; (2) select three of this kind; (3) select another kind; (4) select two of this kind. Recall that there are 13 kinds and 4 of each kind. (b) Three cards of one kind and the other two cards of different kinds (*three of a kind*). *Hint:* There are five tasks to be performed: (1) Select one kind; (2) select three of this kind; (3) select two of the remaining kinds; (4) from the first of these two kinds select one card; (5) from the second of these two kinds select one card.

17 In how many ways can each of the following poker hands be dealt? (a) Two of one kind, two of another kind, and one card of a third kind. Such a hand is called *two pair*. (b) Two cards of one kind and the other three cards of different kinds.

12.8 FURTHER TOPICS IN PROBABILITY

The computations of the probabilities of events which we have made up to now have been accomplished by direct use of the definitions (12.2), (12.3), and (12.6) in Sec. 12.3, the fundamental counting principle, and formulas giving numbers of permutations or combinations. In this section we discuss methods that enlarge the number of types of probability computations that we can make.

We recall that for a given experiment, an event E is a subset of the sample space $S = \{e_1, e_2, e_3, \ldots, e_n\}$ for the experiment, so that E is the union of a number of *simple* events. Also, we recall that the probability $\Pr(E)$ of the event E is the sum of the probabilities of the simple events of which E is the union. Since according to definition (12.2), the sum of the probabilities of all the *simple* events in an experiment is 1, and each of these probabilities is nonnegative, it follows that

for *any* event E
$0 \leq \Pr(E) \leq 1$ (12.26)

If E_1 and E_2 are events with $E_1 \subseteq E_2$, then every sample point that appears in E_1 will also appear in E_2. Thus, if $E_1 = \{e_1, e_2, \ldots, e_k\}$, then E_2 will contain all the sample points $e_1, e_2, \ldots, e_k$ and possibly some other points. Therefore, $\Pr(E_2)$ is *at least as large* as $\Pr(E_1)$ and

if $E_1 \subseteq E_2$, then
$\Pr(E_1) \leq \Pr(E_2)$ (12.27)

For an illustration of result (12.27), consider the experiment of tossing two coins for which the sample space is $S = \{HH, HT, TH, TT\}$. Let $E_1 = \{HH\}$ be the event that both coins are heads and $E_2 = \{HH, HT, TH\}$ be the event that at least one head appears. Here $E_1 \subseteq E_2$, $\Pr(E_1) = \frac{1}{4}$, $\Pr(E_2) = \frac{3}{4}$, and $\Pr(E_1) \leq \Pr(E_2)$.

MUTUALLY EXCLUSIVE EVENTS

Two events E_1 and E_2 are said to be **mutually exclusive** if they cannot both occur on a single trial of a experiment. Recall that an event occurs on a trial of an experiment if the outcome of the trial is an element of E, so if E_1 and E_2 are mutually exclusive, there cannot be an element that belongs to both E_1 and E_2. Therefore,

E_1 and E_2 are mutually exclusive if and only if
$E_1 \cap E_2 = \varnothing$ (12.28)

Similarly, events $E_1, E_2, \ldots, E_k$ are mutually exclusive if and only if

$E_i \cap E_j = \varnothing;\ i, j = 1, 2, \ldots, k,\ i \neq j$

An example of two mutually exclusive events in the experiment of tossing a single die, where the sample space is $\{1, 2, 3, 4, 5, 6\}$, are the events

$E_1 = \{1, 2, 3\}$ of a 1, a 2, or a 3 appearing

and

$E_2 = \{5, 6\}$ of a 5 or a 6 appearing.

Here $E_1 \cap E_2 = \varnothing$, and the events are mutually exclusive. For the same experiment, the events

$E_3 = \{1, 3, 5\}$ of throwing an odd number

and

$E_4 = \{3, 4\}$ of throwing a 3 or a 4

are *not* mutually exclusive since $E_3 \cap E_4 = \{3\}$ is not empty; that is, both events could occur on a single trial of the experiment.

Suppose that we consider an event E which we know to be the *union* of two events E_1 and E_2. Can we determine the probability of E from knowing the probabilities of E_1 and E_2? Since the union of two sets is the set of all elements that are members of *at least one* of the two sets, $E_1 \cup E_2$ is the set of all sample points of the sample space that belong either to E_1 or to E_2 or to both E_1 and E_2. Now $\text{Pr}(E_1) + \text{Pr}(E_2)$ is the sum of the probabilities assigned to the sample points in E_1 plus the sum of the probabilities assigned to the sample points in E_2. Therefore, $\text{Pr}(E_1) + \text{Pr}(E_2)$ includes the probabilities assigned to the sample points in $E_1 \cap E_2$ *twice*. So, if we subtract from $\text{Pr}(E_1) + \text{Pr}(E_2)$ the sum of the probabilities assigned to the sample points in $E_1 \cap E_2$, that is, if we subtract $\text{Pr}(E_1 \cap E_2)$, we have left the sum of the probabilities assigned to the distinct sample points of $E_1 \cup E_2$. This result is symbolized by the formula

$$\text{Pr}(E_1 \cup E_2) = \text{Pr}(E_1) + \text{Pr}(E_2) - \text{Pr}(E_1 \cap E_2) \tag{12.29}$$

In case E_1 and E_2 are mutually exclusive events, $E_1 \cap E_2 = \varnothing$ and we have

$$\text{Pr}(E_1 \cup E_2) = \text{Pr}(E_1) + \text{Pr}(E_2) \quad \text{for mutually exclusive events } E_1 \text{ and } E_2 \tag{12.30}$$

EXAMPLE 1 A single ticket is drawn from a box containing five red tickets, six black tickets, and eight white tickets. What is the probability of the event E that the ticket be red or black?

SOLUTION Consider E to be the union of the events E_1, the ticket is red, and E_2, the ticket is black. E_1 and E_2 are mutually exclusive and we have, by (12.30),

$$\text{Pr}(E) = \text{Pr}(E_1) + \text{Pr}(E_2) = \tfrac{5}{19} + \tfrac{6}{19} = \tfrac{11}{19}$$

EXAMPLE 2 Consider the experiment of tossing a coin and throwing a die once for which $S = \{H1, H2, H3, H4, H5, H6, T1, T2, T3, T4, T5, T6\}$ is a sample space. Let E_1 be the event that the coin falls heads and E_2 be the event that the die turns up even, so

$$E_1 = \{H1, H2, H3, H4, H5, H6\}$$

and

$$E_2 = \{H2, T2, H4, T4, H6, T6\}$$

Then $\Pr(E_1) = \frac{6}{12}$ and $\Pr(E_2) = \frac{6}{12}$. Also,

$$E_1 \cap E_3 = \{H2, H4, H6\} \qquad \text{with} \qquad \Pr(E_1 \cap E_2\} = \tfrac{3}{12}$$

Using (12.29), we get

$$\Pr(E_1 \cup E_2) = \tfrac{6}{12} + \tfrac{6}{12} - \tfrac{3}{12} = \tfrac{9}{12} = \tfrac{3}{4}$$

The result can be obtained directly by observing that

$$E_1 \cup E_2 = \{H1, H2, H3, H4, H5, H6, T2, T4, T6\}$$

from which we see that

$$\Pr(E_1 \cup E_2) = \tfrac{9}{12} = \tfrac{3}{4}$$

Consider again the experiment, described in Example 2, of tossing a coin and throwing a die, for which the sample space S contains 12 sample points. Given that the coin has come up heads, what is the probability that the die shows an even number? Since $E_1 = \{H1, H2, H3, H4, H5, H6\}$ is the event that the coin comes up heads and $E_2 = \{H2, T2, H4, T4, H6, T6\}$ is the event that the die shows an even number, we are asking the question: Given that E_1 has occurred, what is the probability of E_2? Since we are given that E_1 has occurred, the only possible outcomes are the six members of the event E_1. Of these six outcomes, three have the die showing an even number, and so, given that E_1 has occurred, the probability of the die showing an even number is

$$\frac{3}{6} = \frac{1}{2}$$

In this example (with sample space S), we have seen that in finding the probability of an event E_2 given that an event E_1 has occurred, we do not consider S to be the sample space, but consider E_1 (a subset of S) to be the *reduced sample space.*

REDUCED SAMPLE SPACE

We adopt the symbol $\Pr(E_2/E_1)$ for the probability of E_2, given that E_1 has occurred.

EXAMPLE 3 Consider the experiment of throwing two dice. If the sum of the numbers thrown is less than 6, what is the probability that one and only one of the dice shows a 2?

SOLUTION There are $6 \cdot 6 = 36$ possible outcomes of the experiment, and so a sample space consists of 36 sample points (see Exercise 4 in Sec. 12.4). Let E_1 be the event that the sum of the numbers thrown is less than 6, and let E_2 be the event that one and only one of the dice shows a 2. Then

$$E_1 = \{(1, 1), (1, 2), (1, 3), (1, 4), (2, 1), (2, 2), (2, 3), (3, 1), (3, 2), (4, 1)\}$$

and

$$E_2 = \{(2, 1), (2, 3), (2, 4), (2, 5), (2, 6), (1, 2), (3, 2), (4, 2), (5, 2), (6, 2)\}$$

Since we are told that E_1 has occurred, then E_1 is the *reduced sample space*. Of the 10 simple events in the reduced sample space E_1, 4 have one and only one of the dice showing a 2. Thus, the probability of E_2, given that E_1 has occurred, that is $\Pr(E_2/E_1)$, is given by $\Pr(E_2/E_1) = \frac{4}{10} = \frac{2}{5}$.

For the experiment and events given in Example 3, let us consider the probability that the sum of the numbers thrown is less than 6 *and* one and only one of the dice shows a 2. We want the probability of the event $E = E_1 \cap E_2$. We see that

$$E_1 \cap E_2 = \{(1, 2), (2, 1), (2, 3), (3, 2)\}$$

contains four simple events, and therefore

$$\Pr(E_1 \cap E_2) = \frac{4}{36} = \frac{1}{9}$$

Recall that in Example 3,

$$\Pr(E_2/E_1) = \frac{2}{5}$$

and also note that

$$\Pr(E_1) = \frac{10}{36} = \frac{5}{18}$$

Examining these three probabilities, we see that

$$\Pr(E_2/E_1) \cdot \Pr(E_1) = \Pr(E_1 \cap E_2)$$

Thus, *for this example,* the probability $\Pr(E_1 \cap E_2)$ of the joint occurrence of two events and the probability $\Pr(E_2/E_1)$ of event E_2, given that E_1 has occurred, are related by the equation

$$\Pr(E_2/E_1) = \frac{\Pr(E_1 \cap E_2)}{\Pr(E_1)} \qquad \text{if} \qquad \Pr(E_1) \neq 0$$

Similarly, for the experiment of tossing a coin and throwing a die, where

$$S = \{H1, H2, H3, H4, H5, H6, T1, T2, T3, T4, T5, T6\}$$
$$E_1 = \{H1, H2, H3, H4, H5, H6\}$$
$$E_2 = \{H2, T2, H4, T4, H6, T6\}$$

we find that

$$\Pr(E_1 \cap E_2) = \Pr(\{H2, H4, H6\}) = \frac{3}{12} = \frac{1}{4}$$

$$\Pr(E_1) = \frac{6}{12} = \frac{1}{2}$$

$$\Pr(E_2/E_1) = \frac{3}{6} = \frac{1}{2}$$

and again

$$\Pr(E_2/E_1) = \frac{\Pr(E_1 \cap E_2)}{\Pr(E_1)} \qquad \text{if} \qquad \Pr(E_1) \neq 0$$

These two examples (and others that we could examine) lead us to the following definition.

Let E_1 and E_2 be events of a sample space S with the property that $\Pr(E_1) \neq 0$. Let $\Pr(E_2/E_1)$ denote the probability of E_2, given that E_1 has occurred. We define $\Pr(E_2/E_1)$ by

$$\Pr(E_2/E_1) = \frac{\Pr(E_1 \cap E_2)}{\Pr(E_1)} \tag{12.31}$$

PROBABILITY OF E_2, GIVEN E_1

and we call* $\Pr(E_2/E_1)$ the **probability of E_2, given E_1**.

From (12.31), we get

$$\Pr(E_1 \cap E_2) = \Pr(E_2/E_1) \cdot \Pr(E_1) \qquad \text{if} \qquad \Pr(E_1) \neq 0$$

But if $\Pr(E_1) = 0$, then surely $\Pr(E_1 \cap E_2) = 0$, so

$$\Pr(E_1 \cap E_2) = \Pr(E_2/E_1) \cdot \Pr(E_1) \tag{12.32}$$

without restriction of $\Pr(E_1)$.

Similarly, we can show that

$$\Pr(E_2 \cap E_1) = \Pr(E_1/E_2) \cdot \Pr(E_2) \tag{12.33}$$

without restriction on $\Pr(E_2)$. Since $E_1 \cap E_2 = E_2 \cap E_1$, we have $\Pr(E_1 \cap E_2) = \Pr(E_2 \cap E_1)$. Thus, it follows from (12.32) and (12.33) that

$$\Pr(E_1 \cap E_2) = \Pr(E_2/E_1) \cdot \Pr(E_1) = \Pr(E_1/E_2) \cdot \Pr(E_2) \tag{12.34}$$

INDEPENDENT EVENTS

If the occurrence of an event is unaffected by the occurrence or nonoccurrence of another event, we say that the events are **independent.** In other words, E_1 and E_2 are independent events if and only if

$$\Pr(E_1/E_2) = \Pr(E_1) \qquad \text{and} \qquad \Pr(E_2/E_1) = \Pr(E_2) \tag{12.35}$$

If E_1 and E_2 are *independent* events, we can use (12.35) and in (12.34) we can replace $\Pr(E_2/E_1)$ by $\Pr(E_2)$ and $\Pr(E_1/E_2)$ by $\Pr(E_1)$ and obtain the result

if E_1 and E_2 *are independent, then*

$$\Pr(E_1 \cap E_2) = \Pr(E_1) \cdot \Pr(E_2) \tag{12.36}$$

* $\Pr(E_2/E_1)$ is also called the *conditional probability* of E_2, given E_1.

EXAMPLE 4 In the experiment of tossing a coin and throwing a die once, with the sample space of Example 2, let us assume that how the coin falls and how the die falls are independent events. What is the probability $\Pr(E_1 \cap E_2)$ that the coin will fall heads, E_1, *and* the die will fall with a 2, E_2?

SOLUTION Here $\Pr(E_1) = \frac{1}{2}$ and $\Pr(E_2) = \frac{1}{6}$. Using (12.36), we get

$$\Pr(E_1 \cap E_2) = \tfrac{1}{2} \cdot \tfrac{1}{6} = \tfrac{1}{12}$$

This result is readily verified by using the sample space of Example 2, noting that $E_1 \cap E_2 = \{H2\}$, so $\Pr(E_1 \cap E_2) = \frac{1}{12}$.

EXAMPLE 5 If five coins are tossed once, what is the probability of obtaining two heads and three tails?

SOLUTION Five coins may be tossed in $2 \cdot 2 \cdot 2 \cdot 2 \cdot 2 = 2^5 = 32$ ways, so a sample space of the experiment of tossing five coins once has 32 sample points and is the union of 32 simple events.

The number of ways of obtaining two heads and three tails when five coins are tossed once is equal to the number of permutations of five things, two of which are alike in one way (being heads) and the other three of which are alike in a second way (being tails). From Sec. 12.6 we know that the number of such permutations is

$$\frac{5!}{2!3!} = C(5, 2) = \frac{5 \cdot 4 \cdot 3!}{2 \cdot 3!} = 10$$

That is, the event E of throwing two heads and three tails is the union of 10 (mutually exclusive) simple events. Since the sample space has 32 members, we conclude that

$$\Pr(E) = \tfrac{10}{32} = \tfrac{5}{16}$$

Recall that the *complement* of a set R relative to a universe S is the set of all members of S that are *not* members of R. We denote the complement of R by R'. For example, if $S = \{a, b, c, d, e\}$ and $R = \{a, b\}$, then $R' = \{c, d, e\}$.

COMPLEMENTARY EVENTS

Let S be the sample space of an experiment. If E' is the complement of an event E relative to the sample space S, then E and E' are called **complementary events.** An event E *does not occur* on a trial of the experiment if the outcome of the trial is an element of E'. Note that if E and E' are complementary events in a sample space S, then it follows from the definition of the complement of a set that

$$E \cup E' = S \qquad \text{and} \qquad E \cap E' = \varnothing$$

Therefore, from (12.30), $\Pr(E \cup E') = \Pr(E) + \Pr(E')$ and, since $\Pr(E \cup E') = \Pr(S) = 1$, we have

$$\text{Pr}(E) + \text{Pr}(E') = 1 \tag{12.37}$$

This result was previously referred to in Sec. 12.3.

To illustrate, let $S = \{1, 2, 3, 4, 5, 6\}$, $E_1 = \{1, 3, 5\}$, and $E_2 = \{2, 4, 6\}$. Clearly, $E_2 = E_1'$. Here $\text{Pr}(E_1) = \frac{1}{2}$, $\text{Pr}(E_1') = \frac{1}{2}$, and $\text{Pr}(E_1) + \text{Pr}(E_1') = 1$

EXAMPLE 6 Consider the experiment of tossing three coins once, for which we have seen that the sample space contains eight sample points. What is the probability that *at least* one head appears?

SOLUTION Let E denote the event that at least one head appears. Observe that the complementary event E' is the event that no heads appear. So $E' = \{TTT\}$ and $\text{Pr}(E') = \frac{1}{8}$. Now, using (12.37), we obtain

$$\text{Pr}(E) = 1 - \text{Pr}(E') = 1 - \tfrac{1}{8} = \tfrac{7}{8}$$

EXAMPLE 7 If a die is thrown seven times, find the probability that (*a*) a 3 will appear only on the *first two throws;* (*b*) a 3 will appear in *exactly* two of the seven throws.

SOLUTION (*a*) The probability of throwing a 3 the first time is $\frac{1}{6}$ and the second time is $\frac{1}{6}$; the probability of throwing a number other than 3 the third time is $\frac{5}{6}$, the fourth time is $\frac{5}{6}$, the fifth time is $\frac{5}{6}$, the sixth time is $\frac{5}{6}$, and the seventh time is $\frac{5}{6}$. These are independent events, so the probability of throwing a 3 only on the first two throws is the product of their probabilities, and this is

$$\left(\frac{1}{6}\right)^2\left(\frac{5}{6}\right)^5$$

(*b*) The two throws out of the seven throws on which a 3 is to appear may be selected in the number of ways equal to the number of permutations of seven things, two of which are alike in one way (being 3s) and the other five of which are alike in a second way (being a number other than 3). From Sec. 13.5 we know that the number of such permutations is

$$\frac{7!}{2!5!} = C(7, 2) = \frac{7 \cdot 6 \cdot 5!}{2!5!} = \frac{42}{2} = 21 \text{ ways}$$

The event E of throwing a 3 in exactly two of seven throws is therefore the union of 21 mutually exclusive events. Since the probability of each of these 21 mutually exclusive events is $(\frac{1}{6})^2(\frac{5}{6})^5$, we conclude that the required probability is

$$\text{Pr}(E) = 21\left(\frac{1}{6}\right)^2\left(\frac{5}{6}\right)^5$$

EXERCISES

1 A card is drawn from a pack of 52 bridge cards, then replaced, and a second card is drawn. What is the probability that both cards drawn are aces?

2 The probability of A's winning a race is $\frac{1}{3}$, of B's winning $\frac{1}{4}$, and of C's winning $\frac{1}{6}$. What is the probability that none of them will win?

3 One urn contains three white and five red balls, and a second contains four white and three red balls. A ball is drawn from each urn. What is the probability that both are red?

4 If a die is thrown three times, find the probability that (a) each of the throws will be a 1; (b) the first two throws will be 1s, but the third throw will not be a 1; (c) in three throws there will be exactly two 1s.

5 Three coins are tossed, one after the other. Let E_1 be the event "with at least two heads" and E_2 the event "the first coin falls heads." Find the probability of E_1, given E_2 has occurred.

6 In a single throw of an ordinary cubical die, what is the probability of obtaining an even number or a number divisible by 3?

7 Three coins are tossed once. Find the probability that they will all fall heads: (a) by use of an obvious generalization of (12.36); (b) without using (12.36).

8 If $S = \{a, b, c, d, e, f\}$, $R = \{a, c, f\}$, $T = \{e, f\}$, and $C = \varnothing$, find R', T', and C'.

9 If two dice are thrown six times, what is the probability of obtaining (a) a sum of 7 on only the first four throws? (b) A sum of 7 on *exactly* four of the six throws?

10 If 12 men stand in a line, what is the probability that a specified pair stand side by side?

11 The probability that A will solve a problem is $\frac{1}{3}$ and that B will solve the problem is $\frac{2}{3}$. If A and B both try the problem, what is the probability that it will be solved?

12 If the probability that A will die within a year is $\frac{2}{10}$ and the probability that B will die within a year is $\frac{3}{10}$, find the probability that (a) both A and B will die within a year; (b) both A and B will live a year; (c) only one of A and B will die within a year.

12.9 ODDS AND MATHEMATICAL EXPECTATION

Instead of speaking of the probability that a certain event will occur, we frequently speak of the "odds" in favor of a certain event or the odds that a certain event will occur. When we say that the odds are 5 to 2 (or $\frac{5}{2}$) that team A will win the Super Bowl, we are saying that if the same two teams played repeatedly, then, on the average, team A would win 5 out of every 7 games played. That is, the odds in favor of A's winning is the ratio of the number of favorable outcomes (victories) to the number of unfavorable outcomes (losses), and the probability that team A will win is $\frac{5}{7}$. The experiment we have in mind in this example is the playing of a large number of games, say 700, between the two teams. Then there are 700 outcomes of the experiment and the event that A wins would occur 500 times and fail to occur 200 times. The relation between the probability of an event and the odds in favor of that event is formalized by the following definition.

ODDS IN FAVOR OF AN EVENT

We define the **odds in favor of** E as the ratio of the probability $\Pr(E)$ of E to the probability $\Pr(E')$ of the complementary event E'; the odds in favor of E is denoted by $O(E)$. That is,

$$O(E) = \frac{\Pr(E)}{\Pr(E')} \tag{12.38}$$

or, since $\Pr(E') = 1 - \Pr(E)$,

$$O(E) = \frac{\Pr(E)}{1 - \Pr(E)} \tag{12.39}$$

Suppose that a sample space S for a given experiment is the union of n simple events and that an event E is the union of k of these simple events. Then the complementary event E' is the union of $n - k$ simple events and

$$\Pr(E) = \frac{k}{n} \qquad \Pr(E') = \frac{n-k}{n}$$

Therefore, in this situation we can write equation (12.38) in the form

$$O(E) = \frac{\frac{k}{n}}{\frac{n-k}{n}} = \frac{k}{n-k} \tag{12.40}$$

which says that the odds in favor of E will be the quotient of the number of ways E can occur divided by the number of ways E can fail to occur.

EXAMPLE 1 What are the odds in favor of drawing a spade from an ordinary pack of 52 bridge cards?

SOLUTION There are 52 outcomes of the experiment of drawing one card from a pack of 52 cards. There are 13 ways the event E of drawing a spade can occur and 52 − 13 ways the event can fail to occur. So

$$O(E) = \frac{13}{52 - 13} = \frac{13}{39} = \frac{1}{3}$$

EXAMPLE 2 If the probability that team A will win the championship is $\frac{7}{9}$, what are the odds in favor of team A's winning?

SOLUTION Let E be the event that team A wins. Then $\Pr(E) = \frac{7}{9}$ and $\Pr(E') = 1 - \Pr(E) = 1 - \frac{7}{9} = \frac{2}{9}$. From (12.38), we have

$$O(E) = \frac{\Pr(E)}{\Pr(E')} = \frac{\frac{7}{9}}{\frac{2}{9}} = \frac{7}{2}$$

EXAMPLE 3 If the odds in favor of event E are 4 to 3 (or $\frac{4}{3}$), what is the probability of E?

SOLUTION From (12.40) we see that if $O(E) = \frac{4}{3}$, then $\frac{k}{n-k} = \frac{4}{3}$, where n is the number of outcomes of an experiment in which event E can occur in k ways. Thus, setting $k = 4$ and $n - k = 3$, we have $n = 7$ and $\Pr(E) = \frac{4}{7}$.

The result of Example 3 is an example of the general result that

if $O(E) = \dfrac{a}{b}$

then

$$\Pr(E) = \frac{a}{a+b}$$

As an example of another way in which probability comes into play, consider the following situation. Suppose that it is agreed that you win a prize of \$18 if you throw a 5 on a single toss of a die. What should you pay for the privilege of throwing the die? If you were certain of throwing a 5, then the privilege would be worth \$18. However, since the probability of throwing a 5 is $\frac{1}{6}$, the privilege is worth only $\frac{1}{6}(\$18) = \$3$. This example illustrates the concept of mathematical expectation, which we now define.

MATHEMATICAL EXPECTATION

If $\Pr(A)$ is the probability that person A will receive a sum M of money, the **mathematical expectation** of A, denoted by $E(A)$, is defined by

$$E(A) = \Pr(A) \cdot M$$

EXAMPLE 4 Walter will win \$10 if he throws an even number on a single toss of a die. What is Walter's mathematical expectation?

SOLUTION The probability of throwing an even number on the toss of a die is $\frac{3}{6} = \frac{1}{2}$; so Walter's mathematical expectation is $\frac{1}{2}(\$10) = \$5$.

EXERCISES

1 What are the odds in favor of heads appearing when a coin is tossed once?

2 What are the odds that either a 2 or a 3 will appear when a single die is tossed once?

3 If the odds in favor of a certain horse's winning a race are 8 to 5, what is the probability that the horse will win the race?

4 If the odds in a baseball game are quoted as 3 to 1 in favor of team T, what is the probability that team T will win? What is the probability that team T will lose?

5 A man has five tickets in a lottery in which there is one prize and 29 blanks. If the prize is $120, what is his mathematical expectation?

6 A person may choose any one of four envelopes of which one contains $10 and the other three are empty. What is his mathematical expectation?

7 If Robert will win $14,000 if his team wins the Super Bowl, and if the odds in favor of his team's winning are 2 to 5, what is his mathematical expectation?

8 The odds in favor of event E_1 are 4 to 3; the odds in favor of event E_2 are 7 to 1. If *both* events occur, Wesley will receive $2,000. Assuming that the events are independent, what is Wesley's mathematical expectation?

9 If the odds in favor of Jimmy Jones's winning a $10,000 golf prize are 3 to 2, find his mathematical expectation.

10 A bag contains seven white balls and three black balls. A prize of $10 is offered for drawing a black ball. What is the mathematical expectation for a single trial?

11 If one and only one of A and B is alive at the end of a year, he is to receive $12,000. The probability of dying within a year is $\frac{1}{20}$ for A and $\frac{1}{30}$ for B. What is the mathematical expectation for each of A and B?

12 In a certain city it has been found that 9 out of every 2,000 residences are destroyed by fire each year. What premium should an insurance company charge for insuring a house for $12,000 against a total loss, making no allowance for the expense of doing business?

13

Some Topics in Statistics

13.1
INTRODUCTION

Statistics is the science which deals with the collection and study of facts or data. As the name implies, statistics originated as the science concerned with the collection of data pertaining to the state, the word "state" being used here as a unit of government. Federal, state, and local governments continue to make great use of statistics. In addition, businesses, industries, and educational institutions now rely heavily upon statistics for their operation and planning.

Much of the data studied in elementary statistics are collected in the form of a tabulation of corresponding values of two variables. Recall that we discussed in Chap. 7 the four main ways of specifying a relation between two sets of numbers, namely by table, graph, formula, and verbal statement. We may pass from such data arranged in a table to its presentation in a graph, and under suitable circumstances in a formula, and finally in a summarizing verbal statement.

Statistics frequently enables us to get a picture of a problem or a condition which no amount of direct personal observation can give. Usually we are not interested in a particular or isolated fact but in certain representative measures of a large mass of data.

13.2
FREQUENCY DISTRIBUTIONS

Usually a table which presents data in their original form is so extensive that it is not possible to grasp the information which it contains. After data on a situation have been gathered, the next step is to use some scheme in studying the data so that appropriately chosen representative facts will stand out. One such scheme, a *frequency distribution*, is illustrated and described below.

The following table gives the scores made by 100 first-year students in four sections of a mathematics course.

Grades of 100 students in mathematics

88	87	75	77	87	74	81	73	67	62
79	80	55	69	76	74	82	72	75	74
65	70	84	90	76	83	74	73	72	70
80	76	68	70	95	69	75	70	73	85
80	77	79	63	65	58	75	77	72	66
68	73	83	86	55	89	70	72	76	74
90	65	92	75	97	81	63	75	70	80
75	85	70	59	79	75	95	71	77	77
69	67	69	77	60	48	78	84	68	75
85	73	80	81	82	62	74	81	53	71

A set of statistical data usually consists of a set of measurements of a group of objects. The separate items of a set of data are arrived at either by counting or by measurement.

A frequency distribution is formed by separating the data into *classes* and showing the number of items in each class. The classes are determined by choosing a set of equally spaced values which separate the data; these values are called **class boundaries.** The difference between the lower boundary of one class and the lower boundary of the next higher class is the **class width.** Likewise, the class width is the difference between the upper boundaries of two adjacent classes. The number halfway between the upper and lower boundaries of a class is the *class mark*. That is,

CLASS BOUNDARIES

CLASS WIDTH

$$\text{class mark} = \frac{(\text{upper boundary}) + (\text{lower boundary})}{2}$$

Returning to the scores of the preceding table, we note that the lowest score is 48 and the highest score is 97. The grade of 97 might represent any score between 96.5 and 97.5. Similarly, the score of 48 might represent any score between 47.5 and 48.5. Hence the possible range of the scores is from 47.5 to 97.5, or 50. We next choose a class width of 5 and arrange the scores in classes of equal width, with boundaries 47.5 and 52.5, 52.5 and 57.5, . . . , 92.5 and 97.5.

Next the scores are tabulated by classes. Begin with the first item in the original unordered data. Place a tally mark opposite the appropriate interval. Proceed in the same way for all the data. The subsequent counting is facilitated if every fifth mark in a row is made slanting across the preceding four tally marks. In this manner the tally sheet and frequency distribution shown below are prepared from the unorganized data of the preceding table.

Frequency distribution for 100 mathematics grades

CLASS	TALLY	FREQUENCY
92.5 and under 97.5	|||	3
87.5 and under 92.5	𝍸	5
82.5 and under 87.5	𝍸 𝍸	10
77.5 and under 82.5	𝍸 𝍸 𝍸	15
72.5 and under 77.5	𝍸 𝍸 𝍸 𝍸 𝍸 𝍸	30
67.5 and under 72.5	𝍸 𝍸 𝍸 𝍸	20
62.5 and under 67.5	𝍸 |||	8
57.5 and under 62.5	𝍸	5
52.5 and under 57.5	|||	3
47.5 and under 52.5	|	1
Total		100

13.3 HISTOGRAMS AND FREQUENCY POLYGONS

In order to describe or interpret a given frequency distribution, we ask the following questions: What class has the greatest frequency? Are the values evenly distributed? Are they concentrated in certain classes? These and similar questions can be answered by a careful examination and comparison of the class frequencies. A graphical representation of a frequency distribution is often helpful in answering these and related questions.

HISTOGRAM

The simplest form of graphical representation of a frequency distribution is a **histogram**, which is a special kind of bar graph. A histogram is constructed by erecting upon the class intervals rectangles whose altitudes are proportional to the frequencies. Convenient scales on the horizontal and vertical axes should be chosen so that the graph fits the data and is of sufficient size to be readily interpreted. The numbers below the horizontal axis are the midpoints of the classes, or the class marks. The left-hand side of the first rectangle is plotted at the lower boundary of the lowest class, and the right-hand side of the last rectangle is plotted at the upper boundary of the highest class. Figure 13.1 shows the histogram which is constructed from the frequency distribution of the table on page 511.

FREQUENCY POLYGON

Another type of graphical representation of a frequency distribution is a **frequency polygon.** The construction of a frequency polygon is very much like plotting points and graphing curves, which we studied in Chap. 7. To construct a frequency polygon, plot the class marks as abscissas and the corresponding frequencies as ordinates,

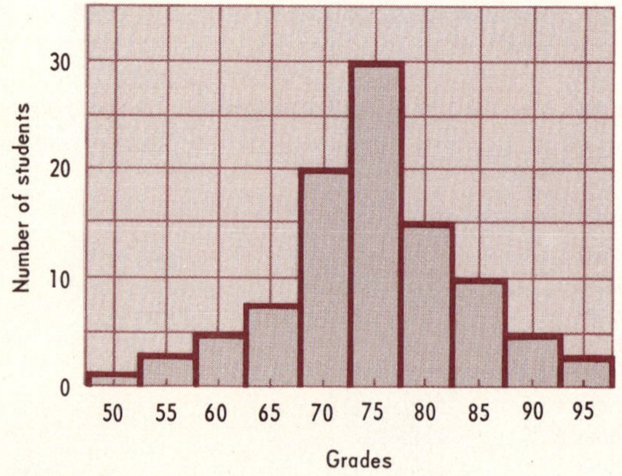

FIGURE 13.1

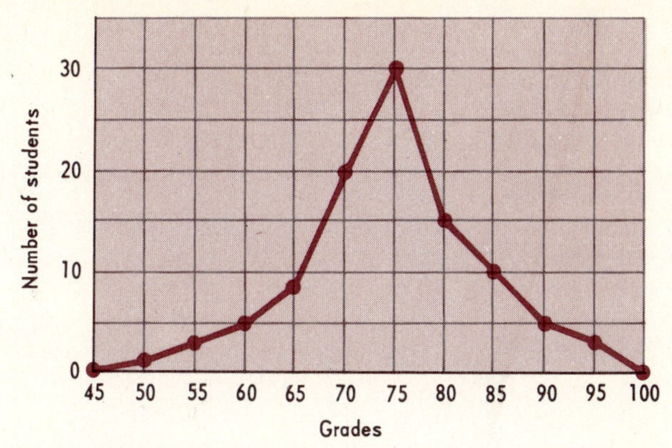

FIGURE 13.2

and then connect the points so obtained by line segments. These line segments make up the frequency polygon. The last point at either end is joined to the base at the center of the next class interval. So the origin O is at the class mark of the class which precedes the lowest class in the frequency distribution. Figure 13.2 shows the frequency polygon for the frequency distribution shown in the table on page 511.

The frequency polygon may be considered to have been derived from the histogram by drawing line segments joining midpoints of the upper bases of adjacent rectangles. The polygon is closed at each end by drawing a line from the midpoint of the top of each of the end rectangles to a point on the base line represented by the class mark of the next outlying class.

If the number of items is increased indefinitely and the class width is decreased indefinitely, the frequency polygon will approach a continuous curve called a *frequency curve*. A frequency curve of special significance, called the *normal frequency curve,* is considered briefly in Sec. 13.10.

EXERCISES

In Exercises 1 to 4, make a histogram and a frequency polygon for the given frequency distribution.

1 The lengths of a sample of 75 beans were measured to the nearest hundredth of a centimeter, and the results recorded in the table on page 514 were obtained:

CLASS, LENGTH	CLASS MARK	FREQUENCY
1.45 and under 1.55	1.50	2
1.55 and under 1.65	1.60	3
1.65 and under 1.75	1.70	7
1.75 and under 1.85	1.80	8
1.85 and under 1.95	1.90	13
1.95 and under 2.05	2.00	19
2.05 and under 2.15	2.10	11
2.15 and under 2.25	2.20	9
2.25 and under 2.35	2.30	2
2.35 and under 2.45	2.40	1
Total		75

2 The following table gives the weights, in pounds, of 1,000 freshmen:

CLASS	CLASS MARK	FREQUENCY
90 and under 100	95	14
100 and under 110	105	28
110 and under 120	115	145
120 and under 130	125	246
130 and under 140	135	242
140 and under 150	145	160
150 and under 160	155	89
160 and under 170	165	47
170 and under 180	175	17
180 and under 190	185	10
190 and under 200	195	2
Total		1,000

3 The following table gives a frequency distribution of data which were obtained by firing 1,000 shots at a target which was divided into 11 horizontal strips, and counting the number of shots in each strip. Regard the number of each strip as a class mark.

STRIP	SHOTS
1	1
2	5
3	9
4	87
5	191
6	212
7	204
8	193
9	79
10	17
11	2
Total	1,000

4 The following table records the grades made by 200 students on a certain test:

CLASS	CLASS MARK	FREQUENCY
136 and under 140	138	3
132 and under 136	134	5
128 and under 132	130	16
124 and under 128	126	23
120 and under 124	122	52
116 and under 120	118	49
112 and under 116	114	27
108 and under 112	110	18
104 and under 108	106	7
Total		200

5 The numbers in the following table give percentages which the 1972 price was of the 1970 price for each of 50 items. We call such numbers price relatives.

109	138	107	111	130
111	127	111	118	108
134	115	101	122	109
109	122	103	99	106
108	133	102	124	146
141	107	106	104	122
138	118	108	113	118
104	112	130	103	124
102	126	99	114	114
114	108	102	100	113

Construct a frequency distribution of these relatives with the class marks at 100.5 (for the class 98 and under 103), 105.5 (for the class 103 and under 108), and so forth, the class width being 5. Using this frequency distribution, construct a histogram and a frequency polygon.

6 For the unorganized data given in Exercise 5 do the following: Construct a frequency distribution with the class marks at 95 (for the class 90 and under 100), 105 (for the class 100 and under 110), and so forth, the class width being 10. Using the frequency distribution you have just prepared, construct a histogram and a frequency polygon.

7 It has been shown that the results given in the following table are to be expected if 1,024 throws are made with 10 coins. Observe that the frequency polygon obtained is symmetrical with respect to the vertical line through the point (5, 0). It is unusual to get such a symmetrical distribution with observed data.

NO. OF HEADS TURNING UP	FREQUENCY
0	1
1	10
2	45
3	120
4	210
5	252
6	210
7	120
8	45
9	10
10	1
Total	1,024

8 Ten coins were actually thrown 1,024 times, and the actual results were as recorded in the following table. When you construct the frequency polygon in this exercise, superimpose it on the frequency polygon of Exercise 7. To construct the two frequency polygons, use a colored pencil or dotted lines for the frequency polygon of this exercise.

NO. OF HEADS TURNING UP	FREQUENCY
0	2
1	9
2	40
3	117
4	204
5	259
6	216
7	128
8	40
9	7
10	2
Total	1,024

13.4 THE ARITHMETIC MEAN

We shall discuss and compare briefly three commonly used averages, the *arithmetic mean,* the *median,* and the *mode.* Many frequency distributions, as those considered in the last section, tend to "pile up" near the center. This behavior is described by saying that such a distribution has a *central tendency.* The three averages we have just mentioned are called *measures of central tendency.*

ARITHMETIC MEAN

The **arithmetic mean** of a set of measures is equal to the sum of the measures divided by the number of measures. If your grades on five tests are 68, 76, 78, 50, and 78, the arithmetic mean M of your grade is given by

$$M = \frac{68 + 76 + 78 + 50 + 78}{5} = \frac{350}{5} = 70 \qquad (13.1)$$

In general, if there are n measures $x_1, x_2, \ldots, x_n$, the arithmetic mean M is given by the formula

$$M = \frac{x_1 + x_2 + x_3 + \cdots + x_n}{n} \qquad (13.2)$$

A measurement may occur more than once, as the score of 78 above. We could write equation (13.1) in the form

$$M = \frac{2 \cdot 78 + 68 + 50 + 76}{5} = \frac{350}{5} = 70$$

Then 78 is said to have a frequency or weight of 2. In general, if a measurement occurs f times in a tabulation, we say that it has a *frequency,* or *weight,* of f. We denote the **weighted arithmetic mean** by M_x and define it by

WEIGHTED ARITHMETIC MEAN

$$M_x = \frac{x_1 f_1 + x_2 f_2 + \cdots + x_n f_n}{f_1 + f_2 + \cdots + f_n} \qquad (13.3)$$

where $f_1, f_2, \ldots, f_n$ denote, respectively, the frequency, or weight, of $x_1, x_2, \ldots, x_n$. The term "weight" originated in experimental work where certain observations are weighted according to their reliability or importance. Weights, chosen in this way, are essentially frequencies.

For an illustration of the weighted arithmetic mean, suppose that the five test grades mentioned above were on courses with the following credits:

68 on a course with 3 semester hours credit

76 on a course with 4 semester hours credit

78 on a course with 3 semester hours credit

50 on a course with 5 semester hours credit

78 on a course with 1 semester hour credit

Clearly the weighted arithmetic mean in this instance is more significant than the ordinary arithmetic mean. We get

$$M_x = \frac{3 \cdot 68 + 4 \cdot 76 + 3 \cdot 78 + 5 \cdot 50 + 1 \cdot 78}{3 + 4 + 3 + 5 + 1} = \frac{1{,}070}{16} = 66\tfrac{7}{8}$$

If the measures have been grouped into a frequency distribution, the arithmetic mean cannot be found by the method just described.

However, it can be closely approximated by finding the weighted mean of the class marks, using the frequencies of the classes as weights. Let $X_1, X_2, \ldots, X_n$ denote the class marks of a frequency distribution, $f_1, f_2, \ldots, f_n$ their respective frequencies, and M_X the weighted arithmetic mean of the class marks. Then

$$M_X = \frac{f_1X_1 + f_2X_2 + \cdots + f_nX_n}{f_1 + f_2 + \cdots + f_n} \tag{13.4}$$

In the following table we give the computation of M_X for the frequency distribution of the 100 mathematics grades of the table on page 511.

Calculation of arithmetic mean

CLASS	CLASS MARK X_i	FREQUENCY f_i	f_iX_i
92.5 and under 97.5	95	3	285
87.5 and under 92.5	90	5	450
82.5 and under 87.5	85	10	850
77.5 and under 82.5	80	15	1,200
72.5 and under 77.5	75	30	2,250
67.5 and under 72.5	70	20	1,400
62.5 and under 67.5	65	8	520
57.5 and under 62.5	60	5	300
52.5 and under 57.5	55	3	165
47.5 and under 52.5	50	1	50
Totals		100	7,470

So

$$M_X = \frac{7{,}470}{100} = 74.70$$

The sum of the original grades in the table on page 510, which gives the ungrouped data, is 7,471. Hence the arithmetic mean M of the ungrouped data is

$$M = \frac{7{,}471}{100} = 74.71$$

If M_X and M are rounded off to one decimal place, they are both equal to 74.7.

EXERCISES

1 The annual salaries paid nine teachers are \$7,000, \$7,100, \$7,400, \$7,600, \$6,950, \$7,150, \$7,300, \$6,800, and \$7,800. What is the arithmetic mean of these salaries?

2 Five plumbers had the following annual earnings: \$11,800, \$13,400, \$12,500, \$15,400, \$12,800. What was the average annual income of the plumbers?

In Exercises 3 to 6, find the arithmetic mean of the given set of numbers.

3 97.0, 90.0, 87.0, 93.0, 96.0, 88.0, 78.0, 95.0, 96.0, 87.0
4 0.653, 0.668, 0.695, 0.679, 0.700, 0.715, 0.738, 0.763, 0.768
5 38.50, 50.75, 55.00, 46.50, 22.75, 34.25, 26.25, 59.00, 20.75
6 3,740, 2,580, 3,273, 4,581, 2,798

7 Compute M_X for the frequency distribution of length of beans given in Exercise 1 of Sec. 13.3.
8 Compute M_X for the frequency distribution of grades of students given in Exercise 4 of Sec. 13.3.
9 Compute M for the 50 items of the ungrouped data given in Exercise 5 of Sec. 13.3.
10 Compute M_X for the frequency distribution which you prepared in Exercise 6 of Sec. 13.3. How do M_X of this exercise and M found in Exercise 9 compare?

In Exercises 11 to 14, find the weighted arithmetic mean of the given set of numbers. The number in parentheses is the weight of the number preceding it; if no weight is indicated, it is 1.

11 351.0 (4), 349.0 (2), 356.0 (7), 343.0 (5), 345.0 (3)
12 72.0 (2), 70.0 (2), 74.0, 69.0, 73.0 (3), 68.0
13 1.42 (50), 1.44 (17), 1.43 (15), 1.47 (7)
14 248 (7), 292 (3), 272 (5), 260 (4)

15 The arithmetic means of the grades of the members of each of four fraternities were 2.02, 2.68, 3.02, and 1.78. If these fraternities have 15, 18, 12, and 32 members, respectively, (a) find the arithmetic mean of the grades of the 77 students by finding the weighted arithmetic mean of the four given grade averages; (b) find the unweighted arithmetic mean of the four grade averages; (c) explain why the result in (b) is larger than the result in (a).
16 A factory owner pays the wages shown in the following table:

POSITION	NO. OF WORKMEN	WEEKLY WAGE
Manager	1	\$900
Foremen	2	300
Skilled workmen	9	220
Unskilled workmen	11	120
Apprentices	3	100

Find the unweighted arithmetic mean of the weekly wages; the weighted arithmetic mean of the weekly wages. Which do you think is the more representative average? Why?

17 Five stores reported sales of eggs as follows:

Store A, 68 dozen at 53 cents per dozen
Store B, 101 dozen at 55 cents per dozen
Store C, 66 dozen at 56 cents per dozen
Store D, 152 dozen at 58 cents per dozen
Store E, 83 dozen at 59 cents per dozen

Find the weighted arithmetic mean price per dozen.

18 The average commission earned in each of six stores in a given week was $180, $244, $140, $156, $124, $168, respectively. If the stores had, respectively, 8, 6, 7, 9, 10, and 11 salesmen, find the weighted arithmetic mean of the commissions.

19 Two mathematics teachers have different ideas as to how test grades should be counted. Each gave four tests during the semester. The first teacher gave these tests the weights 1, 2, 3, 4, respectively, while the second teacher gave these tests the weights 4, 3, 2, 1, respectively. A student had a course with each teacher and made the same grades on the four tests in the two courses, namely 60, 70, 80, and 90. Find the weighted arithmetic mean of the test grades under each of the teachers.

13.5
THE MEDIAN AND THE MODE

If the measures of a set are arranged in order of magnitude, the MEDIAN middle one of the measurements, if it exists, is called the **median.** If the number of measurements is odd, the median always exists. To find the median of the scores 68, 76, 78, 50, and 78, we arrange these scores in order of size, as

50, 68, 76, 78, 78

The score of 76 is the middle one, since there are two scores on either side of it; so 76 is the median score.

If the number of measurements is even, there are two middle terms. In this case the median is defined to be the mean of the two middle terms. To illustrate, the median of the scores

65, 65, 75, 75, 80, 90, 90, 94

is

$$\frac{75 + 80}{2} = 77\tfrac{1}{2}$$

MODE A **mode** is a measure in a set of measurements which occurs most frequently, if such a measure exists. If a student's grades are

60, 62, 62, 62, 70, 80, 92

then 62 is the mode of these grades, for it is the grade which occurs most often. Two or more modes may exist for a given distribution. However, we say that the mode fails to exist when all the frequencies are equal.

The determination of the median and the mode from a frequency distribution requires careful and extended analysis, which we shall not go into here. However, the mode is roughly approximated by the class mark of the class with greatest frequency. That value is appropriately called the **crude mode.** Thus the crude mode for the frequency distribution of the table on page 511 is 75, since this is the class mark of the class with greatest frequency.

CRUDE MODE

It is well that you should be acquainted with the arithmetic mean, the median, and the mode. For when a person speaks of an average, he may mean any one of these (as well as some other averages, such as the harmonic mean and the geometric mean). Usually a particular one of these three averages we have defined will give emphasis to special characteristics of a set of measures.

The arithmetic mean gives emphasis to extreme measurements. For example, the great wealth of one person in a community will unduly affect the arithmetic mean of the wealth of the community. On the other hand, the arithmetic mean is defined in a precise way and is relatively easy to calculate.

The median can be found when exact values of extreme measures are unknown. It can be found readily when different units of measure are used and it is not affected by extreme values.

The crude mode, as we have pointed out before, is the most quickly and easily found of the measures of central tendency.

EXERCISES

1 A student's grades in 10 subjects taken the last two semesters are 94, 90, 86, 86, 86, 75, 73, 70, 60, 54. What is the arithmetic mean of his grades? What is the median grade? What is the mode of the grades?

2 The wages in a certain business are shown in the following table:

POSITION	NO. OF WORKMEN	WEEKLY WAGE
Foremen	2	$220
Skilled workers	22	180
Unskilled workers	16	110
Apprentices	4	80

What is the unweighted arithmetic mean of the wages? What is the weighted arithmetic mean of the wages? What is the median wage? What is the modal wage?

3 The arithmetic mean income of one city is the same as that of another city, namely $5,200 per capita. Does this necessarily mean that the frequency distributions of the income in the two cities are essentially the same? Explain.

4 Give the weighted arithmetic mean and the crude mode for the frequency distribution in the following table.

CLASS MARK	FREQUENCY
30.0	1
41.0	5
52.0	10
63.0	7
74.0	3

In Exercises 5 and 6, find, for the given set of numbers, the arithmetic mean, the median, the mode.

5 7, 15, 31, 31, 31, 43, 65, 72, 79

6 6, 11, 14, 34, 34, 41, 43, 46

7 The following are the prices in dollars charged in each of eight stores for a certain item: 7.20, 7.30, 7.60, 7.80, 8.00, 8.00, 8.00, 8.10. Find the arithmetic mean, the median, and the mode of these prices.

8 What is the crude mode for the frequency distribution given in Exercise 1 of Sec. 13.3?

9 What is the crude mode for the frequency distribution given in Exercise 4 of Sec. 13.3?

13.6
THE STANDARD DEVIATION

The averages which we studied in Secs. 13.4 and 13.5, and which we called *measures of central tendency,* have their value as representative measures of a set of measures. But simply knowing an average is frequently insufficient.

For example, the two sets of numbers

$$8, 10, 12, 14, 16, 18, 20 \tag{13.5}$$

and

$$1, 8, 14, 21, 26 \tag{13.6}$$

have the same arithmetic mean of 14, even though the extreme measures are quite different. The measures of the second set are more widely scattered, or dispersed, from the arithmetic mean than those of the first set.

In this section we study a way of measuring the extent to which the individual measures differ from a measure of central tendency. It is called the *standard deviation* and is the most widely used of several *measures of dispersion.*

DEVIATIONS

The differences between the individual measures and the arithmetic mean of the set are called **deviations.** A deviation is positive if the measure exceeds the arithmetic mean and is negative if the measure is less than the arithmetic mean. It is usual to denote the deviations of the measures $x_1, x_2, \ldots, x_n$ by

$$d_1 = x_1 - M \qquad d_2 = x_2 - M \qquad \cdots \qquad d_n = x_n - M$$

or

$$d_i = x_i - M \qquad \textit{for } i = 1, 2, \ldots, n$$

To illustrate, the deviations for the set of numbers (13.5) are -6, -4, -2, 0, 2, 4, 6, as indicated in the table on the left below. The deviations for the set of numbers (13.6) are -13, -6, 0, 7, and 12, as recorded in the table on the right below. Note that in these two cases the sum of the deviations of a set of measures from the arithmetic mean is zero. This is always true.

STANDARD DEVIATION

The **standard deviation** of a set of measures is defined to be the square root of the arithmetic mean of the squares of the deviations. Let the n measures be $x_1, x_2, \ldots, x_n$, and denote their standard deviation by the Greek letter σ (sigma). Then

$$\sigma = \sqrt{\frac{(x_1 - M)^2 + (x_2 - M)^2 + \cdots + (x_n - M)^2}{n}} \tag{13.7}$$

or

$$\sigma = \sqrt{\frac{d_1^2 + d_2^2 + \cdots + d_n^2}{n}} \tag{13.8}$$

$M = 14$

x_i	$d_i = x_i - M$	$d_i^2 = (x_i - M)^2$
8	-6	36
10	-4	16
12	-2	4
14	0	0
16	2	4
18	4	16
20	6	36
Sums	0	112

$M = 14$

x_i	$d_i = x_i - M$	$d_i^2 = (x_i - M)^2$
1	-13	169
8	-6	36
14	0	0
21	7	49
26	12	144
Sums	0	398

EXAMPLE 1 Find the standard deviation for each of the sets of numbers (13.5) and (13.6).

SOLUTION We first construct the column headed $d_i^2 = (x_i - M)^2$ in each of the tables on page 523.

From formula (13.8) and the table on the left we have, for σ_1, the standard deviation of the first set of numbers,

$$\sigma_1 = \sqrt{\frac{112}{7}} = \sqrt{16} = 4$$

Similarly, from formula (13.8) and the table on the right we have, for σ_2, the standard deviation of the second set of numbers,

$$\sigma_2 = \sqrt{\frac{398}{5}} = \sqrt{79.6} = 8.92 \qquad \textit{approximately}$$

Although the arithmetic mean for both sets of numbers is 14, the larger value of σ for the second set shows that the measures in it are more widely dispersed with respect to the arithmetic mean than are the measures in the first set.

If the measures have been grouped into a frequency distribution, the standard deviation is found approximately by considering all the measures in a class as represented by the class mark. Then the formula for the approximated standard deviation σ is

$$\sigma = \sqrt{\frac{f_1 d_1^2 + f_2 d_2^2 + \cdots + f_n d_n^2}{n}} \tag{13.9}$$

where f_i is the frequency of the class with class mark X_i and

$$d_i = X_i - M_X$$

where M_X is the arithmetic mean given by formula (13.4).

EXAMPLE 2 Compute the standard deviation for the frequency distribution of the 100 mathematics grades in the table on page 511.

SOLUTION Recall that in connection with the table on page 511, we found the value of M_X to be 74.7. The various steps leading to the computation of σ by use of formula (13.9) are shown in the table on page 525. So, by use of formula (13.9), we get

$$\sigma = \sqrt{\frac{7{,}941}{100}} = \sqrt{79.41} = 8.9 \qquad \textit{approximately}$$

Calculation of standard deviation

CLASS MARK X_i	FREQUENCY f_i	DEVIATION $d_i = X_i - M_X$	d_i^2	$f_i d_i^2$
95	3	20.3	412.09	1,236.27
90	5	15.3	234.09	1,170.45
85	10	10.3	106.09	1,060.90
80	15	5.3	28.09	421.35
75	30	0.3	0.09	2.70
70	20	−4.7	22.09	441.80
65	8	−9.7	94.09	752.72
60	5	−14.7	216.09	1,080.45
55	3	−19.7	388.09	1,164.27
50	1	−24.7	610.09	610.09
Totals	100			7,941.00

Since the value of the arithmetic mean M_X is usually a decimal, the method of computing the standard deviation σ which we have just used is ordinarily quite laborious. For this reason special short methods for the calculation of σ are developed in courses in statistics.

It should be realized that some examples, such as Example 1 of this section, have been so chosen to illustrate procedures and at the same time to involve short computations. The values of the measures of central tendency and the standard deviation in such cases have little statistical significance because of the small number of measures involved. Actually a rather large number of measures is necessary in order for statistical measures like the arithmetic mean and standard deviation to have significance.

Many frequency distributions (such as the one in Example 2) that are encountered in applications of statistics are very close to being "normal" distributions. For such distributions, the standard deviation is a particularly useful statistical measure of dispersion. Normal distributions and the significance of the standard deviation for a normal distribution are discussed in Sec. 13.10.

EXERCISES

1 Compute (to tenths) the standard deviations of these two sets of grades: 50, 68, 76, 78, 78 and 35, 52, 67, 75, 77, 79, 94, 98. Compare the results.

2 Compute (to tenths) the standard deviations of the two sets of measures, 48, 49, 50, 51, 52 and 20, 30, 40, 50, 60, 70, 80. Compare the results.

3 Compute (to tenths) the standard deviation of the set of measures 50, 52, 54, 56, 58, 60, 60, 60, 62.

4 Compute σ (to hundredths) for the frequency distribution of Exercise 1 of Sec. 13.3. (You found M_X for that distribution in Exercise 7 of Sec. 13.4.)

5 Compute σ (to hundredths) for the frequency distribution of Exercise 7 of Sec. 13.3.

6 Compute σ (to tenths) for the frequency distribution of Exercise 4 of Sec. 13.3. (You found M_X for that distribution in Exercise 8 of Sec. 13.4.)

7 Two teachers of freshman mathematics gave their students the same examination. The standard deviation of the test scores for one teacher was 6.8 and for the other teacher 8.1. On the basis of these results compare the variability or dispersion in mathematical competency among the students of the two teachers.

8 Forty members of a fraternity made the following grade-point averages during a school term:

2.50	2.40	1.88	3.16	2.30	3.02	2.55	2.91
3.04	2.39	2.24	1.85	1.88	2.12	2.92	3.17
2.88	3.12	2.68	1.94	1.74	2.70	3.09	3.19
1.48	1.92	2.49	2.99	3.27	2.80	2.87	1.75
3.08	2.70	2.64	1.58	1.73	2.54	3.04	3.09

Find the arithmetic mean and the standard deviation of these grade-point averages.

9 Compute the arithmetic mean of the numbers 1, 2, 3, 4, 5, 6, 7, 8, 9, 10 and the arithmetic mean of the numbers 5, 10, 15, 20, 25, 30, 35, 40, 45, 50. Compare these arithmetic means. What is the effect upon the arithmetic mean of the first set of numbers when each member of this set is multiplied by the constant 5?

10 Let M_x be the arithmetic mean of the set of numbers $x_1, x_2, x_3, \ldots, x_n$, and let M_{kx} be the arithmetic mean of the set of numbers $kx_1, kx_2, kx_3, \ldots, kx_n$, where k is a constant. Using the definition of the arithmetic mean given by equation (13.2), prove that $M_{kx} = kM_x$.

11 Compute to three significant figures the standard deviation of the numbers 1, 2, 3, 4, 5, 6, 7, 8, 9, 10 and the standard deviation of the numbers 5, 10, 15, 20, 25, 30, 35, 40, 45, 50. Compare these standard deviations. What is the effect upon the standard deviation of the first set of numbers when each member of this set is multiplied by the constant 5?

12 Let σ_x be the standard deviation of the set of numbers $x_1, x_2, x_3, \ldots, x_n$, and let σ_{kx} be the standard deviation of the set of numbers $kx_1, kx_2, kx_3, \ldots, kx_n$, where k is a constant. Using the definition of the standard deviation given by equation (13.7), prove that $\sigma_{kx} = k\sigma_x$.

13 Compute the arithmetic mean of the numbers 1, 2, 3, 4, 5, 6, 7, 8, 9, 10 and the arithmetic mean of the numbers 21, 22, 23, 24, 25, 26, 27, 28, 29, 30. Compare the arithmetic means. What is the effect upon

the arithmetic mean of the first set of numbers when the constant 20 is added to each member of this set?

14 Let M_x be the arithmetic mean of the set of numbers $x_1, x_2, x_3, \ldots, x_n$, and let M_{x+k} be the arithmetic mean of the set of numbers $x_1 + k, x_2 + k, x_3 + k, \ldots, x_n + k$. Using the definition of the arithmetic mean given by equation (13.2), prove that $M_{x+k} = M_x + k$.

15 Compute to three significant figures the standard deviation of the numbers 1, 2, 3, 4, 5, 6, 7, 8, 9, 10 and the standard deviation of the numbers 21, 22, 23, 24, 25, 26, 27, 28, 29, 30. Compare these standard deviations. What is the effect upon the standard deviation of the first set of numbers when each member of this set is increased by 20?

16 Let σ_x be the standard deviation of the set of numbers $x_1, x_2, x_3, \ldots, x_n$, and let σ_{x+k} be the standard deviation of the set of numbers $x_1 + k, x_2 + k, x_3 + k, \ldots, x_n + k$. Using the definition of the standard deviation given by equation (13.7), prove that $\sigma_{x+k} = \sigma_x$.

17 Let M denote the arithmetic mean and σ denote the standard deviation of the set of numbers $x_1, x_2, x_3, \ldots, x_n$. Similarly, let M' and σ' be the arithmetic mean and the standard deviation, respectively, of the set of numbers $kx_1 + c, kx_2 + c, kx_3 + c, \ldots, kx_n + c$. Express M' and σ' in terms of M and σ and the constants k and c.

13.7 THE BINOMIAL THEOREM

From the definition of a positive integral exponent (Sec. 1.8 and Sec. 3.8) it follows, for any two numbers or algebraic expressions a and b, that

$$(a + b)^1 = a + b \tag{13.10}$$

By multiplication and use of the distributive law, we found in Sec. 3.7 that

$$(a + b)^2 = a^2 + 2ab + b^2 \tag{13.11}$$

In like manner we have

$$\begin{aligned}(a + b)^3 &= (a + b)^2(a + b) = (a^2 + 2ab + b^2)(a + b)\\ &= a^3 + 2aba + b^2a + a^2b + 2ab^2 + b^3\end{aligned}$$

or

$$(a + b)^3 = a^3 + 3a^2b + 3ab^2 + b^3 \tag{13.12}$$

By the same procedures we find that

$$(a + b)^4 = a^4 + 4a^3b + 6a^2b^2 + 4ab^3 + b^4 \tag{13.13}$$

$$(a + b)^5 = a^5 + 5a^4b + 10a^3b^2 + 10a^2b^3 + 5ab^4 + b^5 \tag{13.14}$$

$$\begin{aligned}(a + b)^6 = a^6 + 6a^5b + 15a^4b^2 + 20a^3b^3\\ + 15a^2b^4 + 6ab^5 + b^6\end{aligned} \tag{13.15}$$

The six equalities (13.10) to (13.15) are special cases of the general formula

$$(a+b)^n = a^n + na^{n-1}b + \frac{n(n-1)}{2\cdot 1}a^{n-2}b^2 + \frac{n(n-1)(n-2)}{3\cdot 2\cdot 1}a^{n-3}b^3 + \cdots + \frac{n!}{r!(n-r)!}a^{n-r}b^r + \cdots + b^n \quad (13.16)$$

where n is a positive integer. The equality (13.16) is called the **binomial formula.** The statement that this formula holds for all positive integral values of n is called the **binomial theorem.** The binomial theorem may be proved by use of the principle of mathematical induction. We do not give the proof here. The right side of each of the equalities (13.10) to (13.16) is called the binomial *expansion* of the left side. We have verified the binomial theorem for $n = 1$, 2, and 3, and in subsequent exercises we shall ask the student to verify it for $n = 4$, 5, and 6.

BINOMIAL FORMULA

BINOMIAL THEOREM

If in the expansion of $(a+b)^n$ given by equation (13.16) we number the terms from left to right, then we can make the following observations.

1 The expansion of $(a+b)^n$ has $n+1$ terms.
2 The first term of the expansion is a^n and the last term is b^n.
3 The exponents of a decrease by 1 from term to term, and the exponents of b increase by 1 from term to term.
4 The sum of the exponents of a and b in each term is n.
5 If the coefficient in any term is multiplied by the exponent of a in that term and divided by the number of that term, the result is the coefficient in the next term.

EXAMPLE 1 Find the binomial expansion of $(a+b)^7$.

SOLUTION Applying the binomial theorem as amplified by the observations 1 to 5, we see that the first term is a^7. Note that the exponent of b in the first term is zero, since $a^7 = a^7b^0$. The coefficient in the second term is 7; in the second term the exponent of a is $7-1=6$, and the exponent of b is $0+1=1$. Therefore the second term is $7a^6b$. The coefficient in the third term is $\frac{7\cdot 6}{2} = 21$, and the complete third term is $21a^5b^2$. Continuing this procedure, we obtain

$$(a+b)^7 = a^7 + 7a^6b + 21a^5b^2 + 35a^4b^3 + 35a^3b^4 + 21a^2b^5 + 7ab^6 + b^7$$

as the student should verify.

EXAMPLE 2 Find the binomial expansion of $(x-2y)^4$.

SOLUTION We first write the expansion of $(a+b)^4$, namely,

$$(a+b)^4 = a^4 + 4a^3b + 6a^2b^2 + 4ab^3 + b^4$$

Then in the expansion we let $a = x$ and $b = -2y$. We obtain

$$(x - 2y)^4 = x^4 + 4x^3(-2y) + 6x^2(-2y)^2 + 4x(-2y)^3 + (-2y)^4$$

So

$$(x - 2y)^4 = x^4 - 8x^3y + 24x^2y^2 - 32xy^3 + 16y^4$$

EXAMPLE 3 Use the binomial theorem to find the value of $(1.01)^{10}$ to four decimal places.

SOLUTION We write $(1.01)^{10} = (1 + 0.01)^{10}$, and setting $a = 1$, $b = 0.01$, and $n = 10$ in the binomial formula, obtain

$$\begin{aligned}(1.01)^{10} &= (1 + 0.01)^{10} \\ &= 1^{10} + 10(1)^9(0.01) + 45(1)^8(0.01)^2 + 120(1)^7(0.01)^3 \\ &\qquad + 210(1)^6(0.01)^4 + 252(1)^5(0.01)^5 + \cdots \\ &= 1 + 0.1 + 45(0.0001) + 120(0.000001) + 210(0.00000001) \\ &\qquad + 252(0.0000000001) + \cdots \\ &= 1 + 0.1 + 0.0045 + 0.000120 + 0.00000210 \\ &\qquad + 0.0000000252 + \cdots \\ &= 1.104622 \cdots \\ &= 1.1046 \qquad \text{to four decimal places}\end{aligned}$$

It should be clear that the term $252(1)^5(0.01)^5$ does not affect the result to four decimal places and that any subsequent term will not affect the result to four decimal places.

EXERCISES

1 By multiplication of the factors of $(a + b)^4$, verify the expansion (13.13).
2 By multiplication of the factors of $(a + b)^5$, verify the expansion (13.14).
3 By multiplication of the factors of $(a + b)^6$, verify the expansion (13.15).

In Exercises 4 to 17, find the binomial expansion of the given expression.

4 $(x + 1)^4$
5 $(y - 1)^4$
6 $(y + 2a)^3$
7 $(x + 2)^4$
8 $(x - y)^5$
9 $\left(y + \frac{1}{y}\right)^6$
10 $(3y - 2x)^4$
11 $(x^2 + 3y^2)^5$
12 $(xy^2 - 5)^4$
13 $(2x^2 - y)^5$
14 $(a + b)^{10}$
15 $(x - y)^{11}$
16 $(x + y - 1)^4$
17 $\left(\frac{a}{b} + \frac{b}{a}\right)^3$

In Exercises 18 to 21, find the value of the given expression correct to four decimal places.

18 $(1.02)^7$
19 $(1.04)^6$
20 $(1.06)^4$
21 $(1.05)^6$

22 Simplify $(a+b)^3 - 3b(a+b)^2 + 3b^2(a+b) - b^3$.

23 Simplify $\left(\frac{1}{a}+\frac{1}{b}\right)^3 - \left(\frac{1}{a}-\frac{1}{b}\right)^3$.

13.8 COMBINATIONS AND THE BINOMIAL FORMULA

Recall from Sec. 12.7 that

$$C(n, 0) = 1,\ C(n, 1) = n,\ C(n, 2) = \frac{n(n-1)}{2\cdot 1},$$

$$C(n, 3) = \frac{n(n-1)(n-2)}{3\cdot 2\cdot 1}, \ldots,$$

$$C(n, r) = \frac{n!}{r!(n-r)!}, \ldots, C(n, n) = 1$$

Comparing these results with the binomial formula as given in (13.16) we see that $C(n, 0)$, $C(n, 1)$, $C(n, 2), \ldots, C(n, r), \ldots, C(n, n)$ are the respective coefficients in the binomial formula and that this formula can be written in the form

$$(a+b)^n = C(n, 0)a^n + C(n, 1)a^{n-1}b + C(n, 2)a^{n-2}b^2 + C(n, 3)a^{n-3}b^3 + \cdots + C(n, r)a^{n-r}b^r + \cdots + C(n, n)b^n \tag{13.17}$$

BINOMIAL COEFFICIENTS

The coefficients $C(n, 0)$, $C(n, 1)$, $C(n, 2), \ldots, C(n, n)$ in the binomial formula (13.17) are called the **binomial coefficients** for the binomial expansion of $(a+b)^n$. These coefficients for successive values of n may be displayed as follows:

$$\begin{array}{lccccccccccc}
n=1 & & & & & C(1,0) & & C(1,1) & & & & \\
n=2 & & & & C(2,0) & & C(2,1) & & C(2,2) & & & \\
n=3 & & & C(3,0) & & C(3,1) & & C(3,2) & & C(3,3) & & \\
n=4 & & C(4,0) & & C(4,1) & & C(4,2) & & C(4,3) & & C(4,4) & \\
n=5 & C(5,0) & & C(5,1) & & C(5,2) & & C(5,3) & & C(5,4) & & C(5,5)
\end{array}$$

and so on

PASCAL'S TRIANGLE

Evaluating each of the numbers of combinations in this array will produce the following triangular array known as **Pascal's triangle.**

$$\begin{array}{lccccccccccccc}
n=1 & & & & & & 1 & & 1 & & & & & \\
n=2 & & & & & 1 & & 2 & & 1 & & & & \\
n=3 & & & & 1 & & 3 & & 3 & & 1 & & & \\
n=4 & & & 1 & & 4 & & 6 & & 4 & & 1 & & \\
n=5 & & 1 & & 5 & & 10 & & 10 & & 5 & & 1 & \\
n=6 & 1 & & 6 & & 15 & & 20 & & 15 & & 6 & & 1
\end{array}$$

and so on

In this array we notice that 1 appears as the first entry and as the last entry in each row and that the rth entry in a given row is obtained by adding the $(r-1)$st and rth entries in the preceding row. For example, in the 5th row, the 3d entry, 10, is the sum of 4 and 6, which are the 2d and 3d entries respectively in the 4th row. In this fashion, Pascal's triangle can be extended indefinitely. To illustrate, the 7th row would consist of the eight numbers 1, $1+6=7$, $6+15=21$, $15+20=35$, $20+15=35$, $15+6=21$, $6+1=7$, 1, and from this result we see that the binomial expansion of $(a+b)^7$ is

$$a^7 + 7a^6b + 21a^5b^2 + 35a^4b^3 + 35a^3b^4 + 21a^2b^5 + 7ab^6 + b^7$$

The symmetry of Pascal's triangle follows from the fact that $C(n, r) = C(n, n-r)$, so that, for example, the 3d entry from the *left* in the 6th row, $C(6, 2)$, is the same as the 3d entry from the *right*, $C(6, 4)$.

Besides being used to find the binomial coefficients in a binomial expansion, Pascal's triangle is also useful in another connection (which is actually closely related to the binomial expansion, as we shall see in the next section). Suppose that four fair coins are tossed and we are interested in how many heads appear. We may find no heads, one head, two heads, three heads, or four heads. In how many ways can each of these results occur? No heads can appear in only one way, the number $C(4, 0)$ of 0-combinations of four objects; the number of ways that one head can appear is the number of ways of selecting one object (the coin on which the head appears) from a set of four objects, namely, $C(4, 1)$; the number of ways that two heads can appear is the number of ways of selecting two objects (the coins on which heads appear) from a set of four objects, namely, $C(4, 2)$; similarly, the number of ways that three heads can appear is $C(4, 3)$, and the number of ways that four heads can appear is $C(4, 4)$. Thus, the numbers of ways of obtaining zero heads, one head, two heads, three heads, and four heads, when four coins are tossed, are given by the successive entries in the 4th row of Pascal's triangle. Further, the number of outcomes of the experiment of tossing four coins is the *sum* of the entries in this row, $1+4+6+4+1=16$. Similarly, the numbers of ways in which no heads, one head, two heads, three heads, four heads, five heads, and six heads can occur when six coins are tossed are given by the successive entries in the 6th row of Pascal's triangle, and the number of outcomes of the experiment of tossing six coins is the sum $1+6+15+20+15+6+1=64$ of the entries in the 6th row.

If, in formula (13.17) we set $a=1$ and $b=1$, we get

$$(1+1)^n = C(n, 0) + C(n, 1) + C(n, 2) + \cdots + C(n, n) \quad (13.18)$$

Recall that an r-combination of the n objects $a_1, a_2, a_3, \ldots, a_n$ is a *subset* of $S = \{a_1, a_2, a_3, \ldots, a_n\}$, which contains r members. So, $C(n, r)$ is the *number* of subsets of S which contain r members, and from (13.18) it follows that

if a set S has n members,
then the number of subsets of S is 2^n (13.19)

Also, from (13.18) and the fact that $C(n, 0) = 1$, it follows that

the **total** *number of r-combinations of n objects*
for $r = 1, 2, 3, \ldots, n$ successively is $2^n - 1$ (13.20)

As illustrations of the use of (13.19), recall from Exercise 3 of Sec. 1.3 that the set $\{a, b\}$ has $2^2 = 4$ subsets, namely, $\{a, b\}$, $\{a\}$, $\{b\}$, $\varnothing$; also, as we saw in Exercise 4 of Sec. 1.3, the set $\{a, b, c\}$ has $2^3 = 8$ subsets, namely, $\{a, b, c\}$, $\{a, b\}$, $\{a, c\}$, $\{b, c\}$, $\{a\}$, $\{b\}$, $\{c\}$, $\varnothing$.

The result (13.20) may be used to solve a problem like Exercise 7 in Sec. 12.7: How many different sums can be made by using a penny, a nickel, a dime, a quarter, a half-dollar, and a dollar? Here we are concerned with the *total* number of r-combinations of six objects for $r = 1, 2, 3, 4, 5, 6$ successively. From (13.20) we see that the answer is $2^6 - 1 = 64 - 1 = 63$ possible sums that can be made from the six coins.

EXERCISES

1 The 7th row of Pascal's triangle is

1 7 21 35 35 21 7 1

Determine the 8th row and the 9th row of Pascal's triangle.

2 Use the results of Exercise 1 to write the binomial expansion of $(a + b)^8$ and the expansion of $(2x + 1)^8$.

3 Use the results of Exercise 1 to write the binomial expansions of $(x + y)^9$ and $(x - y)^9$.

4 If five coins are tossed, find the probability that (a) exactly two heads appear; (b) exactly three heads appear; (c) no heads appear.

5 Find the *total* number of r-combinations of 7 objects for $r = 1, 2, 3, 4, 5, 6, 7$ successively.

6 Find the *total* number of r-combinations of 10 objects for $r = 1, 2, 3, 4, 5, 6, 7, 8, 9, 10$ successively.

7 How many selections can be made from 10 different books if one or more may be chosen?

8 Five paths, which we designate by a, b, c, d, e, meet at a junction, and a hiker comes along each path to this junction. In how many different ways may these hikers continue their walks across the junction?

9 A contractor employs 12 men. In how many ways can he choose one or more men to do a certain job?

10 How many subsets are there of a set with 6 members?

11 How many subsets are there of a set with 20 members?

13.9 BINOMIAL DISTRIBUTION

Consider the experiment $\mathcal{E}$ of tossing a coin two times. Each toss of the coin is itself an experiment which we call a *trial*. We can consider the experiment $\mathcal{E}$ as consisting of two trials, $\mathcal{E}_1$ and $\mathcal{E}_2$, and we can regard

$$S_1 = \{H, T\}$$

as the sample space for each of $\mathcal{E}_1$ and $\mathcal{E}_2$. The sample space S of the experiment $\mathcal{E}$ is given by

$$S = \{(H, H), (H, T), (T, H), (T, T)\} \tag{13.21}$$

Note that S has $2^2 = 4$ sample points and that each of the sample points is an ordered pair with each entry of each ordered pair being an H or a T. We follow the convention of writing the sample space S of (13.21) in the abbreviated manner

$$S = \{HH, HT, TH, TT\} \tag{13.22}$$

If in each of the trials $\mathcal{E}_1$ and $\mathcal{E}_2$, for which the sample space is $S_1 = \{H, T\}$, we are interested in getting heads, we call the simple event $\{H\}$ a *success* and its complementary event $\{T\}$ a *failure*. Since each trial has two outcomes, we call each of the trials a *binomial trial*.

Let us consider the question: What is the probability* P of getting *exactly* two heads in three throws of a coin? Here the experiment $\mathcal{E}$ consists of three binomial trials $\mathcal{E}_1$, $\mathcal{E}_2$, $\mathcal{E}_3$, with each trial having the sample space

$$S_1 = \{H, T\}$$

The sample space of $\mathcal{E}$ is given by

$$S = \{(H, H, H), (H, H, T), (H, T, H), (T, H, H), (H, T, T), (T, H, T), (T, T, H), (T, T, T)\}$$

Note that S has $2^3 = 8$ sample points and that each of the sample points is an ordered triple with each entry of each ordered triple

* In the discussion of binomial experiments and the binomial distribution in this section, we shall use the symbol P rather than Pr to denote the probability of obtaining a certain number of successes in a given number of trials of a binomial experiment.

being an H or a T. We write this sample space in abbreviated form as

$$S = \{HHH, HHT, HTH, THH, HTT, THT, TTH, TTT\} \quad (13.23)$$

With the sample space S displayed in (13.23), we see immediately that the probability of throwing exactly two heads, that is, the probability of the event $E = \{HHT, HTH, THH\}$ is $\frac{3}{8}$. We note that this result could be obtained from the entries in the 3d row of Pascal's triangle in the manner indicated in the preceding section.

Now let us see if we can develop a method for finding P when we do not have a tabulation of the simple events in the sample space of the experiment. In order to apply the method to more general cases, we shall suppose that the probability of getting heads in one toss of the coin is p and the probability of getting tails is q. In other words p is the probability of success and q is the probability of failure on any one of the binomial trials. Of course $p + q = 1$. Let us consider how many ways we can get two heads and one tail in three tosses of a coin. We can think of this as having three places to fill (the three tosses) and wishing to fill two of these places with an H and the other place with a T. Thus we want to know the number of permutations of three things, two of which are alike. We recall from Sec. 12.6 that this number is $\frac{3!}{2!1!}$. We recall (Sec. 12.7) that $C(3, 2)$, the number of 2-combinations* of three objects, is also given by $\frac{3!}{2!1!}$. Thus the number of ways of obtaining two heads and one tail in three tosses of a coin is $C(3, 2) = 3$. The event (exactly two heads and one tail) whose probability we are seeking is therefore the union of three mutually exclusive simple events. It follows from statement (12.30) in Sec. 12.8 that the sum of the probabilities of these three simple events will be the probability we are seeking. Now what are the probabilities of these simple events? Each of the simple events is the intersection of three *independent* events. To illustrate, the event $\{HHT\}$ is the intersection of the three events

$E_1 = \{HHH, HTH, HHT, HTT\}$	heads on the first throw
$E_2 = \{HHH, THH, THT, HHT\}$	heads on the second throw
$E_3 = \{HHT, HTT, THT, TTT\}$	tails on the third throw

*It happens that the number of permutations of n things, r of which are alike in one way and the remaining $n - r$ of which are alike in a second way, is equal to $C(n, r)$; that is,

$$\frac{n!}{r!(n-r)!} = C(n, r)$$

[see formulas (12.19) and (12.24)]. It is customary to use the symbol $C(n, r)$ in such a case, but we should remember that in this connection $C(n, r)$ denotes the number of permutations of n things, r of which are alike in one way and $n - r$ of which are alike in a second way.

Since E_1, E_2, E_3 are independent (that is, the occurrence of one of the events does not affect the occurrence of either of the other events), it follows from statement (12.36) in Sec. 12.8 that the probability of $\{HHT\}$ is the product of the probabilities of E_1, E_2, and E_3. The probability of E_1 (heads on the first throw) is given to be p; the probability of E_2 (heads on the second throw) is given to be p; the probability of E_3 (tails on the third throw) is given to be q. So

$$\text{Prob } \{HHT\} = p \cdot p \cdot q = p^2q$$

In a similar way it can be established that the probability of getting two heads and one tail *in any order* is $p \cdot p \cdot q = p^2q$.

Thus we can see that the probability P of throwing two heads and one tail in three tosses of a coin is given by

$$P = p^2q + p^2q + p^2q = C(3, 2)p^2q$$

If we assume (as is customary) that the probability of throwing a head is $\frac{1}{2}$, then

$$P = C(3, 2)\left(\frac{1}{2}\right)^2\frac{1}{2} = 3 \cdot \frac{1}{8} = \frac{3}{8}$$

which is the same result we obtained by tabulating the sample space of the experiment.

In formula (13.17) in Sec. 13.8 we gave one form of the binomial formula. By making use of (12.25), which states that $C(n, r) = C(n, n - r)$, we can write the binomial formula in the following way:

$$(a + b)^n = a^n + C(n, n - 1)a^{n-1}b + C(n, n - 2)a^{n-2}b^2 + \cdots + C(n, n - r)a^{n-r}b^r + \cdots + C(n, 1)ab^{n-1} + b^n \quad (13.24)$$

In this section we shall use the binomial formula in the form (13.24).

Using (13.24), we can write the expansion of $(\frac{1}{2} + \frac{1}{2})^3$ as follows,

$$\left(\frac{1}{2} + \frac{1}{2}\right)^3 = \left(\frac{1}{2}\right)^3 + C(3, 2)\left(\frac{1}{2}\right)^2\left(\frac{1}{2}\right) + C(3, 1)\left(\frac{1}{2}\right)\left(\frac{1}{2}\right)^2 + \left(\frac{1}{2}\right)^3 \quad (13.25)$$

In our discussion of the experiment of tossing a coin three times we identified the second term on the right side of (13.25) as the probability of getting exactly two heads when a coin is tossed three times. The student is asked to verify that the first term on the right side of (13.25) is the probability of getting *exactly* three heads, that the third term is the probability of getting *exactly* one head, and that the fourth term is the probability of getting *exactly* no heads. Therefore, when a coin is tossed three times, the probabilities of getting exactly three heads, two heads, one head, and no heads are, respectively, the successive coefficients in the binomial expansion of $(\frac{1}{2} + \frac{1}{2})^3$ as given by (13.25). We shall be concerned in the remainder of this section with a generalization of this result.

BINOMIAL EXPERIMENT

A **binomial experiment** consists of n independent binomial trials with one of the two simple events in each trial being called a success and with the probability of a success being the same for each trial. We let p denote this probability of a success on any given trial. Note that $n \in N$ and $0 \leq p \leq 1$.

In a binomial experiment with n trials, we can speak of the probability of obtaining a specified number x of successes, where x is one of the integers 0, 1, 2, 3, . . . , n. Thus, paired with each member x of the set $\{0, 1, 2, 3, \ldots, n\}$, there is a number, denoted by $P\{x; n, p)$, which is the probability of obtaining x successes in a binomial experiment with n trials, each trial having probability p of yielding a success. This *set of ordered pairs,* $\{[x, P(x; n, p)]\}$ for $x \in \{0, 1, 2, 3, \ldots, n\}$, is called the **binomial distribution** for the given values of n and p.

BINOMIAL DISTRIBUTION

For the binomial experiment of tossing a coin three times (which we have analyzed above), in which the appearance of a head is considered a success, we can construct the following table showing the binomial distribution for $n = 3$ and $p = \frac{1}{2}$.

x (NUMBER OF HEADS)	3	2	1	0
$P(x; n, p)$	$\frac{1}{8}$	$\frac{3}{8}$	$\frac{3}{8}$	$\frac{1}{8}$

Thus the binomial distribution for $n = 3$, $p = \frac{1}{2}$ is

$$\{(3, \tfrac{1}{8}), (2, \tfrac{3}{8}), (1, \tfrac{3}{8}), (0, \tfrac{1}{8})\} \tag{13.26}$$

In connection with this distribution we observe the following.

1 For each $x \in \{0, 1, 2, 3\}$, $0 \leq P(x; 3, \frac{1}{2}) \leq 1$.
2 $P(3; 3, \frac{1}{2}) + P(2; 3, \frac{1}{2}) + P(1; 3, \frac{1}{2}) + P(0; 3, \frac{1}{2}) = 1$.
3 The denominator in the second entry in each of the ordered pairs is the number $8 = 2^3$ of sample points in the sample space of the experiment.
4 The numerator in the second entry in each ordered pair is the frequency of the appearance of sample points with exactly the number of successes given by the first entry of the pair.
5 Another way of looking at the result is that if we were to toss three coins eight times and on each toss count the number of heads that appear, the "expected" frequency distribution would be as tabulated on page 537.

NUMBER OF HEADS	FREQUENCY
3	1
2	3
1	3
0	1

The graph of the binomial distribution is shown as the points A, B, C, and D in Fig. 13.3*a*. If we join the points A, B, C, and D by line segments as shown in Fig. 13.3*b* we have a curve that is called (by analogy to the situation in Sec. 13.3) the frequency polygon for the binomial distribution. The bar graph shown in Fig. 13.3*b* is the histogram for the binomial distribution.

If a coin is tossed 10 times and if we are interested in the appearance of heads, we have a binomial experiment with 10 trials in which the appearance of a head is considered a success and in which the

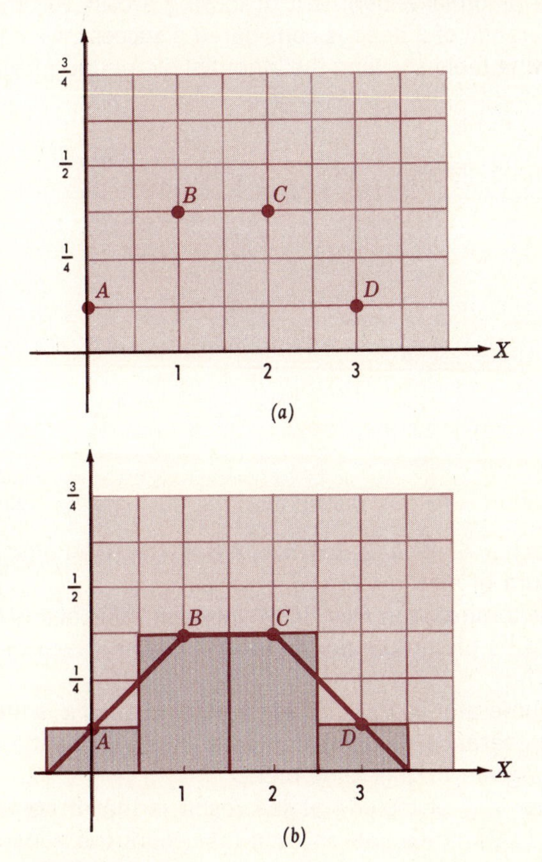

FIGURE 13.3

probability p of a success on any one trial is $\frac{1}{2}$. The student should verify that the successive terms of the binomial expansion,

$$\left(\frac{1}{2}+\frac{1}{2}\right)^{10} = \left(\frac{1}{2}\right)^{10} + C(10,9)\left(\frac{1}{2}\right)^{9}\left(\frac{1}{2}\right) + C(10,8)\left(\frac{1}{2}\right)^{8}\left(\frac{1}{2}\right)^{2}$$

$$+ C(10,7)\left(\frac{1}{2}\right)^{7}\left(\frac{1}{2}\right)^{3} + \cdots + C(10,1)\left(\frac{1}{2}\right)\left(\frac{1}{2}\right)^{9} + \left(\frac{1}{2}\right)^{10}$$

$$\left(\frac{1}{2}+\frac{1}{2}\right)^{10} = \frac{1}{1{,}024} + \frac{10}{1{,}024} + \frac{45}{1{,}024} + \frac{120}{1{,}024} + \frac{210}{1{,}024} + \frac{252}{1{,}024}$$

$$+ \frac{210}{1{,}024} + \frac{120}{1{,}024} + \frac{45}{1{,}024} + \frac{10}{1{,}024} + \frac{1}{1{,}024} \qquad (13.27)$$

are the probabilities of getting exactly 10 heads, 9 heads, 8 heads, 7 heads, 6 heads, 5 heads, 4 heads, 3 heads, 2 heads, 1 head, 0 heads.

For the binomial experiment of tossing a coin 10 times in which the appearance of a head is considered a success, we can construct the following table showing the binomial distribution for $n = 10$ and $p = \frac{1}{2}$.

x (NUMBER OF HEADS)	10	9	8	7	6
$P(x; n, p)$	$\frac{1}{1{,}024}$	$\frac{10}{1{,}024}$	$\frac{45}{1{,}024}$	$\frac{120}{1{,}024}$	$\frac{210}{1{,}024}$

x (NUMBER OF HEADS)	5	4	3	2	1	0
$P(x; n, p)$	$\frac{252}{1{,}024}$	$\frac{210}{1{,}024}$	$\frac{120}{1{,}024}$	$\frac{45}{1{,}024}$	$\frac{10}{1{,}024}$	$\frac{1}{1{,}024}$

In connection with this distribution we observe the following:

1 For each $x \in \{0,1,2,3,4,5,6,7,8,9,10\}$, $0 \leq P(x;\ 10, \frac{1}{2}) \leq 1$.
2 The sum of all the $P(x;\ 10, \frac{1}{2})$ is 1.
3 The denominator in $P(x;\ 10, \frac{1}{2})$ for each value of x is the number ($1{,}024 = 2^{10}$) of sample points in the sample space of the experiment.
4 The numerator in $P(x;\ 10, \frac{1}{2})$ for each value of x is the frequency of the appearance of sample points with exactly the number of successes given by the value of x.
5 Another way of looking at the result is that if we were to toss 10 coins 1,024 times and on each toss count the number of heads that appeared, the "expected" frequency distribution would be as shown in the table on page 539:

NUMBER OF HEADS	FREQUENCY
10	1
9	10
8	45
7	120
6	210
5	252
4	210
3	120
2	45
1	10
0	1

The graph of the binomial distribution for $n = 10$ and $p = \frac{1}{2}$ is shown as the heavy dots in Fig. 13.4. If these points are joined by line segments, as shown in Fig. 13.4, we have the frequency polygon for the binomial distribution.

Consider a binomial experiment $\mathcal{E}$ which consists of n independent binomial trials, each trial having sample space $\{S, F\}$, S denoting success, F denoting failure, the probability of S being p, and the probability of F being $q = 1 - p$. By the fundamental counting principle (Sec. 12.4) the experiment $\mathcal{E}$ has $2 \cdot 2 \cdot 2 \cdot \cdots \cdot 2$ (n factors), or 2^n, different outcomes. Therefore the sample space of $\mathcal{E}$ has 2^n sample points, and each of the sample points is an n-tuple with each entry of the n-tuple an S or an F.

Proceeding in a manner analogous to that used in discussing the experiment of tossing a coin three times, we can see that the probability that the first x trials produce S and the remaining $n - x$ trials produce F is

$$p^x \cdot q^{n-x} \tag{13.28}$$

In (13.28), p is used as a factor x times, and q is used as a factor $n - x$ times.

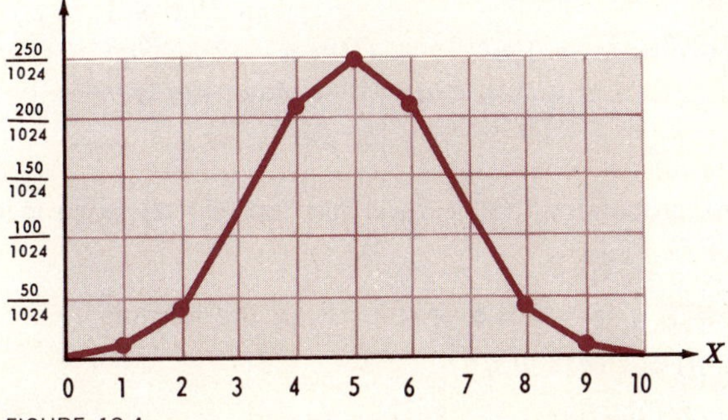

FIGURE 13.4

Suppose that in n trials of the binomial experiment $\mathcal{E}$ we want *exactly* x S's and $(n - x)$ F's, where $x \in \{0, 1, 2, 3, \ldots, n\}$. Let us denote the probability of obtaining x S's and $(n - x)$ F's by

$$P(x; n, p)$$

In the same way in which we obtained (13.28), we conclude that the probability of obtaining x S's and $(n - x)$ F's in *any one given order* is $p^x q^{n-x}$. In how many ways can we get x S's and $(n - x)$ F's in n trials? This number of ways is the number of permutations of n things, x of which are alike in one way, x S's, and $n - x$ of which are alike in a second way, $(n - x)$ F's. This number is $\dfrac{n!}{x!(n-x)!}$, which is also the number of x-combinations of n distinct things and which is denoted* by $C(n, x)$:

$$\frac{n!}{x!(n-x)!} = C(n, x) \tag{13.29}$$

Thus the probability $P(x; n, p)$ is the probability of an event which is the union of $C(n, x)$ sample points of the sample space of $\mathcal{E}$, with each of these sample points having probability $p^x q^{n-x}$. Consequently

$$P(x; n, p) = C(n, x)p^x q^{n-x} \tag{13.30}$$

and we record this result as follows:

If p is the probability of success in a single trial and q is the probability of failure in a single trial $(q = 1 - p)$, then the probability $P(x; n, p)$ that there will be exactly *x successes in n trials is given by* (13.31)

$$P(x; n, p) = C(n, x)p^x q^{n-x}$$

By using (13.31), the student is asked to prove that

The probability of at least *x successes in n trials is*

$$\begin{aligned} &P(n; n, p) + P(n-1; n, p) + P(n-2; n, p) + \cdots \\ &\qquad + P(x+1; n, p) + P(x; n, p) \\ &= p^n + C(n, n-1)p^{n-1}q + C(n, n-2)p^{n-2}q^2 + C(n; n-3)p^{n-3}q^3 \\ &+ \cdots + C(n, x+1)p^{x+1}q^{n-x-1} + C(n, x)p^x q^{n-x} \end{aligned} \tag{13.32}$$

EXAMPLE 1 Find the probability that a 3 will turn up *exactly* twice in five throws of a die.

SOLUTION Here $p = \frac{1}{6}$, $q = \frac{5}{6}$, $n = 5$, $x = 2$. Using the formula

$$P(x; n, p) = C(n, x)p^x q^{n-x}$$

*See the footnote on page 534.

we get

$$P\left(2; 5, \frac{1}{6}\right) = C(5, 2)\left(\frac{1}{6}\right)^2\left(\frac{5}{6}\right)^3 = \frac{5 \cdot 4}{2 \cdot 1} \cdot \frac{1}{6^2} \cdot \frac{125}{6^3} = \frac{1,250}{6^5} = \frac{625}{3,888}$$

EXAMPLE 2 Find the probability that a 3 will turn up *at least* twice in five throws of a die.

SOLUTION The event will occur at least twice if it occurs five times, four times, three times, or two times. Either directly or by use of (13.32), we have that the desired probability is

$$P\left(5; 5, \frac{1}{6}\right) + P\left(4; 5, \frac{1}{6}\right) + P\left(3; 5, \frac{1}{6}\right) + P\left(2; 5, \frac{1}{6}\right)$$

$$= \left(\frac{1}{6}\right)^5 + C(5, 4)\left(\frac{1}{6}\right)^4\left(\frac{5}{6}\right) + C(5, 3)\left(\frac{1}{6}\right)^3\left(\frac{5}{6}\right)^2 + C(5, 2)\left(\frac{1}{6}\right)^2\left(\frac{5}{6}\right)^3$$

$$= \frac{1 + 5 \cdot 5 + 10 \cdot 25 + 10 \cdot 125}{6^5} = \frac{1,526}{6^5} = \frac{763}{3,888}$$

Alternatively, by use of (12.37) in Sec. 12.8,

$$\text{desired probability} = 1 - [P(1; 5, \tfrac{1}{6}) + P(0; 5, \tfrac{1}{6})]$$

$$= 1 - [C(5, 1)(\tfrac{1}{6})(\tfrac{5}{6})^4 + (\tfrac{5}{6})^5]$$

$$= 1 - \frac{5(625)}{6^5} - \frac{5(625)}{6^5} = 1 - \frac{10(625)}{6(6^4)}$$

$$= 1 - \frac{3,125}{3,888} = \frac{763}{3,888}$$

Using the binomial formula as given by (13.24) in conjunction with (13.31), we have the following statement

If p is the probability of success in a single trial and q is the probability of failure ($p + q = 1$), then the successive terms of the binomial expansion of $(p + q)^n$,

$$(p + q)^n = p^n + C(n, n-1)p^{n-1}q + C(n, n-2)p^{n-2}q^2 + \cdots + C(n, x)p^x q^{n-x} + \cdots + q^n \quad (13.33)$$

give the respective probabilities that in n trials there will be exactly *$n, n-1, n-2, \ldots, x, \ldots, 1, 0$ successes.*

In each ordered pair of a binomial distribution $\{[x, P(x; n, p)]\}$, the second entry $P(x; n, p)$ can be considered to be the frequency of the number of successes denoted by the first entry x. Let M_x be the weighted mean of the number of successes in a binomial experiment of n trials in which the weight attached to each number x of successes is the probability $P(x; n, p)$ of that number of successes. Then, in accordance with (13.4), we have [writing $P(x)$ for $P(x; n, p)$]

$$M_x = \frac{nP(n) + (n-1)P(n-1) + (n-2)P(n-2) + \cdots + 1P(1) + 0}{P(n) + P(n-1) + P(n-2) + \cdots + P(1) + 0}$$

To aid in computing M_x, we construct the following table by the use of formula (13.30).

x	$P(x;n,p) = C(n,x)p^x q^{n-x}$	$xP(x;n,p)$
n	p^n	np^n
$n-1$	$np^{n-1}q$	$n(n-1)p^{n-1}q$
$n-2$	$\frac{n(n-1)}{2}p^{n-2}q^2$	$\frac{n(n-1)(n-2)}{2}p^{n-2}q^2$
. . . . . .	. . . .	. . . .
3	$\frac{n(n-1)(n-2)}{3\cdot 2\cdot 1}p^3q^{n-3}$	$\frac{n(n-1)(n-2)}{2}p^3q^{n-3}$
2	$\frac{n(n-1)}{2}p^2q^{n-2}$	$n(n-1)p^2q^{n-2}$
1	npq^{n-1}	npq^{n-1}
0	q^n	0
Totals	$(p+q)^n = 1$	$np(p+q)^{n-1} = np$

The total of the second column was found by use of the binomial expansion given in (13.33) and by use of the fact that $p + q = 1$. The total in the third column was found by noting that np is a factor of each entry in the column. Hence the sum of the entries in the column is given by

$$np\left[p^{n-1} + (n-1)p^{n-2}q + \frac{(n-1)(n-2)}{2}p^{n-3}q^2 + \cdots + \frac{(n-1)(n-2)}{2}p^2q^{n-3} + (n-1)pq^{n-2} + q^{n-1}\right]$$

The expression in the brackets is seen (by the use of the binomial expansion) to be $(p + q)^{n-1}$. So the sum of the entries in the third column is

$$np(p+q)^{n-1}$$

and since $p + q = 1$, this sum becomes np. From these results we obtain

$$M_x = \frac{np}{1} = np \tag{13.34}$$

In a similar way it can be shown that the standard deviation σ of the binomial distribution for n trials with probability of success p is given by

$$\sigma = \sqrt{npq} \tag{13.35}$$

EXPECTED NUMBER OF OCCURRENCES

The **expected number of occurrences** of a success in n trials is $np = M_x$, where p is the probability of success in a single trial.

For the experiment of tossing a coin 10 times, where $n = 10$ and $p = \frac{1}{2}$, the expected number of heads is [by use of (13.34)]

$$M_x = 10(\tfrac{1}{2}) = 5$$

Using (13.35), we find that

$$\sigma = \sqrt{10(\tfrac{1}{2})(\tfrac{1}{2})} = \tfrac{1}{2}\sqrt{10} = 1.58$$

is the standard deviation for the distribution of the number of heads that occur in the experiment.

If each side of (13.33) is multiplied by a suitable positive integer, then the terms on the right side of the resulting equality will represent frequencies. To illustrate, if each side of the equality (13.27) is multiplied by 1,024, we obtain

$$1{,}024(\tfrac{1}{2} + \tfrac{1}{2})^{10} = 1 + 10 + 45 + 120 + 210 + 252 + 210 + 120 + 45 + 10 + 1 \tag{13.36}$$

and the terms on the right side of (13.36) are the number of times we should expect 10, 9, 8, 7, 6, 5, 4, 3, 2, 1, 0 heads, respectively, if 10 coins are tossed 1,024 times.

Let us consider an experiment in which a coin is tossed four times with the probability p of obtaining a head on each toss given as $\frac{1}{3}$ and the probability q of obtaining a tail on each toss given as $\frac{2}{3}$. The probability $P(x; 4, \frac{1}{3})$ of obtaining x heads is given by

$$P(x; 4, \tfrac{1}{3}) = C(4, x)(\tfrac{1}{3})^x(\tfrac{2}{3})^{4-x}$$

and the probabilities of obtaining four heads, three heads, two heads, one head, no heads are, respectively, the first, second, third, fourth, and fifth terms on the right side of the binomial expansion

$$\begin{aligned}(\tfrac{1}{3} + \tfrac{2}{3})^4 &= (\tfrac{1}{3})^4 + C(4, 3)(\tfrac{1}{3})^3(\tfrac{2}{3}) + C(4, 2)(\tfrac{1}{3})^2(\tfrac{2}{3})^2 + C(4, 1)(\tfrac{1}{3})(\tfrac{2}{3})^3 + (\tfrac{2}{3})^4 \\ &= \tfrac{1}{81} + \tfrac{8}{81} + \tfrac{24}{81} + \tfrac{32}{81} + \tfrac{16}{81}\end{aligned} \tag{13.37}$$

The binomial distribution for $n = 4$ and $p = \frac{1}{3}$ is

$$\{(4, \tfrac{1}{81}), (3, \tfrac{8}{81}), (2, \tfrac{24}{81}), (1, \tfrac{32}{81}), (0, \tfrac{16}{81})\}$$

and the graph of this distribution is shown by the heavy dots in Fig. 13.5. The frequency polygon for this distribution is obtained,

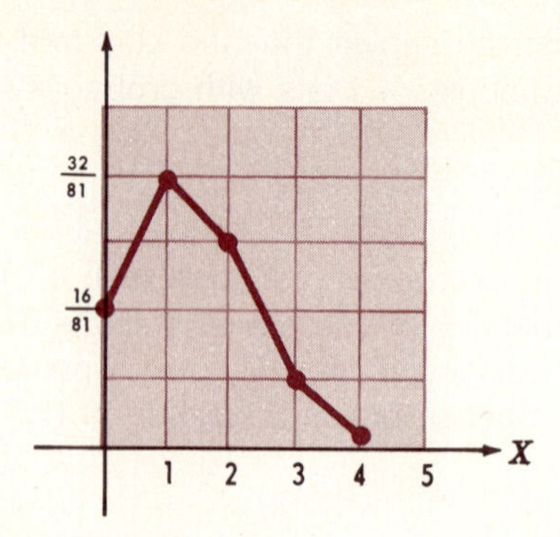

FIGURE 13.5

as shown in Fig. 13.5, by connecting these dots by line segments.

If each side of (13.37) is multiplied by 81 we obtain

$$81(\tfrac{1}{3} + \tfrac{2}{3})^4 = 1 + 8 + 24 + 32 + 16 \qquad (13.38)$$

and the terms on the right side of (13.38) are the number of times we should expect 4, 3, 2, 1, 0 heads, respectively, if four coins are tossed 81 times with the probability of a head on each coin being $\frac{1}{3}$ ("loaded" or lopsided coins).

Comparing the frequency polygons shown in Figs. 13.3, 13.4, and 13.5 for three different binomial distributions, we observe that in Figs. 13.3 and 13.4 the polygons are symmetric about the line which is the graph of $x = M_x$. [For Fig. 13.3, $M_x = 3(\frac{1}{2}) = \frac{3}{2}$; for Fig. 13.4, $M_x = 10(\frac{1}{2}) = 5$.] However, the frequency polygon in Fig. 13.5 is not symmetric about the graph of $x = M_x$ [here $M_x = 4(\frac{1}{3}) = \frac{4}{3}$] but is "skewed" to one side. The symmetry in Figs. 13.3 and 13.4 is a consequence of the fact that for those binomial distributions, $p = q$; the nonsymmetry in Fig. 13.5 is a consequence of the fact that for this distribution $p \neq q$.

For the binomial experiment of tossing a coin 1,024 times (with probability of heads on any one toss being $\frac{1}{2}$), the expected number of heads is $M_x = 1{,}024(\frac{1}{2}) = 512$. The standard deviation for this binomial distribution is $\sigma = \sqrt{1{,}024(\frac{1}{2})(\frac{1}{2})} = \frac{32}{2} = 16$. It is shown in statistics that since this distribution is symmetric, the probability is approximately $\frac{99}{100}$ that the actual number of heads appearing will be between $M_x - 3\sigma$ and $M_x + 3\sigma$, that is, between 464 and 560.

Suppose that upon tossing a coin 1,024 times one were to obtain 700 heads. In this case it is reasonable to suppose that factors other than mere chance were present.

In general, if a binomial distribution is symmetric or only slightly skewed, then the probability that the number of successes obtained is between $M_x - 3\sigma$ and $M_x + 3\sigma$ is approximately $\frac{99}{100}$.

EXERCISES

1 (*a*) If $p = \frac{1}{4}$ is the probability of the success of an event and $q = \frac{3}{4}$ is the probability of its failure, find the binomial expansion of $(\frac{1}{4} + \frac{3}{4})^4$ and interpret its terms in accordance with the statement concerning (13.33). (*b*) Construct the frequency polygon for the binomial distribution with $n = 4$ and $p = \frac{1}{4}$ as we did in Fig. 13.4 for the distribution determined by the expansion (13.27). Is your polygon symmetric? Explain.

2 Find the probability that a 1 will turn up exactly twice in four throws of a die.

3 A coin is tossed seven times. Find the probability of exactly (*a*) no heads; (*b*) one head; (*c*) two heads.

4 Seven coins are tossed 128 times. Using the results of Exercise 3, find the expected number of (*a*) no heads; (*b*) one head; (*c*) two heads.

5 If a die is thrown six times, find the probability of obtaining (*a*) exactly two 4s; (*b*) at least three 4s.

6 Find in two ways the probability of throwing, with a single die, a 2 at least once in five trials. *Hint:* See Example 2.

7 A bag contains four black and two red balls. Five balls are drawn in turn with each being replaced in the bag after each drawing. Find the probability (*a*) that exactly three are black; (*b*) that at least three are red.

8 A man whose batting average is $\frac{3}{10}$ will bat four times in a game. Find the probability that he will get (*a*) exactly two hits; (*b*) at least two hits.

In Exercises 9 to 12, construct a table for the binomial distribution determined by the given expansion $(p + q)^n$, graph the frequency polygon for this distribution, and find M_x and σ for the distribution.

9 $(\frac{5}{6} + \frac{1}{6})^6$ 10 $(\frac{1}{6} + \frac{5}{6})^6$ 11 $(\frac{9}{10} + \frac{1}{10})^4$ 12 $(\frac{3}{5} + \frac{2}{5})^4$

13 *A* reports that upon tossing a coin 400 times, 250 heads appeared. In regard to this result, comment upon the probability that factors other than mere chance were present.

14 A coin is tossed five times. (*a*) Find the probability of getting at least three heads. (*b*) If a prize of $100 is given for getting at least three heads, what is the mathematical expectation?

13.10
NORMAL DISTRIBUTION

Of considerable importance in statistics is the curve having the equation

$$y = Ce^{(-1/2)(x/\sigma)^2}$$

NORMAL FREQUENCY CURVE

where $e = 2.718$ approximately and C is the ordinate for $x = 0$ (the maximum ordinate). This curve is called the **normal frequency curve;**

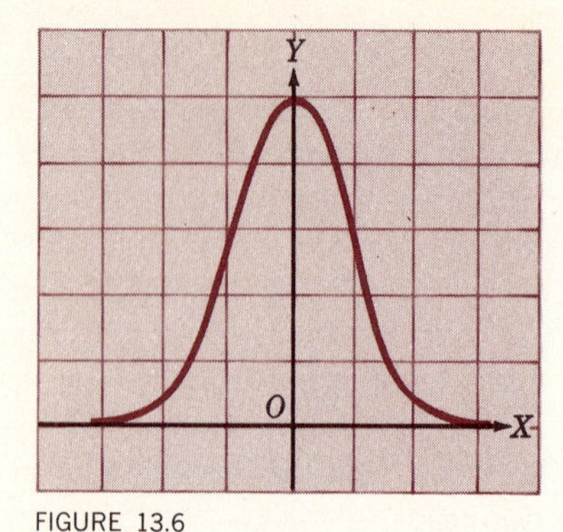

FIGURE 13.6

its graph appears in Fig. 13.6 and is symmetric. The frequency distribution associated with the normal frequency curve is called the **normal distribution.** In the above equation x stands for a deviation from the arithmetic mean of the normal distribution, and σ is the standard deviation of the normal distribution.

NORMAL DISTRIBUTION

Many distributions of measurements of natural objects, as the heights of men and the circumferences of trees, approximate closely the normal distribution. It is helpful, in making conclusions about a set of observations that does not differ too greatly from the normal distribution, to know the following facts, which are proved in statistics. For a normal distribution:

1 The interval $M - \sigma$ to $M + \sigma$ includes approximately $\frac{2}{3}$ of the n measures.
2 The interval $M - 2\sigma$ to $M + 2\sigma$ includes approximately 95% of the n measures.
3 The interval $M - 3\sigma$ to $M + 3\sigma$ includes approximately 99% of the n measures.

Recall that, for the distribution (which is approximately normal) of 100 mathematics grades, which we studied on pages 511, 512, 513, and 518, we found that

$$M = 74.7 \qquad \text{and} \qquad \sigma = 8.9$$

So $3\sigma = 26.7$, and therefore

$$M - 3\sigma = 48 \qquad M + 3\sigma = 101.4$$

From the table on page 510 we see that none of the 100 scores lies outside the interval $M - 3\sigma$ to $M + 3\sigma$.

For the set of numbers (13.5) in Sec. 13.6, recall that

$$M = 14 \qquad \text{and} \qquad \sigma = 4$$

The interval $M - \sigma = 10$ to $M + \sigma = 18$ includes $\frac{5}{7} = 71$ percent, approximately, of the grades.

The interval $M - 2\sigma = 6$ to $M + 2\sigma = 22$ includes all the grades.

The standard deviation is helpful in locating the place of an item in a distribution. To illustrate, if a salesman sells \$400 more than the arithmetic mean of the sales of all the salesmen in his group, and if further we are told that his sales volume is 3 standard deviations above the arithmetic mean, we can conclude (on the assumption that the sales are approximately normally distributed) that he is one of the best salesmen, and probably the best.

In many cases it will be impossible to make measurements on every object that we wish to describe. For example, a manufacturer of light bulbs might be interested in knowing the number of hours that a certain shipment of bulbs will last in continuous use. If measurements are made on all the bulbs, exact information may be obtained but then there will be no bulbs left to sell. At other times, the population, or total number of objects of interest, may be so large that complete enumeration is practically impossible. In such cases, we shall frequently be satisfied to take observations on a sample of objects from the population.

The terms *sample* and *population* introduce two of the most basic ideas of inductive statistics. The normal distribution, and several other common distributions, are extremely useful in making inferences about a population from the partial information that is obtainable from a sample, but these applications are beyond the scope of this book.

EXERCISES

1 For the distribution of 100 mathematics grades which we studied in Example 2 of Sec. 13.6, find the percent of the grades in the interval $M - \sigma$ to $M + \sigma$; the percent of the grades in the interval $M - 2\sigma$ to $M + 2\sigma$.

2 For the set of numbers 1, 8, 14, 21, 26 which we studied in connection with the table on the right on page 523, find the percent of the numbers in the interval $M - \sigma$ to $M + \sigma$; the percent of the numbers in the interval $M - 2\sigma$ to $M + 2\sigma$; the percent of the numbers in the interval $M - 3\sigma$ to $M + 3\sigma$.

For the distributions in Exercises 3 to 5, find the percent of the measures in the intervals $M - \sigma$ to $M + \sigma$, $M - 2\sigma$ to $M + 2\sigma$, $M - 3\sigma$ to $M + 3\sigma$.

3 The distribution of Exercise 4 of Sec. 13.6.

4 The distribution of Exercise 5 of Sec. 13.6.

5 The distribution of Exercise 6 of Sec. 13.6.

6 The arithmetic mean of 1,000 college grades in mathematics is 71, and the standard deviation is 9. Assuming that the distribution of the grades is normal, what range of grades will include approximately $\frac{2}{3}$ of the grades? What is the approximate range of the grades ($M - 3\sigma$ to $M + 3\sigma$)?

7 The arithmetic mean of 600 grades in mathematics is 73, and the standard deviation is 9. Assuming that the distribution of the grades is normal, what range of the grades would include approximately the middle 400 grades? What is the approximate range of the grades?

8 A given frequency distribution contains 10,000 measures which are distributed approximately normally. Approximately how many of the measures are included in the interval $M - \sigma$ to $M + \sigma$; the interval $M - 2\sigma$ to $M + 2\sigma$; the interval $M - 3\sigma$ to $M + 3\sigma$?

9 For a given frequency distribution of 500 measurements which is approximately normal, it is found that $M = 300$, $\sigma = 18$. About how many of the measurements lie between 282 and 318? Between 264 and 336? Between 246 and 354?

10 A manufacturer of men's shoes collected data on sizes of shoes sold and found that the sizes of the shoes closely approximated the normal distribution, that the arithmetic mean of the sizes was 9, and that the standard deviation was 2. Approximately what percent of the sizes were between 7 and 11? Between 5 and 13? Practically all the shoes sold were in what range of sizes?

In Exercises 11 to 14, assume that the given distribution is normal, and make use of the statements 1, 2, and 3 on page 546 pertaining to the normal distribution.

11 The arithmetic mean of the weights of 1,000 men is 156 pounds, and the standard deviation is 8 pounds. Describe the approximate distribution of the weights.

12 The arithmetic mean of the heights of 1,000 men is 69 inches, and the standard deviation is 3 inches. Describe the approximate distribution of the heights.

13 The arithmetic mean of the weights of 1,000 college girls is 121 pounds, and the standard deviation is 9 pounds. Describe the approximate distribution of the weights.

14 The arithmetic mean of the heights of 1,000 college girls is 62.0 inches, and the standard deviation is 2.5 inches. Describe the approximate distribution of the heights.

In Exercises 15 to 18, assume that the population of measures and the sample thereof are both approximately normally distributed.

15 A manufacturer of automobile tires tests a random sample of 200 tires and finds for the useful lives of these tires that $M = 23{,}000$ miles and $\sigma = 3{,}000$ miles. If the manufacturer decides to guarantee the tires to last 20,000 miles, what is the approximate number he should expect to replace out of every 12,000 tires sold?

16 A manufacturer of radio tubes finds that for the lives of 225 tubes selected at random $M = 24$ months and $\sigma = 3$ months. What is the ap-

proximate maximum life for this tube? If the manufacturer decided to guarantee the tube to last 18 months, what is the approximate number he should expect to replace out of every 20,000 tubes sold?

17 A biographical dictionary contains sketches of 42,000 men. In a random sample of 300 sketches, $M = 48$ years and $\sigma = 10$ years for the ages of the persons included. Nearly all the sketches in the book are about men in what range of ages? Approximately how many of the sketches are about men whose ages are 38 to 58 years?

18 For a random sample of 150 pieces of rope, the breaking strength of a certain kind of rope is measured. For these measures it is found that $M = 48$ pounds and $\sigma = 3$ pounds. What weight could the rope be counted on to support in nearly all cases?

REVIEW EXERCISES

Chapter 1

1 Let $A = \{1, 2, 3, 4, 5, 6, 7, 8, 9\}$, $B = \{3, 6, 9, \ldots, 48\}$, $C = \{5, 10, 15, \ldots\}$, $D = \{1, 2, 3, 4\}$.

(*a*) Insert in each of the following blanks the correct symbol from among $\in$, $\notin$, $\subset$, $\not\subset$:

6 ____ A $\quad 6$ ____ B $\quad 6$ ____ C $\quad 6$ ____ D
12 ____ A $\quad 12$ ____ B $\quad D$ ____ A $\quad D$ ____ B

(*b*) How many members are in each of the sets A, B, C, and D?

2 Which of the following are correct statements?

$a \in \{a, b, c, d\}$ $\quad \{a, b\} \subset \{a, b, c, d\}$ $\quad \{a, e\} \subset \{a, b, c, d\}$
$\{a, d\} \in \{a, b, c, d\}$ $\quad \{a\} \in \{a, b, c, d\}$ $\quad \{a\} \subset \{a, b, c, d\}$

3 For the sets A, B, C, and D of Exercise 1 above, tabulate each of the following:

$A \cap B \quad A \cup B \quad A \cap D \quad A \cup D$

4 What do we mean by a one-to-one correspondence between sets? Give three illustrations.

5 We denote the set with members a and b by $\{a, b\}$. Give schematic diagrams to show that the set $\{a, b\}$ can be put into one-to-one correspondence with the set $\{c, d\}$ in two different ways.

6 Give schematic diagrams to show that the set $\{a, b, c\}$ can be put in one-to-one correspondence with the set $\{d, e, f\}$ in six different ways.

Chapter 2 In Exercises 7 to 12, find the solution of the given sentence.

7 $x > -7$ and $x < 4$; the universe is I.
8 $x > -7$ or $x < 4$; the universe is I.
9 $x > -7$ and $x < 4$; the universe is N_0.
10 $x > 2$ and $x < 0$; the universe is I.
11 $x > 2$ or $x < 0$; the universe is N.
12 $-3 < x < 6$; the universe is I.

Chapter 3 In Exercises 13 and 14, find the value of the given expression.

13 $|2(1 - \frac{8}{7}) + 3(-1)|$
14 $|-3(4 - \frac{7}{5}) - 16|$

15 Using the laws of exponents, perform the indicated operations, and simplify:

$\frac{a^3b}{a^2b} \quad a^2 \cdot a^3 \quad (a^2)^3 \quad \left(\frac{a^3}{a^2}\right)^5$

16 Evaluate each of the following: $(25)^{1/2}$; $(27)^{1/3}$; $(\frac{25}{36})^{1/2}$; $(0.25)^{1/2}$.

17 Find the value of the following: 4^{-1}; $(17)^0$; $8^{-1/3}$; $\frac{10^6}{10^4}$; $9 \cdot 4^0$.

18 Express the following in simplest form with positive exponents:

$x^{-1} - y^{-1}$ $\quad (27a^{-3}x^6)^{2/3}$ $\quad (a^1 + b)^2$ $\quad \frac{x^{-3}y^{-5}}{x^{-1}y^2}$

Chapter 4 19 Solve the equation $y - \frac{1}{3}(y + 1) + \frac{1}{5}(y + 3) = \frac{3y - 2}{3}$, and check your solution.

20 Solve the equation $\frac{x}{a} = x - a + \frac{1}{a}$ for x, and check your result.

21 Three children who weigh 62, 75, and 80 pounds, respectively, arrange themselves upon a seesaw. The first sits 4 feet from the fulcrum and the second 5 feet from the fulcrum on the same side. Where must the third sit in order to balance the other two?

22 Airplane A starts from an airport at 8 A.M., flying 150 miles per hour. At 9:30 A.M., B starts after A on the same route, flying 250 miles per hour. How long will it take B to overtake A?

23 A hawk pursues a homing pigeon which has a 5-second start and overtakes it in 25 seconds. If the hawk flew at 75 miles per hour, what was the speed of the pigeon?

24 A travels three times as fast as B. They start toward each other at the same time from towns 320 miles apart and meet in 4 hours. What was the average speed of each?

25 If a boy who weighs 80 pounds sits 7 feet from the fulcrum on the right side of a seesaw, what is the weight of a boy seated 8 feet from the fulcrum on the left side when the seesaw is in balance?

26 A lever is 5 feet long. Where must the fulcrum be placed in order that a force of 50 pounds at one end will lift a weight of 200 pounds at the other end?

27 A merchant mixes 25 pounds of one kind of candy with 15 pounds of another kind to make a mixture which costs 60 cents per pound. If the first kind of candy costs 45 cents per pound, how much per pound does the second kind cost?

28 A boy with a soft-drink stand has 9 quarts of lemonade which cost him 10 cents per quart. How much water should he add in order to reduce the cost of the lemonade to 6 cents per quart, if the water is free?

29 Tea worth \$1.20 per pound is mixed with tea worth \$1 per pound to form a mixture weighing 10 pounds and worth \$1.12 per pound. How many pounds of each kind of tea are used?

30 How many pounds of one kind of candy worth a cents per pound and of another kind worth b cents per pound should be mixed in order to make 100 pounds of candy worth c cents per pound?

31 How many pounds of coffee costing 80 cents per pound should be added to 80 pounds of coffee costing 90 cents per pound to make 100 pounds of coffee costing 88 cents per pound?

32 A merchant mixes 30 pounds of one kind of candy with 50 pounds of another kind to make a mixture which costs 85 cents per pound. If the first kind of candy costs 70 cents per pound, find the cost of the second kind.

33 The length of a rectangle is three times its width, and the perimeter is 96 feet. Find the dimensions.

34 Divide 215 into two parts such that their quotient is $\frac{2}{3}$.

35 A board 62 inches long is sawed into two pieces, so that one piece is 7 inches shorter than twice the length of the other piece. Find the lengths of the two pieces.

36 Twenty-seven coins consisting of quarters and half dollars have a value of $9.25. How many coins are there of each kind?

37 A ship leaves a port at 6 A.M., steaming due east at 25 miles per hour. At 6 A.M. the next day an airplane leaves the same port to overtake the ship. How fast must the airplane fly in order to overtake the ship in 4 hours?

38 A pile of 73 coins, which consists of dimes and quarters, is worth $14.80. Find the number of dimes and the number of quarters.

39 What is the present value of $1,500 due in 9 months if money is worth 4% simple interest?

40 A man can buy a lot for $3,000 cash or $3,100 in 1 year. He has the cash and can invest it at 4% simple interest. Under which plan is it more advantageous for him to buy the lot, and by what amount as of now?

In Exercises 41 and 42, solve the given pair of equations.

41 $2x + 3y = 7$
$3x - 4y = 2$

42 $7x = y + 33$
$12y = x + 19$

43 How much cream containing 25 percent butterfat must be added to 100 pounds of milk containing 3.2 percent butterfat to produce a mixture which contains 3.5 percent butterfat?

44 There are two numbers x and y such that their product is 135, and the difference of their squares is to the square of their difference as 4 is to 1. Find the numbers.

Chapter 5 In Exercises 45 and 46, rationalize the denominator.

45 $\dfrac{\sqrt{3} - 2}{\sqrt{3} + 1}$

46 $\dfrac{\sqrt{7} - 2\sqrt{3}}{\sqrt{7} + 2\sqrt{3}}$

47 Show that $\dfrac{i - 5}{2}$ is a root of the equation $\dfrac{1}{x + 2} + \dfrac{1}{x + 3} = -2i$.

48 Find a number of the form $a + bi$ such that $a + bi = \sqrt{5 - 12i}$. *Hint:* $5 - 12i = a^2 + 2abi - b^2$. Now equate real and imaginary parts, and solve for a and b.

In Exercises 49 and 50, simplify the given expression.

49 $\left(\frac{1}{x}+\frac{1}{y}\right)\left(1+\frac{x+y}{x-y}\right)\left(\frac{1}{x}-\frac{1}{y}\right)\left(1-\frac{y}{x+y}\right)$

50 $\dfrac{\dfrac{a}{b}+\dfrac{b}{a}}{\dfrac{a^2}{b^2}+\dfrac{b^2}{a^2}}\cdot\dfrac{a^6-a^4b^2+a^2b^4-b^6}{a^3+a^2b+ab^2+b^3}$

Chapter 6 Solve each of the following equations, and check your solutions:

51 $2x^2-18=0$ $3x^2-5x=2$ $x^2+6x+4=0$
52 $3x^2-7x=0$ $x^2-6x=8$ $3x^2+1=5x$
53 $4x^2-100=0$ $6y^2+13y+6=0$ $7y^2-7y+1=0$
54 $5x^2-20=0$ $6x^2-3x-45=0$ $3x^2=2(x+3)$
55 $2x^2-11x+12=0$ $4x^2-2x-1=0$
56 $8y^2+6y+1=0$ $y^2-2y+\frac{1}{2}=0$

Chapter 7 57 Define a relation.

58 Define a function. Give two illustrations of a function.

59 Give two illustrations of a nonfunctional relation.

60 Define the domain of a relation. Define the range of a relation. What is the domain and the range of each of the relations which you gave as answers to Exercise 59?

61 What is the general first-degree function? What is the general second-degree function?

62 Referring to a two-dimensional coordinate system, in what quadrant are both the abscissa and the ordinate of a point positive? In what quadrant is the abscissa positive and the ordinate negative? In what quadrant are the abscissa and the ordinate negative?

In Exercises 63 and 64, construct the graph of the given set.

63 $S=\{(x,y)\mid x\geq 0 \text{ and } y\geq 0 \text{ and } 2x+3y\leq 6\}$

64 $S=\{(x,y)\mid x\geq 1 \text{ and } y\geq 0 \text{ and } x\leq 4 \text{ and } x+y<6\}$

Chapter 8 65 z varies jointly with x and y and inversely with the square of w. Also, $z=15$ when $x=12$, $y=20$, and $w=14$. Find a formula for z in terms of x, y, and w.

66 The volume of a gas enclosed in a vessel varies inversely with the pressure. If the volume is 24 cubic inches when the pressure is 100 pounds, what will the volume be when the pressure is 24 pounds?

67 If y varies inversely with x^2 and $y=4$ when $x=2$, find y when $x=\frac{1}{2}$.

68 The formula $E=kIR$ states that the electromotive force E varies directly with the current I and the resistance R. If $E=100$ when $I=0.02$ and $R=5{,}000$, find E when $I=2$ and $R=55$.

69 The illumination of an object from a source of light varies inversely

with the square of the distance of the object from the source of light. If a book now 12 inches away from the light is moved to be 24 inches from the light, what is the ratio of the new illumination to the old?

70 If y varies directly with x and inversely with z and if $y = \frac{5}{2}$ when $x = 5$ and $z = 3$, find the value of x when $y = 6$ and $z = \frac{3}{2}$.

Chapter 9

71 Construct the graph of $\{(x, y) \mid y = \cos x, x \in [-2\pi; 0]\}$.

72 Two similar triangles have corresponding sides which are 15 inches and 12 inches, respectively. If the lengths of the sides of the larger triangle are 20 inches, 15 inches, and 10 inches, what are the lengths of the corresponding sides of the smaller triangle?

73 Find to the nearest degree the angle whose tangent is 0.5.

74 The angle of elevation of the top of a flagpole is 36° at a point which is 25 feet from the pole. Find the height of the pole above the level of the observer's eye.

75 From a lighthouse observation point 250 feet above sea level, the angle of depression of a ship is 52°. How far (to three significant figures) from the lighthouse is the ship?

76 If the specification for a road calls for an inclination of 6° to the horizontal, how much (to three significant figures) does the road rise for each 125 feet along the incline?

Chapter 10

77 Since $5^3 = 125$, the logarithm of 125 to the base ______ is ______.

78 The logarithm of a product is equal to the ______ of the logarithms of the factors; that is, $\log (M \cdot N) =$ ______.

79 The logarithm of a quotient is equal to the ______ of the logarithms of the factors; that is, $\log \frac{M}{N} =$ ______.

80 The logarithm of the pth power of N is equal to the product of ______ and ______; that is, $\log (N)^p =$ ______.

81 Find the following logarithms: $\log_2 8$; $\log_4 2$; $\log_2 \frac{1}{8}$; $\log_{10} 0.001$.

82 Solve each of the following equations for x; $\log_x 9 = 2$; $\log_{10} x = 2$; $\log_x 10 = -1$; $\log_6 x = -2$.

Chapter 11

83 For the arithmetic progression 1, 5, 9, 13, . . . , find a formula in terms of n for each of l (the nth term) and S (the sum of n terms). Also find l and S for $n = 20$.

84 For the geometric progression 8, -4, 2, -1, . . . , find a formula in terms of n for each of l (the nth term) and S (the sum of n terms). Also find l and S for $n = 10$.

85 Give the first five terms in the arithmetic progression for which the nth term is $\frac{3n - 2}{2}$.

86 Give the first five terms in the geometric progression for which the nth term is $(-\frac{1}{2})^{n-1}$.

87 Jack Jones deposited $100 in the Mutual Aid Savings Bank at the end of each year for 10 years, and the bank paid interest at the rate of 4% compounded annually. How much did Jack have on deposit in the bank immediately after his tenth deposit?

88 Find the present value of an annuity of $500 per year for 15 years if the interest rate is 4% compounded annually.

Chapter 12

89 If in the universe in $S = \{1, 2, 3, 4, 5, 6, 7, 8\}$, $R = \{1, 2, 3, 4\}$, and $T = \{4, 6, 8\}$, find R', T', $R' \cup T'$, and $R' \cap T'$.

90 Find the probability that 2 will turn up *exactly* twice in five throws of a die.

91 Find the probability that 2 will turn up *at least* twice in five throws of a die.

92 Prove the so-called Pascal's rule:

$$C(n + 1; r) = C(n, r - 1) + C(n, r)$$

93 If $P(n, r) = 110$ and $C(n, r) = 55$, find n and r.

94 There is a story that an Italian nobleman, who was a professional gambler and an amateur mathematician, had by continued observation with three dice, observed that the sum of 10 appeared more often than the sum of 9 (this sum being the sum of the dots on the upward faces of the three dice for a given throw). He expressed his surprise at this to Galileo, and requested an explanation. Pretend you are Galileo, and give the explanation. *Hint:* Tabulate the simple events each of which yields a sum of 10 and a sum of 9. Then determine the probability of throwing a sum of 10 and the probability of throwing a sum of 9 with three dice.

Chapter 13

95 For the set of numbers 92, 89, 86, 86, 72, 71, 68, 62, 59, what is the arithmetic mean? What is the median? What is the mode?

96 Twenty-five test grades are given in the following table:

CLASS	40–49	50–59	60–69	70–79	80–89	90–99
CLASS MARK	44.5	54.5	64.5	74.5	84.5	94.5
FREQUENCY	1	2	7	9	5	1

Calculate M_x, the weighted arithmetic mean of the class marks given in the table.

97 Calculate the standard deviation for the frequency distribution in the table in Exercise 96.

In Exercises 98 and 99, find the expansion of the given expression by use of the binomial formula.

98 $\left(\frac{x}{y} + \frac{y}{x}\right)^{10}$

99 $\left(\sqrt[3]{a} - \frac{1}{2}b^2\right)^5$

100 If, on the average, one ship in every ten is wrecked, find the probability that at least three ships out of five will arrive safely.

101 How many throws with two dice are required in order that the probability of obtaining a double six at least once will have the value $\frac{1}{2}$? *Hint:* If $\frac{1}{2} = 1 - (\frac{35}{36})^n$, find n.

102 For the 25 grades given in the table in Exercise 96, find the percent of the grades in the interval $M_X - 3\sigma$ to $M_X + 3\sigma$.

Appendix A
The Roman Numeral System

We know from history that the Egyptians and the Babylonians in the East developed their own systems of numerals, and archaeology has established the fact that the Incas and Mayans in the West did likewise. Very probably each ancient civilization had its own system of numerals.

Of all these ancient numeral systems, the only one which is used to any extent at present is the Roman system. The Roman numerals were constructed on the following plan of counting, with its accompanying symbols:

COUNTING METHOD	SYMBOL
By units:	
One	I
Two	II
Three	III
Four, or one from five	IV
Five	V
Six, or five and one	VI
Seven, or five and two	VII
Eight, or five and three	VIII
Nine, or one from ten	IX
Ten	X
By tens:	
One ten	X
Twenty, or two tens	XX
Thirty, or three tens	XXX
Forty, or one ten from five tens	XL
Fifty, or five tens	L
Sixty, or five tens and one ten	LX
Seventy, or five tens and two tens	LXX
Eighty, or five tens and three tens	LXXX
Ninety, or one ten from ten tens	XC
One hundred, or ten tens	C
By hundreds:	
One hundred	C
Two hundreds	CC
Three hundreds	CCC
Four hundreds, or one hundred from five hundreds	CD
Five hundreds	D
And so forth	

The Romans let M stand for 1,000. They wrote 2,000 as MM; 3,000 as MMM; and so forth. In a Roman numeral the thousands are written first, then the hundreds, then the tens, and finally the units. To illustrate, 43 is written XLIII; 632 is written DCXXXII; 2,719 is written MMDCCXIX.

Observe the differences between the Roman system and the numeral system to which we are accustomed. Near the beginning the

Roman system is simpler, for it uses only three symbols to count to ten. In our present-day system we need the 10 symbols 0, 1, 2, 3, 4, 5, 6, 7, 8, 9 to count to ten. As you progress in the Roman system to larger numbers, you need new symbols like L, C, D, and M. In our system you can write larger and larger numbers using only the 10 symbols mentioned above.

EXERCISES

1 Write the Roman numerals from eleven to seventy-one.

2 Express in Roman numerals: 75; 84; 97; 114; 521; 687; 963; 1,492; 2,000; 5,000.

3 Name three ways in which Roman numerals are used today.

4 Express in our present-day numeral system: XXIV; XXVIII; XXXIX; XLI; XC; CX; CCXVI; CDXXVII; MCMLX.

B

Appendix B Some Topics in Logic—Proof that $\sqrt{2}$ is Irrational

In Chap. 2 we described what we meant by the conjunction of two statements and by the disjunction of two statements. When p and q designate statements, the **conjunction** of these two statements is the statement

CONJUNCTION

$$p \text{ and } q \tag{1}$$

DISJUNCTION

The **disjunction** of these two statements is the statement

$$p \text{ or } q \tag{2}$$

The statements (1) and (2) are examples of *compound statements,* and the statements p, q are the *component parts* of these compound statements.

In order to use compound statements such as (1) and (2), we need to know what we mean by the "truth" or "falseness" of the compound statements. We agree that the conjunction

$$p \text{ and } q$$

is *true* when *both* of its component parts are true, and it is *false* when at least one of its component parts is false. We agree that the disjunction

$$p \text{ or } q$$

is *true* when *at least one* of its component parts is true and it is false when both of its component parts are false.

In mathematics we frequently use compound statements of the form

$$\text{if } p \text{ is true, then } q \text{ is true} \tag{3}$$

and statements of the form

$$p \text{ is true if, and only if, } q \text{ is true} \tag{4}$$

We shall discuss what we mean when we say that statement (3) is true and what we mean when we say that statement (4) is true. Statement (3) is usually written more briefly as

$$\text{if } p \text{ then } q$$

and symbolized by writing

$$p \Rightarrow q$$

Statement (4) is usually written more briefly as

$$p \text{ if and only if } q$$

and symbolized by writing

$$p \Leftrightarrow q$$

We agree that the statement symbolized by

$$p \Rightarrow q \tag{5}$$

TRUTH OF $p \to q$

is *true* if it is *impossible* for q to be false when p is true; that is, "if p then q" is true whenever the assumption that p is true *forces* on us the conclusion that q is true. To illustrate,

if $x = 2$, then $x^2 - 4x + 4 = 0$

is *true,* because whenever $x = 2$ is true, $x^2 - 4x + 4 = 0$ *must* be true. However,

if $x^2 = 9$, then $x = 3$

is *false,* because, it is possible for $x = 3$ to be false when $x^2 = 9$ is true.

We agree that the statement symbolized by

TRUTH OF $p \leftrightarrow q$

$$p \Leftrightarrow q \tag{6}$$

is *true* if *both* of the statements

$p \Rightarrow q \qquad q \Rightarrow p$

are true; that is, if it is *impossible* for one of the component parts of (6) to be true and the other false. To illustrate,

$4x - 3 = 0$ if and only if $4x = 3$

is *true,* while

$4x - 3 = 0$ if and only if $4x^2 - 3x = 0$

is *false* because $4x^2 - 3x = 0$ can be true when $4x - 3 = 0$ is false. Further,

$(x - 2)(x + 3) = 0$ if and only if $x = 2$ or $x = -3$

is *true;* also

$y = \sqrt{x}$ if and only if $y \geq 0$ and $y^2 = x$

is *true.*

IMPLICATION

CONDITIONAL

HYPOTHESIS

CONCLUSION

BICONDITIONAL

CONVERSE OF AN IMPLICATION

The compound statement "if p then q" is sometimes called an **implication** or a **conditional.** In the implication $p \Rightarrow q$, the statement p is called the **hypothesis** or *antecedent* and the statement q is called the **conclusion** or *consequent.* The compound statement "p if and only if q" is sometimes called an *equivalence* or a **biconditional.**

If in an implication, we interchange the hypothesis and the conclusion, we produce a compound statement that is called the **converse** of the original implication; that is,

$q \Rightarrow p \qquad$ is the converse of $\qquad p \Rightarrow q$

and *vice versa*. An important fact to note about the converse of an implication is that knowing $p \Rightarrow q$ is true gives us *no* information about the truth or falsity of its converse $q \Rightarrow p$. To illustrate,

if $x = 3$ then $x^2 = 9$

is *true,* and its converse

if $x^2 = 9$ then $x = 3$

is *false;* on the other hand,

if $x = 3$ then $x + 6 = 9$

is *true,* and its converse

if $x + 6 = 9$ then $x = 3$

is *true*. Further,

if $4x^2 - 3x = 0$ then $x = 0$

is *false,* and its converse

if $x = 0$ then $4x^2 - 3x = 0$

is *true*.

NEGATION OF A STATEMENT

Sometimes we wish to consider the denial or the negation of a statement p. The **negation** of p is defined to be a statement that is true when p is false and false when p is true. To illustrate,

the negation of "$x = 2$" is "$x \neq 2$"

The negation of the statement

"all rational numbers are integers"

is the statement

"there is at least one rational number that is not an integer"

Recalling that the implication "if p is true, then q is true" is true if it is *impossible* for q to be false when p is true, we can see that the statement

"p is true and q is false"

is the negation of the implication

"if p is true, then q is true"

For example, the negation of

"if x is an even integer, then x^2 is an even integer"

is

"x is an even integer and x^2 is not an even integer"

Two statements are said to be logically equivalent if it is impossible for one of the statements to be true and the other false. Consider the two statements

if p is true, then q is true (7)
if q is false, then p is false (8)

To say that (7) is true means it is impossible for q to be false when p is true. Thus (7) is true if the falsity of q *forces* on us the falsity of p; but this is just what it means to say (8) is true. Similarly, to say that (7) is false is the same as saying that (8) is false. So the statements (7) and (8) are logically equivalent. The implication

if q is false, then p is false

CONTRAPOSITIVE OF AN IMPLICATION

is called the **contrapositive** of the implication

if p then q

To illustrate, the contrapositive of

if $x = 3$, then $x^2 = 9$

is

if $x^2 \neq 9$, then $x \neq 3$

and both statements are true. The contrapositive of

if x is an odd integer, then x^2 is an odd integer

is

if x^2 is not an odd integer then x is not an odd integer

and (as we shall see in Theorem 2 below) both statements are true.

Let us now apply some of the ideas of the preceding paragraphs to prove that there is no rational number r with the property that $r^2 = 2$. In preparation for this proof we first prove the following three theorems.

THEOREM 1 *If x is an even integer, then x^2 is an even integer.*

PROOF To say that x is an even integer means that there is some integer n with the property that $x = 2n$. So, if x is an even integer, x^2 can be written as $(2n)^2$. But $(2n)^2 = 4n^2 = 2(2n^2)$, and since $2n^2$ is an integer,* $x^2 = (2n)^2 = 2(2n^2)$ is an even integer.

* The student should make sure he understands why this is so.

THEOREM 2 *If x is an odd integer, then x^2 is an odd integer.*

PROOF If x is an odd integer, it can be written as an even integer plus 1. Thus, if x is an odd integer, there is some integer n with the property that $x = 2n + 1$. So x^2 can be written as $(2n + 1)^2$. But

$$(2n + 1)^2 = 4n^2 + 4n + 1 = 2(2n^2 + 2n) + 1$$

where $(2n^2 + 2n)$ is an integer.* So x^2 can be written as an even integer plus 1, and hence x^2 must be odd.

THEOREM 3 *Let x be an integer. If x^2 is even, then x is even.*

PROOF As we have seen, the statement

if x^2 is even, then x is even (9)

and its contrapositive

if x is not even, then x^2 is not even (10)

are logically equivalent. Thus, if we can show that (10) is true, we shall have proved that (9) is true. Now, if an integer is not even, it is odd; so the statement (10) is equivalent to the statement

if x is odd, then x^2 is odd (11)

We recognize (11) as the statement in Theorem 2; therefore (11) is true and (10) is true and (9) is true, and the theorem is proved.

ALTERNATIVE PROOF Since x is an integer, it is either even or it is odd. *Assume* that x^2 is even and x is not even. Then x is odd, and by Theorem 2, x^2 must be odd. Hence if the assumption is true, then it is true that x^2 is even and x^2 is odd. Therefore our assumption cannot be true and so x must be even.

PROOF BY CONTRADICTION

The alternate form of the proof of Theorem 3 just given is an example of a method of proof known as **proof by contradiction.** This method is important, and since we shall use it in the proof of Theorem 4, we shall discuss it briefly.

Suppose that we wish to prove that a statement A is true. The usual procedure in a proof by contradiction is to start with the assumption that A is false, that is, to assume that the *negation of A is true,* and then to show that this assumption leads to a contradiction. From this we conclude that the negation of A *cannot* be true and hence must be false; therefore A is true. Notice that in this procedure we assume as a basic principle of logic that either a statement is true or it is false; there is no third possibility.

*The student should make sure he understands why this is so.

In Theorem 3 we wished to prove that the statement

if x^2 is even, then x is even

is true. The negation of this statement is

x^2 is even, and x is not even (12)

and this is what we assumed to be true in the alternative proof of the theorem. We showed that if (12) is true, then it is true that

x^2 is even and x^2 is odd

This is a contradiction since it says that a statement p (x^2 is even) and the negation of p (x^2 is odd) are both true, contrary to the basic principle that a statement cannot be both true and false. A statement of the form "p is true and p is false" is a very common type of contradiction which occurs in a proof by contradiction. Other types of contradictions which may occur are a denial of a postulate or the negation of a previously proved theorem.

In constructing a proof by contradiction to show that a statement A is true, we must be very careful to start the argument with the negation of A, or with some statement logically equivalent to the negation of A. In this connection, recall that we have seen that the negation of an implication "if p then q" is the statement "p is true and q is false."

$\sqrt{2}$ IS IRRATIONAL

THEOREM 4 *There is no rational number r with the property that* $r^2 = 2$, or equivalently, *if r is a rational number, then* $r^2 \neq 2$.

PROOF We construct a proof by contradiction. The negation of "if r is a rational number, then $r^2 \neq 2$" is

r is a rational number and $r^2 = 2$ (13)

Let us *assume* that (13) is true. Since r is a rational number, it can be represented by $\frac{a}{b}$, where a and b are integers with no common integer factors other than 1 and -1, and $b \neq 0$. If our assumption is true, we have

$$\left(\frac{a}{b}\right)^2 = 2$$

From this equality it follows that

$$a^2 = 2b^2 \quad (14)$$

and since b is an integer, a^2 is an even integer. Therefore, by Theorem 3, *a is an even integer* and there is an integer k with the property that $a = 2k$. Substituting $2k$ in place of a in (14), we have

$$4k^2 = 2b^2 \qquad \text{or} \qquad 2k^2 = b^2$$

from which it follows that b^2 is even, and by Theorem 3, *b is even.* We have now proved that if the assumption

r is a rational number and $r^2 = 2$

is true, then there are integers a and b with the following properties:

1 a and b have no common integer factors except 1 and -1.
2 a and b are both integral multiples of 2.

This of course is a contradiction, and hence the assumption is false and the theorem is proved.

EXERCISES

1 Prove that if x is an even integer, then x^3 is an even integer.

2 Prove that if x is an odd integer, then x^3 is an odd integer.

3 Prove that if x is an integer and x^3 is even, then x is an even integer.

4 Prove that there is no rational number r with the property that $r^3 = 2$. *Hint:* Use Exercise 3 and the method used in the proof of Theorem 4.

5 Prove that if x is an integer which is a multiple of 3 ($x = 3n$, where n is an integer), then x^2 is a multiple of 3.

6 Prove that if x is an integer which is not a multiple of 3 (if x is not a multiple of 3, there is an integer n with the property that either $x = 3n + 1$ or $x = 3n + 2$), then n^2 is not a multiple of 3.

7 Prove that if x is an integer and x^2 is a multiple of 3, then x is a multiple of 3.

8 Prove that there is no rational number r with the property that $r^2 = 3$.

9 Prove that if x is an integer which is a multiple of 5 ($x = 5n$, where n is an integer), then x^2 is a multiple of 5.

10 Prove that if x is an integer that is not a multiple of 5, then x^2 is not a multiple of 5.

11 Prove that if x is an integer and x^2 is a multiple of 5, then x is not a multiple of 5.

12 Prove that there is no rational number r with the property that $r^2 = 5$.

13 Investigate wherein the method used in the proofs of Theorem 4 and Exercises 8 and 12 fails when we try to use that method to prove that there is no rational number r with the property that $r^2 = 4$.

Appendix C
Significant Figures and Rounding Off

SIGNIFICANT FIGURE

In reporting and working with measurements, the concept of significant figures or significant digits is very useful. A **significant figure,** or *significant digit,* is defined to be any digit used in writing a number *except* those zeros which are used only for the purpose of locating the decimal point or those zeros which do not have any nonzero digit on their left. This statement is applicable provided there is at least one nonzero digit in the number. For example, if a measurement is less than 0.005, it would be recorded to the nearest hundredth as 0.00, and the zeros to the right of the decimal point would be significant figures.

Thus, 3.58 has three significant figures, and 0.0025 has two significant figures. The number 35,000 has only two significant figures when the three zeros are used only to fix the position of the decimal point, which would be the case if we are reporting a measurement correct to the nearest 1,000. If, however, 35,000 represents a measurement correct to the nearest 10, then the zeros in the hundreds place and the tens place would have some use other than locating the decimal point. To indicate such a situation we make the following agreement: *A dot placed above a zero means that it and all digits to its left are significant figures.* Thus, $35,\dot{0}00$ has three significant figures (3, 5, and 0), $35,0\dot{0}0$ has four significant figures, and $35,00\dot{0}$ has five significant figures. Note that the dot is not necessary to show that 35,000.0 has six significant figures, since the final zero does not help fix the position of the decimal point and is used only to indicate significant figures. A study of the following table should clarify most questions concerning significant figures.

NUMBER	FIGURES HAVING SIGNIFICANCE	NUMBER OF SIGNIFICANT FIGURES
25.7	2, 5, 7	3
205.7	2, 0, 5, 7	4
2,001	2, 0, 0, 1	4
2.50	2, 5, 0	3
200	2	1
$2,\dot{0}00$	2, 0	2
20.000	2, 0, 0, 0, 0	5
$3,1\dot{0}0$	3, 1, 0, 0	4
0.001	1	1
0.150	1, 5, 0	3

ROUNDING OFF TO k SIGNIFICANT FIGURES

In working with numbers it is frequently convenient to "round off" the numbers to a certain number of significant figures. We say that a number N is **rounded off to k significant figures** when N is replaced by the number which is the closest approximation to N that can be written with k significant figures.

EXAMPLE 1 Round off 27,342 to three significant figures.

SOLUTION We must write a number, using only three significant figures, which is closer to 27,342 than any other number containing only three significant figures. This number will certainly be either 27,300 or 27,400. The correct result is seen to be 27,300 since the difference between 27,300 and 27,342 is 42, while the difference between 27,400 and 27,342 is 58.

EXAMPLE 2 Round off 43.273 to three significant figures.

SOLUTION The number containing only three significant figures which is the closest approximation to 43.273 is surely either 43.2 or 43.3. The correct result is seen to be 43.3 since the difference between 43.3 and 43.273 is 0.027, while the difference between 43.2 and 43.273 is 0.073.

In case the given number is exactly halfway between two numbers with the required number of significant figures, we shall agree to round off so that the last significant figure is *even*.

EXAMPLE 3 Round off 7,250 to two significant figures.

SOLUTION The number 7,250 is exactly halfway between 7,200 and 7,300; so we round off to 7,200.

Study the following table carefully.

	ROUNDED OFF TO			
NUMBER	FOUR SIGNIFICANT FIGURES	THREE SIGNIFICANT FIGURES	TWO SIGNIFICANT FIGURES	ONE SIGNIFICANT FIGURE
73,163	73,160	73,200	73,000	70,000
123.35	123.4	123	120	100
3,451.3	3,451	3,450	3,500	3,000
7,000.5	$7{,}00\dot{0}$	$7{,}0\dot{0}0$	$7{,}\dot{0}00$	7,000
0.00074347	0.0007435	0.000743	0.00074	0.0007

Instead of rounding off a number to a certain number of significant figures, we may wish to round off a number to the nearest hundred, or ten, or unit or to the nearest tenth, or hundredth, and so forth.

We say a number N *is rounded off to the nearest hundred* when we replace N by the number which is the closest approximation to N that can be written with no significant figures to the right of the hundreds place.

Similar definitions can be made for rounding off to the nearest ten, unit, tenth, hundredth, and so forth, by replacing "hundred" by the appropriate word.

For example, the number 4,253.475 rounded off to hundredths becomes 4,253.48 (no significant figure to the right of the hundredths place); rounded off to units, 4,253.475 becomes 4,253 (no significant figures to the right of the units place); and, rounded off to thousands, 4,253.475 becomes 4,000 (no significant figure to the right of the thousands place).

More examples of rounding off are shown in the following table.

	NUMBER ROUNDED OFF TO			
NUMBER	TENTHS	UNITS	TENS	HUNDREDS
167.42	167.4	167	170	200
357.47	357.5	357	360	400
6,891.15	6,891.2	6,891	6,890	6,900
5,955.55	5,955.6	5,956	5,960	$6,\dot{0}00$
4,975.45	4,975.4	4,975	4,980	$5,\dot{0}00$

Decimal numbers are sometimes rounded off to a certain number of decimal places, as illustrated in the following table, with only the desired number of places remaining to the right of the decimal point.

	NUMBER ROUNDED OFF TO		
NUMBER	FOUR DECIMAL PLACES	THREE DECIMAL PLACES	TWO DECIMAL PLACES
0.03517	0.0352	0.035	0.04
6.66666	6.6667	6.667	6.67
3.14549	3.1455	3.145	3.15

EXERCISES

1 Make a table showing the number of significant figures in each of the following numbers: 1.37; 2,050; 3,007; 4,000; 5,702.0; 2,100; 0.0301; $5,\dot{0}00$; 0.001; 42.000; 1.000.

2 Make a table showing the numbers 6,875.1, 4,000.32, 0.0013476, 1,000.0, 96,000.0, 200,000, 96.958 rounded off as follows: to four significant figures; to three significant figures; to two significant figures; to one significant figure.

3 Round off each of the following numbers to tenths: 167.43; 1.056; 0.099; 174.15; 3,214.649; 37.2503; 8.651; 8.65.

4 Round off each of the following numbers to thousands: 3,674.1; 9,901; 47,500.2; 167,099; 6,510; 7,499.9; 19,500; 499,501.

5 Round off each of the following numbers to four decimal places, three decimal places, two decimal places: 2.05655; 0.05173; 41.50505; 0.00493.

6 Round off each of the following numbers to hundredths, to hundreds, to one decimal place; 6,501.372; 176.005; 270.191; 4,685.015; 999.9975; 678.905; 17,000.036; 155.555.

ANSWERS TO ODD-NUMBERED EXERCISES

Sec. 1.3, pp. 11–12

1 (a) T; (b) F; (c) T; (d) F; (e) T; (f) F
3 (a) $\{1, 2\}, \{1\}, \{2\}, \varnothing$; ($b$) $\{1\}, \{2\}, \varnothing$
5 $\{1, 2\}, \{1, 3\}, \{1, 4\}, \{1, 5\}, \{2, 3\}, \{2, 4\}, \{2, 5\}, \{3, 4\}, \{3, 5\}, \{4, 5\}$
7 (a) $\{1, 2, 3, 4, 5\}$; (b) $\{1, 2, 3, 4\}$; (c) $\{2, 3\}$; (d) $\varnothing$
9 $A \cap B = \{b, d\}$; $A \cup B = \{a, b, c, d, e\}$; $A \cap A = A$; $A \cup A = A$; $A \cap \varnothing = \varnothing$; $A \cup \varnothing = A$
11 $A' = \{4, 5, 6\}$, $B' = \{1, 4, 5, 6\}$, $C' = \{1, 2, 4, 5, 6\}$

Sec. 1.5, p. 22

3 2, 3, 5, 7, 11, 13, 17, 19, 23, 29
5 $2 \cdot 2 \cdot 2$; 11; $7 \cdot 2$; $2 \cdot 3 \cdot 3$; $3 \cdot 2 \cdot 2 \cdot 2$; 29 **7** 7
9 1, 3, 31, 93 **11** 1, 3, 5, 7, 9, 15, 21, 35, 45, 63, 105, 315

Sec. 1.6, pp. 24–25

5 45 **7** 29 **9** 31 **11** 78

Sec. 1.7, pp. 28–30

1 $11a$ **3** $2a + 3b$ **5** $8a + 6b + 4c$
7 $12a$ **9** $3a \cdot a + 9a$ **11** $p = 4s$
13 $p = 3s$ **15** $p = 8c$ **17** $d = 7a + 7$
19 $p = 13x$ **21** $a, 3b, 6c; 7a, 2cd; 3abc$
23 $7x, 3y, 2z$; 7, 3, 2 **25** 29; 14 **27** 360

Sec. 1.8, pp. 37

1 $2b + 2c$; $3ac + 4ad$; $8 + 2b + 4a + ab$
3 $7b(y + 3x)$; $b(6a + 1)$; $5(4xy + 3)$
5 $(x + 2)(y + 3)$; $(x + 4)(y + 5)$; $(x + 2r)(y + 3s)$
7 \$3,315 **9** 180 sq ft **11** $3b + 6$ **13** $3x + 6$
15 $a + bc + d$ **17** $5x + 5y$ **19** 13 **21** 30 **23** 24
25 26 **27** $A = 6xy$ **29** $A = 6ac + 3bc$ **31** $V = 24ab$
33 $V = 24ab + 12abc$ **35** $a(b + c + 1)$ **37** $7(x + 1)$
39 $3a(3 + b)$ **41** $V = 5ab$ **43** $D = 275x + 45y$
45 $5(x + y) = 4$ **47** $x + (x + 1) + (x + 2) = 60$
49 $x \cdot x \cdot y \cdot y \cdot y$; $a \cdot a \cdot a \cdot a \cdot b \cdot b \cdot b \cdot b \cdot b$; $2 \cdot a \cdot y \cdot y$; $x \cdot x \cdot x \cdot x \cdot x \cdot y \cdot y \cdot y \cdot y \cdot y$
51 $x^2 + 3x + 2$; $x^2 + 7x + 10$; $2x^2 + 5x + 3$; $x^2 + 3xy + 2y^2$
53 (a) 49 sq in.; (b) 225 sq ft; (c) 841 sq yd; (d) 361 sq mi
55 (a) 125 cu in.; (b) 4,913 cu ft; (c) 9,261 cu yd; (d) 64 cu mi
57 e^2 square units; $6e^2$ square units **59** 35 cu in.
61 36 **63** 25 **65** 98 **67** 576

69 $4ab + 5ac$ **71** $2ab + 2ac + 2bc$
73 $5x^2 + 8x$ **75** $A = 4x^2 + 22x + 10$
77 $A = 12x^2 + 18xy$ **79** $x(2x + 1)$
81 $a(5a^2 + x)$ **83** $(y + 7)(y + 2)$
85 $A = x^2 + 3x$ **87** $S = a^2 + b^2$
89 No **91** $5x^2 + 14x + 9$
93 $8x^2 + 5x + 11$ **95** $5x^3 + 3x^2 + 13$
97 Base x, exponent 3, coefficient 2; base x, exponent 4, coefficient 3; base x, exponent 2, coefficient 4; base x, exponent 3, coefficient 7

99

WHEN $x =$	1	2	3	4	5	6	7	8
THEN $x^2 =$	1	4	9	16	25	36	49	64
AND $x^3 =$	1	8	27	64	125	216	343	512

Sec. 1.9, pp. 41–42

1 $6a^2b$ **3** $3x + 2y$ **5** $7x^2 + x + 2$
7 $ab + 3a$ **9** $5a$ **11** $5a + 9b + c + 3d$
13 Yes, 4 **15** Yes, 8 **17** No
19 Yes, 7 **21** No
23 $a - b$; a must be larger than b. $a \div b$; a must be a factor of a, or a must be a multiple of b
25 29 **27** 12 **29** No; there is no integer c such that $5c = 14$
31 Yes; $3 \cdot 4 = 12$
33 69 **35** 632
37 38 pairs **39** \$64

Sec. 1.10, p. 44

1 (*a*) T; (*b*) F; (*c*) F; (*d*) T; (*e*) T; (*f*) F
3 9 is 4 greater than 5; 5 is 4 less than 9
5 26 is 9 greater than 17; 17 is 9 less than 26
7 6 is greater than 4
9 10 is less than 20
11 (*a*) F; (*b*) T; (*c*) F; (*d*) T

Sec. 1.11, p. 49

1 (*a*) 0, 1, 2, 3, 4, 5, 6; (*b*) 1, 2, 3, 4, 5, 6, 10, 11, 12, 13, 14, 15, 16, 20, 21, 22, 23, 24, 25, 26, 30, 31, 32, 33, 34, 35, 36, 40, 41, 42, 43
3 (*a*) 0, 1, 2, 3, 4, 5, 6, 7, 8, 9, α, β, γ; (*b*) 1, 2, 3, 4, 5, 6, 7, 8, 9, α, β, γ, 10, 11, 12, 13, 14, 15, 16, 17, 18, 19, 1α, 1β, 1γ, 20, 21, 22, 23, 24
5 10100; 1111; 101011; 100101
7 10, 11, 12, 20, 21, 22, 100, 101, 102, 110, 111, 112, 120, 121, 122, 200, 201, 202
9 20110_4; 1024_8; 651_9

Sec. 1.12, p. 55

1

+	0	1	2	3	4	5	6
0	0	1	2	3	4	5	6
1	1	2	3	4	5	6	10
2	2	3	4	5	6	10	11
3	3	4	5	6	10	11	12
4	4	5	6	10	11	12	13
5	5	6	10	11	12	13	14
6	6	10	11	12	13	14	15

3 10; 12; 14

5 100; 83; 65; 486

7

+	0	1	2
0	0	1	2
1	1	2	10
2	2	10	11

·	0	1	2
0	0	0	0
1	0	1	2
2	0	2	11

9 Addition table:

$$\begin{array}{r}1\\+0\\\hline 1\end{array}\quad\begin{array}{r}1\\+1\\\hline 2\end{array}\quad\begin{array}{r}1\\+2\\\hline 3\end{array}\quad\begin{array}{r}1\\+3\\\hline 10\end{array}\quad\begin{array}{r}2\\+0\\\hline 2\end{array}\quad\begin{array}{r}2\\+2\\\hline 10\end{array}\quad\begin{array}{r}2\\+3\\\hline 11\end{array}\quad\begin{array}{r}3\\+0\\\hline 3\end{array}\quad\begin{array}{r}3\\+3\\\hline 12\end{array}$$

Multiplication table:

$$\begin{array}{r}1\\\times 0\\\hline 0\end{array}\quad\begin{array}{r}1\\\times 1\\\hline 1\end{array}\quad\begin{array}{r}1\\\times 2\\\hline 2\end{array}\quad\begin{array}{r}1\\\times 3\\\hline 3\end{array}\quad\begin{array}{r}2\\\times 0\\\hline 0\end{array}\quad\begin{array}{r}2\\\times 2\\\hline 10\end{array}\quad\begin{array}{r}2\\\times 3\\\hline 12\end{array}\quad\begin{array}{r}3\\\times 0\\\hline 0\end{array}\quad\begin{array}{r}3\\\times 3\\\hline 21\end{array}$$

Sec. 2.1, pp. 63–64

1 {9} **3** {6} **5** {1} **7** {3}
9 ∅ **11** {7, 9, 11} **13** {1, 2} **15** {2}
17 {7} **19** {4, 5} **21** {5} **23** ∅
25 {5, 7, 9}

Sec. 2.2, pp. 68–69

1 (*a*) T; (*b*) T; (*c*) F; (*d*) T; (*e*) T; (*f*) T
3 $-6 > -8$ **5** $-4 < 0$
7 13 **9** 5 **11** 5 **13** {1, 2, 3, 4, 5}
15 {−3, −2, −1, 0, 1} **17** {0}
19 32° above zero; 35° below zero; 100° above zero; 15° below zero
21 16°

Sec. 2.3,

1 3 **3** 18 **5** −7 **7** 0
9 −6 **11** 5 **13** 9 **15** 4
17 6 **19** −19 **21** −9 **23** −4
25 −57 **27** 0 **29** −17 **31** 20

33 -18 **35** -56 **37** 0 **39** -100
41 -90 **43** 16 **45** -4 **47** 2
49 -16

Sec. 2.4, pp. 82–83

1 -7 **3** 10 **5** -6 **7** 8
9 5 **11** 4 **13** -7 **15** 33
17 0 **19** $7; 3; 4; 0; 13$ **21** 6 **23** 0
25 $-9; 2; 0; -25; 15$ **27** $-y$
29 $9a$ **31** $10x - 2$
33 $1 - bc$ **35** $10 - 2a$
37 $a(3 - 4c)$ **39** $ax(2 - 1) = ax$
41 $sr - sb + ar - ab$ **43** $x^2 - 4x + 4$
45 $(x - 2)(x - 3)$ **47** $(x - 5)(x - 2)$
49 $8a + 6b$ **51** $9a^2 + 14ab + b^2$
53 $12 - 5x$ **55** $3a + 5b + 7c$
57 5 **59** 4
61 Meaningless **63** -5
65 -3 **67** Yes; $(-4) \cdot (-6) = 24$
69 Yes; $(20 + 8) = 28$; $(-7) \cdot (-4) = 28$

Sec. 2.5, pp. 87–88

1 $\{6\}, \{6\}, \varnothing$ **3** $\{7\}$ **5** $\{7\}$
7 $\{10\}$ **9** $\{7, 9\}$ **11** $\{-1, 0\}$
13 $\{0, 1\}$ **15** $\{7\}$ **17** $\{-3, -2, -1, 0, 1, 2\}$
19 I **21** $\{-3\}$ **23** $\{3, 4, 5, 6, 7\}$
25 (*a*) T; (*b*) T; (*c*) T; (*d*) F; (*e*) T

Sec. 3.1, pp. 94–95

1 4 **3** 21 **5** $6x$ **7** 3 **9** $40x^2$ **11** 16
13 $6, 9, 12, 15, 24, 75$ **15** $6, 9, 12, 15, 30, 60$
17 $\frac{6}{12}; \frac{9}{12}; \frac{-10}{12}; \frac{8}{12}; \frac{-3}{12}$ **19** $\frac{8}{16}; \frac{8}{6}; \frac{8}{-20}; \frac{8}{-56}; \frac{8}{2}$

Sec. 3.2, p. 100

1 $\frac{3}{8}$ **3** $\frac{-14}{3}$ **5** $2a$ **7** $\frac{ab}{d}$
9 $\frac{-1}{b}$ **11** $\frac{3a + 2}{6a}$ **13** $\frac{2 - x}{3}$ **15** $\frac{a + 7}{a}$
17 a **19** No **21** No **23** No **25** Yes

Sec. 3.3, pp. 105–106

1 $\frac{7}{3}$ **3** $\frac{10}{7}$ **5** $\frac{8a}{x}$ **7** 1 **9** $\frac{29}{6}$ **11** $\frac{16y}{5}$ **13** $\frac{11}{2a}$
15 $\frac{bc^2 + 1}{c}$ **17** $\frac{a + 5}{4m}$ **19** $\frac{a + b}{x + y}$ **21** $\frac{2bx + 3ay}{ab}$
23 19 **25** $4\frac{19}{40}$ in.

Sec. 3.4, pp. 112–116

1 $\frac{4}{3}$ **3** 4 **5** $\frac{2}{3}$ **7** $\dfrac{5a^3}{6b^2}$

9 $\dfrac{1}{6x}$ **11** $2a$ **13** 1 **15** $\frac{1}{8}$ **17** $6x$

19 $\dfrac{3b}{14a}$ **21** $5b^2$ **23** $\frac{1}{7}; \frac{1}{10}; \frac{1}{12}; \frac{1}{15}; \frac{1}{18}$ **25** $a + 1$ **27** $a + 1$

29 $7\frac{3}{4}$ **31** $17\frac{5}{14}$ **33** $753\frac{5}{7}$ **35** $76\frac{22}{115}$ **37** 40 **39** $2\frac{3}{4}$ mi

41 (*a*) 154 sq in.; (*b*) 616 sq ft; (*c*) 3,850 sq yd; (*d*) $\frac{891}{14}$ sq ft

43 (*a*) $\frac{528}{7}$ sq in.; (*b*) $\frac{220}{21}$ sq in.; (*c*) 176 sq ft; (*d*) $\frac{440}{63}$ sq ft

45 $\frac{1}{4}$ **47** 5 **49** $\frac{1}{2}, \frac{3}{4}$ **51** (*a*) $\frac{1}{4}$; (*b*) $\frac{4}{9}$; (*c*) $\frac{49}{16}, \frac{9}{16}$

53

WHEN $x =$	$\frac{1}{2}$	$\frac{2}{3}$	$\frac{3}{4}$	$\frac{1}{5}$	$\frac{3}{2}$	$\frac{5}{7}$	$\frac{1}{6}$
THEN $x^2 =$	$\frac{1}{4}$	$\frac{4}{9}$	$\frac{9}{16}$	$\frac{1}{25}$	$\frac{9}{4}$	$\frac{25}{49}$	$\frac{1}{36}$
AND $x^3 =$	$\frac{1}{8}$	$\frac{8}{27}$	$\frac{27}{64}$	$\frac{1}{125}$	$\frac{27}{8}$	$\frac{125}{343}$	$\frac{1}{216}$

55 (*a*) $\frac{5}{6}$ cu ft; (*b*) $\frac{45}{4}$ cu ft; (*c*) $\frac{7}{12}$ cu ft; (*d*) $\frac{9}{4}$ cu ft

57 (*a*) $V = \dfrac{x^2 + 3x}{2}$; (*b*) $V = \dfrac{b^2 + 3b}{6}$; (*c*) $V = \dfrac{x + 3}{6x}$; (*d*) $V = 2a(a - 2b)^2$

59 $A = \dfrac{2x(9x + 5)}{15}$; $A = \dfrac{5(a + 1)(5x + 3)}{3}$; $A = \dfrac{4x(5x + 4)}{5}$; $A = \dfrac{7(2x + 1)(x + 2)}{6}$

Sec. 3.5, pp. 123–124

9 $-\frac{1}{3}$ **11** $-\frac{1}{2}$ **13** $-\frac{1}{2}$ **15** $\frac{5}{6}$ **17** $\frac{1}{30}$ **19** $\frac{7}{30}$ **21** $\frac{35}{22}$

23 $x + y$; xy; $x \div y$ **25** $32\frac{1}{4}$ lb; $87\frac{3}{4}$ lb **27** $5\frac{1}{3}$ h; 5 h and 20 min

29 $6\frac{15}{16}$ in. **31** $1\frac{1}{4}$ h; $2\frac{1}{6}$ h; $1\frac{1}{10}$ h **33** 30

35 $\frac{1}{4}, \frac{9}{28}, \frac{1}{3}, \frac{7}{20}$ **37** $-\frac{4}{35}$ is $\frac{1}{35} > -\frac{1}{7}$ **39** $-\frac{2}{5}$ is $\frac{1}{85} > -\frac{7}{17}$

41 $-, +$

Sec. 3.6, pp. 130–132

1 $\frac{21}{16}$ **2** $\frac{3}{20}$ **5** b

7 $\dfrac{x^2}{x + y}$ **9** $\dfrac{y}{x + y}$ **11** b

13 $\dfrac{a + b}{b}$ **15** $\dfrac{9}{2b}$ **17** $\dfrac{x + 2}{x}$

19 $\dfrac{y}{3}$ **21** $\dfrac{1}{5a}$ **23** $\dfrac{4}{y}$ **25** 12 ft; $\frac{11}{5}$ ft; $\frac{69}{8}$ ft; 20 ft

27 $\frac{50}{27}$ or $1\frac{23}{27}$ **29** $\frac{35}{6}$ or $5\frac{5}{6}$ **31** $\frac{15}{32}$ **33** $\frac{9}{7}$ or $1\frac{2}{7}$

35 $\dfrac{1 + 3x}{4x + 2}$ **37** $\frac{7}{9}$ **39** $\frac{5}{6}$ **41** $\dfrac{40 - 3x^2}{5x}$; exclude $x = 0$

43 $-\dfrac{1}{y+2}$; exclude $y = -2$ **45** 3; exclude $a = \frac{3}{2}$

47 $\dfrac{3}{2m}$; exclude $m = 0$, $m = 4$ **49** 1; exclude $a = -b$

51 $\dfrac{5b^2 - 3a(a-1)}{15ab}$; exclude $a = 0$ and $b = 0$

53 $\dfrac{2}{x-3}$; exclude $x = 3$, $x = -4$ **55** $\dfrac{4(a-1) - a^2}{2a^2}$; exclude $a = 0$

57 $\dfrac{12}{a-b}$; exclude $a = b$ **59** $\dfrac{7}{2x}$; exclude $x = 0$ and $a = -b$

61 $b + a$; exclude $a = 0$, $b = 0$ **63** $\dfrac{x}{y}$; exclude $x = 0$, $y = 0$, $y = \dfrac{1}{x}$

65 q cannot equal 0 for $p \div q$ to be rational

Sec. 3.7, pp. 134–136

1 $14x(x+3)$ **3** $x(x-y+z)$ **5** $5y(2x^2+3x-1)$
7 $2x(x^2+2x-3)$ **9** $4axy - 4x^2y$ **11** $ab^2 - a^2b$ **13** $x^2 - x^2y$
15 $15n^2 - 10n^3 + 5n$ **17** $x^2 + 4x + 4$ **19** $4a^2 - 12ab + 9b^2$
21 $a^2b^2 - 2ab + 1$ **23** $4a^2 + 2a + \frac{1}{4}$ **25** $(x-7)^2$
27 $(y - \frac{1}{3})^2$ **29** $(2b+1)^2$ **31** $(3x-7)^2$ **33** $(x-10)(x+10)$
35 $(x+8)(x-8)$ **37** $(4a - 3xy)(4a + 3xy)$

39 $\left(\dfrac{x}{6} - \dfrac{y}{5}\right)\left(\dfrac{x}{6} + \dfrac{y}{5}\right)$ **41** $(x-10)(x+4)$

43 $(a-5)(a-2)$ **45** $(x+24)(x+2)$ **47** $(x+\frac{1}{2})(x+1)$

49 $\dfrac{3(x+y)}{x-y}$; exclude $x = y$ **51** $10 - a$; exclude $x = -y$, $a = -10$

53 $\dfrac{a+2}{x-9}$; exclude $x = 9$, $a = 3$ **55** $3 - x$; exclude $x = 3$, $x = 1$

57 $\dfrac{4x+13}{x^2-4}$; exclude $x = \pm 2$ **59** $\dfrac{(x+3)(x-2)}{(x+1)^2(x-1)}$; exclude $x = \pm 1$

Sec. 3.8, pp. 140–141

1 16 **3** -27 **5** 11 **7** 248,832 **9** $(-2)^5$ **11** $(0.2)^3$
13 $(\frac{5}{3})^4$ **15** a^6 **17** $5x^7$ **19** a^{10} **21** $5x^{12}$ **23** $125x^{12}$

25 $45b^2$ **27** $\dfrac{a^4}{81}$ **29** $\dfrac{a^2}{9b^2}$ **31** $5y^7$ **33** $\dfrac{a^6}{8b^3}$ **35** 729

37 $512x^6$ **39** $\dfrac{1}{16a^8}$ **41** 27 **43** x^8 **45** $\dfrac{3a^4}{16}$ **47** $19\frac{9}{25}$

49 0.001728 **51** $181\frac{26}{27}$ **53** $5\frac{1}{7}$ **55** $20\frac{1}{4}$ **57** 100,000

59 2^5 **61** 2^8 **63** 2^{12} **65** 3^5 **67** 3^{12} **69** $\dfrac{1}{a^2}$

71 $5a^2$ **73** $\frac{1}{16}$ **75** $\dfrac{1}{a^{18}}$ **77** a^m **79** ab^2 **81** $\dfrac{y^m}{x^{2m-1}}$

83 $\frac{y^2}{x}$ **85** a^{m-1} **87** $\frac{16y^4}{27x}$ **89** 36 **91** 100 **93** $\frac{9}{2}$

Sec. 3.9, pp. 144–146

1 $\frac{1}{25}$ **3** 9 **5** $\frac{3}{2}$ **7** 1 **9** $\frac{25}{27}$ **11** $\frac{5}{26}$ **13** $\frac{1}{3}$ **15** $\frac{81}{16}$
17 $\frac{1}{a^2}$ **19** $\frac{b^3}{a^2}$ **21** $\frac{1}{4x^2}$ **23** $\frac{4}{x^2}$ **25** $\frac{4}{x^2}$ **27** $\frac{b+a}{ab}$
29 x^{-4} **31** $3ac^{-2}d^{-3}$ **33** $a^{-1}b^2$ **35** $3a(4b)^{-1}$ **37** a
39 $(a^2y)^{-1}$ **41** $\frac{9}{4}$ **43** 100 **45** $\frac{4}{11}$

47

WHEN x =	3	2	1	0	−1	−2	−3
THEN 1^x =	1	1	1	1	1	1	1
2^x =	8	4	2	1	$\frac{1}{2}$	$\frac{1}{4}$	$\frac{1}{8}$
3^x =	27	9	3	1	$\frac{1}{3}$	$\frac{1}{9}$	$\frac{1}{27}$
5^x =	125	25	5	1	$\frac{1}{5}$	$\frac{1}{25}$	$\frac{1}{125}$
10^x =	1,000	100	10	1	$\frac{1}{10}$	$\frac{1}{100}$	$\frac{1}{1,000}$
$(\frac{1}{2})^x$ =	$\frac{1}{8}$	$\frac{1}{4}$	$\frac{1}{2}$	1	2	4	8
$(\frac{3}{4})^x$ =	$\frac{27}{64}$	$\frac{9}{16}$	$\frac{3}{4}$	1	$\frac{4}{3}$	$\frac{16}{9}$	$\frac{64}{27}$

49 2^7 **51** 2^1 **53** 2^{-3} **55** 3^{-3} **57** 3^{-12} **59** 3^1
61 10^2 **63** 10^{-8} **65** a^6 **67** 10^3 **69** 10^5 **71** 10^{-2}
73 $x = 5$ **75** $x = -4$ **77** $x = 0$ **79** $\frac{a^2}{b}$ **81** $\frac{1}{xy}$
83 $\frac{ab}{a+b}$

Sec. 3.10, pp. 150–151

1 7 **3** $2a^3$ **5** $-x^3$ **7** 10 **9** 2 **11** 9 **13** 15
15 $9x^2y^3$ **17** 2 **19** 7 **21** 4 **23** $-\frac{1}{27}$ **25** $\frac{1}{625}$
27 24 **29** $\frac{81}{16}$ **31** $\frac{49}{36}$ **33** 10 in.; 11 ft; 9 yd
35 5 ft; 3 ft; 13 yd **37** 5 in. **39** 6 **41** $\frac{1}{2}$ **43** 5
45 $\frac{1}{6}$ **47** $\frac{1}{11}$ **49** 5 **51** 2 **53** 2 **55** 4 **57** −8
59 16 **61** $\frac{1}{4}$ **63** 1,331 **65** 8 **67** $\frac{1}{64}$ **69** 1
71 −1 **73** 40,000 **75** $\frac{1}{2x}$ **77** $-5xy^2$

Sec. 3.11, pp. 157–159

1 $2(10)^1 + 7(10)^0 + 3(10)^{-1} + 4(10)^{-2}$ **3** $6(10)^{-3}$
5 Thirty-one and twenty-five hundredths **7** Five thousandths
9 78.07 **11** 0.00308 **13** 0.8 **15** 0.875 **17** 0.333
19 0.364 **21** $\frac{25}{100}$ **23** $\frac{475}{1,000}$ **25** $\frac{91}{22}$ **27** $\frac{41}{33}$

29 966.21 **31** 3,978.139 **33** 3.39 **35** 8.9835
37 81,955.82628 **39** 0.000202 **41** 28.6 **43** 20
45 $68.25 **47** $2.10; $1.08; $2.08; $5.26
49 $C = 94.20$ in.; $A = 706.50$ sq in.
51 $V = 113.04$ cu in.; $S = 113.04$ sq in.

Sec. 3.12, p. 165
1 No **3** Yes, 0 **5** No **7** Yes; 0, with inverse 0
9 Yes; 1, with inverse 1; -1, with inverse -1 **11** $-\frac{9}{31}$
13 (*a*) T; (*b*) T; (*c*) F; (*d*) T

Sec. 4.2, pp. 176–179
1 x **3** s **5** m **7** x **9** Divide by 8 **11** Subtract 9
13 Multiply by 3 **15** Multiply by 4 **17** Add 52 **19** Add $\frac{5}{3}$
21 Multiply by 2 and add 6 **23** Divide by 5 and add $\frac{8}{5}$ **25** 5
27 $\frac{3}{2}$ **29** $\{3\}$; divide both sides of equation by 3
31 $\{\frac{9}{2}\}$; divide both sides of equation by 4
33 $\{6\}$; multiply both sides of equation by 6
35 $\{\frac{1}{5}\}$; divide both sides of equation by 10
37 $\{6\}$; add 1.5 to both sides of equation
39 $\{4.4\}$; subtract 0.7 from both sides of equation
41 $\{2\}$; multiply both sides of equation by $\frac{2}{9}$
43 $\{2\}$; divide both sides by 6
45 $\{-\frac{3}{7}\}$; divide both sides of equation by 7
47 $\{-6\}$; multiply both sides of equation by $-\frac{3}{17}$
49 $\{-22\}$; multiply both sides of equation by -2
51 $\{200\}$; divide both sides of equation by 1.44
53 $\{-5.13\}$; subtract 5.26 from both sides of equation
57 $\{-10\}$; multiply each side by 2; subtract 18 from each side
59 $\{36\}$; multiply each side by 4; add 8 to each side
61 $\{\frac{13}{5}\}$; subtract 4 from each side; divide each side by 5
63 $\{20\}$; multiply each side by 4; subtract 12 from each side
65 $\{0.2\}$; add 0.34 to each side; divide each side by 21
67 $\{6\}$ **69** $\{3\}$ **71** $\{8\}$ **73** $\{-6\}$ **75** $\{3\}$ **77** $\{-37\}$
79 $\{\frac{5}{4}\}$ **81** $\{-\frac{7}{3}\}$ **83** $\{-1\}$ **85** $\{\frac{5}{3}\}$ **87** $\{\frac{7}{4}\}$ **89** $\{-\frac{3}{4}\}$
91 $\{-2\}$ **93** $\{3\}$ **95** $\frac{46}{3}$, 2, -6 **97** 68°F; 50°F; 32°F; 14°F; -13°F

Sec. 4.3, pp. 181–182
1 $x = \frac{2}{3}$ **3** $x = -\dfrac{2b + 5a}{2a - 3b}$ **5** $x = 4a$ **7** $x = -\dfrac{1}{n}$
9 $x = \dfrac{d - b}{a + c}$ **11** $b = \dfrac{2A}{h}$ **13** $r = \dfrac{I}{Pt}$ **15** $P = \dfrac{I}{rt}$
17 $t = \dfrac{d}{r}$ **19** $t = \dfrac{v - k}{g}$ **21** $a = \dfrac{2s}{t^2}$ **23** $M = \dfrac{Fd^2}{Km}$
25 $w = \dfrac{V}{LH}$ **27** $m = \dfrac{f}{a}$ **29** $b = \dfrac{2A - ch}{h}$

31 $a = l - dn + d$ **33** $d = \frac{l - a}{n - 1}$ **35** $y = \frac{b}{2}\left(\frac{b}{2} - 3\right)$

37 $y = \frac{b^2 + 2ab - a^2}{-2(a + b)}$ **39** $y = \frac{b}{c - a}$ **41** $V = \frac{S - 16t^2}{t}$

43 $r = \frac{C}{2\pi}$ **45** $E = IR$ **47** $a = S(1 - r)$

49 $r = \frac{Cst}{st - Ct - Cs}$ **51** $m = \frac{E}{e^2}$

Sec. 4.4, pp. 188–190

1 10:30 A.M. **3** 24 mi/h and 26 mi/h **5** 4 h
7 5 ft from B **9** 72 lb at 50¢ per lb; 48 lb at 30¢ per lb
11 64 **13** 4, 7; 47 **15** 5, 2; 52 **17** 36 mi/h; 27 mi/h
19 10:18 A.M. **21** A = 157.5 lb; B = 202.5 lb
23 $x = 20$, $x + 1 = 21$ **25** 3, 1, −12 **27** 9 years old
29 Yes; 28, 29, 30 **31** Yes; 24, 26, 28
33 $10\frac{1}{2}$ ft to the right of the fulcrum

Sec. 4.5, pp. 196–200

1 8 **3** 10.5 **5** 37.5% **7** 72% **9** 200 **11** 25
13 510 lb **15** 200 lb; 180 lb; 5,600 lb
17 2% chromium; 3% lead; 12.2% rubber
19 34.5 lb of copper; 161.3 lb approx.
21 30 oz of copper; 6.25% silver **23** 320 gal **25** 37.5% copper
27 81 oz of copper; 204 oz of nickel; 138 oz of copper
29 60 lb of 36% alloy; 240 lb of 41% alloy
31 (Amount of first alloy used): (amount of second alloy used) = 2 : 1
33 $1\frac{3}{5}$ qt **35** 13.95 lb of cream; 586.05 lb of milk **37** 48 qt
39 3.94 qt **41** 15.26 **43** 0.00016 **45** 0.049 **47** 54
49 0.5% **51** 0.9% **53** 6,500% **55** \$6,400
57 150%; 250% **59** $\frac{3}{8}$ qt **61** 54 tons
63 I = \$96.00; A = \$2,496.00 **65** I = \$983.28; A = \$6,767.28
67 I = \$19.50; A = \$319.50 **69** \$30,000.00 **71** $r = 7\%$

73 $t = 25$ yr **75** $t = 12.5$ yr **77** $t = \frac{A - P}{Pr}$

79 $r = \frac{A - P}{Pt}$ **81** A = \$2,200; I = \$200

83 P = \$5,769.23; D_t = \$230.77 **85** P = \$961.54; P = \$952.38; P = \$943.40
89 The offer of \$20,000 now is the better offer; for the present value of \$20,100 two months hence, money being worth 4% simple interest, is only \$19,966.89.
91 P = \$9,076.60

Sec. 4.6, pp. 204–205

1 \$3,000 **3** \$2,004.94 **5** \$8.39; \$2,004.94
7 \$2.79; \$1,002.21 **9** \$14.58; \$485.82 **11** \$6.07; \$368.93

13 \$26.35; \$851.98 **15** \$32; \$800 **17** 7.84%
19 6.03%; 6.06%; 6.09%; 6.12%; 6.19%; 6.38%

Sec. 4.7, pp. 209–210

1 $\{\frac{4}{3}\}$ **3** $\{\frac{6}{5}\}$ **5** $\{5\}$ **7** $\{6\}$ **9** $\varnothing$ **11** $\{-4\}$
13 12 min **15** 6 min **17** $-\frac{1}{2}$ **19** $\dfrac{ab}{a+b}$ days **21** $\{7\}$
23 $\{b\}$ **25** $\{\frac{20}{13}\}$ **27** $\{\frac{4}{7}\}$ **29** -11 **31** 90 min
33 8, 29 **35** 15 min; $\frac{15}{4}$ min

Sec. 4.8, pp. 218–220

1 $3 < 8$; $-1 < 3$; $-2 > -7$; $1.414 > 1.413$ **3** $x > 4$ **5** $x < 5$
7 The set of all rational numbers which are greater than 5
9 The set of all rational numbers which are greater than $\frac{5}{2}$
11 The set of all rational numbers which are greater than -2
13 The set of all rational numbers which are greater than -2
15 The set of all rational numbers which are less than $\frac{12}{5}$
17 The set of all rational numbers which are less than $-\frac{16}{3}$
23 $\{-2, -1, 0, 1, 2\}$ **25** $\{4\}$ **27** $\{4\}$
29 $\{x \in Ra \mid x < -2\} \cup \{x \in Ra \mid x > 2\}$ **31** $\varnothing$

Sec. 4.9, pp. 226–227

1 $\{(\frac{44}{5}, -\frac{61}{35})\}$ **3** $\{(\frac{2}{3}, \frac{1}{9})\}$ **5** $\{(\frac{61}{2}, -\frac{13}{2})\}$ **7** $\{(2.75, 2.25)\}$
9 $\{(5, 2)\}$ **11** $\{(2, 1)\}$ **13** $\{(6, -5)\}$ **15** $\{(-2, -1)\}$
17 $\{(4, 6)\}$ **19** $\{(6, 3)\}$ **21** $\left\{\left(\dfrac{a+b}{2m}, \dfrac{a-b}{2n}\right)\right\}$
23 $\{(m+n, m-n)\}$ **25** 17 and 9 **27** $\dfrac{a+b}{2}$ and $\dfrac{a-b}{2}$
29 Rate in still water = $5\frac{1}{2}$ mi/h; rate of current = $2\frac{1}{2}$ mi/h
31 Airplane, 120 mi/h; wind, 30 mi/h **33** $w = 7.5$ in.; $l = 14.5$ in.
35 Father is 48; son 12
37 Rate of boat in still water = 5 mi/h; rate of current = 1 mi/h

Sec. 5.4, pp. 241–243

1 $10\sqrt{5}$ **3** $6ab^2\sqrt{2ab}$ **5** $x^2\sqrt{1+2x^2y^2}$
7 $4x^2\sqrt[3]{x^2-3y^2}$ **9** $5x\sqrt{3x}$ **11** $3a^{-2}\sqrt{5}$ **13** $3\sqrt[3]{9}$
15 $\dfrac{2a}{y}\sqrt[5]{\dfrac{3a}{y^3}}$ **17** $\sqrt[3]{2xy^2}$ **19** $\sqrt{10}$ **21** $2\sqrt{0.2}$
23 $\sqrt[4]{\dfrac{x+3y}{9}}$ **25** $\sqrt[6]{(x+y)^2}$ **27** $\sqrt[8]{\dfrac{(a-2)^2}{16}}$ **29** $\sqrt[6]{1{,}000}$
31 $\sqrt[10]{243a^5x^5}$, $\sqrt[10]{36a^6x^4}$ **33** $\sqrt[14]{128m^7x^7}$, $\sqrt[14]{16m^{10}x^4}$
35 $\sqrt[8]{\dfrac{16x^4}{81y^4}}$, $\sqrt[8]{\dfrac{x^4}{4y^6}}$, $\sqrt[8]{\dfrac{9x^5}{7y^2}}$ **37** $\sqrt{(x-y)(x+y)^2}$ **39** $\sqrt[3]{125x^4}$
41 $\sqrt{26(2x+1)^2}$ **43** $\frac{1}{10}\sqrt[3]{30}$ **45** $\dfrac{1}{xy}\sqrt{xy(x^2+y^2)}$ **47** $5\sqrt{3}$

49 $\sqrt[3]{5}$ **51** $4\sqrt{3}$ **53** 30 **55** $\sqrt[6]{\dfrac{(x+2)^2}{(x-2)(x-3)^3}}$

57 $3+7\sqrt{15}$ **59** $\sqrt{a^2-x^2}$ **61** $\sqrt[6]{5}$ **63** $-\sqrt[5]{2}$

65 $\dfrac{2(\sqrt{a}-\sqrt{b})}{a-b}$ **67** $\frac{3}{5}(\sqrt{7}+\sqrt{2})$

69 $2+\sqrt{6}-\sqrt{10}-\sqrt{15}$ **71** $\sqrt[6]{48}$

73 $\dfrac{(x+y)^2-2(x+y)\sqrt{x-y}+x-y}{(x+y)^2-x+y}$

75 $\dfrac{3}{2-\sqrt{6}+\sqrt{10}-\sqrt{15}}$ **77** $\dfrac{7}{6-2\sqrt{2}-3\sqrt{3}+\sqrt{6}}$

79 $\dfrac{3}{2(\sqrt{3}+1)}$ **81** $\dfrac{10}{4\sqrt{6}-11}$ **83** $\dfrac{x}{\sqrt{x^2+x}-\sqrt{x}}$

85 $\dfrac{3\sqrt{2}-12}{14}$ **87** $\dfrac{\sqrt[3]{x^4-y^4}\,\sqrt{x^2-y^2}}{x^2-y^2}$ **89** $2\sqrt[3]{5}$

Sec. 5.5, pp. 250–251

1 $6i$ **3** $\frac{2}{3}i$ **5** $\frac{\sqrt{3}}{2}i$ **7** $6\sqrt{2}i$ **9** $x=2, y=-5$

11 $x=8, y=3$ **13** -1 **15** -1 **17** i **19** 1

21 $3+8i$ **23** $\frac{2}{3}-\frac{4}{5}i$ **25** $8+8i$ **27** $-1+3i$

29 $21+\sqrt{3}i$ **31** 0 **33** $\frac{11}{26}+\frac{23}{26}i$ **35** $\frac{8}{3}+\frac{\sqrt{2}}{3}i$ **37** 0

Sec. 5.6, pp. 254–255

1 2 **3** $\sqrt{7}+1$ **5** $1+\sqrt{3}$ **7** $\dfrac{-25}{104}$ **9** $\dfrac{x-y}{a+b}$

11 $\dfrac{ac}{b}$ **13** x^2 **15** $\dfrac{x+y}{x-y}$ **17** $-\dfrac{2xy}{x^2+y^2}$ **19** $\dfrac{ab(a+b)}{a^2+b^2}$

21 $\dfrac{a^2b^2}{a^2+b^2}$ **23** $\dfrac{1}{ab(a+b^2)}$

Sec. 5.7, p. 259

1 63 **3** 1.414 **5** 2.794 **7** 30.708 **9** 218 **11** 8.5

Sec. 5.8, pp. 262–264

1 14.3 ft **3** 44.7 ft **5** 23.2 ft **7** 32.0 mi **9** 11.5 ft **11** 6 ft; 4 yd; 10 in. **13** 13.86 in. **15** $10\sqrt{2}$ in., $a\sqrt{2}$ in.; 14.14 in., 1.414a in. **17** $3\sqrt{91}$ ft, 28.62 ft **19** $6-2\sqrt{3}$ ft; 2.536 ft **23** 6, 8, 10; 5, 12, 13 **25** $8\sqrt{6}$ ft, 19.60 ft

Sec. 6.1, pp. 270–271

1 $8x^2-5x+7$, degree 2 **3** $12x^2-4x-9$, degree 2

5 86, degree 0 **7** $3x^6-8x^3-1$, degree 6

9 $2x^3-x^2$, degree 3 **11** $6x^2+x-2$, degree 2

13 $2x^3 + 5x^2 - 33x + 20$, degree 3
15 $-3x^4 + 3ax^3 + 10a^2x^2 - 14a^3x + 4a^4$, degree 4
17 $6x^4 - 13x^3 - 56x^2 + 30x - 3$, degree 4
19 $x^2 - x + 3 + \frac{-14}{x+3}$, quotient of degree 2
21 $2x^2 - 9 + \frac{-34x + 18}{-2x^2 + 2}$, quotient of degree 2
23 $x^2 + 5x - 4$, degree 2 **25** $3x^2 + 2ax - a^2$, degree 2

Sec. 6.2, pp. 274–275
1 3; 4 **3** 0; -2 **5** 0; 4 **7** 1; 3 **9** -2; -1
11 -3; -4 **13** 3; 6 **15** -8; 5 **17** -3; -9 **19** 5; 6
21 -4; 8 **23** -4; 7 **25** 5 or -3
27 Length = 8 ft, width = 5 ft **29** 4 or -3 **31** 4 or -11
33 $3x^2 - 10x + 3 = 0$, $a = 3$, $b = -10$, $c = 3$
35 $9x^2 + 16 = 0$, $a = 9$, $b = 0$, $c = 16$
37 $5x^2 - 4x = 0$, $a = 5$, $b = -4$, $c = 0$
39 $3x^2 - 11x + 2 = 0$, $a = 3$, $b = -11$, $c = 2$
41 $-8a$; $2a$ **43** $5a$; $-2a$ **45** 8, 9, 10 or -3, -2, -1
47 29 **49** 12 **51** $A = 324$ sq in.

Sec. 6.3, pp. 278–279
1 $18x^2 + 27x - 35$ **3** $2x^2 - 17x - 9$ **5** $(3x + 1)(x + 2)$
7 $(3x - 4)(2x + 5)$ **9** $(2w - 1)(w + 3)$ **11** $(3y - z)(y - z)$
13 1; $\frac{1}{2}$ **15** 7; $-\frac{3}{2}$ **17** $\frac{1}{2}$; -4 **19** 4; $\frac{4}{3}$ **21** $\frac{1}{6}$; $-\frac{4}{3}$
23 0; 12 **25** $-\frac{3}{2}a$; a **27** $-a$; $-2a$ **29** $\frac{2}{5}$ or $\frac{5}{2}$
31 Length = 21 ft, width = 17 ft **33** $t = 15°$
35 Length = $9\frac{1}{2}$ ft; width = 8 ft; perimeter = 35 ft
37 Length = $7\frac{3}{4}$ ft; width = $6\frac{1}{2}$ ft; perimeter = $28\frac{1}{2}$ ft
39 $(4x - 3)(3x - 2)$ **41** $(y + 3)(3y + 2)$ **43** $(4z - 1)(z - 2)$
45 $-\frac{2}{5}$; 3 **47** $-\frac{1}{3}$; 2 **49** $\frac{8}{15}$; -1 **51** $\frac{1}{2}$; $-\frac{3}{2}$ **53** 3; $-\frac{9}{5}$
55 8; -8

Sec. 6.4, pp. 281–283
1 4, -4 **3** $\frac{3}{4}$, $-\frac{3}{4}$ **5** 3.2, -3.2 **7** 9, -9 **9** 4, -4
11 $\frac{12}{11}$, $-\frac{12}{11}$ **13** $\frac{4}{3}$, $-\frac{4}{3}$ **15** $\frac{5}{2}$, $-\frac{5}{2}$ **17** $4a$, $-4a$
19 $\frac{c}{a+b}$, $-\frac{c}{a+b}$ **21** 5 in. and 15 in. **23** 6 ft and 18 ft
25 $\pm\frac{1}{2}\sqrt{2}$; ± 0.71 **27** $3\sqrt{6}$; 7.35 **29** $\pm\frac{1}{3}\sqrt{30}$; ± 1.83
31 $\pm 2\sqrt{2}$; ± 2.83 **33** $\frac{1}{2}\sqrt{10}$, 1.6; $\frac{1}{2}\sqrt{15}$, 1.9; $\sqrt{5}$, 2.2 **35** 25.7 s
37 $e = \sqrt{\frac{A}{6}}$ **39** $r = \sqrt{\frac{V}{\pi h}}$ **41** $s = 2\sqrt{\frac{4}{\sqrt{3}}}$ **43** 2, -2
45 17, -17 **47** $\frac{\sqrt{30}}{2}$, $-\frac{\sqrt{30}}{2}$ **49** $3\sqrt{2}$, $-3\sqrt{2}$

Sec. 6.5, pp. 288–289

1 25 **3** $8x$ **5** $\frac{4}{25}$ **7** 36 **9** $\frac{1}{36}$ **11** $16a^2$ **13** $1, -\frac{1}{2}$

15 $-1, -\frac{1}{2}$ **17** $\frac{3+\sqrt{7}}{2}, \frac{3-\sqrt{7}}{2}$; 2.82, 0.18

19 $\frac{4+\sqrt{11}}{5}, \frac{4-\sqrt{11}}{5}$; 1.47, 0.14 **21** $\frac{3}{2}, -\frac{4}{3}$

23 $\frac{-5+\sqrt{7}}{3}, \frac{-5-\sqrt{7}}{3}$; -0.79, -2.55

25 $\frac{3+\sqrt{33}}{4}, \frac{3-\sqrt{33}}{4}$; 2.19, -0.69 **27** $\frac{3}{2}, -3$

29 $\frac{3+\sqrt{5}}{2}$ or $\frac{3-\sqrt{5}}{2}$

31 Width $= \frac{45-\sqrt{75}\sqrt{11}}{2}$ rods, or 8.14 rods, to two decimal places

33 $\frac{1}{2}$, 2 **35** $\frac{5}{2}$, 15 **37** $1+\sqrt{2}, 1-\sqrt{2}$ **39** $\frac{3}{2}, -\frac{7}{5}$

41 6.97 rods **43** 261 mi/h **45** $-3+2i, -3-2i$

47 $\frac{3+\sqrt{7}\,i}{4}, \frac{3-\sqrt{7}\,i}{4}$

Sec. 6.6, p. 292

1 $-\frac{1}{3}, -\frac{1}{2}$ **3** $\frac{1 \pm i}{3}$ **5** $s_1 + s_2 = \frac{9}{2}$; $s_1 s_2 = -\frac{5}{2}$

7 $s_1 + s_2 = \frac{10}{3}$; $s_1 s_2 = 0$ **9** $x^2 - 5x - 14 = 0$

11 $x^2 - 4x + 13 = 0$ **13** $x^2 - 17 = 0$ **15** $2x^2 + x + 3 = 0$

17 $x^2 - 2x + 4 = 0$

19 $b^2 - 4ac = -8$; solutions are complex numbers; $\frac{-1 \pm i\sqrt{2}}{3}$

Sec. 6.7, pp. 297–298

1 $P_3(5) = 11$ **3** $P_2(k) = ak^2 + bk + c$

5 $P_4(3) = 0$; $x - 3$ is a factor of $P_4(x)$, and $x = 3$ is a solution of $P_4(x) = 0$

11 No; $P_4(-3) = 4$ **13** Yes **15** $(x+1)(x+2)(x+3)$

17 $(x+2)(x+3)(x+4)(x-4)$

19 $x^3 - x^2 - 4x + 4 = (x-1)(x-2)(x+2)$, $x = 1$, $x = 2$, $x = -2$

21 $x^4 - 2x^3 - 7x^2 + 20x - 12 = (x-1)(x-2)(x-2)(x+3)$, $x = 1$, $x = 2$, $x = 2$, $x = -3$

23 3 in., 5 in., 7 in. **25** Width = 4 ft, height = 9 ft, length = 11 ft

Sec. 6.8, pp. 302–303

1 $(x-3y)(x+2y)$ **3** $(3-4x)(2-x)$

5 $(x+y)(x^6 - x^5y + x^4y^2 - x^3y^3 + x^2y^4 - xy^5 + y^6)$

7 $(4+c)(16-4c+c^2)$ **9** $(7x-12y)(3x+2y)$

11 $(2x^3-3y^3)(2x^3+3y^3)$ **13** $0.01(a-b)(a+b)$

15 $3y(y+1)^2$ **17** $(x-2)(x+1)(x^2-x+2)$

19 $(x+y-2)(x-y)$ **21** $(ab+cd)(ac-bd)$

23 $(x-1)^2(x^2+x+1)^2$ **25** $(a-b-5)^2$
27 $(x-3-2y)[(x-3)^2+(x-3)(2y)+(2y)^2]$
29 $2x(x-2)(x+2)(x+3)$ **31** $24(x-2)(x-3)(x+1)$
33 $(x-1)(x-1-\sqrt{3})(x-1+\sqrt{3})$

35 $(x-1)^3(x+2)$ **37** $-\left(\frac{x^2+2x+4}{x^2}\right)$ **39** $\frac{a^2b^4(a^2-b^2)}{3a^2-2b^2}$

41 $(x-\sqrt[3]{9})\left(x+\frac{\sqrt[3]{9}+\sqrt{3}\,\sqrt[3]{9}\,i}{2}\right)\left(x+\frac{\sqrt[3]{9}-\sqrt{3}\,\sqrt[3]{9}\,i}{2}\right)$

43 $(x-2)(x+1)[x-(-1+i)][x-(-1-i)]$
45 $(y-2)[y-(1+\sqrt{2}\,i)][y-(1-\sqrt{2}\,i)]$; $2, 1+\sqrt{2}\,i, 1-\sqrt{2}\,i$
47 $(x+1)(x+1)(x-2)(x+2i)(x-2i)$; $-1, 2, -2i, 2i$

Sec. 6.9, pp. 305–306
1 15 **3** 3 **5** 6 **7** 5 **9** 4 **11** 19
13 493 ft, to the nearest foot

Sec. 7.1, pp. 311–312
1 (1, 1), (1, 2), (2, 1), (2, 2)
3 (1, 1), (1, 2), (1, 3), (1, 4), (2, 1), (2, 2), (2, 3), (2, 4), (3, 1), (3, 2), (3, 3), (3, 4), (4, 1), (4, 2), (4, 3), (4, 4)
7 (1, 1), (1, −1), (4, 2), (4, −2), (9, 3) **9** R_1 and R_4 are functions
11 No

Sec. 7.2, p. 316
1 12, 6, 5, 0
3 $F(-2)=18$; $F(-1)=11$; $F(0)=6$; $F(1)=3$; $F(2)=2$; $F(3)=3$; $F(5)=11$

5

WHEN $x=$	−4	−3	−2	0	1	2	3	4
THEN $y=$	−14	−11	−8	−2	1	4	7	10

When x is increased by 1, y is increased by 3.

7

WHEN $x=$	−4	−3	−2	0	1	2	3	4
THEN $y=$	−8	−6	−4	0	2	4	6	8

When x is increased by 1, y is increased by 2.

9

WHEN $x=$	−4	−3	−2	−1	0	1	2	3	4
THEN $y=$	20	13	8	5	4	5	8	13	20

11

WHEN $x=$	−4	−3	−2	−1	0	1	2	3	4
THEN $y=$	12	5	0	−3	−4	−3	0	5	12

Sec. 7.3, pp.320–323

1 $A(\frac{3}{2}, \frac{5}{2})$; $B(\frac{3}{2}, 0)$; $C(-\frac{5}{2}, 2)$; $D(-2, 0)$; $E(-3, -2)$; $F(0, -3)$; $G(3, -2.5)$

3 A, II; B, I; C, III; D, IV **5** 0 **7** $(-5, 0)$

11 $AB = 2\sqrt{37}$; $BC = 7\sqrt{2}$; $AC = 5\sqrt{2}$ **15** $5\sqrt{5}$ **17** 4.12

Sec. 7.4, pp. 334–335

9 $y = 2x + 2$ **11** $10r = 6s - 1$

17 (*a*) Lowest point $\left(-\frac{1}{2}, -\frac{25}{4}\right)$ (*b*) $(-3, 0)$, $(2, 0)$

19 (*a*) Highest point $\left(-\frac{1}{2}, \frac{13}{2}\right)$ (*b*) $\left(\frac{-1 + \sqrt{13}}{2}, 0\right)$, $\left(\frac{-1 + \sqrt{13}}{2}, 0\right)$

Sec. 7.5, p. 340

1 $\{(3, -4)\}$ **3** $\{(-5, -1)\}$ **5** $\{(\sqrt{2}, 0)\}$ **7** $\{(-5, 7)\}$

9 The solution is the empty set because the lines are parallel.

11 $\{(5, 9), (1, 1)\}$ **13** $\{(4, 4), (4, -4)\}$

Sec. 7.7, p. 355

1 $P_1(3, 0)$, $P_2(5, 0)$, $P_3(0, 12)$, $P_4(0, 4)$; maximum value is 12, minimum value is 3

3 $P_1(-10, 0)$, $P_2(0, 7)$, $P_3(\frac{25}{6}, \frac{17}{6})$; maximum value is $\frac{4}{3}$, minimum value is -10

5 $P_1(0, 8)$, $P_2(5, 8)$, $P_3(5, 4)$, $P_4(2, 4)$, $P_5(0, 6)$; maximum value is 21, minimum value is 10

7 1,000 units to each customer yields $3,960 maximum proceeds

Sec. 8.1, pp. 359–360

1 $\frac{4}{5}$ **3** 25; 5 **5** $3.00; $1.20 **7** $\frac{7}{8}$; $\frac{2}{1}$; $\frac{4}{3}$; $\frac{8y}{z}$ **9** 378 mi

11 60 mi/h **13** 11 ft/min **15** $\frac{4}{5}$ qt/s **17** 45 lb

19 123.2 mi; 1.42 gal to two decimal places **21** $1\frac{2}{3}$ lb **23** $2.97

Sec. 8.2, pp. 361–362

1 2 **3** $\frac{57}{2}$ **5** 5 **7** $-\frac{2}{3}$ **9** $\frac{35}{78}$ **11** 10 **13** No solution

15 $\frac{1}{2}$ **17** 210 lb **19** $4\frac{4}{5}$ lb; $2\frac{2}{5}$ lb

Sec. 8.3, pp. 365–367

1 $k = 4$; $y = 16$ **3** $k = 4$; $Y = 14$ **5** $A = kl$ **7** $N = kd$

9 $y = 24$ **11** $43.50

13 $I = 3$ amperes when $E = 110$ volts; $I = 2\frac{8}{11}$ amperes when $E = 100$ volts. $E = 146\frac{2}{3}$ volts when $I = 4$ amperes; $E = 128\frac{1}{3}$ volts when $I = 3.5$ amperes

Sec. 8.4, pp. 368–369

1 $k = \frac{1}{6}$, $y = 24$

3 $k = \frac{7}{128}$; $P = \$6,835.94$ when $L = 50$ ft; $P = \$54,687.50$ when $L = 100$ ft

5 $\frac{V_1}{V_2} = \frac{r_1^2}{r_2^2}$ **7** $135

Sec. 8.5, pp. 371–372

1 25 ohms **3** 4:1 **5** 178 lb **7** 25.6

Sec. 8.7, pp. 373–374

1 $R = kST$ **3** $H = k\dfrac{m^2}{p^3}$

5 S varies directly with the square of r and the square of h
7 I varies directly with P and inversely with the square of d
9 12,960 lb **11** $z = \frac{8}{3}xy^2$; $z = 128$ when $x = 3$ and $y = 4$

Sec. 9.1, p. 378

1 2 in.; 3 in.; 4 in. **3** 6 units; $\frac{78}{11}$ units; $\frac{90}{11}$ units **5** 135 ft
7 $\frac{40}{47}$ ft **9** 13.6 ft

Sec. 9.2, p. 379

1 475 mi **3** $333\frac{1}{4}$ sq ft **5** 408 sq ft **7** 5 **9** 3.3

Sec. 9.3, pp. 384–386

1 0.404 **3** 1.00 **5** 139 ft
7 186 ft **9** 29,700 ft **11** 205 ft

Sec. 9.4, pp. 389–391

1 0.326 **3** 0.707 **5** 0.857
7 0.707 **11** 218 ft **13** 13.5 ft
15 $\sin A = \frac{4}{5}$, $\cos A = \frac{3}{5}$, $\tan A = \frac{4}{3}$; $\sin B = \frac{3}{5}$, $\cos B = \frac{4}{5}$, $\tan B = \frac{3}{4}$
21 68° **23** 8° **25** 35.9 ft **27** 206 ft **29** 81 ft

Sec. 9.6, pp. 398–399

1 $\dfrac{\pi}{18}$ **3** $\dfrac{2\pi}{9}$ **5** $22\frac{1}{2}°$ **7** −90° **9** 0.985 **11** −0.485

Sec. 9.7, p. 404

1 120° **3** 35°, 86°, 59°
5 $\alpha = 30°$, $\beta = 43°$, $a = 295$ **7** $\gamma = 60°$, $b = 30.6$, $c = 26.9$

Sec. 10.1, p. 413

1 $\log_5 25 = 2$ **3** $\log_7 \frac{1}{49} = -2$ **5** $2^7 = 128$ **7** $10^3 = 1{,}000$
9 0, 1, 2, 3, 4, 5, 6, −1, −2, −3, −5, −6

Sec. 10.2, pp. 420–421

1 $1.73 \cdot 10^8$ **3** $1.55 \cdot 10^6$ **5** $6.9643217 \cdot 10^7$
7 $1.7432 \cdot 10^{-8}$ **9** $2.950 \cdot 10^{-4}$ **11** $1.543798 \cdot 10^{-3}$
13 736,400 **15** 34,700,000,000 **17** 4.6 **19** $1.2 \cdot 10^{17}$
21 $3.1 \cdot 10^8$ **23** $2.0 \cdot 10^6$ **25** $1.9 \cdot 10^9$ **27** $1.7 \cdot 10^{-3}$
29 $4.8 \cdot 10^7$ **31** $2.01 \cdot 10^2$ light-years **33** $2.0 \cdot 10^{27}$ tons
35 0,5843 + 2 **37** 0.6096 + 3 **39** 0.8987 − 2
41 0.7235 + 4 **43** 0.8751 − 1 **45** 0.9881 − 4
47 56,700 **49** 5.67

Sec. 10.3, pp. 422–423

1 0.8974 + 2 **3** 0.5706 + 1 **5** 0.8929 + 5
7 0.7961 + 1 **9** 0.6330 − 3 **11** 0.9530 + 6

Sec. 10.4, p. 424

1 $N = 6{,}940$ **3** 268.3 **5** 26.87 **7** $N = 24{,}700$ **9** 3.659

Sec. 10.5, pp. 425–426

1 log 26 + log 35; log $\frac{2}{3}$ + log $\frac{5}{7}$; log 217 + log 314

3 3 log 17; −3 log 219; 8 log 47
5 $a + b$ **7** $a + 3b$ **9** $2a - 3b$

Sec. 10.6, p. 430
1 29,600 **3** 54,090 **5** 2.119 **7** 6,540 **9** 9,865
11 35,780 cu in. **13** $N = 0.2850$ **15** $A = 61,870$ sq in.

Sec. 10.7, pp. 435–436
5 *Re* **7** *Re*

Sec. 11.1, p. 440
1 \$15.00; \$15.93 **3** \$570.00; \$598.98 **5** 6 months; 4; 3.5%
7 1 month; 12; 0.5% **9** \$1,120.00; \$1,123.60; \$1,125.51
11 \$1,357.40; \$1,360.48; \$1,362.11 **13** 4.04%; 6.14%

Sec. 11.2, pp. 444–446
1 \$1,338.23; \$338.23 **3** \$3,207.14; \$2,207.14
5 \$1,345.87; \$345.87 **7** \$1,485.95; \$485.95
9 \$2,285.61; \$410.61 **11** \$6,976.59; \$6,013.59
13 \$9,004.72
15 *A*, \$3,000.00; *B*, \$4,037.38; *B* has the greater amount by \$1,037.38
17 \$84.37 **19** \$63,814.10; \$81,444.75; \$103,946.40; \$573,370.00
21 \$3,617 **23** 51
25 (*a*) \$10,800.00; (*b*) \$10,816.00; (*c*) \$10,824.32

Sec. 11.3, p. 449
1 Yes; $d = 6$; next two terms are 28 and 34 **3** No
5 Yes; $d = d$; next two terms are $a + 2d$ and $a + 3d$
7 $l = 35, S = 222$ **9** $n = 28$ **11** $S = 5{,}050$ **13** $S = 2{,}550$
19 $n = 16$, interest = \$4,080

Sec. 11.4, pp. 453–454
1 Yes; $r = 2$; next two terms are 48 and 96 **3** No **5** No
7 $l = 118{,}098$; $S = 177{,}146$ **9** $n = 7$ **11** $n = 8, S = \frac{255}{128}$
13 $S = 1.04\dfrac{(1.04)^n - 1}{0.04}$ **15** \$3,001.52 **17** $n = 6, l = 160$

Sec. 11.5, pp. 459–460
1 $A_{20} = \$12{,}148.69$; $P_{20} = \$8{,}175.73$
3 $A_{20} = \$14{,}889.04$; $P_{20} = \$6{,}795.16$
5 $A_{100} = \$273{,}203.27$; $P_{100} = \$37{,}711.06$
7 $A_{50} = \$8{,}457.94$; $P_{50} = \$3{,}142.37$ **9** $P_8 = \$5{,}461.42$
13 \$542.64 **15** \$505.18 **17** \$299.64

Sec. 12.2, p. 465
1 (*a*) $\frac{3}{4}$; (*b*) $\frac{1}{4}$; (*c*) $\frac{1}{4}$
3 (*a*) $\frac{1}{3}$; (*b*) $\frac{1}{6}$; (*c*) $\frac{2}{3}$

Sec. 12.3, pp. 473–474
1 $\{1H, 1T, 2H, 2T, 3H, 3T, 4H, 4T, 5H, 5T, 6H, 6T\}$
3 Yes, each probability is nonnegative and the sum of the probabilities is 1
5 $\Pr(e_4) = \frac{1}{6}$

7 Pr(throwing a number greater than 3) = $\frac{1}{2}$; Pr(throwing a number greater than 4) = $\frac{1}{3}$; Pr(throwing a number greater than 6) = 0
9 Pr(ace) = $\frac{1}{13}$ **11** $\Pr(H) = \frac{1}{4}$, $\Pr(T) = \frac{3}{4}$

Sec. 12.4, pp. 478–479
1 120; 216 **3** 210 **5** 6 **7** 1,728
9 Number of slates with a girl for president = 1,980; number of slates with a boy for president = 2,520; total number of slates = 4,500
11 $\Pr(A) = \frac{4}{7}$ **13** $\frac{1}{9}$ **15** $(a)\ \frac{1}{18}$; $(b)\ \frac{1}{6}$; (c) 7

Sec. 12.5, pp. 484–485
1 5,040; 8; 90; 35 **3** n; $n^2 + n$; $\dfrac{(n+1)(n+2)\cdots(2n)}{n!}$
5 3,024; 2,730; 1,560; 56 **7** 13,824 **9** 120
11 (a) 362,880; (b) 725,760; (c) 725,760; (d) 2,903,040 **13** $n = 8$

Sec. 12.6, pp. 488–489
1 720 **3** 4!2! = 48 **5** 9! **7** 7(8!) **9** $\dfrac{11!}{2!2!2!}$
11 3,360 **13** 12,600

Sec. 12.7, pp. 494–496
1 56; 153; 252; 55 **3** 1,326 **5** $C(6, 4) \cdot C(5, 3) = 150$
7 63 **9** 371 **11** $(a)\ \frac{3}{7}$; $(b)\ \frac{1}{2}$; $(c)\ \frac{1}{14}$; $(d)\ \frac{13}{14}$
13 $\dfrac{1}{635{,}013{,}559{,}600}$; $\dfrac{1}{158{,}753{,}389{,}900}$
15 (a) 4; (b) 40; (c) 10,200; (d) 624
17 (a) 123,552; (b) 1,098,240

Sec. 12.8, p. 504
1 $\frac{4}{52} \cdot \frac{4}{52}$ **3** $\frac{15}{56}$ **5** $\frac{3}{4}$
7 $(a)\ \frac{1}{8}$; $(b)\ \frac{1}{8}$ **9** $(a)\ \dfrac{25}{46{,}656}$; $(b)\ \dfrac{125}{15{,}552}$ **11** $\frac{7}{9}$

Sec. 12.9, pp. 506–507
1 1 to 1, or even **3** $\frac{8}{13}$ **5** \$20
7 \$4,000 **9** \$6,000 **11** A, \$380; B, \$580

Sec. 13.4, pp. 518–520
1 \$7,233.33 **3** 90.7 **5** 39.30 **7** 1.96 **9** 115
11 349.7 **13** 1.43 **15** 2.23; 2.38 **17** 56¢ per dozen
19 80.70

Sec. 13.5, pp. 521–522
1 77.4; 80.5; 86 **5** 41.6; 31; 31 **7** 7.75; 7.90; 8.00 **9** 122

Sec. 13.6, pp. 525–527
1 10.6; 19.5 **3** 3.6 **5** 1.58
9 5.5, 27.5. When each member of the set is multiplied by 5, the arithmetic mean is also multiplied by 5.
11 2.87, 14.4. When each member of the set is multiplied by 5, the standard deviation is also multiplied by 5.
13 5.5, 25.5. When 20 is added to each member of the set, 20 is added to the arithmetic mean.

15 2.87, 2.87. Adding 20 to each member of the set does not affect the standard deviation.
17 $M' = kM + c$, $\sigma' = k\sigma$

Sec. 13.7, pp. 529–530

5 $y^4 - 4y^3 + 6y^2 - 4y + 1$ **7** $x^4 + 8x^3 + 24x^2 + 32x + 16$
9 $y^6 + 6y^4 + 15y^2 + 20 + \frac{15}{y^2} + \frac{6}{y^4} + \frac{1}{y^6}$
11 $x^{10} + 15x^8y^2 + 90x^6y^4 + 270x^4y^6 + 405x^2y^8 + 243y^{10}$
13 $32x^{10} - 80x^8y + 80x^6y^2 - 40x^4y^3 + 10x^2y^4 - y^5$
15 $x^{11} - 11x^{10}y + 55x^9y^2 - 165x^8y^3 + 330x^7y^4 - 462x^6y^5 + 462x^5y^6 - 330x^4y^7 + 165x^3y^8 - 55x^2y^9 + 11xy^{10} - y^{11}$
17 $\frac{a^3}{b^3} + 3\frac{a}{b} + 3\frac{b}{a} + \frac{b^3}{a^3}$ **19** 1.2653
21 1.3401 **23** $\frac{6}{a^2b} + \frac{2}{b^3}$

Sec. 13.8, pp. 532–533

1 1 8 28 56 70 56 28 8 1
1 9 36 84 126 126 84 36 9 1
3 $(x + y)^9 = x^9 + 9x^8y + 36x^7y^2 + 84x^6y^3 + 126x^5y^4 + 126x^4y^5 + 84x^3y^6 + 36x^2y^7 + 9xy^8 + y^9$;
$(x - y)^9 = x^9 - 9x^8y + 36x^7y^2 - 84x^6y^3 + 126x^5y^4 - 126x^4y^5 + 84x^3y^6 - 36x^2y^2 + 9xy^8 - y^9$ **5** 127 **7** 1,023 **9** 4,095
11 1,048,576

Sec. 13.9, p. 545

1 (*a*) $\frac{1}{256}$ is the probability of four successes; $\frac{12}{256}$ is the probability of three successes; $\frac{54}{256}$ is the probability of two successes; $\frac{108}{256}$ is the probability of one success; $\frac{81}{256}$ is the probability of no successes
3 (*a*) $\frac{1}{128}$; (*b*) $\frac{7}{128}$; (*c*) $\frac{21}{128}$
5 (*a*) $\frac{9375}{6^6}$; (*b*) $\frac{2906}{6^6}$ **7** (*a*) $\frac{80}{3^5}$; (*b*) $\frac{51}{3^5}$
9 $M_x = 5$, $\sigma = 0.91$ **11** $M_x = 3.6$, $\sigma = 0.6$
13 $M_x = 200$, $\sigma = 10$; probability of the number of heads being in the interval 170 to 230 is $\frac{99}{100}$. It appears that forces other than mere chance have operated.

Sec. 13.10, pp. 547–549

1 60%; 93% **3** 68%; 96%; 100%
5 76%; 92%; 100% **7** 64 – 82; 46 – 100 **9** 333; 475; 495
11 Approximately 667 weigh 148 to 164 lb, 950 weigh 140 to 172 lb, and all weigh 132 to 180 lb
13 Approximately 667 weigh 112 to 130 lb, 950 weigh 103 to 139 lb, and all weigh 94 to 148 lb
15 2,000 **17** 18 to 78 years; 28,000

Review Exercises, pp. 550–556

1 (*a*) $6 \in A$; $6 \in B$; $6 \notin C$; $6 \notin D$; $12 \notin A$; $12 \in B$; $D \subset A$; $D \not\subset B$
(*b*) 9, 16, infinite, 4

3 $\{3, 6, 9\}$; $\{1, 2, 3, 4, 5, 6, 7, 8, 9, 12, 15, \ldots, 48\}$; D; A
7 $\{-6, -5, -4, -3, -2, -1, 0, 1, 2, 3\}$
9 $\{0, 1, 2, 3\}$ **11** $\{3, 4, 5, 6, \ldots\}$ **13** $\frac{23}{7}$
15 a; a^5; a^6; $-x^{15}y^{12}$; a^5 **17** $\frac{1}{4}$; 1; $\frac{1}{2}$; 100; 9 **19** 7
21 $\frac{623}{80}$, or 7.8 ft (approx.) on the other side of the fulcrum
23 $62\frac{1}{2}$ mi/h **25** 70 lb **27** 85¢ per lb
29 6 lb of $1.20 tea; 4 lb of $1 tea
31 20 lb **33** $w = 12$ ft, $l = 36$ ft
35 23 in., 39 in. **37** 175 mi/h **39** $1,456.31
41 $\{(2, 1)\}$ **43** 20 lb **45** $\frac{1}{2}(5 - 3\sqrt{3})$
49 $-\dfrac{2}{y^2}$ **51** $\{3, -3\}$; $\{-\frac{1}{3}, 2\}$; $\{-3 + \sqrt{5}, -3 - \sqrt{5}\}$
53 $\{5, -5\}$; $\{-\frac{2}{3}, -\frac{3}{2}\}$; $\left\{\dfrac{7 - \sqrt{21}}{14}, \dfrac{7 + \sqrt{21}}{14}\right\}$
55 $\{\frac{3}{2}, 4\}$; $\left\{\dfrac{1 + \sqrt{5}}{4}, \dfrac{1 - \sqrt{5}}{4}\right\}$ **65** $z = \dfrac{49xy}{4w^2}$ **67** 64 **69** $\frac{1}{4}$
73 27° **75** 195.25 **77** 3, 5 **79** Difference; $\log M - \log N$
81 3, $\frac{1}{2}$, -3, -3 **83** $l = 4n - 3$, $S = n(2n - 1)$; 77, 780
85 $\frac{1}{2}$, 2, $\frac{7}{2}$, 5, $\frac{13}{2}$ **87** $1,200.61
89 $R' = \{5, 6, 7, 8\}$; $T' = \{1, 2, 3, 5, 7\}$; $R' \cup T' = \{1, 2, 3, 5, 6, 7, 8\}$; $R' \cap T' = \{5, 7\}$
91 $\dfrac{763}{3{,}888}$ **93** $m = 11, n = 2$
95 Arithmetic mean = 77; median = 79; mode = 86 **97** $\sigma = 11.14$
99 $\sqrt[3]{a^5} - \frac{5}{2}\sqrt[3]{a^4}\,b^2 + \frac{5}{2}ab^4 - \frac{5}{4}\sqrt[3]{a^2}\,b^6 + \frac{5}{16}\sqrt[3]{a}\,b^8 - \frac{1}{32}b^{10}$
101 $m = 24.6$

Appendix A, p. 539 **1** XI; XII; XIII; XIV; XV; XVI; XVII; XVIII; XIX; XX; XXI; XXII; XXIII; XXIV; XXV; XXVI; XXVII; XXVIII; XXIX; XXX; XXXI; XXXII; XXXIII; XXXIV; XXXV; XXXVI; XXXVII; XXXVIII; XXXIX; XL; XLI; XLII; XLIII; XLIV; XLV; XLVI; XLVII; XLVIII; XLIX; L; LI; LII; LIII; LIV; LV; LVI; LVII; LVIII; LIX; LX; LXI; LXII; LXIII; LXIV; LXV; LXVI; LXVII; LXVIII; LXIX; LXX; LXXI

Appendix C, p. 572 **1**

NUMBER	1.37	2,050	3,007	4,000	5,702.0	$2,1\dot{0}0$
NUMBER OF SIGNIFICANT FIGURES	3	3	4	1	5	3
NUMBER	0.0301	$5,0\dot{0}0$	0.001	42.000	1.000	
NUMBER OF SIGNIFICANT FIGURES	3	3	1	5	4	

3 167.4; 1.1; 0.1; 174.2; 3,214.6; 37.3; 8.7; 8.7

5

	NUMBER ROUNDED OFF TO		
NUMBER	4 DECIMAL PLACES	3 DECIMAL PLACES	2 DECIMAL PLACES
2.05655	2.0566	2.057	2.06
0.05173	0.0517	0.052	0.05
41.50505	41.5050	41.505	41.51
0.00493	0.0049	0.005	0.00

INDEX